T0214976

Partial Differential Equations
Equations
Analytical Methods and Applications

Textbooks in Mathematics

Series editors:
Al Boggess and Ken Rosen

CRYPTOGRAPHY: THEORY AND PRACTICE, FOURTH EDITION
Douglas R. Stinson and Maura B. Paterson

GRAPH THEORY AND ITS APPLICATIONS, THIRD EDITION
Jonathan L. Gross, Jay Yellen and Mark Anderson

COMPLEX VARIABLES: A PHYSICAL APPROACH WITH APPLICATIONS, SECOND EDITION
Steven G. Krantz

GAME THEORY: A MODELING APPROACH
Richard Alan Gillman and David Housman

FORMAL METHODS IN COMPUTER SCIENCE
Jiacun Wang and William Tepfenhart

AN ELEMENTARY TRANSITION TO ABSTRACT MATHEMATICS
Gove Effinger and Gary L. Mullen

ORDINARY DIFFERENTIAL EQUATIONS: AN INTRODUCTION TO THE FUNDAMENTALS, SECOND EDITION
Kenneth B. Howell

SPHERICAL GEOMETRY AND ITS APPLICATIONS
Marshall A. Whittlesey

COMPUTATIONAL PARTIAL DIFFERENTIAL PARTIAL EQUATIONS USING MATLAB®, SECOND EDITION
Jichun Li and Yi-Tung Chen

AN INTRODUCTION TO MATHEMATICAL PROOFS
Nicholas A. Loehr

DIFFERENTIAL GEOMETRY OF MANIFOLDS, SECOND EDITION
Stephen T. Lovett

MATHEMATICAL MODELING WITH EXCEL
Brian Albright and William P. Fox

THE SHAPE OF SPACE, THIRD EDITION
Jeffrey R. Weeks

CHROMATIC GRAPH THEORY, SECOND EDITION
Gary Chartrand and Ping Zhang

PARTIAL DIFFERENTIAL EQUATIONS: ANALYTICAL METHODS AND APPLICATIONS
Victor Henner, Tatyana Belozerova, and Alexander Nepomnyashchy

https://www.crcpress.com/Textbooks-in-Mathematics/book-series/CANDHTEXBOOMTH

Partial Differential Equations

Analytical Methods and Applications

Victor Henner
Tatyana Belozerova
Alexander Nepomnyashchy

CRC Press
Taylor & Francis Group
Boca Raton London New York

CRC Press is an imprint of the
Taylor & Francis Group, an **informa** business

A CHAPMAN & HALL BOOK

CRC Press
Taylor & Francis Group
6000 Broken Sound Parkway NW, Suite 300
Boca Raton, FL 33487-2742

First issued in paperback 2022

© 2020 by Taylor & Francis Group, LLC
CRC Press is an imprint of Taylor & Francis Group, an Informa business

No claim to original U.S. Government works

ISBN 13: 978-1-03-247508-0 (pbk)
ISBN 13: 978-1-138-33983-5 (hbk)

DOI: 10.1201/9780429440908

Visit the Taylor & Francis Web site at
http://www.taylorandfrancis.com

and the CRC Press Web site at
http://www.crcpress.com

*In memory of the wonderful scientist
and friend Sergei Shklyaev*

Contents

Preface   **xi**

1   Introduction   **1**
   1.1   Basic Definitions 1
   1.2   Examples 4

2   First-Order Equations   **7**
   2.1   Linear First-Order Equations 7
      2.1.1   General Solution 7
      2.1.2   Initial Condition 10
   2.2   Quasilinear First-Order Equations 12
      2.2.1   Characteristic Curves 12
      2.2.2   Examples 14

3   Second-Order Equations   **21**
   3.1   Classification of Second-Order Equations 21
   3.2   Canonical Forms 22
      3.2.1   Hyperbolic Equations 23
      3.2.2   Elliptic Equations 24
      3.2.3   Parabolic Equations 26

4   The Sturm-Liouville Problem   **29**
   4.1   General Consideration 29
   4.2   Examples of Sturm-Liouville Problems 34

5   One-Dimensional Hyperbolic Equations   **43**
   5.1   Wave Equation 43
   5.2   Boundary and Initial Conditions 45
   5.3   Longitudinal Vibrations of a Rod and Electrical Oscillations 48
      5.3.1   Rod Oscillations: Equations and Boundary Conditions . . . 48
      5.3.2   Electrical Oscillations in a Circuit 50
   5.4   Traveling Waves: D'Alembert Method 52
   5.5   Cauchy Problem for Nonhomogeneous Wave Equation 57
      5.5.1   D'Alembert's Formula 57
      5.5.2   Green's Function 58
      5.5.3   Well-Posedness of the Cauchy Problem 59
   5.6   Finite Intervals: The Fourier Method for Homogeneous Equations 60
   5.7   The Fourier Method for Nonhomogeneous Equations 71
   5.8   The Laplace Transform Method: Simple Cases 76
   5.9   Equations with Nonhomogeneous Boundary Conditions 78
   5.10   The Consistency Conditions and Generalized Solutions 83
   5.11   Energy in the Harmonics 84

5.12 Dispersion of Waves 88
    5.12.1 Cauchy Problem in an Infinite Region 88
    5.12.2 Propagation of a Wave Train 91
5.13 Wave Propagation on an Inclined Bottom: Tsunami Effect 93

6 One-Dimensional Parabolic Equations   **99**
6.1 Heat Conduction and Diffusion: Boundary Value Problems 99
    6.1.1 Heat Conduction 99
    6.1.2 Diffusion Equation 100
    6.1.3 One-dimensional Parabolic Equations and Initial and Boundary
        Conditions 101
6.2 The Fourier Method for Homogeneous Equations 103
6.3 Nonhomogeneous Equations 111
6.4 Green's Function and Duhamel's Principle 114
6.5 The Fourier Method for Nonhomogeneous Equations with
    Nonhomogeneous Boundary Conditions 118
6.6 Large Time Behavior of Solutions 126
6.7 Maximum Principle 129
6.8 The Heat Equation in an Infinite Region 131

7 Elliptic Equations   **139**
7.1 Elliptic Differential Equations and Related Physical Problems 139
7.2 Harmonic Functions 141
7.3 Boundary Conditions 142
    7.3.1 Example of an Ill-posed Problem 142
    7.3.2 Well-posed Boundary Value Problems 143
    7.3.3 Maximum Principle and its Consequences 144
7.4 Laplace Equation in Polar Coordinates 146
7.5 Laplace Equation and Interior BVP for Circular Domain 147
7.6 Laplace Equation and Exterior BVP for Circular Domain 151
7.7 Poisson Equation: General Notes and a Simple Case 151
7.8 Poisson Integral 154
7.9 Application of Bessel Functions for the Solution of Poisson
    Equations in a Circle 156
7.10 Three-dimensional Laplace Equation for a Cylinder 160
7.11 Three-dimensional Laplace Equation for a Ball 164
    7.11.1 Axisymmetric Case 164
    7.11.2 Non-axisymmetric Case 165
7.12 BVP for Laplace Equation in a Rectangular Domain 167
7.13 The Poisson Equation with Homogeneous Boundary Conditions 169
7.14 Green's Function for Poisson Equations 171
    7.14.1 Homogeneous Boundary Conditions 171
    7.14.2 Nonhomogeneous Boundary Conditions 175
7.15 Some Other Important Equations 176
    7.15.1 Helmholtz Equation 177
    7.15.2 Schrödinger Equation 180

8 Two-Dimensional Hyperbolic Equations   **187**
8.1 Derivation of the Equations of Motion 187
    8.1.1 Boundary and Initial Conditions 189
8.2 Oscillations of a Rectangular Membrane 191

8.2.1   The Fourier Method for Homogeneous Equations with
        Homogeneous Boundary Conditions 192
8.2.2   The Fourier Method for Nonhomogeneous Equations with
        Homogeneous Boundary Conditions 199
8.2.3   The Fourier Method for Nonhomogeneous Equations with
        Nonhomogeneous Boundary Conditions 203
8.3   Small Transverse Oscillations of a Circular Membrane 205
8.3.1   The Fourier Method for Homogeneous Equations with
        Homogeneous Boundary Conditions 206
8.3.2   Axisymmetric Oscillations of a Membrane 209
8.3.3   The Fourier Method for Nonhomogeneous Equations with
        Homogeneous Boundary Conditions 214
8.3.4   Forced Axisymmetric Oscillations 216
8.3.5   The Fourier Method for Equations with Nonhomogeneous
        Boundary Conditions 218

9   Two-Dimensional Parabolic Equations   227
9.1   Heat Conduction within a Finite Rectangular Domain 227
9.1.1   The Fourier Method for the Homogeneous Heat Equation
        (Free Heat Exchange) 230
9.1.2   The Fourier Method for Nonhomogeneous Heat Equation with
        Homogeneous Boundary Conditions 233
9.2   Heat Conduction within a Circular Domain 237
9.2.1   The Fourier Method for the Homogeneous Heat Equation 238
9.2.2   The Fourier Method for the Nonhomogeneous Heat Equation 241
9.2.3   The Fourier Method for the Nonhomogeneous Heat Equation
        with Nonhomogeneous Boundary Conditions 246
9.3   Heat Conduction in an Infinite Medium 248
9.4   Heat Conduction in a Semi-Infinite Medium 250

10 Nonlinear Equations   261
10.1 Burgers Equation 261
10.1.1 Kink Solution 261
10.1.2 Symmetries of the Burger's Equation 262
10.2 General Solution of the Cauchy Problem 264
10.2.1 Interaction of Kinks 265
10.3 Korteweg-de Vries Equation 267
10.3.1 Symmetry Properties of the KdV Equation 267
10.3.2 Cnoidal Waves 268
10.3.3 Solitons 270
10.3.4 Bilinear Formulation of the KdV Equation 271
10.3.5 Hirota's Method 272
10.3.6 Multisoliton Solutions 274
10.4 Nonlinear Schrödinger Equation 277
10.4.1 Symmetry Properties of NSE 277
10.4.2 Solitary Waves 278

A   Fourier Series, Fourier and Laplace Transforms   283
A.1   Periodic Processes and Periodic Functions 283
A.2   Fourier Formulas 284
A.3   Convergence of Fourier Series 286

A.4 Fourier Series for Non-periodic Functions 288
A.5 Fourier Expansions on Intervals of Arbitrary Length 289
A.6 Fourier Series in Cosine or in Sine Functions 290
A.7 Examples 292
A.8 The Complex Form of the Trigonometric Series 294
A.9 Fourier Series for Functions of Several Variables 295
A.10 Generalized Fourier Series 296
A.11 The Gibbs Phenomenon 298
A.12 Fourier Transforms 299
A.13 Laplace Transforms 303
A.14 Applications of Laplace Transform for ODE 306

B Bessel and Legendre Functions   **309**
B.1 Bessel Equation 309
B.2 Properties of Bessel Functions 312
B.3 Boundary Value Problems and Fourier-Bessel Series 315
B.4 Spherical Bessel Functions 320
B.5 The Gamma Function 322
B.6 Legendre Equation and Legendre Polynomials 324
B.7 Fourier-Legendre Series in Legendre Polynomials 328
B.8 Associated Legendre Functions 331
B.9 Fourier-Legendre Series in Associated Legendre Functions 334
B.10 Airy Functions 335

C Sturm-Liouville Problem and Auxiliary Functions for One
and Two Dimensions   **341**
C.1 Eigenvalues and Eigenfunctions of 1D Sturm-Liouville Problem for Different
Types of Boundary Conditions 341
C.2 Auxiliary Functions 343

D The Sturm-Liouville Problem for Circular and Rectangular Domains   **349**
D.1 The Sturm-Liouville Problem for a Circle 349
D.2 The Sturm-Liouville Problem for the Rectangle 352

E The Heat Conduction and Poisson Equations for Rectangular
Domains – Examples   **355**
E.1 The Laplace and Poisson Equations for a Rectangular Domain with
Nonhomogeneous Boundary Conditions – Examples 355
E.2 The Heat Conduction Equations with Nonhomogeneous Boundary
Conditions – Examples 366

Bibliography   **375**

Index   **377**

Preface

This book covers all the basic topics of a PDE course for undergraduate students and a beginner course for graduate students. The book is written in simple language, it is easy to read, and it provides qualitative physical explanations of mathematical results. At the same time, we keep all the rigor of mathematics (but without the emphasis on proofs – some simple theorems are proved; for more sophisticated ones we discuss only the key steps of the proofs). In our best judgment, a balanced presentation has been achieved, one which is as informative as possible at this level, and introduces and practices all necessary problem-solving skills, yet is concise and friendly to the reader. A part of the philosophy of the book is 'teaching-by-example', and thus we provide numerous carefully chosen examples that guide step-by-step learning of concepts and techniques, and problems to be solved by students. The level of presentation and the book structure allow its use in the engineering, physics and applied mathematics departments as the main text for the entire course.

The primary motivation for writing this book was as follows. Traditionally, textbooks written for students studying engineering and physical sciences are concentrated on the three basic second-order PDEs, the wave equation, the heat equation, and the Laplace/Poisson equation. A typical example of such an approach is the popular book by J.W. Brown and R.V. Churchill [1]. As a rule, only the method of separation of variables is presented and only the simplest problems are discussed. On the other hand, complete books in that field (for instance, the excellent book by R. Haberman [2]) contain too much information and therefore they are difficult for beginner reading or for a one-semester course. In the presented book, we tried to get rid of those shortcomings. The book contains a simple and concise description of the main types of PDEs and presents a wide collection of analytical methods for their solutions, from the method of characteristics for quasilinear first-order equations to sophisticated methods for solving integrable nonlinear equations. The text is written in such a way that an instructor can easily include definite chapters of the book and omit some other chapters, depending on the structure of a particular course syllabus. Numerous examples, applications, and graphical material make the book advantageous for students taking a distance learning course and studying the subject by themselves.

The structure of the book is as follows.

Chapter 1 presents a general introduction. Examples of physical systems described by basic PDEs are given, and general definitions (order of equation, linear, quasilinear and nonlinear equations, initial and boundary condition, general and particular solutions) are presented. Also, we consider some problems that can be solved without knowing any special methods.

Chapter 2 is devoted to first-order PDEs. The consideration starts with linear PDEs, where the problem can be reduced to ODEs by a coordinate transformation, and then the approach is generalized so that it includes quasilinear equations. We consider the formation of a shock wave and explain its geometrical and physical meanings, and discuss some other nontrivial examples. The conditions of existence and uniqueness of the solution are also explained.

In Chapter 3, we provide the general classification of second-order PDEs, show its invariance to transformations of coordinates, and consider the transformation of equations to their canonical forms.

The subject of Chapter 4 is the Sturm-Liouville problem, which gives the basis for the development of the classical technique of variables' separation in the consequent chapters, where three types of second-order PDEs are considered, hyperbolic (Chapter 5), parabolic (Chapter 6), and elliptic (Chapter 7). In each case, physical systems described by the corresponding type of equations are presented. Besides the method of separation of variables, Green's function method is described. These chapters cover the full range of relevant subjects including the notion of well-posedness, consequences of wave dispersion, Duhamel's principle, maximum principle, etc.

Chapter 8 is devoted mostly to multidimensional hyperbolic equations, with oscillations of an elastic two-dimensional membrane as a primary example. In Chapter 9, the method of separation of variables is applied to the two-dimensional heat equation in a rectangular domain and in a circular domain.

The last chapter of the book, Chapter 10, is devoted to three famous nonlinear PDEs, which have numerous applications in physics. First, we consider the Burgers equation, obtain the solution of the Cauchy problem for that equation using the Hopf-Cole transformation, and consider the interaction of shock waves. Further, the analysis of the Korteweg-de Vries equation gives us the opportunity to present an example of an integrable equation. By means of the Hirota transformation, we obtain its multisolitons solutions. Finally, the nonlinear Schrödinger equation is considered. We discuss its symmetry properties and obtain some basic particular solutions.

The appendices are an important part of the book. Appendices A and B are written in the same style as the chapters of the main text, with reading exercises, examples, and problems for solving. The subjects of these appendices (Fourier series, Fourier and Laplace transforms, Bessel and Legendre functions) do not belong formally to the field of PDEs, but they give mathematical tools, which are crucial for solving boundary-value problems for different kinds of second-order partial differential equations. Therefore, they can be included in the course syllabus if the class time allows. Appendices C, D, and E contain mathematical materials related to the Sturm-Liouville problems which are used in several chapters of the book.

We are grateful to Profs. Kyle Forinash and Mikhail Khenner, our former coauthors, for sharing their experience in writing books.

1

Introduction

1.1 Basic Definitions

Let us start with the discussion of the basic notion of *partial differential equation*. In a contradistinction to an algebraic equation, e.g., a quadratic equation

$$ax^2 + bx + c = 0,$$

where one has to find the unknown *number* x, and to an ordinary differential equation (ODE), e.g., that corresponding to Newton's second law,

$$m\frac{d^2 x}{dt^2} = F(x, t),$$

where the unknown object is *the function $x(t)$ of one variable*, a partial differential equation (PDE) is an equation that determines a function $u(x_1, x_2, \ldots, x_n)$ of *several variables*. Because we live in a four-dimensional world, which has three spatial dimensions and one temporal dimension, the natural phenomena are described by fields of physical quantities (velocity, temperature, electric and magnetic fields etc.) that depend on spatial variables and evolve in time. Therefore, it is not surprising that partial differential equations provide the language for the formulation of the laws of nature.

Let us consider the following simple example. A metal rod of length L is subject to an external heating (that may be a barbecue skewer), see Figure 1.1.

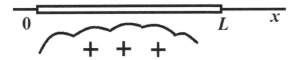

FIGURE 1.1
Heating a rod.

The temperature of the rod, $u(x, t)$, is governed by the *heat equation*,

$$\frac{\partial u}{\partial t} = \kappa \frac{\partial^2 u}{\partial x^2} + q(x, t), \tag{1.1}$$

where x is the coordinate along the rod, t is time, κ is thermal diffusivity, and $q(x, t)$ is the rate of heating. It is assumed that κ and $q(x, t)$ are known, while $u(x, t)$ is the unknown function that has to be found. When dealing with that equation, we have to indicate the region Ω where that equation is defined, say,

$$\Omega: \ 0 < x < L, \quad 0 < t < T. \tag{1.2}$$

Any function $u(x,t)$ satisfying Equation (1.1), is a *solution* (or *particular solution*) of that equation. It is assumed that the derivatives $\partial u/\partial t$ and $\partial^2 u/\partial x^2$, which appear in the equation, really exist and are continuous in the whole region Ω; in that case we call $u(x,t)$ a *classical* or *strong solution* (in Chapter 5 we shall extend the notion of solution). Note that there are infinitely many solutions of Equation (1.1); for instance, if $u(x,t)$ is a solution, then any function $u(x) + ax + b$, where a and b are constant, is also a solution. The set of all the particular solutions of a PDE form its *general solution*.

When we consider a definite real physical process, we are interested to describe it in a deterministic, unique way; hence finding the general solution of the corresponding PDE is not the final goal. In the case of the heat equation (1.1), as it will be shown in Chapter 6, for getting a unique solution, it is sufficient to add the following data.

First, we have to present an *initial condition*, i.e., the distribution of temperature over the whole rod including its ends in the very beginning of the process, i.e., at $t = 0$:

$$u(x,0) = u_0(x), \quad 0 \le x \le L. \tag{1.3}$$

Secondly, we have to indicate *boundary conditions*, i.e., the law of heating or cooling at the ends of the rod, $x = 0$ and $x = L$. For instance, as it will be shown in Chapter 6, if the rod ends are thermally insulated, the appropriate boundary conditions are

$$\frac{\partial u}{\partial x}(0,t) = 0, \quad \frac{\partial u}{\partial x}(L,t) = 0, \quad 0 \le t \le T. \tag{1.4}$$

More precisely, it is assumed that the functions that appear in the boundary conditions are continuous by approaching the boundaries, i.e., relations (1.3) and (1.4) are actually understood as

$$\lim_{t \to 0} u(x,t) = u_0(x), \quad 0 \le x \le L; \tag{1.5}$$

$$\lim_{x \to 0} \frac{\partial u}{\partial x}(x,t) = 0, \quad \lim_{x \to L} \frac{\partial u}{\partial x}(x,t) = 0, \quad 0 \le t \le T. \tag{1.6}$$

The full formulation of the problem, which includes Equation (1.1), (1.2), initial condition (1.3) and boundary conditions (1.4), is called *initial-boundary value problem* (see Figure 1.2). Under a natural assumption that $u_0(x)$ is a continuous function, that problem has a unique solution that allows us to predict the process of heating

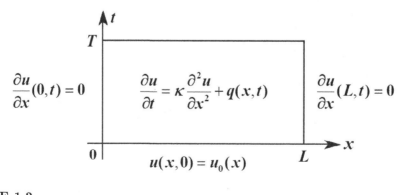

FIGURE 1.2
Initial-boundary value problem.

There is one more subtle point that should be kept in mind. It is clear that the initial temperature distribution, $u_0(x)$, and the rate of heating, $q(x,t)$, cannot be measured absolutely precisely. There is always a certain error in the measurement data. If an arbitrary small error in those data leads to a drastic change of the solution, than the mathematical model is of no use. Fortunately, as we shall see in Chapter 6, problem (1.1)-(1.4) is *stable* with respect to the measurement error: using that model, one can obtain the solution with arbitrary precision, if the measurement data are precise correspondingly. Existence and uniqueness of the solution, and its continuous dependence of the initial data and the heating rate mean that model (1.1)-(1.4) is physically sound, or *well-posed*. An example of an unsatisfactorily formulated, ill-posed, problem will be given in Chapter 7.

Equation (1.1) describes many other *dissipative* processes (e.g., diffusion and viscosity) where initial inhomogeneities tend to be smoothed and exterminated in an irreversible way. It is one of the three *basic equations of the mathematical physics*. Let us mention the other two equations.

The *wave equation*

$$\frac{\partial^2 u}{\partial t^2} - a^2 \frac{\partial^2 u}{\partial x^2} = 0, \quad 0 < x < L, \quad 0 < t < T \tag{1.7}$$

describes the propagation of waves of different physical nature, including sound waves, oscillations of a string, electromagnetic waves etc. We will discuss the initial and boundary conditions for that equation, which make the initial-boundary value problem well-posed, in Chapter 5.

Static distributions of the temperature $u(x, y)$ in a two-dimensional body Ω are governed by the *Poisson equation*

$$\frac{\partial^2 u}{\partial x^2} + \frac{\partial^2 u}{\partial y^2} = -\frac{q(x, y)}{\kappa}, \quad (x, y) \in \Omega; \tag{1.8}$$

here the meaning of variables is the same as in (1.1) but now the temperature depends on two spatial variables. This equation is valid for electrical potential, gravitational potential etc. The question about the adequate boundary conditions is not trivial, and we will discuss it in Chapter 7. If $q(x, y) = 0$, Equation (1.8) is called the *Laplace equation*.

Generally, a PDE for the function $u(x_1, x_2, \ldots, x_n)$ can be written as

$$F\left(x_1, \ldots, x_n, \frac{\partial u}{\partial x_1}, \ldots, \frac{\partial u}{\partial x_n}, \ldots\right) = 0. \tag{1.9}$$

The *order* of a PDE is the highest order of derivatives that appear in that equation. Thus, Equations (1.1), (1.7) and (1.8) are of the second order.

If function F in (1.9) is a linear function of u and its derivatives (while the dependence on $x_1, \ldots, x_n$ can be arbitrary), the PDE is called a *linear equation*. For instance, the most general form of the second-order linear PDE for function $u(x, y)$ is

$$a(x, y)\frac{\partial^2 u}{\partial x^2} + 2b(x, y)\frac{\partial^2 u}{\partial x \partial y} + c(x, y)\frac{\partial^2 u}{\partial y^2} + d(x, y)\frac{\partial u}{\partial x}$$
$$+ e(x, y)\frac{\partial u}{\partial y} + f(x, y)u = g(x, y), \tag{1.10}$$

where $a(x, y)$, $b(x, y), \ldots, g(x, y)$ are known functions. The second coefficient is written as $2b(x, y)$ in order to simplify some formulas that will be obtained in Chapter 3. All the Equations (1.1), (1.7), (1.8) are linear. If $g(x, y) \neq 0$, the equation is *inhomogeneous*; otherwise it is *homogeneous*.

As an example of a nonlinear equation, let us present the *non-viscous Burgers equation*

$$\frac{\partial u}{\partial t} + u\frac{\partial u}{\partial x} = 0, \tag{1.11}$$

which describes the propagation of an intense sound wave (as well as many other phenomena, including traffic waves). We shall consider that equation in detail in Chapter 2. Note that the expression in the left-hand side of (1.11) is linear with respect to highest order derivatives, $\partial u/\partial t$ and $\partial u/\partial x$. In that case, the nonlinear equation is called *a quasilinear equation*.

Later on, we will use often the following compact form of the notation for partial derivatives:

$$u_x = \frac{\partial u}{\partial x}, \quad u_y = \frac{\partial u}{\partial y}, \quad u_{xx} = \frac{\partial^2 u}{\partial x^2}, \quad u_{xy} = \frac{\partial^2 u}{\partial x \partial y}, \quad u_{yy} = \frac{\partial^2 u}{\partial y^2} \qquad (1.12)$$

etc.

1.2   Examples

Let us consider some simple example of PDEs that can be solved without knowing any special methods.

Example 1.1 Let us consider equation

$$\frac{\partial u}{\partial x} = a(x, y), \quad -\infty < x < \infty, \quad -\infty < y < \infty \qquad (1.13)$$

for a function $u(x, y)$ of two variables.

If u and a were functions of one variable x, the general solution of the ODE

$$\frac{du}{dx} = a(x) \qquad (1.14)$$

would be

$$u(x) = \int_0^x a(z)dz + C,$$

where C is an *arbitrary constant*. In the case of a PDE, one can see that the expression

$$u(x, y) = \int_0^x a(z, y)dz + f(y) \qquad (1.15)$$

where $f(y)$ is an *arbitrary function*, presents the general solution of (1.13). Thus, in a contradistinction to general solutions of ODEs that contain arbitrary numbers, general solutions of PDEs contain arbitrary functions.

Generally, if a PDE contains only derivatives with respect to one variable, it can be solved by the same methods which are used for solving ODEs, and the solution looks like a solution of an ODE with constants replaced by functions. Let us consider the following example that we will use below in Chapter 2 when studying the methods for solving linear equations of the first order.

Example 1.2 Find the general solution $u(x, y)$ of PDE

$$\frac{\partial u}{\partial x} + c(x, y)u = d(x, y), \quad -\infty < x < \infty, \quad -\infty < y < \infty. \qquad (1.16)$$

Like in the case of an ODE, Equation (1.16) can be solved by means of an *integrating factor*. Let us multiply both sides of (1.16) by a still unknown function $\mu(x, y)$,

$$\mu(x, y)\frac{\partial u}{\partial x} + \mu(x, y)c(x, y)u = \mu(x, y)d(x, y),$$

and demand that the left-hand side will be a full derivative:

$$\mu(x, y)\frac{\partial u}{\partial x} + \mu(x, y)c(x, y)u = \frac{\partial}{\partial x}[\mu(x, y)u].$$

That gives the relation

$$\frac{\partial \mu(x, y)}{\partial x} = \mu(x, y)c(x, y),$$

thus, we can take

$$\mu(x, y) = \exp\left[\int_0^x c(t, y)dt\right]. \tag{1.17}$$

Solving the obtained equation

$$\frac{\partial}{\partial x}[\mu(x, y)u] = \mu(x, y)d(x, y) \tag{1.18}$$

in the same way as in Example 1.1, we find:

$$\mu(x, y)u = \int_0^x \mu(z, y)d(z, y)dz + f(y),$$

where $f(y)$ is an arbitrary function, thus

$$u(x, y) = \frac{1}{\mu(x, y)}\left[\int_0^x \mu(z, y)d(z, y)dz + f(y)\right]. \tag{1.19}$$

Formulas (1.17) and (1.19) together give the general solution of Equation (1.16).

In some cases a PDE can be reduced to "the form of an ODE" by a transformation of independent variables.

Example 1.3 Let us consider equation

$$\frac{\partial u}{\partial x} - \frac{\partial u}{\partial y} = 0, \quad -\infty < x < \infty, \quad -\infty < y < \infty. \tag{1.20}$$

This equation contains both derivatives with respect to x and y. Let us perform the following change of variables:

$$s = x + y, \quad t = x - y, \quad u(x, y) = v(s(x, y), t(x, y)). \tag{1.21}$$

Using the chain rule, we find:

$$\frac{\partial u}{\partial x} = \frac{\partial v}{\partial s}\frac{\partial s}{\partial x} + \frac{\partial v}{\partial t}\frac{\partial t}{\partial x} = \frac{\partial v}{\partial s} + \frac{\partial v}{\partial t}, \quad \frac{\partial u}{\partial y} = \frac{\partial v}{\partial s}\frac{\partial s}{\partial y} + \frac{\partial v}{\partial t}\frac{\partial t}{\partial y} = \frac{\partial v}{\partial s} - \frac{\partial v}{\partial t}. \tag{1.22}$$

Substituting (1.22) into (1.20), we obtain:

$$2\frac{\partial v}{\partial t} = 0.$$

Thus, v is an arbitrary function of t;

$$v = f(s).$$

Returning to $u(x, y)$, we find that

$$u = f(x + y), \tag{1.23}$$

where f is an arbitrary differentiable function, is the general solution of Equation (1.20).

Let us consider one more example, that we will use below in Chapter 5, when considering the propagation of waves.

Example 1.4 Find the general solution of equation

$$\frac{\partial^2 u}{\partial x \partial y} = 0, \quad -\infty < x < \infty, \quad -\infty < y < \infty. \tag{1.24}$$

We will apply the reduction of the given equation to ODEs twice. First, define

$$v = \frac{\partial u}{\partial y} \tag{1.25}$$

and rewrite (1.24) as

$$\frac{\partial v}{\partial x} = 0. \tag{1.26}$$

The general solution of Equation (1.26) is

$$v = f(y), \tag{1.27}$$

where $h(y)$ is an arbitrary function of y. Let us consider now (1.25) as an equation for u:

$$\frac{\partial u}{\partial y} = f(y),$$

thus

$$u(x, y) = \int_0^y f(z)dz + g(x), \tag{1.28}$$

where $g(x)$ is an arbitrary function of x. Because the integral of an arbitrary function is also an arbitrary function, we can write the general solution of Equation (1.24) as

$$u(x, y) = h(y) + g(x), \tag{1.29}$$

where both $f(y)$ and $g(x)$ are arbitrary differentiable functions.

Applying the change of independent variables, as in Example 1.3, and/or the dependent variable, as in Example 1.4, one can find general solutions of the following problems.

Problems

Find general solutions $u = u(x, y)$ of the following PDEs.

1. $4u_x + u_y = 0, -\infty < x, y < \infty.$
2. $yu_{xx} = \cos x, -\infty < x < \infty, y > 0.$
3. $x^2 u_{xx} + xu_x - u = 0, x > 0, -\infty < x < \infty.$
4. $y^2 u_{xx} - u = 0, -\infty < x < \infty, y > 0.$
5. $u_{xyy} = 0, -\infty < x, y < \infty.$
6. $u_{xy} = x^2 e^y, -\infty < x, y < \infty.$
7. $u_{xy} + u_x + u_y + u = 0, -\infty < x, y < \infty.$
8. $u_{xy} + yu_x + u_y + yu = 0, -\infty < x, y < \infty.$

Find solutions $u = u(x, y)$ of the following initial value problems.

9. $u_x + 2u_y = 0, -\infty < x, y < \infty; u(x, x) = x.$
10. $u_{xy} - u_y = y, -\infty < x, y < \infty; u(x, 0) = 0, u(0, y) = 0.$

2

First-Order Equations

In the present chapter, we consider basic methods for solving first-order partial differential equations for an unknown function $u(x, y)$ of two independent variables.

2.1 Linear First-Order Equations

2.1.1 General Solution

We have acquired already some experience in dealing with linear first-order equations (see Section 1.2, Examples 1.1-1.3.) Our next step is solving the general linear first-order equations,

$$a(x,y)u_x + b(x,y)u_y + c(x,y)u = d(x,y); \tag{2.1}$$

here and below, we use notation (1.12),

$$u_x = \frac{\partial u}{\partial x}, \quad u_y = \frac{\partial u}{\partial y}.$$

All the coefficients in (2.1) are assumed to be continuous functions. The approach that we describe in this section (*method of variable transformation*), has actually been applied in Example 1.3: we will construct a transformation of variables,

$$s = s(x,y), \quad t = t(x,y), \quad u(x,y) = v(s(x,y), t(x,y)), \tag{2.2}$$

which allows us to obtain a solvable equation of Example 1.2.

Using the chain rule,

$$u_x = v_s s_x + v_t t_x, \quad u_y = v_s s_y + v_t t_y, \tag{2.3}$$

we rewrite (2.1) as

$$(as_x + bs_y)v_s + (at_x + bt_y)v_t + cv = d. \tag{2.4}$$

Thus, if we find a function $s(x, y)$ such that

$$as_x + bs_y = 0, \tag{2.5}$$

we shall obtain an equation that contains only one partial derivative v_t, i.e., an equation of the type of (1.16).

Equation (2.5) has a simple geometric meaning. Define a vector field

$$\boldsymbol{\tau}(x,y) = (a(x,y), b(x,y)); \tag{2.6}$$

then Equation (2.5) can be written as

$$\boldsymbol{\tau} \cdot \nabla s = 0. \tag{2.7}$$

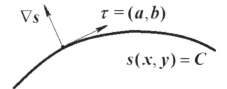

FIGURE 2.1
Characteristic curve.

The vector ∇s is *normal* to the level curves of function $s(x, y)$ determined by the relation

$$s(x, y) = C, \tag{2.8}$$

where C is a constant (see Figure 2.1). Therefore, vector $\tau(x, y)$ is a *tangent* vector to curve (2.8), which passes through the point (x, y). Thus, we arrive at the following geometric problem: for a given vector field (2.6), find the family of *characteristic curves*

$$y = y(x; C), \tag{2.9}$$

such that in each point, $\tau(x, y)$ is a tangent vector to a certain curve of that family.

Differentiating (2.8) along a level curve, we obtain,

$$\frac{ds(x, y(x))}{dx} = s_x + s_y \frac{dy}{dx} = 0,$$

hence

$$\frac{dy}{dx} = -\frac{s_x}{s_y}.$$

Using (2.5), we find:

$$\frac{dy}{dx} = \frac{b(x, y)}{a(x, y)}. \tag{2.10}$$

Thus, in order to find the appropriate function $s = s(x, y)$, we have to solve the ordinary differential equation (2.10) (the *characteristic equation*) and present the solution in the implicit form (2.8).

The choice of $t = t(x, y)$ is almost arbitrary, but we have to take care of a one-to-one correspondence between coordinates (s, t) and (x, y). According to the Inverse Function Theorem of calculus, the Jacobian determinant of the transformation (2.2),

$$J(x, y) = \frac{\partial(s, t)}{\partial(x, y)} = \begin{vmatrix} s_x & s_y \\ t_x & t_y \end{vmatrix} = s_x t_y - s_y t_x, \tag{2.11}$$

has to be non-vanishing. That allows us to invert the transformation (2.2) and find relations

$$x = x(s, t), \quad y = y(s, t), \tag{2.12}$$

that will allow us to present the coefficients in Equation (2.4) as functions of s, t.

When coordinates (s, t) are used, the first term in (2.4) vanishes; thus we obtain equation

$$(at_x + bt_y) v_t + cv = d, \tag{2.13}$$

which contains only a derivative with respect to t, while s appears just as a parameter. We can integrate the linear equation (2.13) keeping s constant, i.e., *along the characteristic curves*, as it was done in Section 1.2, Example 1.3. That integration gives us the solution in the form

$$v = v(s, t). \tag{2.14}$$

At the last stage, we obtain the solution

$$u(x, y) = v(s(x, y), t(x, y)). \tag{2.15}$$

Note that $a(x, y)$ can vanish at a certain point, which creates a problem with using Equation (2.10) near that point. If $b(x, y) \neq 0$ at that point, one can use equation

$$\frac{dx}{dy} = \frac{a(x, y)}{b(x, y)}$$

instead of (2.10). A simultaneous vanishing of $a(x, y)$ and $b(x, y)$, is a true difficulty: different characteristic curves cross at that point.

Let us apply the approach described above to the following problem.

Example 2.1 Consider equation

$$xu_x + yu_y + u = 1. \tag{2.16}$$

The characteristic equation is

$$\frac{dy}{dx} = \frac{y}{x}. \tag{2.17}$$

Let us assume that $x \neq 0$ for now. The family of solutions of Equation (2.17),

$$y(x) = Cx,$$

can be presented as

$$\frac{y}{x} = C,$$

thus we find that

$$s(x, y) = \frac{y}{x}. \tag{2.18}$$

Let us choose

$$t(x, y) = x. \tag{2.19}$$

The Jacobian (2.11) of the transformation (2.18), (2.19) is

$$J = -\frac{1}{x},$$

which is acceptable if $x \neq 0$. The chain rule (2.3) gives

$$u_x = -v_s \frac{y}{x^2} + v_t, \quad u_y = v_s \frac{1}{x};$$

the inverse transformation of variables is

$$x = t, \quad y = st, \tag{2.20}$$

hence

$$u_x = -v_s \frac{s}{t} + v_t, \quad u_y = \frac{v_s}{t}. \tag{2.21}$$

Substituting (2.20) and (2.21) into (2.16), we obtain

$$tv_t + v = 1. \tag{2.22}$$

Solving (2.22), we find

$$v(s,t) = \frac{f(s)}{t} + 1, \qquad (2.23)$$

where $f(s)$ is an arbitrary differentiable function. Substituting (2.18), (2.19) into (2.23), we obtain the general solution of Equation (2.16):

$$u(x,y) = \frac{f(y/x)}{x} + 1, \quad x \neq 0. \qquad (2.24)$$

Alternatively, we can rewrite the characteristic Equation (2.17) as

$$\frac{dx}{dy} = \frac{x}{y}, \quad y \neq 0$$

and use the transformation

$$s(x,y) = \frac{x}{y}, \quad t(x,y) = y. \qquad (2.25)$$

That will lead us to the following presentation of the solution:

$$u(x,y) = \frac{g(x/y)}{y} + 1, \quad y \neq 0, \qquad (2.26)$$

where g is an arbitrary differentiable function. In the region $x \neq 0$, $y \neq 0$, both forms (2.24) and (2.26) are equivalent, with

$$g\left(\frac{x}{y}\right) \equiv \frac{y}{x} f\left(\frac{y}{x}\right).$$

In the point $(0,0)$, where the characteristic lines intersect, all the solutions of Equation (2.16) are singular except $u = 1$, which corresponds to $f = 0$ or $g = 0$.

2.1.2 Initial Condition

As expected, the general solution of a first-order equation includes an arbitrary function. For getting a unique solution, we have to formulate an *initial value problem*, i.e., impose an *initial condition* that assigns some definite values to function $u(x,y)$ on a certain *initial curve* Γ. Let us consider several examples.

Example 2.2 Find the solution of Equation (2.16) satisfying the additional condition,

$$u(1,y) = 2, \quad -\infty < y < \infty. \qquad (2.27)$$

In this example, the initial curve Γ is the line $x = 1$. Substituting (2.24) into (2.27), we find $f(y) = 1$, hence we get a unique solution

$$u(x,y) = \frac{1}{x} + 1. \qquad (2.28)$$

This solution diverges at $x = 0$. Note that though formula (2.28) describes a certain solution of Equation (2.16) in the region $x < 0$, it *does not* present a solution of the *initial value problem* (2.16), (2.27) in that region. Indeed, the classical solution of an initial value problem for a first-order PDE has to be *continuously differentiable* in the whole region where it is defined, including the line where the initial condition is imposed. The solution of the initial value problem, (2.28), cannot be continued from the region $x > 0$ into the region $x < 0$ in a continuously differentiable way, because of its singularity on the line $x = 0$. Therefore, the definition region of the solution of the initial value problem (2.16), (2.27) is $x > 0$.

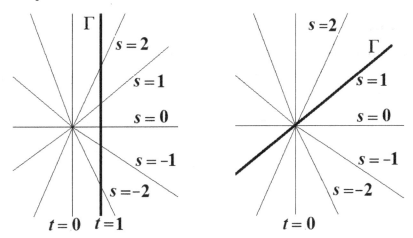

FIGURE 2.2
(a) Example 2.1; (b) Examples 2.2 and 2.3.

Example 2.3 Find the solution of Equation (2.16) satisfying the additional condition,

$$u(x, x) = 2, \quad -\infty < x < \infty. \tag{2.29}$$

In this case, the initial curve Γ is the line $y = x$. Substituting (2.24) into (2.29), we find the relation $f(1)/x = 1$, which cannot be satisfied. Thus, the initial value problem (2.24), (2.29) has no solutions.

Example 2.4 Let us take the same initial curve $y = x$ but demand

$$u(x, x) = 1, \quad -\infty < x < \infty. \tag{2.30}$$

Substituting (2.24) into (2.29), we find that the initial condition is satisfied if

$$f(1) = 0. \tag{2.31}$$

Thus, there are infinitely many solutions corresponding to different functions $f(y/x)$ in (2.24), which satisfy condition (2.31).

What is the origin of the difference between these three cases? In Example 2.2, the characteristic curves (actually, straight lines) $y = sx$, $s = const$ pass through the initial line $x = 1$ *transversally*, i.e., forming a non-zero angle with it (see Figure 2.2a). Thus, when definite values of $u(x, y)$ are assigned on the initial curve, the value of $u(x, y)$ in any other point with $x > 0$ is determined in the unique way by integrating along the characteristic line connecting that point with a certain point on the initial curve Γ. In the point $x = 0$ the solution diverges, and one cannot extend the classical solution beyond that point.

In Examples 2.3 and 2.4, the initial line $y = x$ is a characteristic line itself (see Figure 2.2b). Thus, we get two prescriptions for the calculation of $u(x, y)$ on Γ: one of them is given by Equation (2.22), and another one by the initial condition. If these two prescriptions contradict each other, there are no solutions of the initial value problem. If they match each other, there are infinitely many solutions, because the initial condition given on the initial curve Γ does not influence the solution outside that line.

We will formulate the conditions for the existence and uniqueness of solutions of initial value problems in the next section in a more general context.

2.2   Quasilinear First-Order Equations

In the present section we develop an approach for solving initial value problems for more general, quasilinear, first-order equations,

$$a(x, y, u)u_x + b(x, y, u)u_y = c(x, y, u) \qquad (2.32)$$

That approach is based on a generalization of the notion of *characteristic curve* defined in the previous section.

2.2.1   Characteristic Curves

Our success in solving the linear equation, (2.1), was based on finding the curves

$$x = x(s, t), \quad y = y(s, t) \qquad (2.33)$$

such that the transformed equation contains only a *directional derivative* in the direction $\tau = (a(x, y), b(x, y))$ along that curve. Now we will apply the same idea to a more general equation, (2.32).

Consider a particular solution of Equation (2.32),

$$u = f(x, y), \qquad\qquad (2.34)$$

which satisfies equation

$$a\left(x, y, f(x, y)\right) f_x + b\left(x, y, f(x, y)\right) f_y - c\left(x, y, f(x, y)\right) = 0. \qquad (2.35)$$

The plot of this solution is a surface in the three-dimensional space (x, y, u), which is called the *integral surface* of the partial differential equation (2.32). Let us present relation (2.34) as

$$F(x, y, u) \equiv f(x, y) - u = 0. \qquad (2.36)$$

Thus, the integral surface of the equation is the surface where a certain function $F(x, y, u)$ vanishes. Note that

$$\nabla F = (f_x, f_y, -1) \qquad (2.37)$$

determines the direction of the normal vector in every point of the surface (2.36). Define the vector field

$$\tau(x, y, u) = (a(x, y, u), \ b(x, y, u), \ c(x, y, u)). \qquad (2.38)$$

On the integral surface (2.36),

$$\tau = (a\left(x, y, f(x, y)\right), \ b\left(x, y, f(x, y)\right), \ c\left(x, y, f(x, y)\right)). \qquad (2.39)$$

Obviously, Equation (2.35) can be written as

$$\tau \cdot \nabla F = 0. \qquad (2.40)$$

The geometric meaning of that relation is quite clear: τ is *tangent* to surface (2.36), therefore the directional derivative of F along the direction τ vanishes (see Figure 2.3).

We come to the following geometric reformulation of the partial differential equation (2.32): for a given vector field (2.38), find a smooth surface $u = f(x, y)$ such that vector (2.39) is tangent in every point of that surface. The obtained surface will be an integral surface of equation (2.32).

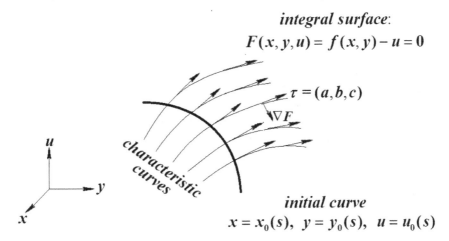

FIGURE 2.3
Integral surface, characteristic curves, and initial curve.

As the first step for solving that problem, let us consider the following system of ordinary differential equations:

$$\frac{dx(t)}{dt} = a(x(t), y(t), u(t)), \quad \frac{dy(t)}{dt} = b(x(t), y(t), u(t)),$$

$$\frac{du(t)}{dt} = c(x(t), y(t), u(t)), \tag{2.41}$$

which is solved with some initial conditions

$$x(0) = x_0, \quad y(0) = y_0, \quad u(0) = u_0 \tag{2.42}$$

(t is changed in both positive and negative directions). The solution of the problem (2.41), (2.42), $(x(t), y(t), u(t)$, determines a *curve* (called the *characteristic curve* of the PDE) in the three-dimensional space (x, y, u). According to (2.41), vector $\tau = (a, b, c)$ is tangent to that curve at each of its points. Changing the initial condition (2.42), we can obtain many curves of that kind. Taking a *one-parametric family* of such curves, we can obtain a *surface* such that vector τ is tangent at every point of that surface. That will solve the geometric problem formulated above, which is equivalent to finding a certain integral surface of Equation (2.32).

For instance, if we take initial conditions corresponding to a certain parametrically defined curve,

$$x_0 = x_0(s), \quad y_0 = y_0(s), \quad u_0 = u_0(s), \tag{2.43}$$

called the *initial curve* (see Figure 2.3), then solving the system of Equations (2.41) with initial conditions (2.43) will give us an integral surface, which is the plot of the solution $u = u(x, y)$ of Equation (2.32) with the initial condition (2.43). The solution is obtained in the parametric form:

$$x = x(t, s), \quad y = y(t, s), \tag{2.44}$$

$$u = u(t, s). \tag{2.45}$$

In order to get the solution $u = u(x, y)$ in the explicit form, we have to invert relations (2.44),

$$t = t(x, y), \quad s = s(x, y), \tag{2.46}$$

and substitute (2.46) into (2.45):

$$u = u\left(t(x, y), s(x, y)\right).$$

Thus, we have developed the following algorithm for solving quasilinear first-order equation (2.32) with definite initial conditions.

1. Present initial condition in the parametric form (2.43).

2. Solve ordinary differential equations

$$\frac{dx(t, s)}{dt} = a\left(x(t, s), y(t, s), u(t, s)\right), \quad \frac{dy(t, s)}{dt} = b\left(x(t, s), y(t, s), u(t, s)\right),$$
$$\frac{du(t, s)}{dt} = c\left(x(t, s), y(t, s), u(t, s)\right)$$

with initial conditions

$$x(0, s) = x_0(s), \quad y(0, s) = y_0(s), \quad u(0, s) = u_0(s). \tag{2.47}$$

That will give the solution in the parametric form,

$$x = x(t, s), \quad y = y(t, s), \quad u = u(t, s).$$

The existence and uniqueness of that solution in a certain interval of t around 0 is guaranteed if functions $a(x, y)$, $b(x, y)$ and $c(x, y)$ are smooth.

3. Determine $t = t(x, y)$, $s = s(x, y)$ and find the solution in the explicit form,

$$u = u\left(t(x, y), s(x, y)\right).$$

The general solution of Equation (2.32) is found as a set of all particular solutions with arbitrary functions (2.43).

2.2.2   Examples

In order to find some "underwater rocks" in the way described above, let us consider the following examples.

First, let us apply the approach described above to the *linear* equation,

$$xu_x + yu_y = 1 - u, \tag{2.48}$$

formerly solved in Section 2.1 with different initial conditions (Examples 2.2-2.4).

Example 2.5 In Example 2.2, we have applied the initial condition

$$u(1, y) = 2, \quad -\infty < y < \infty. \tag{2.49}$$

This initial condition determines the *initial curve* that can be presented in the parametric form as

$$x_0(s) = 1, \quad y_0(s) = s, \quad u_0(s) = 2; \quad -\infty < s < \infty. \tag{2.50}$$

In order to find the set of characteristic curves crossing different points of the initial curve (2.50), we solve the system of ordinary differential equations

$$\frac{dx}{dt}(t, s) = x, \quad \frac{dy}{dt}(t, s) = y, \quad \frac{du}{dt}(t, s) = 1 - u \tag{2.51}$$

with initial conditions

$$x(0, s) = 1, \quad y(0, s) = s, \quad u(0, s) = 2. \tag{2.52}$$

The general solution of the equation for x is $x(t, s) = C_1(s) \exp(t)$; from the initial condition $x(0, s) = 1$ we find $C_1(s) = 1$. Similarly, we get $y(t, s) = C_2(s) \exp(t)$ while solving the equation for y, with $C_2(s) = s$. Finally, solving the equation for u gives $u(t, s) = C_3(s) \exp(-t) + 1$, with $C_3(s) = 1$. Relations

$$x(t, s) = \exp(t), \quad y(t, s) = s \exp(t), \tag{2.53}$$

$$u(t, s) = \exp(-t) + 1 \tag{2.54}$$

determine the solution of the initial value problem (2.51), (2.52) *in the parametric form*. To obtain the solution in its explicit form, $u = u(x, t)$, we have to invert relations (2.53):

$$t(x, y) = \ln x, \quad s(x, y) = y/x, \tag{2.55}$$

and substitute expressions (2.55) into (2.54). We get

$$u(x, y) = \frac{1}{x} + 1, \tag{2.56}$$

which coincides with the result obtained in Example 2.2.

The inverse transformation (2.55) is successful, because the projections of characteristic curves on the plane (x, y) cross the projection of the initial curve (2.50) *transversally*, i.e., neither of the characteristic curves is *tangent* to the initial curve anywhere. Indeed, the projection of the tangent vector to (2.50) is

$$x_0'(s) = x_s(0, s) = 0, \quad y_0'(s) = y_s(0, s) = 1, \tag{2.57}$$

while the projection of the tangent vector to the characteristic curve on the line $t = 0$, according to Equations (2.53), is

$$x_t(0, s) = x(0, s) = 1, \quad y_t(0, s) = y(0, s) = s. \tag{2.58}$$

To check whether vectors (2.57) and (2.58) are nonparallel, we have to calculate the determinant

$$J(0, s) \equiv \begin{vmatrix} x_t(0, s) & y_t(0, s) \\ x_s(0, s) & y_s(0, s) \end{vmatrix} = \begin{vmatrix} 1 & s \\ 0 & 1 \end{vmatrix} = -1.$$

Because that determinant is different from 0 at the initial curve, the possibility of the inversion of relations (2.53) is granted in a certain vicinity of the initial curve. For arbitrary t, the Jacobian of transformation (2.52)

$$J(t, s) \equiv \begin{vmatrix} x_t(t, s) & y_t(t, s) \\ x_s(t, s) & y_s(t, s) \end{vmatrix} = \begin{vmatrix} \exp(t) & s \exp(t) \\ 0 & \exp(t) \end{vmatrix} = \exp(2t) = x^2.$$

Thus, we can expect that something wrong may happen at $x = 0$, where the Jacobian vanishes. Indeed, the construction of the solution following the characteristic curves, which starts at $x = 1$, cannot be continued below $x = 0$. When $t \to -\infty$, then $x \to 0$ and $y \to 0$ *for all* s. Thus, all the characteristic curves merge at the point $x = 0$, $y = 0$, and they cannot be continued further.

In the region $x > 0$, the solution is defined in a unique way.

Example 2.6 Let us consider now the same equation (2.48) with boundary conditions of Example 2.3,

$$u(x, x) = 2, \quad -\infty < x < \infty. \tag{2.59}$$

Presenting the initial curve in the parametric form,

$$x_0(s) = s, \quad y_0(s) = s, \quad u_0(s) = 2; \quad -\infty < s < \infty, \tag{2.60}$$

we formulate the problem for finding the characteristic curves that consists of the same system of ordinary differential equations, (2.51), and initial conditions

$$x(0, s) = s, \quad y(0, s) = s, \quad u(0, s) = 2. \tag{2.61}$$

First, let us check the transversality of the characteristic curves to the initial curve:

$$J(0, s) \equiv \begin{vmatrix} x_t(0, s) & y_t(0, s) \\ x_s(0, s) & y_s(0, s) \end{vmatrix} = \begin{vmatrix} s & s \\ 1 & 1 \end{vmatrix} = 0.$$

Thus, the projection of the tangent vector to the initial curve, $(1, 1)$, is parallel to the projection of the tangent vector of the characteristic curve, (s, s), hence our approach fails: instead of leaving the initial curve and forming the integral surface, the characteristic curves go along the initial curve. Moreover, we get a contradiction between the prescriptions of (2.51) and (2.61). According to (2.51), the tangent vector to the integral surface with the projection (s, s) is

$$(x_t(0, s), y_t(0, s), u_t(0, s)) = (x, y, 1 - u)|_{t=0} = (s, s, -1).$$

At the same time, the initial condition demands that the tangent vector with the projection in that direction has to be

$$(x_s(0, s), y_s(0, s), u_s(0, s)) = (1, 1, 0).$$

These two vectors are not parallel; thus it is impossible to simultaneously satisfy the equations and the boundary conditions: the solution of (2.51) and (2.61) does not exist.

Example 2.7 Let us return to Example 2.4, i.e., consider Equation (2.48) with the initial condition

$$u(x, x) = 1, \quad -\infty < x < \infty.$$

The initial curve is

$$x_0(s) = s, \quad y_0(s) = s, \quad u_0(s) = 1,$$

hence we have to consider system (2.50) with initial conditions

$$x(0, s) = s, \quad y(0, s) = s, \quad u(0, s) = 1. \tag{2.62}$$

Again,

$$J(0, s) \equiv \begin{vmatrix} x_t(0, s) & y_t(0, s) \\ x_s(0, s) & y_s(0, s) \end{vmatrix} = \begin{vmatrix} s & s \\ 1 & 1 \end{vmatrix} = 0.$$

This time, however, the tangent vector to the characteristic curve, $(x_t(0, s), y_t(0, s), u_t(0, s)) = (s, s, 0)$, and that to the initial curve, $(x_s(0, s), y_s(0, s), u_s(0, s)) = (1, 1, 0)$, are parallel *for every* s. Thus, we do not get any contradictions: the characteristic curve just coincides with the initial curve. But our approach does not work: instead of the set of characteristic curves forming the integral surface, we obtain only one characteristic curve which coincides with the initial curve. Thus, though the initial condition (2.62) is not harmful, it is also not helpful in finding the solution.

In order to construct the solution in the whole region of its existence, let us return to the general solution of (2.51),

$$x(t,s) = C_1(s)\exp(t), \quad y(t,s) = C_2(s)\exp(t), \quad u(t,s) = C_3\exp(-t) + 1. \qquad (2.63)$$

It is useless to substitute that solution into the initial condition (2.62), because that gives us the solution only at the line $y = x$. Let us apply the following trick. Instead of the useless initial condition along the line $y = x$, we take a curve that is transversal to projections of the characteristic curves, e.g., the line $x = 1$ of Example 2.5. Impose artificial initial conditions,

$$x_0(s) = 1, \quad y_0(s) = s, \quad u_0(s) = h(s),$$

where $h(s)$ is a certain unknown function, and substitute (2.63) into those conditions:

$$C_1(s) = 1, \quad C_2(s) = s, \quad C_3(s) = h(s) - 1,$$

Hence

$$x(t,s) = \exp(t), \quad y(t,s) = s\exp(t), \qquad (2.64)$$

$$u(t,s) = (h(s) - 1)\exp(-t) + 1. \qquad (2.65)$$

Inversion of relations (2.64) and substitution into (2.65) gives the solution,

$$u(x,y) = \frac{h(y/x) - 1}{x} + 1. \qquad (2.66)$$

The original initial condition (2.62) determines the value of h only in one point:

$$u(x,x) = h(1) = 1.$$

The obtained solution coincides with (2.24), (2.31), where $f(y/x) = h(y/x) - 1$.

Examples 2.5-2.7 allow us to come to the following conclusions (the proof can be found elsewhere, e.g., in [3]).

1. If the determinant

$$J(0,s) \equiv \begin{vmatrix} x_t(0,s) & y_t(0,s) \\ x_s(0,s) & y_s(0,s) \end{vmatrix} \neq 0$$

   everywhere, the unique solution exists near the initial curve.

2. In the points where $J(t,s) = 0$, the solution may cease to exist.

3. If $J(0,s) = 0$ everywhere along the initial curve, then it is necessary to check whether the vectors

$$(x_t(0,s), y_t(0,s), u_t(0,s)) = (a,b,c) \quad \text{and}$$

$$(x_s(0,s), y_s(0,s), u_s(0,s)) = (x_0'(s), y_0'(s), u_0'(s))$$

   are parallel. If they are parallel everywhere, there are infinitely many solutions. Otherwise, the solution does not exist.

Example 2.8 Let us consider equation

$$u_y + uu_x = 0, \quad -\infty < x < \infty, \quad y > 0 \tag{2.67}$$

with initial condition

$$u(x,0) = p(x), \quad -\infty < x < \infty, \tag{2.68}$$

where $p(x)$ is a known continuously differentiable function. As mentioned in Section 1, this equation is called the non-viscous Burgers equation.

Equation (2.67) describes the propagation of a strong sound wave in a gas. The physical meaning of x is the coordinate along the direction of the wave propagation; the variable y denotes time. The wave propagation is described in the reference frame moving at the speed of sound. Function u has the meaning of pressure.

Applying the approach described above, we present the initial curve in the parametric form

$$x_0(s) = s, \quad y_0(s) = 0, \quad u_0(s) = p(s) \tag{2.69}$$

and consider the system of *ordinary* differential equations depending on the parameter s

$$\frac{dx}{dt}(t,s) = u(t,s), \tag{2.70}$$

$$\frac{dy}{dt}(t,s) = 1, \tag{2.71}$$

$$\frac{du}{dt}(t,s) = 0, \tag{2.72}$$

with initial conditions

$$x(0,s) = s, \tag{2.73}$$

$$y(0,s) = 0, \tag{2.74}$$

$$u(0,s) = p(s), \tag{2.75}$$

which determine the characteristic curves.

First, let us calculate the value of the Jacobian $J(0,s)$ of the transformation $(x,y) \to (t,s)$ on the initial curve $t = 0$:

$$J(0,s) = \begin{vmatrix} x_t(0,s) & y_t(0,s) \\ x_s(0,s) & y_s(0,s) \end{vmatrix} = \begin{vmatrix} p(s) & 1 \\ 1 & 0 \end{vmatrix} = -1 \neq 0.$$

Therefore, we expect that the solution of the problem (2.70)-(2.75) exists at least for values of y sufficiently close to 0, and it is unique.

Solving Equation (2.72) with initial condition (2.75), we find

$$u(t,s) = p(s). \tag{2.76}$$

Solution of Equation (2.72) with initial condition (2.74) is

$$y(t,s) = t. \tag{2.77}$$

Finally, substituting (2.76) into (2.70) and solving the latter equation with initial condition (2.72), we find

$$x(t,s) = s + p(s)t. \tag{2.78}$$

Formulas (2.76)-(2.78) describe the solution of the initial value problem (2.70)-(2.75) in a parametric way. We cannot write that solution in an explicit way $u = u(x,y)$ for an

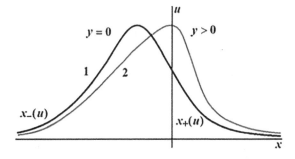

FIGURE 2.4
Example 2.8.

arbitrary function $p(s)$. However, the evolution of the spatial pressure distribution $u = u(x)$ with time $y = t$ can be easily understood. For sake of simplicity, assume that the plot of function $u(x,0) = p(x)$ has a shape shown in Figure 2.4 (line 1): $p(x)$ grows from 0 to p_* for $-\infty < x < x_*$ and decreases from p_* to 0 for $x_* < x < \infty$. In the intervals of the monotonicity of function $p(x)$, we can invert the relation $u = p(x)$ and present $x(u)$ as $x_-(u)$ in the region where $p'(x) > 0$ (hence $x'_-(u) > 0$) and $x_+(u)$ in the region where $p'(x) < 0$ (hence $x'_+(u) < 0$) (see Figure 2.4).

For $y > 0$, the solution in both regions can be written in implicit form as

$$x = x(y, u) = x_\pm(u) + uy. \tag{2.79}$$

That means that with the growth of $y = t$, according to (2.76), each point of the plot $u = u(x)$ keeps its height. According to (2.78), that point moves horizontally with the velocity $u = p(s)$ corresponding to its height: the higher the point, the faster it moves (see line 2 in Figure 2.4). One can see that the curve $u(x)$ becomes more gently sloping with time in the region where $\partial u/\partial x > 0$ and steeper in the region where $\partial u/\partial x < 0$. Indeed, differentiating (2.79) with respect to u, we find that

$$\frac{\partial x}{\partial y} = x'_\pm(u) + y. \tag{2.80}$$

Because $x'_-(u) > 0$, $\partial x/\partial u > 0$ grows with y, i.e., $\partial u/\partial x > 0$ decreases. But $x'_+(u) < 0$, therefore with the growth of y, $|\partial x/\partial u|$ decreases, and $|\partial u/\partial x|$ grows. In the latter case, according to (2.80), $\partial x/\partial u$ becomes equal to zero when $y = -x'_\pm(u)$. First it happens at the time instant

$$y = y_m = \min_u \left[-x'_+(u)\right] = \frac{1}{\max_s \left[-p'(s)\right]}. \tag{2.81}$$

At $y = y_m$, the derivative $\partial u/\partial x$ diverges at a certain point. The classical solution cannot be extended beyond that time instant.

Let us calculate the Jacobian of the transformation $(x, y) \to (t, s)$ for the obtained solution (2.76)-(2.78):

$$J(t, s) = \begin{vmatrix} x_t(t, s) & y_t(t, s) \\ x_s(t, s) & y_s(t, s) \end{vmatrix} = \begin{vmatrix} p(s) & 1 \\ 1 + p'(s)t & 0 \end{vmatrix} = -(1 + p'(s)t).$$

One can see that $J(t, s)$ first becomes equal to zero exactly at the time instant $t = y = y_m$ determined by relation (2.81), which is the formal reason for the failure of the approach.

In the case of gas motion, the divergence of $\partial u/\partial x$ corresponds to the appearance of a *shock wave*. For the description of the further evolution of the physical system, a modification of model (2.67) is needed (see Chapter 10).

Problems

Find the general solution of the equation and solutions satisfying initial conditions.

1. Equation: $u_x = 1$, $-\infty < x < \infty$, $y > 0$; initial conditions:

   (a) $u(x,0) = 0$, $-\infty < x < \infty$;
   (b) $u(x,0) = x$, $-\infty < x < \infty$;
   (c) $u(x,x^2) = 0$, $-\infty < x < \infty$.

2. Equation: $u_x + u_y = 1$, $-\infty < x < \infty$, $-\infty < y < \infty$; initial conditions:

   (a) $u(x,0) = v(x)$, $-\infty < x < \infty$;
   (b) $u(x,x) = v(x)$, $-\infty < x < \infty$;
   (c) $u(x,x^2) = 0$, $-\infty < x < 1/2$.

3. Equation: $u_y + (axu)_x = 0$, $a = const$, $-\infty < x < \infty$, $y > 0$; initial conditions:

   (a) $u(x,0) = v(x)$, $-\infty < x < \infty$;
   (b) $u(0,y) = v(y)$, $y \geq 0$.

4. Equation: $xu_x - yu_y = 0$, $x > 0$, $y > 0$; initial conditions:

   (a) $u(x,x) = h(x)$, $x \geq 0$;
   (b) $u(x,k/x) = h(x)$, $k = const$, $k > 0$, $x > 0$;
   (c) $u(x,1-x) = h(x)$, $1/2 < x < 1$.

5. Equation: $xu_y - yu_x = 0$, $x > 0$, $y > 0$; initial conditions:

   (a) $u(x,0) = v(x)$, $0 \leq x < \infty$;
   (b) $u(x,\sqrt{1-x^2} = v(x)$, $0 \leq x \leq 1$;
   (c) $u(x,1) = x$, $0 \leq x < \infty$;
   (d) $u(x,1) = x^2$, $0 \leq x < \infty$.

   Solve the following initial value problems.

6. $u_y + u^2 u_x = 0$, $-\infty < x < \infty$, $y > 0$; $u(x,0) = p(x)$; $p(x)$ is a bounded function. Find the solution in the implicit form. Find its existence region.

7. $u_y + uu_x = -au$, $-\infty < x < \infty$, $y > 0$; $a = const$, $a > 0$; $u(x,0) = p(x)$; $p(x)$ is a bounded function. Find the solution in the implicit form. What is the condition for the existence of the solution for any y?

8. $u_x + (x-u)u_y = 0$, $x > 0$, $-\infty < y < \infty$; $u(x,0) = x$.

3

Second-Order Equations

Now we start studying the most important class of PDEs, the linear equations of the second order.

3.1 Classification of Second-Order Equations

In the Introduction, we already formulated the most general form of the second-order linear PDE for function $u(x, y)$,

$$a(x, y)\frac{\partial^2 u}{\partial x^2} + 2b(x, y)\frac{\partial^2 u}{\partial x \partial y} + c(x, y)\frac{\partial^2 u}{\partial y^2} + d(x, y)\frac{\partial u}{\partial x}$$

$$+ e(x, y)\frac{\partial u}{\partial y} + f(x, y)u = g(x, y), \tag{3.1}$$

and considered three basic examples, the *wave equation*

$$\frac{\partial^2 u}{\partial y^2} - c_0^2 \frac{\partial^2 u}{\partial x^2} = 0, \tag{3.2}$$

the *heat equation*

$$\frac{\partial u}{\partial y} - \kappa \frac{\partial^2 u}{\partial x^2} = q(x, y) \tag{3.3}$$

(in order to unify notations, we denote time as y in the examples above), and the *Poisson equation*

$$\frac{\partial^2 u}{\partial x^2} + \frac{\partial^2 u}{\partial y^2} = g(x, y). \tag{3.4}$$

It has been announced that these three equations correspond to three types of problems, namely description of wave propagation, dissipative processes and static fields. What kind of solution behavior is expected for Equation (3.1) with a definite set of functions $a(x, y), \ldots, g(x, y)$?

Let us calculate *the discriminant*

$$\delta(x, y) = b^2(x, y) - a(x, y)c(x, y) \tag{3.5}$$

for each of Equations (3.2)-(3.4).

For the wave equation, $a = -c_0^2$, $b = 0$, $c = 1$, hence $\delta = c_0^2$ is positive. We shall call the equation *hyperbolic* in that case. As we shall see in Chapter 5, equations of this kind describe waves.

For the heat equation, $a = -\kappa$, $b = c = 0$, hence $\delta = 0$. Equations of this kind are called *parabolic*. They describe a non-wave evolution of the system (like heat conductivity and diffusion).

For the Poisson equation, $a = c = 1$, $b = 0$, hence $\delta = -1 < 0$. The negative values of δ correspond to the *elliptic* type of equations. Equations of this kind are used for description of static physical fields (e.g., temperature, concentration, pressure, gravity potential, electric potential etc.).

Note that the equation type is determined by the first three terms of (3.1), which contain derivatives of the second order. These terms form the *principal part* of the equation.

It is crucial that the type of the equation is not changed by transformation of coordinates

$$s = s(x, y), \quad t = t(x, y), \quad u(x, y) = v(s(x, y), t(x, y)); \tag{3.6}$$

it is assumed that the Jacobian of the transformation,

$$J(x, y) = \frac{\partial(s, t)}{\partial(x, y)} = \begin{vmatrix} s_x & s_y \\ t_x & t_y \end{vmatrix} = s_x t_y - s_y t_x, \tag{3.7}$$

is different from 0 everywhere. Using the chain rule,

$$u_x = v_s s_x + v_t t_x, \quad u_y = v_s s_y + v_t t_y, \tag{3.8}$$

we find that

$$u_{xx} = v_{ss} s_x^2 + 2v_{st} s_x t_x + v_{tt} t_x^2 + v_s s_{xx} + v_t t_{xx}, \tag{3.9}$$

$$u_{xy} = v_{ss} s_x s_y + v_{st} (s_x t_y + s_y t_x) + v_{tt} t_x t_y + v_s s_{xy} + v_t t_{xy}, \tag{3.10}$$

$$u_{yy} = v_{ss} s_y^2 + 2v_{st} s_y t_y + v_{tt} t_y^2 + v_s s_{yy} + v_t t_{yy}. \tag{3.11}$$

Substituting (3.9)-(3.11) into (3.1), we obtain:

$$Av_{ss} + 2Bv_{st} + Cv_{tt} + g(s, t, v, v_s, v_t) = 0, \tag{3.12}$$

where

$$A = as_x^2 + 2bs_x s_y + cs_y^2, \tag{3.13}$$

$$B = as_x t_x + b(s_x t_y + s_y t_x) + cs_y t_y, \tag{3.14}$$

$$C = at_x^2 + 2bt_x t_y + ct_y^2, \tag{3.15}$$

and g does not contain derivatives of the second order of v.

It can be shown directly that

$$\begin{pmatrix} A & B \\ B & C \end{pmatrix} = \begin{pmatrix} s_x & s_y \\ t_x & t_y \end{pmatrix} \begin{pmatrix} a & b \\ b & c \end{pmatrix} \begin{pmatrix} s_x & t_x \\ s_y & t_y \end{pmatrix}. \tag{3.16}$$

Calculating the determinants of the both parts of (3.16), we find that the discriminant characterizing the type of Equation (3.12),

$$\delta' = B^2 - AC = J^2(b^2 - ac) = J^2 \delta.$$

Because $J^2 > 0$, the type of the equation is not changed by transformation (3.16).

3.2   Canonical Forms

By means of a definite transformation (3.16), one can transform the principal part of Equation (3.1) to a certain standard form called *canonical form*. That transformation standardizes and simplifies the further analysis of the equation.

3.2.1 Hyperbolic Equations

Let us start with hyperbolic equations, $\delta = b^2 - ac < 0$. The canonical form for those equations is defined as

$$v_{st} + G\left(s, t, v, v_s, v_t\right) = 0, \tag{3.17}$$

where G does not contain derivatives of the second order.

Obviously, in order to obtain the form of (3.17), it is necessary to make A and C in (3.12) equal to zero:

$$as_x^2 + 2bs_x s_y + cs_y^2 = 0, \quad at_x^2 + 2bt_x t_y + ct_y^2 = 0. \tag{3.18}$$

The transformation is needed if at least one of the coefficients a and c is nonzero. Assume that $a \neq 0$ in the whole region of the definition of the equation. Solving quadratic equations (3.18), we find that

$$s_x = \frac{-b \pm \sqrt{b^2 - ac}}{a} s_y, \quad t_x = \frac{-b \pm \sqrt{b^2 - ac}}{a} t_y.$$

Because we are interested in finding a transformation with $s \neq t$, let us take different signs of the root for s_x and t_x, e.g.,

$$s_x + \frac{b - \sqrt{b^2 - ac}}{a} s_y = 0, \tag{3.19}$$

$$t_x + \frac{b + \sqrt{b^2 - ac}}{a} t_y = 0. \tag{3.20}$$

For solving first-order partial differential equations (3.19), (3.20), we can apply the approach described in Section 2.1.1. For (3.19), we obtain the characteristic equation

$$\frac{dy}{dx} = \frac{b - \sqrt{b^2 - ac}}{a}. \tag{3.21}$$

If the general solution of (3.21) is found in the implicit form $f_-(x, y) = \text{const}$, then we can choose the appropriate change of variable as

$$s = f_-(x, y). \tag{3.22}$$

Similarly, finding the general solution of the characteristic equation

$$\frac{dy}{dx} = \frac{b + \sqrt{b^2 - ac}}{a}. \tag{3.23}$$

in the form $f_+(x, y) = \text{const}$, we obtain the corresponding variable

$$t = f_+(x, y). \tag{3.24}$$

Using transformation (3.22), (3.24), we obtain the equation

$$B v_{st} + g = 0,$$

where B is determined by formula (3.14) and g does not contain derivatives of the second order. Dividing by B, we get an equation in the canonical form,

$$v_{st} + G = 0, \quad G = g/B.$$

Instead expressions (3.13) and (3.15), one can use the following simple mnemonic rule. The characteristic curves (3.21) and (3.23) satisfy the following quadratic equation:

$$a \left(\frac{dy}{dx} \right)^2 - 2b \frac{dy}{dx} + c = 0 \tag{3.25}$$

that can be written immediately while looking at Equation (3.1) (just do not forget to change the sign in the term with b). Below we shall see that this rule is useful also for other types of equations.

Example 3.1 The wave equation

$$u_{yy} - c_0^2 u_{xx} = 0 \tag{3.26}$$

is given. Let us transform it to the canonical form and find its general solution.

In this example, $a = -c_0^2$, $b = 0$, and $c = 1$; hence, as shown above, $\delta = b^2 - ac = c_0^2 > 0$. This, Equation (3.26) is of the hyperbolic type. The equation (3.25) for the characteristic curve is

$$-c_0^2 \left(\frac{dy}{dx} \right)^2 + 1 = 0,$$

hence

$$\frac{dy}{dx} = \pm \frac{1}{c_0}. \tag{3.27}$$

Solving (3.27) gives

$$s = x - c_0 y, \quad t = x + c_0 y. \tag{3.28}$$

Using formulas (3.8)-(3.11), we obtain equation

$$v_{st} = 0. \tag{3.29}$$

As shown in Example 1.4, the general solution of Equation (3.29) is the sum of two arbitrary functions (with continuous second derivatives); each of them depends only on one variable:

$$v(s, t) = F_+(s) + F_-(t).$$

Returning to original variables according to (3.28), we find:

$$u(x, y) = F_+(x - c_0 y) + F_-(x + c_0 y). \tag{3.30}$$

The physical meaning of Equation (3.26) and the obtained solution (3.30) will be discussed in Chapter 5.

3.2.2 Elliptic Equations

In the case of elliptic equations, $\delta = b^2 - ac < 0$, the canonical form is

$$v_{ss} + v_{tt} + G(s, t, v, v_s, v_t) = 0, \tag{3.31}$$

where G does not contain derivatives of the second order.

As the first step, let us apply exactly the same transformation as in the hyperbolic case. Consider the *complex* transformation

$$\tilde{s} = \tilde{s}(x, y), \quad \tilde{t} = \tilde{t}(x, y), \quad u(x, y) = \tilde{v}\left(\tilde{s}(x, y), \tilde{t}(x, y)\right), \tag{3.32}$$

determined by equations

$$\tilde{s}_x + \frac{b - i\sqrt{ac - b^2}}{a}\tilde{s}_y = 0, \tag{3.33}$$

$$\tilde{t}_x + \frac{b + i\sqrt{ac - b^2}}{a}\tilde{t}_y = 0. \tag{3.34}$$

Note that $\tilde{t}(x, y) = \bar{\tilde{s}}(x, y)$, where a bar is the sign of complex conjugation. Following the derivation given in Section 3.2.1, we obtain equation

$$\tilde{v}_{\tilde{s}\tilde{t}} + \tilde{G}\left(\tilde{s}, \tilde{t}, \tilde{v}, \tilde{v}_{\tilde{s}}, \tilde{v}_{\tilde{t}}\right) = 0. \tag{3.35}$$

At the second step, we introduce real variables

$$s = \tilde{s} + \tilde{t}, \quad t = \left(\tilde{s} - \tilde{t}\right)/i, \quad v(s, t) = \tilde{v}(\tilde{s}, \tilde{t}). \tag{3.36}$$

According to the chain rule,

$$\frac{\partial}{\partial \tilde{t}} = \frac{\partial}{\partial s} + i\frac{\partial}{\partial t}, \quad \frac{\partial}{\partial \tilde{s}} = \frac{\partial}{\partial s} - i\frac{\partial}{\partial t}, \tag{3.37}$$

hence we obtain Equation (3.31). Note that Equation (3.25) can be used in the case of an elliptic equation for finding characteristic curves in a complex space.

Example 3.2 Let us consider equation

$$au_{xx} + 2bu_{xy} + cu_{yy} = 0 \tag{3.38}$$

with constant coefficients.

Following the algorithm described above, we write the characteristic Equation (3.25) and obtain its solutions,

$$\frac{dy}{dx} = \frac{b \pm i\sqrt{ac - b^2}}{a}. \tag{3.39}$$

Integrating Equations (3.39), we find the transformation

$$\tilde{s} = ay - \left(b - i\sqrt{ac - b^2}\right)x, \quad \tilde{t} = ay - \left(b + i\sqrt{ac - b^2}\right)x, \tag{3.40}$$

which brings Equation (3.38) to the form

$$\tilde{v}_{\tilde{s}\tilde{t}} = 0; \tag{3.41}$$

thus, the general solution is given by the formula

$$\tilde{v}(\tilde{s}, \tilde{t}) = F_+(\tilde{s}) + F_-(\tilde{t}),$$

or

$$u(x, y) = F_+\left(ay - \left(b - i\sqrt{ac - b^2}\right)x\right) + F_-\left(ay - \left(b + i\sqrt{ac - b^2}\right)x\right), \tag{3.42}$$

where F_+ and F_- are arbitrary continuously differentiable functions.

Generally, solution (3.42) is complex, but it can be used for construction of the real general solution. Note that if $u_1(x, y)$ and $u_2(x, y)$ are solutions of a linear homogeneous equation (e.g., Equation (3.38)), then any linear combination of those solutions, $c_1u_1(x, y) + c_2u_2(x, y)$ is also a solution of that equation (that feature of solutions of linear homogeneous equations is called the *superposition principle*). Also, for equations with real coefficients, if

$u(x, y)$ is a solution, then the complex conjugate function, $\bar{u}_1(x, y)$, is a solution. Therefore, starting with (3.42), one can construct two *real* solutions of Equation (3.38),

$$u_1(x, y) = \operatorname{Re} u(x, y) = \frac{1}{2}\left[u(x, y) + \bar{u}(x, y)\right] \tag{3.43}$$

and

$$u_2(x, y) = \operatorname{Re} u(x, y) = \frac{1}{2i}\left[u(x, y) - \bar{u}(x, y)\right] \tag{3.44}$$

The set of linear combinations of (3.43) and (3.44) with real coefficients form the real general solution of (3.38).

According to formulas (3.36), the transformation

$$s = 2(ay - bx), \quad t = 2\sqrt{ac - b^2}x, \quad v(s, t) = \tilde{v}(\tilde{s}, \tilde{t})$$

leads to the canonical form

$$v_{ss} + v_{tt} = 0,$$

which is *the Laplace equation.*

3.2.3 Parabolic Equations

In the case of a parabolic equation, $\delta = b^2 - ac = 0$, the canonical form is

$$w_{ss} + G\left(s, t, w, w_s, w_t\right) = 0, \tag{3.45}$$

where G does not contain derivatives of the second order. This time, the characteristic equation (3.25) gives only one equation for a characteristic curve,

$$\frac{dy}{dx} = \frac{b}{a}, \tag{3.46}$$

which corresponds to the equation for the change of only one variable, say,

$$t = t(x, y); \tag{3.47}$$

that equation is

$$t_x + \frac{b}{a}t_y = 0. \tag{3.48}$$

As the second variable, let us choose

$$s = x. \tag{3.49}$$

According to (3.14), (3.15),

$$B = at_x + bt_y = 0,$$

$$C = at_x^2 + 2bt_xt_y + \frac{b^2}{a}t_y^2 = a\left(t_x + \frac{b^2}{a}t_y\right)^2 = 0,$$

hence the transformation (3.47), (3.49) brings Equation (3.1) to the canonical form (3.45).

Example 3.3 Let us transform equation

$$au_{xx} + 2bu_{xy} + cu_{yy} = 0, \quad b^2 - ac = 0 \tag{3.50}$$

with constant coefficients a, b and c to the canonical form, and find its general solution.

Following (3.46), we apply the transformation

$$t = ay - bx, \quad s = x$$

and obtain the canonical form of the equation,

$$v_{ss} = 0.$$

Its general solution is

$$v(s, t) = f_1(t)s + f_2(t),$$

hence

$$u(x, y) = x f_1(ay - bx) + f_2(ay - bx),$$

where f_1 and f_2 are arbitrary functions, which are twice continuously differentiable.

Problems

Transform the equation to the canonical form and find its general solution.

1. $u_{xx} + 2u_{xy} + u_x = 0$, $\quad -\infty < x < \infty$, $\quad -\infty < y - \infty$.

2. $u_{xx} - 4u_{xy} + 4u_{yy} = 1$, $\quad -\infty < x < \infty$, $\quad -\infty < y - \infty$.

3. $xu_{xx} + 2x^2 u_{xy} - u_x = 1$, $\quad 0 < x < \infty$, $\quad -\infty < y - \infty$.

4. $x^2 u_{xx} - 2xy u_{xy} + y^2 u_{yy} + x u_x + y u_y = 0$, $\quad 0 < x < \infty$, $\quad -\infty < y - \infty$.

Transform the equation to the canonical form.

5. $u_{xx} - x u_{yy} = 0$, $\quad 0 < x < \infty$, $\quad -\infty < y - \infty$.

6. $u_{xx} + x^2 u_{yy} = 0$, $\quad 0 < x < \infty$, $\quad -\infty < y - \infty$.

7. $y^2 u_{xx} - x^2 u_{yy} = 0$, $\quad 0 < x < \infty$, $\quad 0 < y < \infty$.

8. $\sin^2 x u_{xx} + \sin(2x) u_{xy} + \cos^2 x u_{yy} = 0$, $\quad 0 < x < \pi$, $\quad 0 < y < \infty$.

4

The Sturm-Liouville Problem

4.1 General Consideration

In the following chapters we carry out a systematic study of methods for solving second order linear PDEs. We will develop a powerful approach, the method of separation of variables (or Fourier expansion method), which is efficient for solving all three types of PDE equations. That method includes solution of a certain kind of problems for ordinary differential equations, the Sturm-Liouville problem, which is not always included in a standard course of ODEs. Therefore, we devote the present chapter to consideration of that problem.

The Sturm-Liouville problem consists of a linear, second order ordinary differential equation containing a parameter whose value is determined by the condition that there exists a nonzero solution to the equation which satisfies a given boundary conditions. The set of orthogonal functions generated by the solutions to such problems gives the base functions for the Fourier expansion method of solving partial differential equations.

Consider a general linear second order ordinary differential equation

$$a(x)y''(x) + b(x)y'(x) + c(x)y(x) + \lambda d(x)y(x) = 0 \tag{4.1}$$

with a *parameter* λ, generally a complex number, multiplied by the function $y(x)$, which is generally a complex function, while the functions $a(x)$, $b(x)$, $c(x)$ and $d(x)$ are real.

This equation can be written in the form

$$\frac{d}{dx}\left[p(x)y'(x)\right] + \left[q(x) + \lambda r(x)\right]y(x) = 0, \tag{4.2}$$

where

$$p(x) = e^{\int \frac{b(x)}{a(x)}dx}, \quad q(x) = \frac{p(x)c(x)}{a(x)}, \quad r(x) = \frac{p(x)d(x)}{a(x)}. \tag{4.3}$$

Reading Exercise: Verify that substitution of Equations (4.3) into Equation (4.2) gives Equation (4.1).

As we will see studying PDE, many physical problems result in the linear ordinary Equation (4.2) where the function $y(x)$ is defined on an interval $[a, b]$ and obeys *homogeneous boundary conditions* of the form

$$\begin{aligned}\alpha_1 y' + \beta_1 y|_{x=a} &= 0, \\ \alpha_2 y' + \beta_2 y|_{x=b} &= 0.\end{aligned} \tag{4.4}$$

This kind of condition, imposed in two different points, strongly differs from the set of initial conditions $y(a) = y_0$, $y'(a) = y_1$ imposed at the same point, which most often are considered in ODE problems.

It is clear that the constants α_1 and β_1 cannot both be zero simultaneously, nor the constants α_2 and β_2. The constants α_k and β_k, which are determined by physical laws, are real. We note also that the relative signs for α_k and β_k are not arbitrary; we generally must have $\beta_1/\alpha_1 < 0$ and $\beta_2/\alpha_2 > 0$. This choice of signs (details of which are discussed in books [7, 8]) is necessary in setting up boundary conditions for various classes of physical problems. The very rare cases when the signs are different occur in problems where there is explosive behavior, such as an exponential temperature increase. Everywhere in the book we consider "normal" physical situations in which processes occur smoothly and thus the parameters in boundary conditions are restricted as above.

Equations (4.2) and (4.4) define a Sturm-Liouville problem. Solving this problem involves determining the values of the constant λ for which nontrivial solutions $y(x)$ exist. If $\alpha_k = 0$ the boundary condition simplifies to $y = 0$ (known as the *Dirichlet boundary condition*), if $\beta_k = 0$ the boundary condition is $y' = 0$ (called the *Neumann boundary condition*), otherwise the boundary condition is referred to as a *mixed boundary condition*.

Notice that for the function $y(x)$ defined on an infinite or semi-infinite interval, the conditions (4.4) may not be specified and are often replaced by the condition of regularity or physically reasonable behavior as $x \to \pm\infty$, for example that $y(\infty)$ be finite.

Later on we let $p(x)$, $q(x)$, $r(x)$ and $p'(x)$ be continuous, real functions on the interval $[a, b]$ and let $p(x) > 0$ and $r(x) > 0$ on the interval $[a, b]$. The coefficients α_k and β_k in Equations (4.4) are assumed to be real and independent of λ.

The differential Equation (4.2) and boundary conditions (4.4) are *homogeneous* which is essential for the subsequent development. The trivial solution $y(x) = 0$ is always possible for homogeneous equations but we seek special values of λ (called *eigenvalues*) for which there are nontrivial solutions (called *eigenfunctions*) that depend on λ.

If we introduce the differential operator (called the *Sturm-Liouville operator*)

$$Ly(x) = -\frac{d}{dx}\left[p(x)y'(x)\right] - q(x)y(x) = -p(x)y''(x) - p'(x)y'(x) - q(x)y(x) \qquad (4.5)$$

or

$$L = -p(x)\frac{d^2}{dx^2} - p'(x)\frac{d}{dx} - q(x),$$

then Equation (4.2) becomes

$$Ly(x) = \lambda r(x)y(x). \qquad (4.6)$$

As it is seen from Equation (4.5), $Ly(x)$ is a real *linear operator*. When $r(x) = const.$ (in this case one can take $r(x) = 1$) this equation appears as an ordinary *eigenvalue problem*, $Ly(x) = \lambda y(x)$, for which we have to determine λ and $y(x)$. For $r(x) \neq 1$ we have a modified problem where the function $r(x)$ is called a *weight function*. As we stated above, the only requirement on $r(x)$ is that it is real and chosen to be nonnegative. Equations (4.2) and (4.6) are different ways to specify the same boundary value problem.

Now we discuss the properties of eigenvalues and eigenfunctions of a Sturm-Liouville problem. Let us write Equation (4.6) for two eigenfunctions, $y_n(x)$ and $y_m(x)$, and take the complex conjugate of the equation for $y_m(x)$. Notice that in spite of the fact that $p(x)$, $q(x)$, and $r(x)$ are real, we cannot assume from the very beginning that λ and $y(x)$ are real: that has to be checked. We have

$$Ly_n(x) = \lambda_n r(x)y_n(x)$$

and

$$Ly_m^*(x) = \lambda_m^* r(x)y_m^*(x).$$

Multiplying the first of these equations by $y_m^*(x)$ and the second by $y_n(x)$, we then integrate both from a to b and subtract the two results to obtain

$$\int_a^b y_m^*(x)Ly_n(x)dx - \int_a^b y_n(x)Ly_m^*(x)dx = (\lambda_n - \lambda_m^*)\int_a^b r(x)y_m^*(x)y_n(x)dx. \qquad (4.7)$$

Using the definition of L given by Equation (4.5), the left side of Equation (4.7) is

$$\left\{ p(x) \left[\frac{dy_m^*}{dx} y_n(x) - y_m^*(x) \frac{dy_n}{dx} \right] \right\}_a^b. \tag{4.8}$$

Reading Exercise: Verify the previous statement.

Then, using the boundary conditions (4.4), it can be easily proved that the expression (4.8) equals to zero.

Reading Exercise: Verify that the expression in Equation (4.8) equals to zero.

Thus, we are left with

$$\int_a^b y_m^*(x) L y_n(x) dx = \int_a^b y_n(x) L y_m^*(x) dx. \tag{4.9}$$

An operator, L, that satisfies Equation (4.9), is named a *Hermitian* or *self-adjoint operator*. Thus, we may say that *the Sturm-Liouville linear operator* satisfying homogeneous boundary conditions is *Hermitian*. Many important operators in physics, especially in quantum mechanics, are Hermitian.

Let us show that Hermitian operators have *real eigenvalues* and their eigenfunctions are *orthogonal*. The right side of Equation (4.7) gives

$$(\lambda_n - \lambda_m^*) \int_a^b r(x) y_m^*(x) y_n(x) dx = 0. \tag{4.10}$$

When $m = n$ the integral cannot be zero (recall that $r(x) > 0$); thus $\lambda_n^* = \lambda_n$ and we have proved that *the eigenvalues of a Sturm-Liouville problem are real*.

Then, for $\lambda_m \neq \lambda_n$, Equation (4.10) is

$$\int_a^b r(x) y_m^*(x) y_n(x) dx = 0 \tag{4.11}$$

and we conclude that *the eigenfunctions corresponding to different eigenvalues of a Sturm-Liouville problem are orthogonal (with the weight function $r(x)$).*

The squared *norm* of the eigenfunction $y_n(x)$ is defined to be

$$\|y_n\|^2 = \int_a^b r(x) |y_n(x)|^2 dx. \tag{4.12}$$

Note that the eigenfunctions of Hermitian operators always *can be chosen to be real*. This can be done by using some linear combinations of the functions $y_n(x)$, for example, choosing $\sin x$ and $\cos x$ instead of $\exp(\pm ix)$ for the solutions of the equation $y'' + y = 0$. Real eigenfunctions are more convenient to work with because it is easier to match them to boundary conditions which are intrinsically real since they represent physical restrictions.

The above proof fails if $\lambda_m = \lambda_n$ for some $m \neq n$ (in other words there exist different eigenfunctions belonging to the same eigenvalue) in which case we cannot conclude that the corresponding eigenfunctions, $y_m(x)$ and $y_n(x)$, are orthogonal (although in some cases they are). If there are f eigenfunctions that have the same eigenvalue, we have an *f-fold degeneracy* of the eigenvalue. In general, a degeneracy reflects a symmetry of the underlying physical system (examples can be found in the text). For a Hermitian operator it is always possible to construct linear combinations of the eigenfunctions belonging to the same eigenvalue so that these new functions are orthogonal.

If $p(a) \neq 0$ and $p(b) \neq 0$ then $p(x) > 0$ on the closed interval $[a, b]$ (which follows from $p(x) > 0$) and we have the so-called *regular Sturm-Liouville problem*. If $p(a) = 0$ then we do not impose the first of the boundary conditions in Equations (4.4), instead we require $y(x)$ and $y'(x)$ to be finite at $x = a$. Similar situations occur if $p(b) = 0$, or if both $p(a) = 0$ and $p(b) = 0$. All these cases correspond to the so-called *singular Sturm-Liouville problem*.

Beside problems with boundary conditions (4.4) imposed separately in points $x = a$ and $x = b$, we will also consider *the periodic Sturm-Liouville problem* with periodic boundary conditions $y(a) = y(b)$, $y'(a) = y'(b)$. (In that case, it is assumed also that $p(a) = p(b)$, $q(a) = q(b)$, and $r(a) = r(b)$.)

The following summarizes the types of Sturm-Liouville problems:

   i. For $p(x) > 0$ and $r(x) > 0$ we have the *regular* problem;

   ii. For $p(x) \geq 0$ and $r(x) \geq 0$ we have the *singular* problem;

   iii. For $p(a) = p(b)$ and $r(x) > 0$ we have the *periodic* problem.

Notice that if the interval (a, b) is infinite, the Sturm-Liouville problem is also considered as singular.

The following theorem gives a list of several important *properties of the Sturm-Liouville problem*:

Theorem 1

   i) Each regular and each periodic Sturm-Liouville problem has an infinite number of non-negative, discrete eigenvalues $0 \leq \lambda_1 < \lambda_2 < ... < \lambda_n < ...$ such that $\lambda_n \to \infty$ as $n \to \infty$. All eigenvalues are real numbers.

   ii) For each eigenvalue of a regular Sturm-Liouville problem there is only one eigenfunction; for a periodic Sturm-Liouville problem this property does not hold.

   iii) For each of the types of Sturm-Liouville problems the eigenfunctions corresponding to different eigenvalues are linearly independent.

   iv) For each of the types of Sturm-Liouville problems the set of eigenfunctions is orthogonal with respect to the weight function $r(x)$ on the interval $[a, b]$.

   v) If $q(x) \leq 0$ on $[a, b]$ and $\beta_1/\alpha_1 < 0$ and $\beta_2/\alpha_2 > 0$, then all $\lambda_n \geq 0$.

Some of these properties have been proven previously, such as property *iv)* and part of property *i)*. The remaining part of property *i)* will be shown in several examples below, as well as property *v)*. Property *ii)* can be easily proved when there are two eigenfunctions corresponding to the same eigenvalue. We then apply Equations (4.2) and (4.4) to show that these two eigenfunctions coincide or differ at most by some multiplicative constant. We leave this proof to the reader as a *Reading Exercise*. Similarly, can be proven property *iii)*.

Solutions of some important Sturm-Liouville problems will be considered in subsequent chapters. Many of them cannot be written using elementary functions. They form new classes of functions called special functions. Specifically, Bessel functions and the orthogonal polynomials, such as Legendre polynomials, arise from singular Sturm-Liouville problems; thus the first statement in the above theorem is not directly applicable to these important cases. In spite of that, singular Sturm-Liouville problems may also have an infinite sequence of discrete eigenvalues which we will later verify directly for Bessel functions and for the orthogonal polynomials (see Appendix B).

Since Equation (4.11) is satisfied, eigenfunctions $y_n(x)$ form a *complete orthogonal set* on $[a, b]$. This means that any reasonable well-behaved function, $f(x)$, defined on $[a, b]$ can be expressed as a series (called a *generalized Fourier series*) of eigenfunctions of a Sturm-Liouville problem in which case we may write

$$f(x) = \sum_{n}^{\infty} c_n y_n(x), \tag{4.13}$$

where it is convenient, in some cases, to start the summation with $n = 1$, in other cases with $n = 0$. An expression for the coefficients c_n can be found by multiplying both sides of Equation (4.13) by $r(x)y_n^*(x)$ and integrating over $[a, b]$ to give

$$c_n = \frac{\int_a^b r(x)f(x)y_n^*(x)dx}{\|y_n\|^2}. \tag{4.14}$$

Let us substitute (4.14) into (4.13). Changing the order of summation and integration (we can do it due to uniform convergence), we obtain:

$$f(x) = \sum_{n}^{\infty} \frac{1}{\|y_n\|^2} \int_a^b r(\xi)f(\xi)y_n(x)y_n^*(\xi)d\xi.$$

That means that

$$\sum_{n}^{\infty} \frac{1}{\|y_n\|^2} \int_a^b r(\xi)y_n(x)y_n^*(\xi)d\xi = \delta(x - \xi).$$

This is *the completeness property* for eigenfunctions of a Sturm-Liouville problem.

The Sturm-Liouville theory provides a theorem for convergence of the series in Equation (4.13) at every point x of $[a, b]$:

Theorem 2

Let $\{y_n(x)\}$ be the set of eigenfunctions of a regular Sturm-Liouville problem and let $f(x)$ and $f'(x)$ be piecewise continuous on a closed interval. Then the series expansion (4.13) converges to $f(x)$ at every point where $f(x)$ is continuous and to the value $[f(x_0 + 0) + f(x_0 - 0)]/2$ if x_0 is a point of discontinuity.

The theorem is also valid for the orthogonal polynomials and Bessel functions related to singular Sturm-Liouville problems. This theorem, which is extremely important for applications, is similar to the theorem for trigonometric Fourier series.

Formulas in Equations (4.11) through (4.14) can be written in a more convenient way if we define a *scalar product of* functions φ and ψ as the number given by

$$\varphi \cdot \psi = \int_a^b r(x)\varphi(x)\psi^*(x)dx \tag{4.15}$$

(the other notation of the scalar product is (φ, ψ)).

This definition of the scalar product has properties identical to those for vectors in linear Euclidian space, a result which can be easily proved:

$$\varphi \cdot \psi = (\psi \cdot \varphi)^*,$$
$$(a\varphi) \cdot \psi = a\varphi \cdot \psi, \qquad \text{(where } a \text{ is a number)}$$
$$\varphi \cdot (a\psi) = a^*\varphi \cdot \psi, \tag{4.16}$$
$$\varphi \cdot (a\psi + b\phi) = a^*\varphi \cdot \psi + b^*\varphi \cdot \phi,$$
$$\varphi \cdot \varphi = |\varphi|^2 \geq 0.$$

The last property relies on the assumption made for the Sturm-Liouville equation that $r(x) \geq 0$. If φ is continuous on $[a, b]$, then $\varphi \cdot \varphi = 0$ only if φ is zero.

Reading Exercise: Prove the relations given in Equations (4.16).

In terms of the scalar product, the orthogonality of eigenfunctions (defined by Equation (4.11)) means that

$$y_n \cdot y_m = 0, \quad \text{if} \quad n \neq m \tag{4.17}$$

and the formula for the Fourier coefficients in Equation (4.14) becomes

$$c_n = \frac{f \cdot y_n^*}{y_n \cdot y_n} \tag{4.18}$$

Functions satisfying the condition

$$\varphi \cdot \varphi = \int_a^b r(x) |\varphi(x)|^2 dx < \infty \tag{4.19}$$

belong to a *Hilbert space*, L^2, which is infinite dimensional. The complete orthogonal set of functions $\{y_n(x)\}$ serves as the orthogonal basis in L^2.

4.2   Examples of Sturm-Liouville Problems

Example 4.1 Solve the equation

$$y''(x) + \lambda y(x) = 0 \tag{4.20}$$

on the interval $[0, l]$ with boundary conditions

$$y(0) = 0 \quad \text{and} \quad y(l) = 0. \tag{4.21}$$

Solution. First, comparing Equation (4.20) with Equations (4.5) and (4.6), it is clear that we have a Sturm-Liouville problem with linear operator $L = -d^2/dx^2$, i.e. functions $q(x) = 0$ and $p(x) = r(x) = 1$. As a *Reading Exercise*, verify that L is Hermitian.

Let us discuss the cases $\lambda = 0$, $\lambda < 0$ and $\lambda > 0$ separately. If $\lambda = 0$, then a general solution to Equation (4.20) is

$$y(x) = C_1 x + C_2$$

and from boundary conditions (4.21) we have $C_1 = C_2 = 0$, i.e. there exists only the trivial solution $y(x) = 0$. If $\lambda < 0$, then

$$y(x) = C_1 e^{\sqrt{-\lambda} x} + C_2 e^{-\sqrt{-\lambda} x}$$

and the boundary conditions (4.21) again give $C_1 = C_2 = 0$ and therefore the trivial solution $y(x) = 0$. Thus we have only the possibility $\lambda > 0$, in which case we write $\lambda = \mu^2$ with μ real and we have a general solution of Equation (4.20) given by

$$y(x) = C_1 \sin \mu x + C_2 \cos \mu x.$$

The boundary condition $y(0) = 0$ requires that $C_2 = 0$ and the boundary condition $y(l) = 0$ gives $C_1 \sin \mu l = 0$. From this we must have $\sin \mu l = 0$ and $\mu_n = \frac{n\pi}{l}$ since the choice $C_1 = 0$ again gives the trivial solution. Thus, the eigenvalues are

$$\lambda_n = \mu_n^2 = \left(\frac{n\pi}{l}\right)^2, \quad n = 1, 2, \ldots \tag{4.22}$$

and the eigenfunctions are $y_n(x) = C_n \sin \frac{n\pi x}{l}$, where for $n = 0$ we have the trivial solution $y_0(x) = 0$. It is obvious that we can restrict ourselves to positive values of n since negative values do not give new solutions. These eigenfunctions are orthogonal over the interval $[0, l]$ since we can easily show that

$$\int_0^l \sin \frac{n\pi x}{l} \sin \frac{m\pi x}{l} dx = 0 \quad \text{for} \quad m \neq n. \tag{4.23}$$

The orthogonality of eigenfunctions follows from the fact that the Sturm-Liouville operator, L, is Hermitian for the boundary conditions given in Equation (4.21).

The eigenfunctions may be normalized by writing

$$C_n^2 \int_0^1 \sin^2 \frac{n\pi x}{l} dx = C_n^2 \cdot \frac{l}{2} = 1,$$

which results in the orthonormal eigenfunctions

$$y_n(x) = \sqrt{\frac{2}{l}} \sin n\pi x, \quad n = 1, 2, \ldots. \tag{4.24}$$

Thus, we have shown that the boundary value problem consisting of Equations (4.20) and (4.21) has eigenfunctions that are *sine* functions. It means that *the expansion in eigenfunctions of the Sturm-Liouville problem for solutions to Equations (4.20) and (4.21) is equivalent to the trigonometric Fourier sine series.*

Reading Exercise: Suggest alternatives to boundary conditions (4.21) which will result in cosine functions as the eigenfunctions for Equation (4.20).

Example 4.2 Determine the eigenvalues and corresponding eigenfunctions for the Sturm-Liouville problem

$$y''(x) + \lambda y(x) = 0, \tag{4.25}$$

$$y'(0) = 0, \quad y(l) = 0. \tag{4.26}$$

Solution. As in the previous example, the reader may check as a *Reading Exercise* that the parameter λ must be positive in order to have nontrivial solutions. Thus, we may write $\lambda = \mu^2$, so that we have oscillating solutions given by

$$y(x) = C_1 \sin \mu x + C_2 \cos \mu x.$$

The boundary condition $y'(0) = 0$ gives $C_1 = 0$ and the boundary condition $y(l) = 0$ gives $C_2 \cos \mu l = 0$. If $C_2 = 0$ we have a trivial solution; otherwise we have $\mu_n = (2n+1)\pi/2l$, for $n = 0, 1, 2, \ldots$. Therefore, the eigenvalues are

$$\lambda_n = \mu_n^2 = \left[\frac{(2n-1)\pi}{2l}\right]^2, \quad n = 1, 2, \ldots \tag{4.27}$$

and the eigenfunctions are

$$y_n(x) = C_n \cos \frac{(2n-1)\pi x}{2l}, \quad n = 1, 2, \ldots, \quad n = 1, 2, \ldots. \tag{4.28}$$

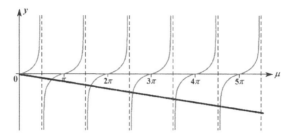

FIGURE 4.1
The functions $\tan \mu_n l$ and $-\mu_n/h$ (for $h = 5$) plotted against μ. The eigenvalues of the Sturm-Liouville problem in Example 4.3 are given by the intersections of these lines.

We leave it to the reader to prove that the eigenfunctions in Equation (4.28) are orthogonal on the interval $[0, l]$. The reader may also normalize these eigenfunctions to find the normalization constant C_n which is equal to $\sqrt{2/l}$.

Example 4.3 Determine the eigenvalues and eigenfunctions for the Sturm-Liouville problem

$$y''(x) + \lambda y(x) = 0, \tag{4.29}$$

$$y(0) = 0, \quad y'(l) + h y(l) = 0. \tag{4.30}$$

Solution. As in the previous examples, nontrivial solutions exist only when $\lambda > 0$ (the reader should verify this as a *Reading Exercise*). Letting $\lambda = \mu^2$ we obtain a general solution as

$$y(x) = C_1 \sin \mu x + C_2 \cos \mu x.$$

From the boundary condition $y(0) = 0$ we have $C_2 = 0$. The other boundary condition gives $\mu \cos \mu l + h \sin \mu l = 0$. Thus, the eigenvalues are given by the equation

$$\tan \mu_n l = -\mu_n/h. \tag{4.31}$$

We can obtain these eigenvalues by plotting $\tan \mu_n l$ and $-\mu_n/h$ on the same graph as in Figure 4.1. The graph is plotted for positive μ, because negative μ do not bring new solutions.

From the figure it is directly seen that there is an infinite number of discrete eigenvalues. The eigenfunctions

$$y_n(x) = C_n \sin \mu_n x, \quad n = 1, 2, \ldots \tag{4.32}$$

are orthogonal so that

$$\int_0^l \sin \mu_n x \cdot \sin \mu_m x\, dx = 0 \quad \text{for} \quad m \neq n. \tag{4.33}$$

The orthogonality condition shown in Equation (4.33) follows from the general theory as a direct consequence of the fact that the operator $L = -\frac{d^2}{dx^2}$ is Hermitian for the boundary conditions (4.30). We leave it to the reader to verify the previous statement as a *Reading Exercise*.

The normalized eigenfunctions are

$$y_n(x) = \sqrt{\frac{2(\mu_n^2 + h^2)}{l(\mu_n^2 + h^2) + h}} \sin \mu_n x. \tag{4.34}$$

Example 4.4 Solve the Sturm-Liouville problem

$$y'' + \lambda y = 0, \quad 0 < x < l, \tag{4.35}$$

on the interval $[0, l]$ with periodic boundary conditions

$$y(0) = y(l), \quad y'(0) = y'(l). \tag{4.36}$$

Solution. Again, verify as a *Reading Exercise* that nontrivial solutions exist only when $\lambda > 0$ (for which we will have oscillating solutions as before). Letting $\lambda = \mu^2$ we can write a general solution in the form

$$y(x) = C_1 \cos \mu x + C_2 \sin \mu x.$$

The boundary conditions in Equations (4.36) give

$$\begin{cases} C_1(\cos \mu l - 1) + C_2 \sin \mu l = 0, \\ -C_1 \sin \mu l + C_2(\cos \mu l - 1) = 0. \end{cases} \tag{4.37}$$

This system of homogeneous algebraic equations for C_1 and C_2 has a nontrivial solution only when its determinant is equal to zero:

$$\begin{vmatrix} \cos \mu l - 1 & \sin \mu l \\ -\sin \mu l & \cos \mu l - 1 \end{vmatrix} = 0, \tag{4.38}$$

which yields

$$\cos \mu l = 1. \tag{4.39}$$

The roots of Equation (4.39) are

$$\lambda_n = \left(\frac{2\pi n}{l}\right)^2, \quad n = 0, 1, 2, \ldots \tag{4.40}$$

With these values of λ_n, Equations (4.37) for C_1 and C_2 have two linearly independent nontrivial solutions given by

$$C_1 = \begin{pmatrix} 1 \\ 0 \end{pmatrix} \quad \text{and} \quad C_2 = \begin{pmatrix} 0 \\ 1 \end{pmatrix}. \tag{4.41}$$

Substituting each set into the general solution we obtain the eigenfunctions

$$y_n^{(1)}(x) = \cos \sqrt{\lambda_n} x \quad \text{and} \quad y_n^{(2)}(x) = \sin \sqrt{\lambda_n} x. \tag{4.42}$$

Therefore for the eigenvalue $\lambda_0 = 0$ we have the eigenfunction $y_0(x) = 1$ (and a trivial solution $y(x) \equiv 0$). Each nonzero eigenvalue λ_n has two linearly independent eigenfunctions so that for this example we have two-fold degeneracy.

Collecting the above results we have that this boundary value problem with periodic boundary conditions has the following eigenvalues and eigenfunctions:

$$\lambda_n = \left(\frac{2\pi n}{l}\right)^2, \quad n = 0, 1, 2, \ldots \tag{4.43}$$

$$y_0(x) = 1, \quad y_n(x) = \begin{cases} \cos(2\pi n x/l), \\ \sin(2\pi n x/l), \end{cases} \tag{4.44}$$

$$\|y_0\|^2 = l, \quad \|y_n\|^2 = \frac{l}{2} \quad (\text{for} \quad n = 1, 2, \ldots).$$

In particular, when $l = 2\pi$ we have

$$\lambda_n = n^2, \quad y_0(x) \equiv 1, \quad y_n(x) = \begin{cases} \cos nx, \\ \sin nx. \end{cases}$$

From this we see that the boundary value problem consisting of Equations (4.35) and (4.36) results in eigenfunctions for this Sturm-Liouville problem which allows *an expansion of the solution equivalent to the trigonometric Fourier series expansion.*

Example 4.5 In the book we meet a number of two-dimensional Sturm-Liouville problems. Here we present a simple example.

Consider the equation

$$\frac{\partial^2 u}{\partial x^2} + \frac{\partial^2 u}{\partial y^2} + k^2 u = 0, \tag{4.45}$$

where k is a real constant which determines the function $u(x, y)$ with independent variables in domain $0 \leq x \leq l, 0 \leq y \leq h$. Define the two-dimensional Sturm-Liouville operator in a similar fashion as was done for the one-dimensional case.

Solution. We have

$$L = -\frac{d^2}{dx^2} - \frac{d^2}{dy^2}. \tag{4.46}$$

Let the boundary conditions be Dirichlet type so that we have

$$u(0, y) = u(l, y) = u(x, 0) = u(x, h) = 0. \tag{4.47}$$

Reading Exercise: By direct substitution into Equation (4.45) and using boundary conditions (4.47) check that this Sturm-Liouville problem has the eigenvalues

$$k_{nm}^2 = \frac{1}{\pi^2} \left(\frac{n^2}{l^2} + \frac{m^2}{h^2} \right) \tag{4.48}$$

and the corresponding eigenfunctions

$$u_{nm} = \sin\frac{n\pi x}{l} \sin\frac{m\pi y}{h}, \quad n, m = 1, 2, \ldots \tag{4.49}$$

In the case of a square domain, $l = h$, the eigenfunctions u_{nm} and u_{mn} have the same eigenvalues, $k_{nm} = k_{mn}$, which is a degeneracy reflecting *the symmetry* of the problem with respect to x and y.

Example 4.6 Obtain the expansion of the function $f(x) = x^2(1 - x)$ using the eigenfunctions of the Sturm-Liouville problem

$$y'' + \lambda y = 0, \quad 0 \leq x \leq \pi/2, \tag{4.50}$$

$$y'(0) = y'(\pi/2) = 0. \tag{4.51}$$

Solution. First, prove as a *Reading Exercise* that the eigenvalues and eigenfunctions of this boundary value problem are

$$\lambda = 4n^2, \quad y_n(x) = \cos 2nx, \quad n = 0, 1, 2, \ldots \tag{4.52}$$

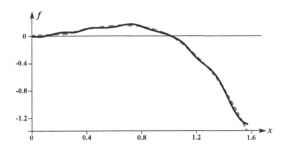

FIGURE 4.2
Graphs of the function $f(x) = x^2(1 - x)$ (dashed line) and partial sum with $n = 10$ of the Fourier expansion of $f(x)$ (solid line).

A Fourier series expansion, given in Equation (4.13), of the function $f(x)$ using the eigenfunctions above is

$$x^2(1 - x) = \sum_{n=0}^{\infty} c_n y_n(x) = \sum_{n=0}^{\infty} c_n \cos 2nx. \tag{4.53}$$

Since f and f' are continuous functions, this expansion will converge to $x^2(1 - x)$ for $0 < x < \pi/2$ as was stated previously. In Equation (4.50) we see that the function $r(x) = 1$; thus the coefficients of this expansion obtained from Equation (4.14) are

$$c_0 = \frac{\int_0^{\pi/2} x^2(1 - x)dx}{\int_0^{\pi/2} dx} = \frac{\pi^2}{4}\left(\frac{1}{3} - \frac{\pi}{8}\right),$$

$$c_n = \frac{\int_0^{\pi/2} x^2(1 - x) \cos 2nx \, dx}{\int_0^{\pi/2} \cos^2 2nx \, dx}$$

$$= \frac{(-1)^n}{n^4}\left[1 - \frac{3\pi}{4} + \frac{3}{2\pi n^2}\right] - \frac{3}{2\pi n^4}, \quad n = 1, 2, 3, \ldots$$

Figure 4.2 shows the partial sum ($n = 10$) of this series, compared with the original function $f(x) = x^2(1 - x)$.

Two important *special functions*, Legendre and Bessel functions, are discussed in Appendix B. In the two following examples they serve simply as illustrations of Sturm-Liouville problems.

Example 4.7 (*Fourier-Legendre Series*).
The Legendre equation is

$$\frac{d}{dx}\left[\left(1 - x^2\right)y'\right] + \lambda y = 0 \tag{4.54}$$

for x on the closed interval $[-1, 1]$. There are no boundary conditions in a straight form because $p(x) = 1 - x^2$ vanishes at the endpoints. However, we seek a finite solution, a condition which in this case acts as a boundary condition.

Solution. The Legendre polynomials, $P_n(x)$, are the only solutions of Legendre's equation that are bounded on the closed interval $[-1, 1]$. The set of functions $\{P_n(x)\}$, where $n = 0, 1, 2, \ldots$, is orthogonal with respect to the weight function $r(x) = 1$ on the interval $[-1, 1]$

in which case the orthogonality relation is

$$\int_{-1}^{1} P_n(x)P_m(x)dx = 0 \quad \text{for} \quad m \neq n. \tag{4.55}$$

The eigenfunctions for this problem are thus $P_n(x)$ with eigenvalues $\lambda = n(n+1)$ for $n = 0, 1, 2, \ldots$ (see Appendix B).

If $f(x)$ is a piecewise smooth function on $[-1, 1]$, the series

$$\sum_{n=0}^{\infty} c_n P_n(x) \tag{4.56}$$

converges to

$$\frac{1}{2}[f(x_0 + 0) + f(x_0 - 0)] \tag{4.57}$$

at any point x_0 on $(-1, 1)$. Because $r(x) = 1$ in Equation (4.14) the coefficients c_n are

$$c_n = \frac{\int_{-1}^{1} f(x)P_n(x)dx}{\int_{-1}^{1} P_n^2(x)dx}, \tag{4.58}$$

or written in terms of the scalar product,

$$c_n = \frac{f(x) \cdot P_n(x)}{P_n(x) \cdot P_n(x)}. \tag{4.59}$$

Example 4.8 (*Fourier-Bessel Series*).

Consider the Sturm-Liouville problem

$$y'' + \frac{y'}{x} + \left(\lambda - \frac{\nu^2}{x^2}\right)y = 0, \quad 0 \leq x \leq 1 \tag{4.60}$$

with boundary conditions such that $y(0)$ is finite and $y(1) = 0$. Here ν is a constant.

Solution. The eigenvalues for this problem are $\lambda = j_n^2$ for $n = 1, 2, \ldots$, where j_1, j_2, j_3, ... are the positive zeros of the functions $J_\nu(x)$ which are Bessel functions of order ν (see Appendix B). If $f(x)$ is a piecewise smooth function on the interval $[0, 1]$, then for $0 < x < 1$ it can be resolved in the series

$$\sum_{n=1}^{\infty} c_n J_\nu(j_n x), \tag{4.61}$$

which converges to

$$\frac{1}{2}[f(x_0 + 0) + f(x_0 - 0)]. \tag{4.62}$$

Since, in this Sturm-Liouville problem, $r(x) = x$, the coefficients c_n are

$$c_n = \frac{\int_0^1 xf(x)J_\nu(j_n x)dx}{\int_0^1 xJ_\nu^2(j_n x)dx}, \tag{4.63}$$

or in terms of the scalar product,

$$c_n = \frac{f(x) \cdot J_\nu(j_n x)}{J_\nu(j_n x) \cdot J_\nu(j_n x)}. \tag{4.64}$$

Problems

In problems 1 through 7, find eigenvalues and eigenfunctions of the Sturm-Liouville problem for the equation

$$y''(x) + \lambda y(x) = 0$$

with the following boundary conditions:

1. $y(0) = 0$, $\quad y'(l) = 0$;
2. $y'(0) = 0$, $\quad y(\pi) = 0$;
3. $y'(0) = 0$, $\quad y'(1) = 0$;
4. $y'(0) = 0$, $\quad y'(1) + y(1) = 0$;
5. $y'(0) + y(0) = 0$, $\quad y(\pi) = 0$;
6. $y(-l) = y(l)$, $\quad y'(-l) = y'(l)$ (periodic boundary conditions). As noted above, if the boundary conditions are periodic then the eigenvalues can be degenerate. Show that in this problem two linearly independent eigenfunctions exist for each eigenvalue.
7. For the operator $L = -d^2/dx^2$ acting on functions $y(x)$ defined on the interval $[0,1]$, find its eigenvalues and eigenfunctions (assume $r(x) = 1$):

   (a) $y(0) = 0$, $\quad y'(1) + y(1) = 0$;
   (b) $y'(0) - y(0) = 0$, $\quad y(1) = 0$;
   (c) $y(0) = y(1)$, $\quad y'(0) = y'(1)$.

5

One-Dimensional Hyperbolic Equations

5.1   Wave Equation

In Chapter 3 (Example 3.1) we introduced the wave equation as an example of hyperbolic PDE and obtained its general solution. In the present chapter, we study that equation in more detail. Let us start with the following physical example. Consider the problem of small transverse oscillations of a thin, stretched string. The transverse oscillations mean that the movement of each point of the string is perpendicular to the x axis with no displacements or velocities along this axis. Let $u(x, t)$ represent displacements of the points of the string from the equilibrium at location x and time t (u plays the role of the y coordinate, see Figure 5.1). Small oscillations mean that the displacement amplitude $u(x, t)$ is small relative to the string length, and what is important is our assumption that the partial derivative $u_x(x, t)$ is small for all values of x and t (i.e. the slope is small everywhere during the string's motion), and its second power can be neglected: $(u_x)^2 \ll 1$ (u_x has no dimension). With these assumptions which are justified in many applications, the equations that we will derive will be *linear* partial differential equations.

Consider an interval $(x, x + \Delta x)$ in Figure 5.1. Points on the string move perpendicularly to the x direction; thus the sum of the x components of the tension forces at points x and $x + \Delta x$ equals zero. Because the tension forces are directed along tangent lines, we have:

$$-T(x) \cos \alpha(x) + T(x + \Delta x) \cos \alpha(x + \Delta x) = 0.$$

Clearly, $\tan \alpha = u_x$ and for small oscillations $\cos \alpha = 1/\sqrt{1 + \tan^2 \alpha} = 1/\sqrt{1 + u_x^2} \approx 1$; thus

$$T(x) \approx T(x + \Delta x),$$

that is, the value of tension T does not depend on x, and for all x and t it approximately equals its value in the equilibrium state.

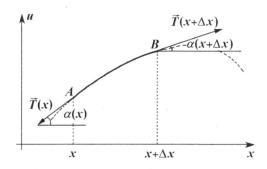

FIGURE 5.1
Small oscillations of a string.

In the situation shown in Figure 5.1, at point x the vertical component of force of tension is

$$T_{vert} = -T \sin \alpha(x).$$

The same expression with a positive sign holds at the point $x + \Delta x$: $T_{vert} = T \sin \alpha(x + \Delta x)$. The signs of T_{vert} at x and $x + \Delta x$ depend on the orientation of the string segment and are opposite for the two ends of the segment.

For small oscillations $\sin \alpha = \tan \alpha / \sqrt{1 + \tan^2 \alpha} = u_x / \sqrt{1 + u_x^2} \approx u_x$, so that

$$T_{vert} = -T u_x(x,t) \tag{5.1}$$

and the sum of vertical components of tension forces at points x and $x + \Delta x$ is

$$T_{vert}^{net} = T \left[\sin \alpha(x + \Delta x) - \sin \alpha(x) \right] = T \left[u_x(x + \Delta x, t) - u_x(x,t) \right].$$

As $\Delta x \to 0$ we arrive at

$$T_{vert}^{net} = T \frac{\partial^2 u}{\partial x^2} dx.$$

On the other hand, the force T_{vert}^{net} acting on segment Δx is equal to the mass of this segment, $\rho(x) dx$ (where $\rho(x)$ is a linear mass density of the string) times acceleration:

$$T_{vert}^{net} = \rho(x) \frac{\partial^2 u}{\partial t^2} dx.$$

If there is also an additional external force $F(x,t)$ per unit length acting on the string perpendicular to the x axis (for small oscillations the force should be small with respect to tension, T), we obtain the equation for *forced transverse oscillations of a string*:

$$\rho(x) \frac{\partial^2 u}{\partial t^2} = T \frac{\partial^2 u}{\partial x^2} + F(x,t). \tag{5.2}$$

For the case of a constant mass density, $\rho = \text{const}$, i.e. for a uniform string, this equation can be written as

$$\frac{\partial^2 u}{\partial t^2} = a^2 \frac{\partial^2 u}{\partial x^2} + f(x,t), \tag{5.3}$$

where $a = \sqrt{T/\rho} = \text{const}$, $f(x,t) = F(x,t)/\rho$. For instance, if the weight of the string cannot be neglected and the gravity force is directed down perpendicularly to the x axis, we have $f(x,t) = -mg/l\rho = -g$. If there is no external force, $F(x,t) \equiv 0$, we have the equation for free oscillations of a string

$$\frac{\partial^2 u}{\partial t^2} = a^2 \frac{\partial^2 u}{\partial x^2} \tag{5.4}$$

which is referred to as the *homogeneous wave equation*; Equation (5.3) is called the *nonhomogeneous wave equation*. With subscripts for derivatives, Equation (5.4) is

$$u_{tt}(x,t) = a^2 u_{xx}(x,t).$$

Equation (5.4) was considered in Section 3.2.1 (Example 3.1), up to a change of notations. We have found (see Equation (3.30)) that the general solution of that equation is the sum of two solutions,

$$u(x,t) = f_1(x - at) \quad \text{and} \quad u(x,t) = f_2(x + at),$$

where f_1 and f_2 are arbitrary, twice differentiable functions. Each of these solutions has a simple physical interpretation. In the first case, the displacement $u = f_1$ at point x and time t is the same as that at point $x + a\Delta t$ at time $t + \Delta t$. Thus, the disturbance moves in the direction of increasing x with velocity a. The quantity a is therefore the "velocity of propagation" of the disturbance, or the *wave speed*. In the second case, the displacement $u = f_2$ at point x at time t is found at the point with coordinate $x - a\Delta t$ at a later time $t + \Delta t$. This disturbance therefore travels in the direction of decreasing x with velocity a. The general solution of Equation (5.4) can be written as the sum of f_1 and f_2:

$$u(x,t) = f_1(x - at) + f_2(x + at). \tag{5.5}$$

This solution is considered with more details in the section devoted to the method of D'Alembert.

Equation (5.4) describes the simplest situation with no external forces (including string's weight) and no dissipation. For a string vibrating in an elastic medium when the force on the string from the medium is proportional to the string's deflection, $F = -\alpha u$ (that is Hooke's law; α is a coefficient with the dimension of force per length squared, F is a force per unit length) we have the wave equation in the form

$$\rho \frac{\partial^2 u}{\partial t^2} = T \frac{\partial^2 u}{\partial x^2} - \alpha u. \tag{5.6}$$

When a string oscillates in a medium with force of friction proportional to the speed, the force per unit length, F, is given by $F = -ku_t$, where k is a coefficient of friction. In that case, the equation contains the time derivative $u_t(x,t)$:

$$\frac{\partial^2 u}{\partial t^2} = a^2 \frac{\partial^2 u}{\partial x^2} - 2\kappa \frac{\partial u}{\partial t}, \tag{5.7}$$

where $2\kappa = k/\rho$ (κ has the dimension of inverse time).

All Equations (5.4), (5.6), and (5.7) are hyperbolic type (as we discussed in Chapter 3).

Next, consider the homogeneous wave equation with constant coefficients in the general form

$$u_{tt} - a^2 u_{xx} + b_1 u_t + b_2 u_x + cu = 0. \tag{5.8}$$

If we introduce a new function, $v(x,t)$, using the substitution

$$u(x,t) = e^{\lambda x + \mu t} v(x,t), \tag{5.9}$$

with $\lambda = b_1/2$ and $\mu = b_2/2$ Equation (5.8) reduces to a substantially simpler form

$$v_{tt} - a^2 v_{xx} + cv = 0. \tag{5.10}$$

Exercise: Prove the statement above.

It is immediately seen that if $c \neq 0$ Equation (5.10) does not allow a solution in the form $f(x \pm at)$. Physically this result is related to the phenomena of wave dispersion which will be discussed in Section 5.12.

5.2 Boundary and Initial Conditions

As we have seen in Chapter 2, where first order PDEs were considered, for a problem described by a differential equation additional conditions are needed to find a particular solution describing the behavior of the system. These additional conditions are determined by the nature of the system and should obey the following demands:

i) They should guarantee the *uniqueness* of the solution, i.e. there should not be two different functions satisfying the equation and additional conditions;

ii) They should guarantee the *stability* of the solution, i.e. any small variation of these additional conditions or the coefficients of the differential equation results in only insignificant variations in the solution. In other words, the solution should depend *continuously* on these additional conditions and the coefficients of the equation.

These additional conditions may be categorized as two distinct types: initial conditions and boundary conditions.

Initial conditions characterize the function satisfying the equation at the initial moment $t = 0$. Equations that are of the second order in time have two initial conditions. For example, in the problem of transverse oscillations of a string, the initial conditions define the string's shape and speed distribution at zero time:

$$u(x,0) = \varphi(x) \quad \text{and} \quad \frac{\partial u}{\partial t}(x,0) = \psi(x), \tag{5.11}$$

where $\varphi(x)$ and $\psi(x)$ are given functions of x.

Boundary conditions characterize the behavior of the function satisfying the equation at the boundary of the physical region of interest for all moments of time t. In most cases, the boundary conditions for partial differential equations give the function $u(x,t)$ or $u_x(x,t)$, or their combination at the boundary.

Let us consider various boundary conditions for transverse oscillations of a string over the finite interval $0 \le x \le l$ from a physical point of view.

1. If the left end of the string, located at $x = 0$, is rigidly fixed, the boundary condition at $x = 0$ is

$$u(0,t) = 0. \tag{5.12}$$

    A similar condition exists for the right end at $x = l$, if it is fixed. These are called *fixed end* boundary conditions.

2. If the motion of the left end of the string is driven with the function $g(t)$, then

$$u(0,t) = g(t) \tag{5.13}$$

    in which case we have *driven end* boundary conditions.

3. If the end at $x = 0$ can move and experiences a force $f(t)$ perpendicular to the x axis (e.g., a string is attached to a ring which is driven up and down on a vertical rod), then from Equation (5.1) we have

$$-Tu_x(0,t) = f(t). \tag{5.14}$$

    If this boundary condition is applied to the right end of the string at $x = l$, the left-hand side of this formula will have a positive sign. These are called *forced end* boundary conditions.

4. If the end at $x = 0$ moves freely, but is still attached (e.g., a string is attached to a ring which can slide up and down on a vertical rod with no friction), then the slope at the end will be zero. In this case, the last equation gives

$$u_x(0,t) = 0 \tag{5.15}$$

    with similar equations for the right end. The conditions in this case are called *free end* boundary conditions.

5. If the left end is attached to a surface that can stretch, we have an *elastic boundary*, in which case a vertical component of elastic force is $T_{vert} = -\alpha u(0, t)$. Together with Equation (5.1) this results in the boundary condition

$$u_x(0, t) - hu(0, t) = 0, \quad h = \alpha/T > 0. \tag{5.16}$$

For the right end we have

$$u_x(l, t) + hu(l, t) = 0. \tag{5.17}$$

If the point to which the string is elastically attached is also moving and its deviation from the initial position is described by the function $g(t)$, replacing $u(0, t)$ in Equation (5.17) by $u(0, t) - g(t)$ leads to the boundary condition

$$u_x(0, t) - h\left[u(0, t) - g(t)\right] = 0. \tag{5.18}$$

It can be seen that for stiff attachment (large h) when even a small shift of the end causes strong tension ($h \to \infty$), the boundary condition (5.18) becomes $u(0, t) = g(t)$. For weak attachment ($h \to 0$, weak tensions), this condition (5.18) becomes the condition for a free end, $u_x(0, t) = 0$.

In general, for one-dimensional problems, the boundary conditions at the ends $x = 0$ and $x = l$ can be summarized in the form

$$\alpha_1 u_x + \beta_1 u|_{x=0} = g_1(t), \quad \alpha_2 u_x + \beta_2 u|_{x=l} = g_2(t), \tag{5.19}$$

where $g_1(t)$ and $g_2(t)$ are given functions, and α_1, β_1, α_2, β_2 are real constants. As is discussed in Chapter 4, due to physical constraints the normal restrictions on these constants are $\beta_1/\alpha_1 < 0$ and $\beta_2/\alpha_2 > 0$.

When functions on the right-hand sides of Equation (5.19) are zero (i.e. $g_{1,2}(t) \equiv 0$), the boundary conditions are said to be *homogeneous*. In this case, if $u_1(x, t), u_2(x, t), \ldots, u_n(x, t)$ satisfy these boundary conditions, then any linear combination of these functions

$$C_1 u_1(x, t) + C_2 u_2(x, t) + \ldots + C_n u_n(x, t)$$

(where $C_1, \ldots, C_n$ are constants) also satisfies these conditions. This property will be used frequently in the following discussion.

We may classify the above physical notions of boundary conditions as formally belonging to one of three main types:

1. *Boundary conditions of the first kind (Dirichlet boundary condition).* For this case we are given $u|_{x=A} = g(t)$, where here and below $A = 0$ or l. This describes a given boundary regime; for example, if $g(t) \equiv 0$ we have fixed ends.

2. *Boundary conditions of the second kind (Neumann boundary condition).* In this case, we are given $u_x|_{x=A} = g(t)$, which describes a given force acting at the ends of the string; for example, if $g(t) \equiv 0$ we have free ends;

3. *Boundary conditions of the third kind (mixed boundary condition).* Here we have $u_x \pm hu|_{x=A} = g(t)$ (minus sign for $A = 0$, plus sign for $A = l$); for example, an elastic attachment for the case $h = $ const.

Applying these three conditions alternately to the two ends of the string results in nine types of boundary problems. A list classifying all possible combinations of boundary conditions can be found in Appendix C, part 1.

As mentioned previously, the initial and boundary conditions completely determine the solution of the wave equation. It can be proved that under certain conditions of smoothness of the functions $\varphi(x)$, $\psi(x)$, $g_1(t)$, and $g_2(t)$ defined in Equations (5.11) and (5.19), a unique solution always exists. The following sections investigate many examples of the dependence of the solutions on the boundary conditions.

In some physical situations, either the initial conditions or the boundary conditions may be ignored, leaving only one condition to determine the solution. For instance, suppose the point M_0 is rather distant from the boundary and the boundary conditions are given such that the influence of these conditions at M_0 is exposed after a rather long time interval. In such cases, if we investigate the situation for a relatively short time interval, one can ignore the boundaries and study the *initial value problem* (or *the Cauchy problem*). These solutions for an infinite region can be used for a finite one for times short enough that the boundary conditions have not had time to have an effect. For short time periods, we may ignore the boundary conditions and search for the solution of the equation

$$u_{tt} = a^2 u_{xx} + f(x,t) \quad \text{for} \quad -\infty < x < \infty, \ t > 0$$

with the initial conditions

$$\left. \begin{array}{l} u(x,0) = \varphi(x) \\ u_t(x,0) = \psi(x) \end{array} \right\} \quad \text{for} \quad -\infty < x < \infty.$$

Similarly, if we study a process close enough to one boundary (at one end for the one-dimensional case) and rather far from the other boundary for some characteristic time of that process the boundary condition at the distant end may be insignificant. For the one-dimensional case, we arrive at a boundary value problem for a semi-infinite region, $0 \leq x < \infty$, where in addition to the differential equation we have initial conditions

$$\left. \begin{array}{l} u(x,0) = \varphi(x) \\ u_t(x,0) = \psi(x) \end{array} \right\}, \quad 0 \leq x < \infty,$$

and a boundary condition

$$u(0,t) = g(t), \quad t > 0.$$

Here as well as in the previous situation, other kinds of boundary conditions described above can be applied.

5.3   Longitudinal Vibrations of a Rod and Electrical Oscillations

In this section we consider other boundary value problems which are similar to the vibrating string problem. The intent here is to show the similarity in the approach to solving physical problems, which on the surface appear quite different but in fact have a similar mathematical structure.

5.3.1   Rod Oscillations: Equations and Boundary Conditions

Consider a thin elastic rod of cylindrical, rectangular or other uniform cross section. In this case forces applied along the axis, perpendicular to the (rigid) cross section, will cause

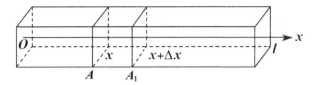

FIGURE 5.2
Arbitrary segment of a rod of length Δx.

changes in the length of the rod. We will assume that the forces act along the rod axis and each cross-sectional area can move only along the rod axis. Such assumptions can be justified if the transverse dimensions are substantially smaller compared to the length of the rod and the forces acting on the rod are comparatively weak.

If a force compresses the rod along its axis and is then released, the rod will begin to vibrate along this axis – contrary to transverse string oscillations, such considered rod oscillations are *longitudinal*. Let the ends of the rod be located at the points $x = 0$ and $x = l$ when it is at rest. The location of some cross-section at rest will be given by x (Figure 5.2). Let the function $u(x, t)$ be the longitudinal shift of this cross-section from equilibrium at time t. The force of tension T is proportional to $u_x(x, t)$ – the relative length change at location x, and the cross-sectional area of the rod (Hook's law); thus, $T(x, t) = EAu_x(x, t)$, where E is the elasticity modulus.

Consider the element of the rod between two cross-sections A and A_1 with coordinates at rest of x and $x + dx$. For small deflections, the resultant of two forces of tension at these cross-sections is

$$T_{x+dx} - T_x = EA \left.\frac{\partial u}{\partial x}\right|_{x+dx} - EA \left.\frac{\partial u}{\partial x}\right|_x \approx EA \frac{\partial^2 u}{\partial x^2} dx.$$

The acceleration of this element is $\partial^2 u/\partial t^2$ in the direction of the resultant force. Together, these two equations give the equation of longitudinal motion of a cross-sectional element as

$$\rho A dx \frac{\partial^2 u}{\partial t^2} = EA \frac{\partial^2 u}{\partial x^2} dx,$$

where ρ is the rod density. Using the notation

$$a = \sqrt{E/\rho} \tag{5.20}$$

we obtain the differential equation for *longitudinal free oscillations of a uniform rod*:

$$\frac{\partial^2 u}{\partial t^2} = a^2 \frac{\partial^2 u}{\partial x^2}. \tag{5.21}$$

As we already know, solutions of such hyperbolic equations have a wave character with the speed of wave propagation, a, given by (5.20).

If there is also an external force per unit volume, $F(x, t)$, we obtain instead the equation

$$\rho A dx \frac{\partial^2 u}{\partial t^2} = EA \frac{\partial^2 u}{\partial x^2} dx + F(x, t) A dx,$$

or

$$\frac{\partial^2 u}{\partial t^2} = a^2 \frac{\partial^2 u}{\partial x^2} + \frac{1}{\rho} F(x, t). \tag{5.22}$$

This is the equation for *forced oscillations of a uniform rod*. Note the similarity of Equation (5.22) to Equation (5.3) for a string under forced oscillations: the two equations are equivalent.

For longitudinal waves there is an important physical restriction: the derivative $u_t(x, t)$ must be small in comparison to the speed of wave propagation: $u_t \ll a$.

The initial conditions are similar to those for a string:

$$u(x, 0) = \varphi(x), \quad \frac{\partial u}{\partial t}(x, 0) = \psi(x),$$

which are initial deflection and initial speed of points of a rod, respectively.

Now consider boundary conditions for a rod.

1. If the left end (here and below similarly for the right end) at $x = 0$ is *fixed*, the boundary condition is
$$u(0, t) = 0.$$

2. If the left end at $x = 0$ is *driven*, and its displacement is determined by the function $g_1(t)$, then
$$u(0, t) = g_1(t),$$
   where $g_1(t)$ is a given function of t.

3. If the end at $x = 0$ can move and experiences a force $f(t)$ along the rod axis, then from equation $T(x, t) = EAu_x(x, t)$ we have
$$-u_x(0, t) = f(t)/EA.$$

   If a force $f(t)$ is applied, instead, to the right end at $x = l$, the left-hand side of this formula will have an opposite sign. These are *forced end* boundary conditions.

4. If the left end is *free*, the tension at that location is zero, $T(a, t) = 0$. Then, from $T(x, t) = EAu_x(x, t)$, the condition follows that
$$u_x(0, t) = 0.$$

5. If the either end of the rod is attached to an elastic material (a wall that "gives" in the horizontal direction) obeying Hook's law, $F = -\alpha u$, it is an *elastic end* boundary condition. For the left end of the rod at $x = 0$, similarly to (5.16) we have
$$u_x(0, t) - hu(0, t) = 0.$$
   For the right end of the rod, located at $x = l$, similarly to (5.17) we have
$$u_x(l, t) + hu(l, t) = 0,$$
   where $h = \alpha/EA > 0$ is the elasticity coefficient.

5.3.2   Electrical Oscillations in a Circuit

Let us briefly set up the boundary value problem for a current and a voltage in a circuit which contains resistance, capacitance and inductance R, C, and L as well as the possibility of leakage, G. For simplicity, these quantities are considered uniformly distributed in a wire placed along the x-axis and defined to be per unit length. The functions $i(x, t)$ and $V(x, t)$ represent current and voltage at a location x along the wire and at time t. Applying Ohm's law for a circuit with nonzero self-inductance and using charge conservation, the so-called *telegraph equations* can be obtained (for more details see books [7, 8]):

$$i_{xx}(x,t) = LCi_{tt}(x,t) + (RC + LG)i_t(x,t) + RGi(x,t) \tag{5.23}$$

and

$$V_{xx}(x,t) = LCV_{tt}(x,t) + (RC + LG)V_t(x,t) + RGV(x,t). \tag{5.24}$$

These equations are similar to the equations of string oscillations; therefore they describe electrical oscillations in an RCL circuit, which can be considered as a longitudinal wave along a conductor. When $R = 0$ and $G = 0$, the equations have the simplest form. For instance, for current

$$\frac{\partial^2 i}{\partial t^2} = a^2 \frac{\partial^2 i}{\partial x^2},$$

where $a^2 = 1/LC$, once again we have the wave Equation (5.4). If $G = 0$, the equation is similar to Equation (5.7) describing oscillations in a medium with the force of resistance proportional to the speed:

$$\frac{\partial^2 i}{\partial t^2} = a^2 \frac{\partial^2 i}{\partial x^2} - 2\kappa \frac{\partial i}{\partial t},$$

where $a^2 = 1/LC$, $2\kappa = (R/L + G/C)$.

Consider initial and boundary conditions for a current and voltage. Let initially, at $t = 0$, the current be $\varphi(x)$, and the voltage along the wire be $\psi(x)$.

The equation for current contains the second-order time derivative; thus, we need two initial conditions. One initial condition is the initial current in the wire

$$i(x, 0) = \varphi(x). \tag{5.25}$$

The second initial condition for the current is:

$$i_t(x, 0) = -\frac{1}{L}[\psi'(x) - R\varphi(x)]. \tag{5.26}$$

Two initial conditions for $V(x, t)$ are:

$$V(x, 0) = \psi(x), \tag{5.27}$$

$$V_t(x, 0) = -\frac{1}{C}[\varphi'(x) - G\psi(x)]. \tag{5.28}$$

For the details on how to obtain conditions (5.26) and (5.28) see [7, 8].

Let us give two examples of the boundary conditions.

1. One end of the wire of length l is attached to a source of electro-motive force (*emf*) $E(t)$, and the other end is grounded. The boundary conditions are

$$V(0, t) = E(t), \quad V(l, t) = 0.$$

2. A sinusoidal voltage with frequency ω is applied to the left end of the wire, and the other end is insulated. The boundary conditions are

$$V(0, t) = V_0 \sin \omega t, \quad i(l, t) = 0.$$

One-dimensional wave equations describe many other periodic phenomena, among others the acoustic longitudinal waves propagating in different materials, and transverse waves in fluids in shallow channels – for details see [7, 8].

5.4   Traveling Waves: D'Alembert Method

In this section, we study waves propagating along an infinite $(-\infty < x < \infty)$ interval. In this case, a physically intuitive way to solve the wave equation is D'Alembert's method. Our physical model will be waves on a string; however, as shown in the previous sections, the same wave equation describes many different physical phenomena and the results derived here apply to those cases as well. The material related to semi-infinite intervals, $0 \leq x < \infty$ and $-\infty < x \leq 0$ is discussed in books [7, 8].

It is clear that oscillations in the parts of a very long string very distant from its ends do not depend on the behavior of the ends. (This is correct during the time interval needed for a wave to arrive from the ends.) Recall that for oscillations we consider, there should be $(u_x)^2 \ll 1$; also $u(x,t)$ is supposed to be small relative to some characteristic scale (instead of the string's length) which always appears in scientific problems. Also, in most physical applications u_t should be small compared to the speed of propagation of a wave, $u_t \ll a$. Physically this is equivalent to the condition $u_x^2 \ll 1$ and $u_x \sim u/L$, where L is the characteristic scale of the string deformation.

The general solution of the equation

$$\frac{\partial^2 u}{\partial t^2} - a^2 \frac{\partial^2 u}{\partial x^2} = 0 \tag{5.29}$$

can be presented, as we discussed earlier, in *D'Alembert's* form as a sum of two twice differentiable functions

$$u = f_1(x - at) + f_2(x + at), \tag{5.30}$$

where f_1 and f_2 represent the waves moving with constant speed a to the right (along the x-axis), and to the left, respectively.

In order to describe the solution for free oscillations in a particular physical case we should find the functions f_1 and f_2 for that particular situation. Since we are considering boundaries at infinity, these functions will be determined only by the initial conditions of the string. Actually, the solution of (5.29) has form (5.30) also when the ends are considered; just f_1 and f_2 are not determined by initial conditions but also by boundary conditions.

The Cauchy problem for the infinite string is defined as the search for a solution of Equation (5.29) which satisfies the initial conditions

$$u|_{t=0} = \varphi(x) \quad \text{and} \quad \frac{\partial u}{\partial t}\bigg|_{t=0} = \psi(x). \tag{5.31}$$

Substituting Equation (5.30) into Equation (5.31) gives

$$f_1(x) + f_2(x) = \varphi(x),$$

$$-af_1'(x) + af_2'(x) = \psi(x).$$

From the second relation

$$-af_1(x) + af_2(x) = \int_0^x \psi(x)dx + aC,$$

(where C is an arbitrary constant), thus

$$f_1(x) = \frac{1}{2}\left[\varphi(x) - \frac{1}{a}\int_0^x \psi(x)dx - C\right],$$

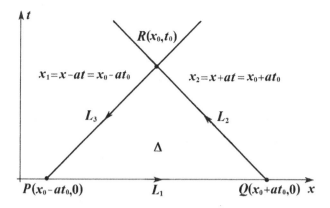

FIGURE 5.3
The characteristic triangle with vertex at point (x_0, t_0).

$$f_2(x) = \frac{1}{2}\left[\varphi(x) + \frac{1}{a}\int_0^x \psi(x)dx + C\right].$$

Therefore

$$u(x,t) = \frac{1}{2}\left[\varphi(x - at) - \frac{1}{a}\int_0^{x-at}\psi(x)dx - C\right.$$
$$\left. + \varphi(x + at) + \frac{1}{a}\int_0^{x+at}\psi(x)dx + C\right],$$

which finally gives

$$u(x,t) = \frac{\varphi(x - at) + \varphi(x + at)}{2} + \frac{1}{2a}\int_{x-at}^{x+at}\psi(x)dx. \tag{5.32}$$

If the function $\psi(x)$ is differentiable and $\varphi(x)$ twice differentiable, this solution satisfies Equation (5.29) and initial conditions (5.31). (The reader may check this as a *Reading Exercise*.) The method of construction of the solution given by Equation (5.29) proves its uniqueness. The solution is stable if functions $\varphi(x)$ and $\psi(x)$ depend continuously on x.

Another way to write the solution of Equation (5.29) is to use *the characteristic triangle* shown in Figure 5.3. Suppose we want to find the solution at some point (x_0, t_0). The vertex of this triangle is the point (x_0, t_0) and the two sides are given by the equations $x - at = x_0 - at_0$ and $x + at = x_0 + at_0$. The base of this triangle, i.e. the line between the points $P(x_0 - at_0, 0)$ and $Q(x_0 + at_0, 0)$, determines the wave amplitude at point (x_0, t_0); all other points beyond this base do not contribute to the solution at this point, as follows from Equation (5.32):

$$u(x_0, t_0) = \frac{\varphi(P) + \varphi(Q)}{2} + \frac{1}{2a}\int_P^Q \psi(x)dx. \tag{5.33}$$

We now discuss, in more detail, two physical situations which are typically encountered: waves created by a displacement and waves created by a pulse.

a) *Waves created by a displacement*. Let the initial speeds of points on the string be zero, but the initial displacements are not zero. The solution given by Equation (5.32) in this case is

$$u(x,t) = \frac{1}{2}\left[\varphi(x - at) + \varphi(x + at)\right]. \tag{5.34}$$

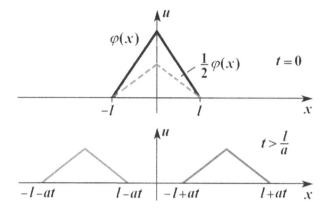

FIGURE 5.4
Propagation of an initial displacement.

The first term, $\frac{1}{2}\varphi(x - at)$, is a constant shape disturbance propagating with speed a in the positive x direction, and the term $\frac{1}{2}\varphi(x + at)$ is the same shaped disturbance moving in the opposite direction. Suppose that the initial disturbance exists only on a limited interval $-l \leq x \leq l$. This kind of disturbance is shown schematically in Figure 5.4, where the function $\varphi(x)$ is plotted with a solid line in the upper part of the figure. The dashed line shows the function $\frac{1}{2}\varphi(x)$. We can consider $u(x, t = 0)$ given by Equation (5.34) as two independent disturbances, $\frac{1}{2}\varphi(x)$, each propagating in opposite directions with unchanged amplitude. Initially, at $t = 0$, the profiles of both waves coincide, after which they separate and the distance between them increases. The bottom part of Figure 5.4 shows these waves after some time $t > l/a$. As they pass a given section of the string, this part returns to rest. Only two intervals of the string of length $2l$ each are deflected at any instant t.

Note that the function (5.34), which corresponds to the initial condition shown in Figure 5.4, is not differentiable in some points; hence (5.34) is not a classical solution of Equation (5.29) in that case (the *classical solution* assumes that the derivatives of the unknown function $u(x, t)$ exist and are continuous). However, it can be understood in the following way.

We can change the non-smooth function $\varphi(x)$ a little to make it smooth, i.e., consider $\varphi(x)$ as a limit of a sequence of smooth, twice differentiable, functions $\varphi_n(x)$, $n = 1, 2, \ldots$ such that

$$\lim_{n \to \infty} \varphi_n(x) = \varphi(x) \tag{5.35}$$

for any x. For any finite n, we can construct a classical solution of Equation (5.29) satisfying the initial condition $u_n(x, 0) = \varphi_n(x)$,

$$u_n(x, t) = \frac{1}{2}\left[\varphi_n(x - at) + \varphi_n(x + at)\right].$$

Obviously,

$$\lim_{n \to \infty} u_n(x, t) = u(x, t) \tag{5.36}$$

of classical solutions, which slightly differ from the formal solution (5.34) and converge to it as $n \to \infty$. In that case, we say that $u(x, t)$ is a generalized solution of the wave equation (5.29) with the initial condition $\varphi(x)$.

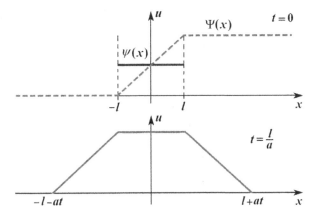

FIGURE 5.5
Propagation of an initial pulse.

Reading Exercise: Let the initial displacement be $\varphi(x) = \exp(-x^2/b^2)$; b is a constant. Use Equation (5.31) to obtain

$$u(x,t) = \frac{1}{2}\left[\exp\left(-(x-at)^2/b^2\right) + \exp\left(-(x+at)^2/b^2\right)\right].$$

Notice, that introducing constant b (as well as other constants to keep proper dimension in the Examples and Problems everywhere in the book) is not necessary if x is dimensionless – it is very common that problems' parameters are used in dimensionless form.

b) *Waves created by a pulse.* In this case we let the initial displacement be zero, $\varphi(x) = 0$, but the initial velocity is given as a function of position, $\psi(x)$. In other words, points on the string are given some initial velocity by an external agent. An example of a distribution of velocities is schematically shown in Figure 5.5, where, to simplify the plot, a constant function $\psi(x)$ on $-l \leq x \leq l$ was chosen. As in the case considered above, the solution corresponding to Figure 5.5 is a generalized solution. The function $\psi(x)$ is plotted with a solid line in the upper part of Figure 5.5. Consider the integral

$$W(x) = \frac{1}{2a}\int_{-\infty}^{x}\psi(x)dx,$$

which is zero on the interval $-\infty < x \leq -l$; for $x \geq l$ the function $W(x)$ is constant and equal to

$$\frac{1}{2a}\int_{-l}^{l}\psi(x)dx.$$

The graph for $W(x)$ is shown with a dashed line in Figure 5.5. From Equation (5.31) we have

$$u(x,t) = \frac{1}{2a}\int_{x-at}^{x+at}\psi(x)dx = W(x+at) - W(x-at). \qquad (5.37)$$

Equation (5.37) means that we again have two waves moving in opposite directions, but now the waves have opposite signs.

At some location x for large enough time, we have $x + at > l$ and the wave $W(x+at)$ becomes a constant and at the same instant of time $x - at < -l$ and the wave $W(x-at)$ is equal to zero. As a result, the perturbation propagates in both directions but, contrary to the case of a wave created by a displacement, none of the elements of the string return to the initial position existing at $t = 0$. The bottom part of Figure 5.5 illustrates this situation for an instant of time $t = l/a$. When $t > l/a$ we have a similar trapezoid for $u(x, t)$ (with the same height as at $t = l/a$), which expands in both directions uniformly with time.

Example 5.1 Let $\varphi(x) = 0$, $\psi(x) = \psi_0 = $ const for $x_1 < x < x_2$ and $\psi(x) = 0$ outside of this interval.

In this case we have

$$W(x) = \frac{1}{2a} \int_{x_1}^{x} \psi(x) dx = \begin{cases} 0, & x < x_1 \\ \dfrac{\psi_0}{2a}(x - x_1), & x_1 < x < x_2 \\ \dfrac{\psi_0}{2a}(x_2 - x_1) = \text{const}, & x > x_2. \end{cases}$$

Reading Exercise: Solve the problem in Example 5.1 and generate the wave propagation with the parameters $a = 0.5$, $l = 1$, $\psi_0 = 1$. Explain why the maximum deflection is $u_{max} = 2$, and why, for the point $x = 0$, this maximum deflection value is reached at time $t = l/a = 2$.

Example 5.2 An infinite stretched string is excited by the initial deflection

$$\varphi(x) = \frac{A}{x^2 + B}$$

with no initial velocities. Find the vibrations of the string. Write an analytical solution representing the motion of the string and illustrate the spatial-time-dependent solution $u(x, t)$ with an animation sequence including snapshots at times $t = 0, 1, \ldots, n$.

Solution. The initial speeds of points on the string are zero ($\psi(x) \equiv 0$), so the solution given by Equation (5.34) is

$$u(x, t) = \frac{\varphi(x - at) + \varphi(x + at)}{2} = \frac{A}{2} \left[\frac{1}{(x - at)^2 + B} + \frac{1}{(x + at)^2 + B} \right].$$

Figure 5.6 shows the solution for the case when $a^2 = 0.25$, $A = B = 1$. The black bold line represents the initial deflection and the bold gray line is the string profile at time $t = 12$.

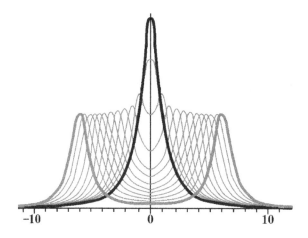

FIGURE 5.6
Graph of the solution to Example 5.2.

The gray lines show the evolution of the string profile within the period of time from $t = 0$ to $t = 12$.

Notice, that for a proper dimension of $\varphi(x)$, the values of A and B should have dimensions of length cubed and squared, correspondingly (if we consider string oscillations). In the problems following this and other chapters, in most cases we use dimensionless parameters and leave it to the reader to discuss particular physical situations.

5.5 Cauchy Problem for Nonhomogeneous Wave Equation

5.5.1 D'Alembert's Formula

Let us consider now the *nonhomogeneous* wave equation,

$$\frac{\partial^2 u}{\partial t^2} - a^2 \frac{\partial^2 u}{\partial x^2} = f(x, t), \quad -\infty < x < \infty, \ 0 < t < T \tag{5.38}$$

with the same initial conditions,

$$u|_{t=0} = \varphi(x), \quad \left.\frac{\partial u}{\partial t}\right|_{t=0} = \psi(x), \quad -\infty < x < \infty. \tag{5.39}$$

In the previous section, we defined the *characteristic triangle* Δ that contains all the points (x, t) that can influence the value $u(x_0, t_0)$ in the world where the signals propagate with velocity a. Let us integrate Equation (5.38) over that triangle:

$$\iint_\Delta (a^2 u_{xx} - u_{tt}) dx dt = - \iint_\Delta f(x, t) dx dt$$

Recall *Green's theorem* from calculus: for any differentiable functions $p(x, t)$ and $q(x, t)$, the following relation exists between the double integral over a two-dimensional region Δ and the integral along its boundary L,

$$\iint_\Delta [q_x(x, t) - p_t(x, t)] \, dx dt = \oint_L [p(x, t) dx - q(x, t) dt]. \tag{5.40}$$

The integration has to be carried out counterclockwise.

In our case the integration path consists of three sides of the triangle, L_1, L_2 and L_3 (see Figure 5.3). Substituting $q = a u_x$ and $p = u_t$ into (5.40), we find that

$$\iint_\Delta (a^2 u_{xx} - u_{tt}) \, dx dt = \oint_L (u_t dx + a^2 u_x dt), \tag{5.41}$$

where $L = L_1 \cup L_2 \cup L_3$.

Along the base of the triangle L_1, $t = const$, hence $dt = 0$ and

$$\int_{L_1} (u_t dx + a^2 u_x dt) = \int_{x_0 - a t_0}^{x_0 + a t_0} u_t(x, 0) dx = \int_{x_0 - a t_0}^{x_0 + a t_0} \psi(x) dx.$$

Along the side L_2, $x + at = const$, hence $dx + a dt = 0$. Replacing dx with $-a dt$ and dt with $-dx/a$, we find:

$$\int_{L_2} (u_t dx + a^2 u_x dt) = -a \int_{L_2} (u_t dt + u_x dx) = -a \int_Q^R du$$

$$= -a [u(x_0, t_0) - u(x_0 + a t_0, 0)] = -a [u(x_0, t_0) - \varphi(x_0 + a t_0)].$$

Similarly, along the side L_3 $x - at = const$, hence $dx - adt = 0$, $dx = adt$ and $dt = dx/a$, thus we get

$$\int_{L_3} \left(u_t dx + a^2 u_x dt \right) = a \int_{L_3} \left(u_t dt + u_x dx \right) = a \int_R^P du$$
$$= a \left[u(x_0 - at_0, 0) - u(x_0, t_0) \right] = a \left[\varphi(x_0 - at_0) - u(x_0, t_0) \right].$$

Summating the expressions obtained above, we find:

$$\int_L \left(u_t dx + a^2 u_x dt \right) = \int_{x_0 - at_0}^{x_0 + at_0} \psi(x) dx$$
$$+ a \left[\varphi(x_0 - at_0) + \varphi(x_0 + at_0) \right] - 2au(x_0, t_0). \tag{5.42}$$

Substituting (5.42) into (5.41), we find:

$$u(x_0, t_0) = \frac{\varphi(x_0 - at_0) + \varphi(x_0 + at_0)}{2}$$
$$+ \frac{1}{2a} \int_{x_0 - at_0}^{x_0 + at_0} \psi(x) dx + \frac{1}{2a} \iint_\Delta f(x, t) dx dt. \tag{5.43}$$

To unify the notations with D'Alembert's formula (5.31) obtained for the homogeneous wave equation, let us replace (x_0, t_0) with (x, t) and (x, t) with (ξ, τ):

$$u(x, t) = \frac{\varphi(x - at) + \varphi(x + at)}{2} + \frac{1}{2a} \int_{x - at}^{x + at} \psi(\xi) d\xi + \frac{1}{2a} \iint_\Delta f(\xi, \tau) d\xi d\tau. \tag{5.44}$$

5.5.2 Green's Function

Solution (5.44) is a sum of two functions,

$$u(x, t) = v(x, t) + w(x, t),$$

where

$$v(x, t) = \frac{\varphi(x - at) + \varphi(x + at)}{2} + \frac{1}{2u} \int_{x - at}^{x + at} \psi(\xi) d\xi$$

is the contribution of nonzero initial conditions and

$$w(x, t) = \frac{1}{2a} \iint_\Delta f(\xi, \tau) d\xi d\tau \tag{5.45}$$

is produced by the inhomogeneity. This expression can be rewritten as

$$w(x, t) = \int_{-\infty}^\infty d\xi \int_0^T d\tau G(x, t; \xi, \tau) f(\xi, \tau), \tag{5.46}$$

where the kernel $G(x, t; \xi, \tau)$ of the linear transformation (5.46), which is called *Green's function*, is equal to $1/2a$ for (ξ, τ) inside the characteristic triangle of the point (x, t) and equal to zero outside it. Green's function can be written in a compact form using the Heaviside step function $H(y)$,

$$H(y) = 1, \quad y \geq 0; \quad H(y) = 0, \quad y < 0. \tag{5.47}$$

Indeed,

$$G(x, t; \xi, \tau) = \frac{1}{2a} H \left(t - \tau - \frac{|x - \xi|}{a} \right). \tag{5.48}$$

Note that Green's function depends only on differences of its arguments, $t - \tau$ and $x - \xi$. It satisfies Equation (5.38) with the nonhomogeneous term $f(x,t) = \delta(x - \xi)\delta(t - \tau)$,

$$\frac{\partial^2 G(x,t;\xi,\tau)}{\partial t^2} - a^2 \frac{\partial^2 G(x,t;\xi,\tau)}{\partial x^2} = \delta(x - \xi)\delta(t - \tau); \tag{5.49}$$

$$-\infty < x < \infty, \quad 0 < t < T; \quad -\infty < \xi < \infty, \quad 0 < \tau < T,$$

and zero initial conditions:

$$G(x,0;\xi,\tau) = \frac{\partial G(x,0;\xi,\tau)}{\partial t} = 0, \tag{5.50}$$

$$-\infty < x < \infty; \quad -\infty < \xi < \infty, \quad 0 < \tau < T.$$

Example 5.3 Using Green's function (5.48), find the solution of the following initial value problem:

$$\frac{\partial^2 u}{\partial t^2} - a^2 \frac{\partial^2 u}{\partial x^2} = \delta(x), \quad -\infty < x < \infty, \quad t > 0.$$

with initial conditions

$$u(x,0) = 0, \quad u_t(x,0) = 0, \quad -\infty < x < \infty.$$

Because of zero initial conditions, the solution is determined by expression (5.46). It is convenient to change the order of integrations,

$$u(x,t) = \frac{1}{2a} \int_0^\infty d\tau \int_{-\infty}^\infty d\xi \, \delta(\xi) H\left(t - \tau - \frac{|x - \xi|}{a}\right) = \frac{1}{2a} \int_0^\infty d\tau \, H\left(t - \tau - \frac{|x|}{a}\right).$$

The argument of the Heaviside function decreases monotonically from $t - |x|/a$ at $\tau = 0$ to $-\infty$ as $\tau \to \infty$. Thus, it is always negative, if $|x| > at$, hence $u(x,t) = 0$ in that region. If $|x| < at$, then

$$u(x,t) = \frac{1}{2a} \int_0^{t - |x|/a} d\tau = \frac{1}{2a}\left(t - \frac{|x - \xi|}{a}\right).$$

5.5.3   Well-Posedness of the Cauchy Problem

Equation (5.44) gives the explicit solution of the Cauchy problem (5.38), (5.39). Substituting (5.44) into (5.38), one can show that solution (5.44) is the *classical solution*; i.e., the derivatives of $u(x,t)$ which appear in (5.38) exist and are continuous, if (i) $\varphi(x)$ is twice continuously differentiable, (ii) $\psi(x)$ is continuously differentiable, and (iii) $f(x,t)$, $f_x(x,t)$ are continuous. Also, this solution is unique.

There is one more property of the solution that has to be checked. It is clear that in reality the functions $\varphi(x)$ and $\psi(x)$, which describe the initial conditions, and the external load $f(x,t)$ cannot be *measured* precisely. There is always a certain difference between the actual functions $\varphi_1(x)$, $\psi_1(x)$, $f_1(x,t)$ and the measured function $\varphi_2(x)$, $\psi_2(x)$, $f_2(x,t)$; the corresponding solutions $u_1(x,t)$ and $u_2(x,t)$ are not equal. It is the crucial question whether both solutions are arbitrarily close to each other if the measurements are sufficiently precise. If that is not the case, the solution obtained with not perfectly measured data is of no use. That argument leads to the following mathematical question: can we get $|u_1(x,t) - u_2(x,t)| < \varepsilon$, where ε is an arbitrary small number, by means of sufficiently precise measurements, i.e., under conditions $|\varphi_1(x) - \varphi_2(x)| < \delta(\varepsilon)$, $|\psi_1(x) - \psi_2(x)| < \delta(\varepsilon)$, $|f_1(x,t) - f_2(x,t)| < \delta(\varepsilon)$?

To estimate the difference $|u_1(x,t) - u_2(x,t)|$, we shall use the following inequalities that can be easily checked.

1. For any real a and b,

$$|a + b| \leq |a| + |b|.$$

2. A similar inequality is correct for an arbitrary number of terms:

$$a_1 + a_2 + \ldots a_n| \leq |a_1| + |a_2| + \ldots |a_n|.$$

3. The further generalization is for integrals:

$$\left| \int_a^b F(x)dx \right| \leq \int_a^b |F(x)|\, dx.$$

4. If $F_1(x) \leq F_2(x)$, then

$$\int_a^b F_1(x)dx \leq \int_a^b F_2(x)dx.$$

Using the inequalities listed above, we get the following sequence of inequalities for $|u_1(x,t) - u_2(x,t)|$:

$$
\begin{aligned}
|u_1(x,t) - u_2(x,t)| &= \left| \frac{1}{2} \left[\varphi_1(x - at) - \varphi_2(x - at) \right] + \frac{1}{2} \left[\varphi_1(x + at) - \varphi_2(x + at) \right] \right. \\
&\quad \left. + \frac{1}{2a} \int_{x-at}^{x+at} \left[\psi_1(\xi) - \psi_2(\xi) \right] d\xi + \frac{1}{2a} \iint_\Delta \left[f_1(\xi, \tau) - f_2(\xi, \tau)) \right] d\xi d\tau \right| \\
&\leq \frac{1}{2} |\varphi_1(x - at) - \varphi_2(x - at)| + \frac{1}{2} |\varphi_1(x + at) - \varphi_2(x + at)| \\
&\quad + \frac{1}{2a} \int_{x-at}^{x+at} |\psi_1(\xi) - \psi_2(\xi)|\, d\xi + \frac{1}{2a} \iint_\Delta |f_1(\xi, \tau) - f_2(\xi, \tau)|\, d\xi d\tau \\
&< \frac{1}{2}\delta + \frac{1}{2}\delta + \frac{\delta}{2a} \cdot 2aT + \frac{\delta}{2a} \cdot \frac{1}{2}(2aT) \cdot T = \delta \left(1 + T + \frac{1}{2}T^2 \right).
\end{aligned}
$$

Thus, for any ε, if

$$\delta < \frac{\varepsilon}{1 + T + T^2/2},$$

then

$$|u_1(x,t) - u_2(x,t)| < \varepsilon.$$

The problem, which has a unique classical solution and depends continuously on initial conditions, is *well-posed*.

In Chapter 7 we shall see that the Cauchy problem for an *elliptic* equation is *ill-posed*, i.e., not well-posed.

5.6 Finite Intervals: The Fourier Method for Homogeneous Equations

In this and the following sections, we introduce a powerful Fourier method for solving partial differential equations for *finite intervals*. The Fourier method, or the method of separation of variables, is one of the most widely used methods for an analytical solution of boundary value problems in mathematical physics. The method gives a solution in terms of a series of eigenfunctions of the corresponding Sturm-Liouville problem (discussed in Chapter 4).

Let us apply the Fourier method in the case of the general one-dimensional homogeneous hyperbolic equation

$$\frac{\partial^2 u}{\partial t^2} + 2\kappa \frac{\partial u}{\partial t} - a^2 \frac{\partial^2 u}{\partial x^2} + \gamma u = 0, \tag{5.51}$$

where a, κ, γ are constants. As we discussed when deriving Equations (5.6) and (5.7) in Section 5.1, for physical situations the requirement is: $\kappa \geq 0, \gamma \geq 0$[1]. Here we will work on a finite interval, $0 \leq x \leq l$, and obviously $t \geq 0$.

To obtain a unique solution of Equation (5.51), boundary and initial conditions must be imposed on the function $u(x,t)$. Some of them can be homogeneous, some not. Initially, we will search for a solution of Equation (5.51) satisfying the *homogeneous* boundary conditions

$$P_1[u] \equiv \alpha_1 u_x + \beta_1 u|_{x=0} = 0, \quad P_2[u] \equiv \alpha_2 u_x + \beta_2 u|_{x=l} = 0, \tag{5.52}$$

with constant $\alpha_1, \beta_1, \alpha_2$ and β_2, and initial conditions

$$u(x,t)|_{t=0} = \varphi(x), \quad u_t(x,t)|_{t=0} = \psi(x), \tag{5.53}$$

where $\varphi(x)$ and $\psi(x)$ are given functions. As we discussed in Section 5.2, normally there are physical restrictions on the signs of the coefficients in Equation (5.52) so that we have $\alpha_1/\beta_1 < 0$ and $\alpha_2/\beta_2 > 0$. Obviously, only these ratios are significant in Equations (5.52), but to formulate a solution of the Sturm-Liouville problem for functions $X(x)$ it is more convenient to keep all the constants $\alpha_1, \beta_1, \alpha_2$ and β_2.

We begin by assuming that a nontrivial (non-zero) solution of Equation (5.51) can be found that is a product of two functions, one depending only on x, another depending only on t:

$$u(x,t) = X(x)T(t). \tag{5.54}$$

Substituting Equation (5.54) into Equation (5.51), we obtain

$$X(x)T''(t) + 2\kappa X(x)T'(t) - a^2 X''(x)T(t) + \gamma X(x)T(t) = 0$$

or, by rearranging terms,

$$\frac{T''(t) + 2\kappa T'(t) + \gamma T(t)}{a^2 T(t)} = \frac{X''(x)}{X(x)}, \tag{5.55}$$

where primes indicate derivatives with respect to t or x. The left-hand side of this equation depends only on t, and the right-hand side only on x, which is possible only if each side equals a constant. By using the notation $-\lambda$ for this constant, we obtain

$$\frac{T''(t) + 2\kappa T'(t) + \gamma T(t)}{a^2 T(t)} \equiv \frac{X''(x)}{X(x)} = -\lambda$$

(it is seen from this relation that λ has dimension of inversed length squared).

Thus, Equation (5.55) gives two ordinary second-order linear homogeneous differential equations:

$$T''(t) + 2\kappa T'(t) + \left(a^2\lambda + \gamma\right)T(t) = 0, \tag{5.56}$$

---

[1]Typical wave problems do not contain terms like bu_x, but if included in Equation (5.51), the substitution $u(x,t) = e^{bx/2a^2}\nu(x,t)$ leads to the equation for function $\nu(x,t)$ without ν_x term.

and

$$X''(x) + \lambda X(x) = 0. \qquad (5.57)$$

Therefore, we see that we have successfully separated the variables, resulting in separate equations for functions $X(x)$ and $T(t)$. These equations share the common parameter λ.

To find λ, we apply the boundary conditions. The homogenous boundary condition of Equation (5.52), imposed on $u(x, y)$, gives the homogeneous boundary conditions on the function $X(x)$:

$$\alpha_1 X' + \beta_1 X|_{x=0} = 0, \qquad \alpha_2 X' + \beta_2 X|_{x=l} = 0 \qquad (5.58)$$

with restrictions $\alpha_1/\beta_1 < 0$ and $\alpha_2/\beta_2 > 0$.

This result therefore leads to the *Sturm-Liouville boundary value problem*, which may be stated in the present case as the following:

Find values of the parameter λ (eigenvalues) for which nontrivial (not identically equal to zero) solutions to Equation (5.57), $X(x)$ (eigenfunctions), satisfying boundary conditions (5.58) exist.

Let us briefly recall the main properties of eigenvalues and eigenfunctions of the Sturm-Liouville problem given in Equations (5.57) and (5.58) (see Chapter 4).

1. There exists an *infinite set* of real nonnegative discrete eigenvalues $\{\lambda_n\}$ and corresponding eigenfunctions $\{X_n(x)\}$. The eigenvalues increase as the number n increases:

$$0 \le \lambda_1 < \lambda_2 < \lambda_3 < ... < \lambda_n < \quad ... \quad (\lim \lambda_n = +\infty).$$

2. Eigenfunctions corresponding to different eigenvalues are *linearly independent and orthogonal*:

$$\int_0^l X_i(x)X_j(x)dx = 0, \quad i \ne j. \qquad (5.59)$$

3. *The completeness property* states that any function $f(x)$ which is twice differentiable on $(0, l)$ and satisfies the homogeneous boundary conditions in Equations (5.58) can be resolved in an absolutely and uniformly converging series with eigenfunctions of the boundary value problem given in Equations (5.57) and (5.58):

$$f(x) = \sum_{n=1}^{\infty} f_n X_n(x), \qquad f_n = \frac{1}{\|X_n\|^2} \int_0^l f(x)X_n(x)dx, \qquad (5.60)$$

where $\|X_n\|^2 = \int_0^1 X_n^2 dx$.

Eigenvalues and eigenfunctions of a boundary value problem depend on the type of the boundary conditions: Dirichlet, Neumann or mixed. The values of the constants α_i and β_i in Equation (5.58) determine one of these three possible types. All possible variants are presented in Chapter 4. For all of these variants, the solution for the problem given in Equations (5.57) and (5.58) is

$$X(x) = C_1 \cos \sqrt{\lambda} x + C_2 \sin \sqrt{\lambda} x. \qquad (5.61)$$

The coefficients C_1 and C_2 are determined from the system of Equations (5.58):

$$\begin{cases} C_1 \beta_1 + C_2 \alpha_1 \sqrt{\lambda} = 0, \\ C_1 \left[-\alpha_2 \sqrt{\lambda} \sin \sqrt{\lambda} l + \beta_2 \cos \sqrt{\lambda} l \right] + C_2 \left[\alpha_2 \sqrt{\lambda} \cos \sqrt{\lambda} l + \beta_2 \sin \sqrt{\lambda} l \right] = 0. \end{cases} \qquad (5.62)$$

This system of linear homogeneous algebraic equations has a nontrivial solution only when its determinant equals zero:

$$(\alpha_1\alpha_2\lambda + \beta_1\beta_2)\tan\sqrt{\lambda}l - \sqrt{\lambda}(\alpha_1\beta_2 - \beta_1\alpha_2) = 0. \tag{5.63}$$

It is easy to determine (for instance by using graphical methods) that Equation (5.63) has an infinite number of roots $\{\lambda_n\}$, which conforms to the general Sturm-Liouville theory. For each root λ_n, we obtain a nonzero solution of Equations (5.62).

It is often convenient to present the solution in a form that allows us to consider in a unified way a mixed boundary condition and two other kinds of boundary conditions as well, when some of the constants α_i or β_i may be equal to zero. We will do this in the following way. Using the first expression in Equations (5.62), we represent C_1 and C_2 as

$$C_1 = C\alpha_1\sqrt{\lambda_n}, \quad C_2 = -C\beta_1,$$

where $C \neq 0$ is an arbitrary constant (because the determinant is equal to zero, the same C_1 and C_2 satisfy the second equation of the system of Equations (5.62)). For these constraints the choice

$$C = 1/\sqrt{\lambda_n\alpha_1^2 + \beta_1^2}$$

(with positive square root and $\alpha_1^2 + \beta_1^2 \neq 0$) allows us to obtain a simple set of coefficients C_1 and C_2. For Dirichlet boundary conditions, we assign $\alpha_1 = 0$, $\beta_1 = -1$ (we may also use $\beta_1 = 1$ because the overall sign of $X_n(x)$ is not important) so that $C_1 = 0$ and $C_2 = 1$. For Neumann boundary conditions $\beta_1 = 0$, $\alpha_1 = 1$, and we have $C_1 = 1$, $C_2 = 0$.

With this choice of C, the functions $X_n(x)$ are *bounded by the values* ± 1. The alternative often used for the coefficients C_1 and C_2 corresponds to the normalizations $\|X_n\|^2 = 1$. Here and elsewhere in the book we will use the first choice because in this case the graphs for $X_n(x)$ are easier to plot.

From the above discussion, the eigenfunctions of the Sturm-Liouville problem given by the equations

$$X'' + \lambda X = 0,$$

$$\alpha_1 X' + \beta_1 X|_{x=0} = 0, \quad \alpha_2 X' + \beta_2 X|_{x=l} = 0$$

can be written as

$$X_n(x) = \frac{1}{\sqrt{\alpha_1^2\lambda_n + \beta_1^2}}\left[\alpha_1\sqrt{\lambda_n}\cos\sqrt{\lambda_n}x - \beta_1\sin\sqrt{\lambda_n}x\right]. \tag{5.64}$$

The orthogonality property in Equation (5.59) can be easily verified by the reader as a *Reading Exercise*. The square norms of eigenfunctions are

$$\|X_n\|^2 = \int_0^l X_n^2(x)dx = \frac{1}{2}\left[l + \frac{(\beta_2\alpha_1 - \beta_1\alpha_2)(\lambda_n\alpha_1\alpha_2 - \beta_1\beta_2)}{(\lambda_n\alpha_1^2 + \beta_1^2)(\lambda_n\alpha_2^2 + \beta_2^2)}\right]. \tag{5.65}$$

The eigenvalues are $\lambda_n = \left(\frac{\mu_n}{l}\right)^2$, where μ_n is the nth root of the equation

$$\tan\mu = \frac{(\alpha_1\beta_2 - \alpha_2\beta_1)l\mu}{\mu^2\alpha_1\alpha_2 + l^2\beta_1\beta_2}. \tag{5.66}$$

This equation remains unchanged when the sign of μ changes, which indicates that positive and negative roots are placed symmetrically on the μ axis. Because the eigenvalues λ_n do not depend on the sign of μ, it is *enough to find only positive roots* μ_n of Equation (5.66) since negative roots do not give new values of λ_n. Clearly, μ has no dimension.

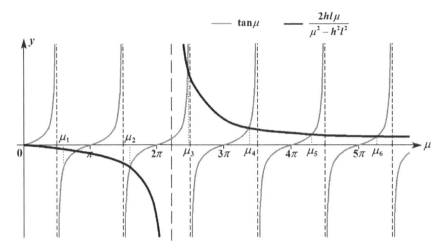

FIGURE 5.7
A graphical solution of Equation (5.64) for $l = 100$ and $h = 0.07$.

It is not difficult to demonstrate (for instance using a graphical solution of this equation) that $\mu_{n+1} > \mu_n$, again in accordance with the Sturm-Liouville theory. In the cases when either both boundary conditions are of Dirichlet type, or Neumann type, or when one is Dirichlet type and the other is Neumann type, a graphical solution of Equation (5.66) is not needed and we leave to the reader as the *Reading Exercises* to obtain analytical expressions for $X_n(x)$. The obtained results can be compared with those collected in Chapter 4.

Let us find the roots of Equation (5.66) for mixed boundary conditions

$$u_x(0,t) - hu(0,t) = 0 \quad \text{and} \quad u_x(l,t) + hu(l,t) = 0. \tag{5.67}$$

Such a particular case with equal values of h in both Equations (5.67) is rather common.

The eigenvalues are determined from Equation (5.66), where $\alpha_1 = 1$, $\beta_1 = -h$, $\alpha_2 = 1$, $\beta_2 = h$:

$$\tan \mu = \frac{2hl\mu}{\mu^2 - h^2 l^2}. \tag{5.68}$$

This equation has an infinite number of roots, μ_n. Figure 5.7 shows curves of the two functions $y = \tan \mu$ and $y = \frac{2hl\mu}{\mu^2 - h^2 l^2}$, plotted on the same set of axes. The values of μ at the intersection points of these curves are the roots of the Equation (5.68).

For the values $l = 100$ and $h = 0.07$ the first six positive roots of this equation are:

$$\mu_1 = 2.464 \in \left(\frac{\pi}{2}, \frac{3\pi}{2}\right), \quad \mu_2 = 5.036 \in \left(\frac{3\pi}{2}, \frac{5\pi}{2}\right), \quad \mu_3 = 7.752 \in \left(\frac{3\pi}{2}, \frac{5\pi}{2}\right),$$

$$\mu_4 = 10.59 \in \left(\frac{5\pi}{2}, \frac{7\pi}{2}\right), \quad \mu_5 = 13.25 \in \left(\frac{7\pi}{2}, \frac{9\pi}{2}\right), \quad \mu_6 = 16.51 \in \left(\frac{9\pi}{2}, \frac{11\pi}{2}\right).$$

The line $\mu = hl$ (dashed line in Figure 5.7) is the asymptote of the graph $y = 2hl\mu/(\mu^2 - h^2 l^2)$. In our example $\mu = 7 \in (3\pi/2, 5\pi/2)$, and we notice that this interval includes *two roots*, μ_2 and μ_3, whereas each of the other intervals, $((2k-1)\pi/2, (2k+1)\pi/2)$, contains *one root* of Equation (5.68). This is a commonly met situation.

Each eigenvalue

$$\lambda_n = \left(\frac{\mu_n}{l}\right)^2, \quad n = 1, 2, 3, \ldots \tag{5.69}$$

corresponds to an eigenfunction

$$X_n(x) = \frac{1}{\sqrt{\lambda_n + h^2}} \left[\sqrt{\lambda_n} \cos \sqrt{\lambda_n} x + h \sin \sqrt{\lambda_n} x \right]. \tag{5.70}$$

The norms of these eigenfunctions are

$$\|X_n\|^2 = \frac{l}{2} + \frac{h}{\lambda_n + h^2}. \tag{5.71}$$

The next step of the solution is to *find the function* $T(t)$. Equation (5.56) is an ordinary linear second-order homogeneous differential equation. For $\lambda = \lambda_n$

$$T_n''(t) + 2\kappa T_n''(t) + (a^2\lambda_n + \gamma)T_n(t) = 0 \tag{5.72}$$

and a general solution of this equation is

$$T_n(t) = a_n y_n^{(1)}(t) + b_n y_n^{(2)}(t), \tag{5.73}$$

where a_n and b_n are arbitrary constants. Two particular solutions, $y_n^{(1)}(t)$ and $y_n^{(2)}(t)$, are

$$y_n^{(1)}(t) = \begin{cases} e^{-\kappa t} \cos \omega_n t, & \text{if } \kappa^2 < a^2\lambda_n + \gamma, \\ e^{-\kappa t} \cosh \omega_n t, & \text{if } \kappa^2 > a^2\lambda_n + \gamma, \\ e^{-\kappa t}, & \text{if } \kappa^2 = a^2\lambda_n + \gamma, \end{cases} \tag{5.74}$$

and

$$y_n^{(2)}(t) = \begin{cases} e^{-\kappa t} \sin \omega_n t, & \text{if } \kappa^2 < a^2\lambda_n + \gamma, \\ e^{-\kappa t} \sinh \omega_n t, & \text{if } \kappa^2 > a^2\lambda_n + \gamma, \\ te^{-\kappa t}, & \text{if } \kappa^2 = a^2\lambda_n + \gamma, \end{cases} \tag{5.75}$$

where

$$\omega_n = \sqrt{|a^2\lambda_n + \gamma - \kappa^2|}$$

(obviously, ω_n has the dimension of inversed time).

Reading Exercise: Verify the above expressions.

It is clear that each function

$$u_n(x,t) = T_n(t)X_n(x) = \left[a_n y_n^{(1)}(t) + b_n y_n^{(2)}(t) \right] X_n(x) \tag{5.76}$$

is a solution of Equation (5.51) and satisfies boundary conditions in Equation (5.52).

Then we compose the series

$$u(x,t) = \sum_{n=1}^{\infty} \left[a_n y_n^{(1)}(t) + b_n y_n^{(2)}(t) \right] X_n(x), \tag{5.77}$$

which can be considered as the expansion of the unknown function $u(x,t)$ into a Fourier series using an orthogonal system of functions $\{X_n(x)\}$.

The superposition (5.77) allows us to satisfy the initial conditions (5.53). The first one gives

$$u|_{t=0} = \varphi(x) = \sum_{n=1}^{\infty} a_n X_n(x). \tag{5.78}$$

If the series (5.77) converges uniformly, it can be differentiating with respect to t and the second initial condition gives

$$\frac{\partial u}{\partial t}\Big|_{t=0} = \psi(x) = \sum_{n=1}^{\infty} \left[\omega_n b_n - \kappa a_n\right] X_n(x) \tag{5.79}$$

(here and in (5.81) we replace ω_n by 1 when $\kappa^2 = a^2\lambda_n + \gamma$ to consider all three cases for κ^2 in (5.75) simultaneously).

For "reasonably smooth", like piecewise continuous, initial conditions, the series (5.78) and (5.79) converge uniformly which allows us to find coefficients a_n and b_n. By multiplying both sides of these equations by $X_n(x)$, integrating from 0 to l, and using the orthogonality condition defined in Equation (5.59), we obtain

$$a_n = \frac{1}{||X_n||^2} \int_0^l \varphi(x) X_n(x) dx \tag{5.80}$$

and

$$b_n = \frac{1}{\omega_n} \left[\frac{1}{||X_n||^2} \int_0^l \psi(x) X_n(x) dx + \kappa a_n \right]. \tag{5.81}$$

The series (5.77) with these coefficients gives the solution of the boundary value problem given in Equations (5.51) through (5.53).

Recall that the success of this method is based on the following details: the functions $\{X_n(x)\}$ are orthogonal to each other and form a complete set (i.e., a basis for an expansion of $u(x,t)$); the functions $\{X_n(x)\}$ satisfy the same boundary conditions as the solutions, $u(x,t)$; and solutions to linear equations obey the superposition principle (i.e., sums of solutions are also solutions).

The obtained solution describes *free oscillations*. Processes with not very big damping, i.e., $\kappa^2 < a^2\lambda_n + \gamma$, are periodic (or quasi-periodic) and have a special physical interest. For $\kappa = 0$, the motion is purely periodic; thus $\omega_n = a\sqrt{\lambda_n}$ are the frequencies. The partial solutions $u_n(x,t) = T_n(t)X_n(x)$ are called *normal modes*. The first term, $u_1(x,t)$, called the *first (or fundamental) harmonic*, has time dependence with frequency ω_1 and period $2\pi/\omega_1$. The *second harmonic* (or *first overtone*), $u_2(x,t)$, oscillates with greater frequency ω_2 (for Dirichlet or Neumann type boundary conditions and $\kappa = \gamma = 0$, it is twice ω_1); the *third harmonic* is called *second overtone*, etc. The points where $X_n(x) = 0$ are not moving are called *nodes* of the harmonic $u_n(x,t)$. Between the nodes the string oscillates up and down. The waves $u_n(x,t)$ are also called *standing waves* because the position of the nodes are fixed in time. The general solution, $u(x,t)$, is a *superposition of standing waves*; thus, any oscillation can be presented in this way.

Clearly, $\omega_n = a\sqrt{\lambda_n}$ increase with the tension and decrease with the length and density: tuning any stringed instrument is based on changing the tension, and the bass strings are longer and heavier. The loudness of a sound is characterized by the energy or amplitude of the oscillations; tone by the period of oscillations; timbre by the ratio of energies of the main mode and overtones. The presence of high overtones destroys the harmony of a sound producing dissonance. Low overtones, in contrast, give a sense of completeness to a sound.

We give two examples for homogeneous wave equations with homogeneous boundary conditions. Note, that to simplify the formulas and the graphs for $X_n(x)$, we take them in dimensionless form; in particular $X_n(x)$ are bounded by the values ± 1. From that it follows that the coefficients a_n and b_n have dimension of function $u(x,t)$ which, for a case of string oscillations, is a length.

Example 5.4 The ends of a uniform string of length l are fixed and all external forces including the gravitational force can be neglected. Displace the string from equilibrium by

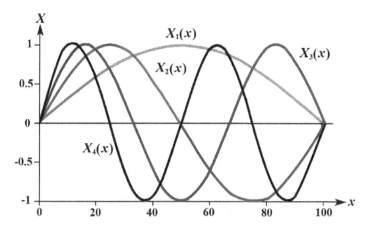

FIGURE 5.8
Eigenfunctions $X_1(x)$ through $X_4(x)$ for Example 5.4.

shifting the point $x = x_0$ by distance A at time $t = 0$ and then release it with zero initial speed. Find the displacements $u(x, t)$ of the string for times $t > 0$.

Solution. The boundary value problem is

$$u_{tt} - a^2 u_{xx} = 0, \quad 0 < x < l, t > 0,$$

$$u(x, 0) = \begin{cases} \dfrac{A}{x_0} x, & 0 < x \le x_0, \\ \dfrac{A(l - x)}{l - x_0}, & x_0 < x < l, \end{cases} \qquad \dfrac{\partial u}{\partial t}(x, 0) = 0, \quad u(0, t) = u(l, t) = 0.$$

This initial condition is not differentiable in the point $x = x_0$; it means we are searching for a generalized solution.

The eigenvalues and eigenfunctions are those of the Dirichlet problem on $0 < x < l$:

$$\lambda_n = \left(\frac{n\pi}{l}\right)^2, \quad X_n(x) = \sin \frac{n\pi x}{l}, \quad \|X_n\|^2 = \frac{l}{2}, \quad n = 1, 2, 3, \ldots.$$

Figure 5.8 shows the first four eigenfunctions for $l = 100$.
Using Equations (5.80) and (5.77), we obtain

$$a_n = \frac{2Al^2}{\pi^2 x_0(l - x_0)n^2} \sin \frac{n\pi x_0}{l}, \quad b_n = 0.$$

Therefore, string vibrations are given by the series

$$u(x, t) = \frac{2Al^2}{\pi^2 x_0(l - x_0)} \sum_{n=1}^{\infty} \frac{1}{n^2} \sin \frac{n\pi x_0}{l} \sin \frac{n\pi x}{l} \cos \frac{n\pi a t}{l}.$$

Figure 5.9 shows the space-time-dependent solution $u(x, t)$ for Example 5.4 (for the case when $a^2 = 1$, $l = 100$, $A = 6$, and $x_0 = 25$). The animation sequence in Figure 5.9 shows snapshots of the animation at times $t = 0, 1, \ldots, 12$. Because there is no dissipation, it is sufficient to run the simulation until the time is equal to the period of the main harmonic (until $2l/a = 200$ in this Example).

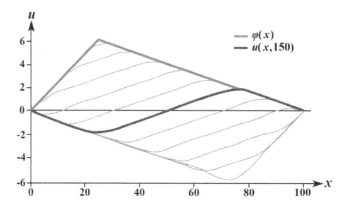

FIGURE 5.9
Solutions, $u(x,t)$, at various times for Example 5.4.

For $t = 0$ the obtained solution gives

$$u(x,0) = \sum_{n=1}^{\infty} a_n \sin \frac{n\pi x}{l}$$

–this is the Fourier expansion of the initial condition function.

In the solution $u(x,t)$ the terms for which $\sin(n\pi x_0/l) = 0$ vanish from the series, that is, the solution does not contain overtones for which $x = x_0$ is a node. For instance, if x_0 is at the middle of the string, the solution does not contain harmonics with even numbers.

The formula for $u(x,t)$ can be presented in a more physically intuitive form. Let us denote $\sqrt{\lambda_n}$ as k_n, where $k_n = \frac{n\pi}{l}$ which are called *wave numbers*. The frequencies $\omega_n = \frac{n\pi a}{l}$ give *the frequency spectrum* of the problem. The solution to the problem now can be written as

$$u(x,t) = \sum_{n=1}^{\infty} A_n \cos \omega_n t \cdot \sin k_n x,$$

where the amplitude $A_n = \frac{D}{n^2} \sin \frac{n\pi x_0}{l}$ with $D \equiv \frac{2Al^2}{\pi^2 x_0(l-x_0)}$.

The first harmonic (or mode), $u_1(x,t)$, is zero for all x when $\omega_1 t = \pi/2$, that is, when $t = l/2a$. It becomes zero again for $t = 3l/2a$, $t = 5l/2a$, etc. The second harmonic $u_2(x,t)$ is zero for all x for the moments of time $t = l/4a$, $t = 3l/4a$, etc.

The bar chart in Figure 5.10 represents $|V_n(t)| = |A_n| \cos \omega_n t$ (in units of D) for the case $x_0 = l/4 = 25$ and $t = 0$ (i.e. $|V_n(t)| = |A_n|$). The terms with numbers $n = 4k$ vanish from the series. Note that the amplitudes A_n decrease as $1/n^2$.

Example 5.5 A homogeneous rod of length l elastically fixed at the end $x = l$ is stretched by a longitudinal force $F_0 = \text{const}$, applied to the end at $x = 0$. At time $t = 0$ the force F_0 stops acting. Find the longitudinal oscillations of the rod if initial velocities are zero; the resistance of a medium as well as external forces are absent.

Solution. Let us first find initial displacements of locations along the rod, $u|_{t=0} = \varphi(x)$. Because in each cross section the force of tension T is constant and equals F_0 we have $\varphi'(x) = -F_0/EA$, where E is Young's modulus and A is the cross-sectional area of the rod. Negative $\varphi'(x)$ corresponds to a decrease of the longitudinal shift from the left end to the right. If F_0 is a compressing force, the expression for $\varphi'(x)$ will have the opposite

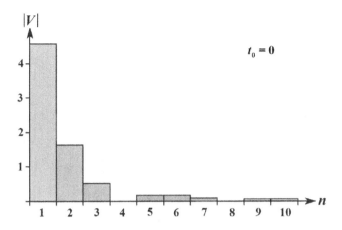

FIGURE 5.10
Bar charts of $|V_n(t)|$ for Example 5.4.

sign. Integrating and using $\varphi'(l) + h\varphi(l) = 0$ (since the rod is elastically fixed at $x = l$) we obtain

$$\varphi(x) = \frac{F_0}{EA}\left(l + \frac{1}{h} - x\right).$$

Here, h is the elasticity coefficient. Thus, we have the equation

$$u_{tt} = a^2 u_{xx}, \quad 0 < x < l, \quad t > 0,$$

with initial and boundary conditions

$$u(x, 0) = \frac{F_0}{EA}\left(l + \frac{1}{h} - x\right), \quad u_t(x, 0) = 0,$$

$$u_x(0, t) = 0, \quad u_x(l, t) + hu(l, t) = 0.$$

Figure 5.11 shows graph curves of the two functions $\tan\mu$ and hl/μ, plotted on the same set of axes. The roots μ_n are at the intersection points of these curves.

The boundary conditions are the Neumann type at $x = 0$ (the free end) and mixed type at $x = l$ (an elastic connection). The eigenvalues are easily obtained and are:

$$\lambda_n = \left(\frac{\mu_n}{l}\right)^2, \quad n = 1, 2, 3, \ldots,$$

where μ_n is the nth root of the equation $\tan\mu = \frac{hl}{\mu}$. Each eigenvalue corresponds to an eigenfunction $X_n(x) = \cos\sqrt{\lambda_n}x$ with the norm

$$\|X_n\|^2 = \frac{1}{2}\left(l + \frac{h}{\lambda_n + h^2}\right).$$

Figure 5.12 shows first four eigenfunctions for $l = 100$, $h = 0.1$.

Because the initial speeds are zero, all coefficients, b_n, are zero. Coefficients a_n are found using Equation (5.76):

$$a_n = \frac{1}{\|X_n\|^2}\int_0^l \varphi(x)\cos(\sqrt{\lambda_n}x)dx = \frac{2F_0}{EA}\frac{1 - \cos\sqrt{\lambda_n}l + (\sqrt{\lambda_n}/h)\sin\sqrt{\lambda_n}l}{\lambda_n[l + h/(\lambda_n + h^2)]}.$$

The results of the computation of λ_n, $\|X_n\|^2$ and a_n for the first ten eigenfunctions are summarized in the table presented in Figure 5.13.

n	μ_n
1	1.4288
2	4.3058
3	7.2281
4	10.2003
5	13.2142
6	16.2594
7	19.3270
8	22.4108
9	25.5064
10	28.6106

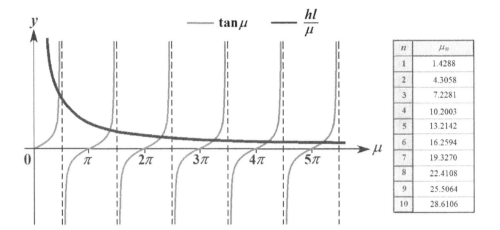

FIGURE 5.11
Graphical solution of the eigenvalue equation for $l = 100$, $h = 0.1$.

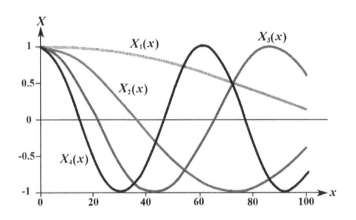

FIGURE 5.12
Eigenfunctions $X_1(x)$ through $X_4(x)$ for Example 5.5.

n	λ_n	$\|X_n\|^2$	a_n
1	0.0002042	54.9000	4.460800
2	0.0018540	54.2180	0.497415
3	0.0052247	53.2842	0.179607
4	0.0104045	52.4504	0.091622
5	0.0174615	51.8207	0.055257
6	0.0264367	51.3722	0.036816
7	0.0373534	51.0559	0.026218
8	0.0502246	50.8302	0.019585
9	0.0650576	50.6662	0.015169
10	0.0818565	50.5443	0.012085

FIGURE 5.13
Eigenvalues, norms, $\|X_n\|^2$, coefficients a_n for Example 5.5.

Factor $\cos\sqrt{\lambda_n}at$ describing the time dependence of the solution, can be written as $\cos\omega_n t$, thus harmonics' periods are $2\pi/\omega_n = 2\pi/\sqrt{\lambda_n}a$.

Finally, we have the series that describes the rod's oscillations:

$$u(x,t) = \frac{2F_0}{EA}\sum_{n=1}^{\infty}\frac{1-\cos\sqrt{\lambda_n}l + (\sqrt{\lambda_n}/h)\sin\sqrt{\lambda_n}l}{\lambda_n\left[l+h/(\lambda_n+h^2)\right]}\cos\sqrt{\lambda_n}at\cdot\cos\sqrt{\lambda_n}x.$$

5.7 The Fourier Method for Nonhomogeneous Equations

Consider the nonhomogeneous linear equation

$$\frac{\partial^2 u}{\partial t^2} + 2\kappa\frac{\partial u}{\partial t} - a^2\frac{\partial^2 u}{\partial x^2} + \gamma u = f(x,t). \tag{5.82}$$

As we will show below, generally a solution has a form of a sum of the infinite number of terms of the (5.54) with $X_n(x)$ being the eigenfunctions of the same Sturm-Liouville problem as in the case of a homogeneous equation.

First, let the function $u(x,t)$ satisfy *homogeneous boundary conditions* given in Equation (5.52) and initial conditions given in Equation (5.53). The approach is to search for the solution as a sum of two functions

$$u(x,y) = u_1(x,y) + u_2(x,y), \tag{5.83}$$

where $u_1(x,t)$ is the solution of the homogeneous equation satisfying given boundary and initial conditions

$$\frac{\partial^2 u_1}{\partial t^2} + 2\kappa\frac{\partial u_1}{\partial t} - a^2\frac{\partial^2 u_1}{\partial x^2} + \gamma u_1 = 0, \tag{5.84}$$

$$u_1(x,t)|_{t=0} = \varphi(x), \qquad \frac{\partial u_1}{\partial t}(x,t)\bigg|_{t=0} = \psi(x), \tag{5.85}$$

$$\alpha_1 u_{1x} + \beta_1 u_1|_{x=0} = 0, \qquad \alpha_2 u_{1x} + \beta_2 u_1|_{x=l} = 0, \tag{5.86}$$

and $u_2(x,t)$ is the solution of a nonhomogeneous equation satisfying the same boundary conditions and *zero initial conditions*

$$\frac{\partial^2 u_2}{\partial t^2} + 2\kappa\frac{\partial u_2}{\partial t} - a^2\frac{\partial^2 u_2}{\partial x^2} + \gamma u_2 = f(x,t), \tag{5.87}$$

$$u_2(x,t)|_{t=0} = 0, \qquad \frac{\partial u_2}{\partial t}(x,t)\bigg|_{t=0} = 0, \tag{5.88}$$

$$\alpha_1 u_{2x} + \beta_1 u_2|_{x=0} = 0, \qquad \alpha_2 u_{2x} + \beta_2 u_2|_{x=l} = 0. \tag{5.89}$$

Clearly, the function $u_1(x,t)$ represents *free oscillations* (i.e., the oscillations due to initial perturbation only) and the function $u_2(x,t)$ represents *forced oscillations* (i.e., the oscillations due to an external force when initial disturbances are zero).

The methods for finding the solution in the case of free oscillations, $u_1(x,t)$, have been discussed in the previous section; therefore we turn our attention in this section to finding the forced oscillation solutions, $u_2(x,t)$. Similar to the case of free oscillations, let us search the solution for the function $u_2(x,t)$ as the series

$$u_2(x,t) = \sum_{n=1}^{\infty} T_n(t)X_n(x), \tag{5.90}$$

where $X_n(x)$ are eigenfunctions of the corresponding boundary value problem for $u_1(x,t)$, and $T_n(t)$ are unknown functions of t (they are different from the functions $T_n(t)$ in Equation (5.73) obtained for the homogeneous wave equation).

We choose the functions $T_n(t)$ in such a way that the series (5.90) satisfies the nonhomogeneous Equation (5.87) and *zero initial conditions* given in Equation (5.88). Substituting the series (5.90) into Equation (5.87), we obtain (assuming that this series can be differentiated the necessary number of times)

$$\sum_{n=1}^{\infty} \left\{ [T_n''(t) + 2\kappa T_n'(t) + \gamma T_n(t)] X_n(x) - a^2 T_n(t) X_n''(x) \right\} = f(x,t).$$

Because $X_n(x)$ are the eigenfunctions of the corresponding homogeneous boundary value problem (5.84) – (5.86), they satisfy equation

$$X_n''(x) + \lambda_n X_n(x) = 0.$$

Using this, we obtain

$$\sum_{n=1}^{\infty} \left[T_n''(t) + 2\kappa T_n'(t) + (a^2 \lambda_n + \gamma) T_n(t) \right] X_n(x) = f(x,t). \tag{5.91}$$

Because of the completeness of the set of functions $\{X_n(x)\}$, we can expand the function $f(x,t)$ on the interval $(0,l)$ into a Fourier series of the functions $X_n(x)$ so that

$$f(x,t) = \sum_{n=1}^{\infty} f_n(t) X_n(x), \tag{5.92}$$

where

$$f_n(t) = \frac{1}{||X_n||^2} \int_0^l f(x,t) X_n(x) dx. \tag{5.93}$$

Comparing the series in Equation (5.91) and that in Equation (5.92), we obtain an ordinary second order linear differential equation with constant coefficients for each function $T_n(t)$:

$$T_n''(t) + 2\kappa T_n'(t) + (a^2 \lambda_n + \gamma) T_n(t) = f_n(t), \quad n = 1, 2, 3, \ldots. \tag{5.94}$$

In order that the function $u_2(x,t)$ represented by the series given in Equation (5.90) satisfies initial conditions in Equation (5.88), it is clear that functions $T_n(t)$ should satisfy the conditions

$$T_n(0) = 0, \quad T_n'(0) = 0, \quad n = 1, 2, 3, \ldots. \tag{5.95}$$

Solutions of linear equation (5.94) can be easily obtained in a standard way, but here we give the solution of the Cauchy problem given in Equations (5.94) and (5.95) for functions $T_n(t)$ as an integral representation which is very convenient for our further purposes:

$$T_n(t) = \int_0^t f_n(\tau) Y_n(t - \tau) d\tau. \tag{5.96}$$

Here

$$Y_n(t) = \begin{cases} \dfrac{1}{\omega_n} e^{-\kappa t} \sin \omega_n t, & \text{if } \kappa^2 < a^2 \lambda_n + \gamma, \\[2mm] \dfrac{1}{\omega_n} e^{-\kappa t} \sinh \omega_n t, & \text{if } \kappa^2 > a^2 \lambda_n + \gamma, \\[2mm] t e^{-\kappa t}, & \text{if } \kappa^2 = a^2 \lambda_n + \gamma, \end{cases} \tag{5.97}$$

where $\omega_n = \sqrt{|a^2 \lambda_n + \gamma - \kappa^2|}$.

To prove that the representation in Equation (5.96) yields Equation (5.94), we differentiate Equation (5.96) with respect to t twice to get

$$T_n'(t) = \int_0^t f_n(\tau) \frac{\partial}{\partial t} Y_n(t - \tau) d\tau + Y_n(0) f_n(t)$$

and

$$T_n''(t) = \int_0^t f_n(\tau) \frac{\partial^2}{\partial t^2} Y_n(t - \tau) d\tau + Y_n'(0) f_n(t).$$

By using these formulas, we can prove that Equation (5.94) and the initial conditions in Equation (5.95) are satisfied.

Reading Exercise: Verify this result for $\kappa = 0$ (i.e., for $Y_n(t) = \sin(\omega_n t)/\omega_n$).

With expression (5.96) for $T_n(t)$ the series (5.90) gives, assuming that it converges uniformly as well as the series obtained by differentiating by x and t up to two times, the solution of the boundary value problem given in Equations (5.87)-(5.89).

Combining the results for the functions $u_1(x,t)$ and $u_2(x,t)$, the solution of the forced oscillations problem with initial conditions (5.53) and homogeneous boundary conditions (5.52) is

$$u(x,t) = u_1(x,t) + u_2(x,t)$$
$$= \sum_{n=1}^{\infty} \left\{ \left[a_n y_n^{(1)}(t) + b_n y_n^{(2)}(t) \right] + T_n(t) \right\} X_n(x), \qquad (5.98)$$

where the $T_n(t)$ are defined by Equation (5.92) and $y_n^{(1)}(t)$, $y_n^{(2)}(t)$, a_n, and b_n are given by the formulas (5.74), (5.75), (5.80) and (5.81).

Let us consider two examples of nonhomogeneous problems with homogeneous boundary conditions.

Example 5.6 Consider a homogeneous string of mass density ρ with rigidly fixed ends. Starting at time $t = 0$ a uniformly distributed harmonic force with linear density

$$F(x,t) = F_0 \sin \omega t$$

acts on the string. The initial deflection and speed are zero. Neglecting friction find the resulting oscillations and investigate the resonance behavior.

Solution. The boundary value problem modeling this process is

$$\frac{\partial^2 u}{\partial t^2} - a^2 \frac{\partial^2 u}{\partial x^2} = \frac{F_0}{\rho} \sin \omega t, \quad 0 < x < l, \quad t > 0, \qquad (5.99)$$

$$u(x,0) = 0, \quad u_t(x,0) = 0,$$
$$u(0,t) = u(l,t) = 0.$$

The boundary conditions here are homogeneous Dirichlet conditions, thus the eigenvalues and eigenfunctions are

$$\lambda_n = \left(\frac{n\pi}{l} \right)^2, \quad X_n(x) = \sin \frac{n\pi x}{l}, \quad \|X_n\|^2 = \frac{l}{2}, \quad n = 1, 2, 3, \ldots.$$

Because the initial conditions are $\varphi(x) = \psi(x) = 0$, the solution of the homogeneous wave equation, $u_1(x,t) = 0$, and the only nonzero contribution is the solution of the nonhomogeneous equation, $u_2(x,t)$, which is given by the series

$$u(x,t) = \sum_{n=1}^{\infty} T_n(t) \sin \frac{n\pi x}{l}. \tag{5.100}$$

We can find $T_n(t)$ by using Equation (5.96):

$$T_n(t) = \int_0^t f_n(\tau) Y_n(t - \tau) d\tau,$$

where $\omega_n = a\sqrt{\lambda_n} = \frac{n\pi a}{l}$ are the natural frequencies and $Y_n(t) = \frac{1}{\omega_n} \sin \frac{n\pi a t}{l}$. We then have

$$f_n(t) = \frac{2}{l} \int_0^l f(x,t) \sin \frac{n\pi x}{l} dx = \frac{2}{l} \frac{F_0}{\rho} \sin \omega t \int_0^l \sin \frac{n\pi x}{l} dx,$$

which gives

$$f_{2n-1}(t) = \frac{4F_0}{(2n-1)\pi\rho} \sin \omega t, \quad f_{2n}(t) = 0.$$

Then, for $T_n(t)$ we have

$$T_{2n-1}(t) = \frac{l}{(2n-1)\pi a} \frac{4F_0}{(2n-1)\pi\rho} \int_0^t \sin \omega\tau \, \sin\left[\omega_{2n-1}(t-\tau)\right] d\tau$$

$$= \frac{4F_0 l}{a\pi^2(2n-1)^2\rho} \frac{\omega_{2n-1} \sin \omega t - \omega \sin \omega_{2n-1} t}{\omega_{2n-1}^2 - \omega^2},$$

$$T_{2n}(t) = 0.$$

These expressions apply when, for any n, the frequency of the external force, ω, is not equal to any of the natural frequencies of the string. Substituting $T_n(t)$ into Equation (5.100) we obtain the solution

$$u(x,t) = \frac{4F_0 l}{\rho a\pi^2} \sum_{n=1}^{\infty} \frac{1}{(2n-1)^2} \frac{\omega_{2n-1} \sin \omega t - \omega \sin \omega_{2n-1} t}{\omega_{2n-1}^2 - \omega^2} \sin \frac{n\pi x}{l}. \tag{5.101}$$

Because ω_n is proportional to n, terms in (5.101) with $\sin \omega t$, which are purely periodic with frequency ω, decrease as $1/n$, whereas terms with $\sin \omega_{2n-1} t$, which represent oscillations (eigenmodes) with different frequencies, decrease as $1/n^2$.

Figure 5.14 shows the spatial-time-dependent solution $u(x,t)$ for Example 5.6 (for the case when $a^2 = 1$, $l = 100$, $\rho = 1$, $F_0 = 0.025$, $\omega = 0.3$). With these parameters the period of the main harmonic is $2l/a = 200$. The animation sequence in Figure 5.14 shows snapshots of the animation at times $t = 0, 1, \ldots, n$.

If for some $n = k$, the frequency of the external force $\omega = \omega_{2k-1}$, we have a case of *resonance*. To proceed, we apply L'Hôpital's rule to this term, taking the derivatives of the numerator and denominator with respect to ω_{2k-1} to yield

$$-\frac{2F_0 l}{\rho a\pi^2(2k-1)^2} \frac{\omega_{2n-1} t \cos \omega_{2k-1} t - \sin \omega_{2n-1} t}{\omega_{2k-1}}$$

$$= \frac{2F_0 l^2}{\rho a^2\pi^3(2k-1)^3} \left[\sin \omega_{2n-1} t - \omega_{2n-1} t \cos \omega_{2k-1} t\right].$$

In this case, one can rewrite the solution in the form

$$u(x,t) = \frac{4F_0 l}{\rho a \pi^2} \sum_{\substack{n=1 \\ n \neq k}}^{\infty} \frac{1}{(2n-1)^2} \frac{\omega_{2n-1} \sin \omega t - \omega \sin \omega_{2n-1} t}{\omega_{2n-1}^2 - \omega^2} \sin \frac{n\pi x}{l}$$

$$+ \frac{2F_0 l^2}{\rho a^2 \pi^3 (2k-1)^3} [\sin \omega t - \omega t \cos \omega t] \sin \frac{(2k-1)\pi x}{l}. \tag{5.102}$$

Let us obtain a periodic solution for Example 5.6. Contrary to the solution obtained above in the form of an infinite series, a steady-state solution is described by a finite formula which is more convenient for the analysis of the properties of the solution. Let us write the equation as

$$\frac{\partial^2 u}{\partial t^2} - a^2 \frac{\partial^2 u}{\partial x^2} = A \sin \omega t \tag{5.103}$$

and seek a steady-state solution in the form

$$u(x,t) = X(x) \sin \omega t. \tag{5.104}$$

For $X(x)$ we obtain an ordinary linear differential equation

$$X''(x) + \left(\frac{\omega}{a}\right)^2 X(x) = -\frac{A}{a^2}, \tag{5.105}$$

which has a general solution

$$X(x) = c_1 \cos kx + c_2 \sin kx - \frac{A}{\omega^2} \tag{5.106}$$

where $k = \omega/a$. The boundary conditions $X(0) = X(l) = 0$ give $c_1 = A/\omega^2$ and $c_2 = \frac{A}{\omega^2} \frac{1-\cos kl}{\sin kl}$. Then, using standard trigonometric formulas for double angles we obtain

$$u(x,t) = \frac{2A}{\omega^2} \frac{\sin(kx/2) \sin[k(l-x)/2]}{\cos(kl/2)} \sin \omega t. \tag{5.107}$$

Notice, that for this solution the initial speed is not zero.

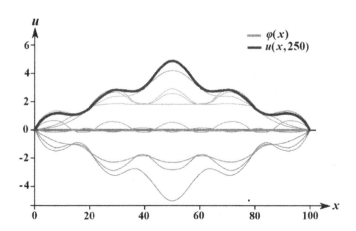

FIGURE 5.14
Graph of solutions, $u(x,t)$, at various times for Example 5.6.

Example 5.7 The upper end of an elastic homogeneous heavy rod is rigidly attached to the ceiling of free falling elevator. When the elevator reaches the speed v_0 it stops instantly. Set up the boundary value problem for vibrations of the rod.

Solution. The mathematical model for this problem is:

$$u_{tt} - a^2 u_{xx} = -g, \quad 0 < x < l, \quad t > 0,$$

$$u(x,0) = 0, \quad u_t(x,0) = v_0,$$

$$u(0,t) = 0, \quad u_x(l,t) = 0,$$

where g is the acceleration of gravity.

The boundary conditions are of Dirichlet type at $x = 0$ (the fixed end) and Neumann type at $x = l$ (the free end). From our previous work, we have eigenvalues and eigenfunctions for the problem given by

$$\lambda_n = \left[\frac{(2n-1)\pi}{2l} \right]^2, \quad X_n(x) = \sin \frac{(2n-1)\pi x}{2l}, \quad n = 1, 2, 3, \dots.$$

From Equations (5.76), (5.77), (5.89), and (5.92), we have

$$a_n = 0, \quad b_n = \frac{8v_0 l}{a\pi^2(2n-1)^2},$$

$$f_n(t) = -\frac{4g}{\pi(2n-1)}, \quad T_n(t) = -\frac{16gl^2}{a^2\pi^3(2n-1)^3} \left[1 - \cos \frac{(2n-1)a\pi t}{2l} \right].$$

Thus, a general solution of this boundary value problem is

$$u(x,t) = \frac{8l}{a\pi^2} \sum_{n=1}^{\infty} \frac{1}{(2n-1)^2}$$

$$\times \left\{ \frac{-2gl}{a\pi(2n-1)} \left[1 - \cos \frac{(2n-1)a\pi t}{2l} \right] + v_0 \sin \frac{(2n-1)\pi a t}{2l} \right\} \sin \frac{(2n-1)\pi x}{2l}.$$

5.8   The Laplace Transform Method: Simple Cases

In this section we demonstrate how Laplace transform (LT) can be applied to solve heat, or generally a PDE equation. The LT allows us to obtain an ODE from PDE and then an algebraic equation from an ODE.

As a simple illustration, let us consider oscillations of a semi-infinite string governed by the wave equation

$$u_{tt} - a^2 u_{xx} = 0, \quad x > 0, \quad -\infty < t < \infty. \tag{5.108}$$

Initially, the string is at rest, $u(x,t) = 0$ for $t \le 0$ (hence also $u_t(x,0) = 0$ for $t \le 0$). At $t = 0$, the left end of the string starts moving according to a certain law $u(0,t) = f(t)$, while at infinity, function $u(\infty, t)$ has to be bounded.

Let us find the solution of the problem at $t > 0$ using the Laplace transform with respect to t.

Denote

$$\hat{u}(x,p) = \int_0^\infty e^{-pt} u(x,t) dt$$

and

$$\hat{f}(p) = \int_0^\infty e^{-pt} f(t) dt.$$

Because $u(x,0) = 0$, $u_t(x,0) = 0$ for any $x > 0$, the Laplace transform of Equation (5.108) leads to the ODE with respect to x,

$$p^2 \hat{u}(x,p) - a^2 \hat{u}_{xx}(x,p) = 0$$

with the boundary condition

$$\hat{u}(0,p) = \hat{f}(p).$$

The solution of this problem bounded on infinity is

$$\hat{u}(x,p) = \hat{f}(p) e^{-px/a}, \quad x > 0.$$

Using the Delay Theorem, we find that $u(x,t) = 0$ for $t < x/a$, i.e., $x > at$, and $u(x,t) = f(x - t/a)$ for $t > x/a$, i.e., $x < at$.

Example 5.8 Solve the nonhomogeneous equation,

$$u_{tt} - a^2 u_{xx} = \sin \frac{\pi x}{l}, \quad t > 0, \tag{5.109}$$

with zero initial and boundary conditions

$$u(x,0) = 0, \quad u_t(x,0) = 0, \tag{5.110}$$

$$u(0,t) = u(l,t) = 0, \quad t > 0. \tag{5.111}$$

Let us apply the Laplace transform to Equation (5.109) in order to obtain an ordinary differential equation. Taking into account the initial conditions (5.110), we obtain the ODE with respect to x:

$$p^2 \hat{u}(x,p) - a^2 \hat{u}_{xx}(x,p) = \frac{1}{p} \sin \frac{\pi x}{l},$$

or

$$\hat{u}_{xx}(x,p) - \frac{p^2}{a^2} \hat{u}(x,p) = -\frac{1}{pa^2} \sin \frac{\pi x}{l}. \tag{5.112}$$

Solving the ordinary differential equation (5.112) with boundary conditions

$$\hat{u}(0,p) = \hat{u}(l,p) = 0, \tag{5.113}$$

we obtain

$$\hat{u}(x,p) = \frac{1}{p[p^2 + (\pi a/l)^2]} \sin \frac{\pi x}{l} = \left(\frac{l}{\pi a}\right)^2 \left[\frac{1}{p} - \frac{p}{p^2 + (\pi a/l)^2}\right] \sin \frac{\pi x}{l}.$$

The inverse Laplace transform (see the table in Appendix A) gives

$$u(x,t) = \left(\frac{l}{\pi a}\right)^2 \left(1 - \cos \frac{\pi a t}{l}\right) \sin \frac{\pi x}{l}.$$

Let us present some general consideration for the Cauchy problem for function $T_n(t)$ satisfying the nonhomogeneous ODE (5.94),

$$T_n''(t) + 2\kappa T_n'(t) + (a^2 \lambda_n + \gamma) T_n(t) = f_n(t), \quad n = 1, 2, 3, \ldots.$$

The solution has an especially simple form when the initial conditions are zero, $T_n(0) = 0$, $T_n'(0) = 0$.

Defining

$$\hat{T}_n(p) = L[T_n(t)] = \int_0^\infty e^{-pt} T_n(t) dt,$$

$$\hat{f}_n(p) = L[f_n(t)] = \int_0^\infty e^{-pt} f_n(t) dt,$$

we obtain the following algebraic equation for $\hat{T}_n(p)$:

$$(p^2 + 2\kappa p + a^2\lambda_n + \gamma)\hat{T}_n(p) = \hat{f}_n(p),$$

hence

$$\hat{T}_n(p) = \frac{\hat{f}_n(p)}{p^2 + 2\kappa p + a^2\lambda_n + \gamma}.$$

Using the convolution property, we find that

$$T_n(t) = \int_0^t f_n(\tau) Z_n(t - \tau) d\tau,$$

where

$$Z_n(t) = L^{-1}\left[\frac{1}{p^2 + 2\kappa p + a^2\lambda_n + \gamma}\right] = L^{-1}\left[\frac{1}{(p + \kappa)^2 + (a^2\lambda_n + \gamma - \kappa^2)}\right].$$

To find the inverse Laplace transform, we can use the table given in Appendix A. If $a^2\lambda_n + \gamma - \kappa^2 \equiv \omega_n^2 > 0$, we can see that

$$Z_n(t) = \frac{1}{\omega_n} e^{-\kappa t} \sin \omega_n t.$$

If $a^2\lambda_n + \gamma - \kappa^2 \equiv -\omega_n^2 < 0$, we can use the Shift Theorem:

$$Z_n(t) = L^{-1}\left[\frac{1}{(p + \kappa)^2 - \omega_n^2}\right] = e^{-\kappa t} L^{-1}\left[\frac{1}{p^2 - \omega_n^2}\right] = \frac{1}{\omega_n} e^{-\kappa t} \sinh \omega_n t.$$

If $a^2\lambda_n + \gamma - \kappa^2 = 0$, we find that

$$Z_n(t) = L^{-1}\left[\frac{1}{(p + \kappa)^2}\right] = e^{-\kappa t} L^{-1}\left[\frac{1}{p^2}\right] = t e^{-\kappa t}.$$

This is the same result as obtained in formulas (5.96), (5.97) with the replacement $Y_n(t)$ by $Z_n(t)$.

Notice, that the LT is not really useful for PDEs without formulas for the inverse Laplace transforms – obtaining the final results needs the technique of integration in the complex plane.

5.9   Equations with Nonhomogeneous Boundary Conditions

With the preceding development, we are prepared to study a general boundary value problem for the nonhomogeneous equation with nonhomogeneous boundary conditions defined by

$$\frac{\partial^2 u}{\partial t^2} + 2\kappa \frac{\partial u}{\partial t} - a^2 \frac{\partial^2 u}{\partial x^2} + \gamma u = f(x, t), \tag{5.114}$$

$$u(x,t)|_{t=0} = \varphi(x), \quad u_t(x,t)|_{t=0} = \psi(x), \tag{5.115}$$

$$\alpha_1 u_x + \beta_1 u|_{x=0} = g_1(t), \quad \alpha_2 u_x + \beta_2 u|_{x=l} = g_2(t). \tag{5.116}$$

The Fourier method cannot be applied to this problem directly because the boundary conditions are nonhomogeneous. However, we can reduce this problem to the previously investigated case with homogeneous boundary conditions.

To proceed, let us search for the solution as a sum of two functions

$$u(x,t) = v(x,t) + w(x,t), \tag{5.117}$$

where $v(x,t)$ is a new, unknown function and the function $w(x,t)$ is chosen in a way that satisfies the given nonhomogeneous boundary conditions

$$\alpha_1 w_x + \beta_1 w|_{x=0} = g_1(t), \quad \alpha_2 w_x + \beta_2 w|_{x=l} = g_2(t). \tag{5.118}$$

The function $w(x,t)$ should also have the necessary number of continuous derivatives in x and t.

For the function $v(x,t)$, we obtain the boundary value problem with *homogeneous boundary conditions*

$$\frac{\partial^2 v}{\partial t^2} + 2\kappa \frac{\partial v}{\partial t} - a^2 \frac{\partial^2 v}{\partial x^2} + \gamma v = \tilde{f}(x,t), \tag{5.119}$$

$$v(x,t)|_{t=0} = \tilde{\varphi}(x), \quad v_t(x,t)|_{t=0} = \tilde{\psi}(x), \tag{5.120}$$

$$\alpha_1 v_x + \beta_1 v|_{x=0} = 0, \quad \alpha_2 v_x + \beta_2 v|_{x=l} = 0, \tag{5.121}$$

where

$$\tilde{f}(x,t) = f(x,t) - \frac{\partial^2 w}{\partial t^2} - 2\kappa \frac{\partial w}{\partial t} + a^2 \frac{\partial^2 w}{\partial x^2}, \tag{5.122}$$

$$\tilde{\varphi}(x) = \varphi(x) - w(x,0), \tag{5.123}$$

$$\tilde{\psi}(x) = \psi(x) - w_t(x,0). \tag{5.124}$$

The solution to this problem has been described above.

Reading Exercise: Verify Equations (5.119)-(5.124).

For the auxiliary function, $w(x,t)$, a number of choices are possible. One criterion is to simplify the form of the equation for the function, $v(x,t)$. To this end, we search for $w(x,t)$ in the form

$$w(x,t) = P_1(x)g_1(t) + P_2(x)g_2(t), \tag{5.125}$$

where $P_1(x)$ and $P_2(x)$ are polynomials; we will show that in some cases we need polynomials of the first order, and in some of the second order. Coefficients of these polynomials should be chosen in such a way that the function $w(x,t)$ satisfies the boundary conditions (5.118); therefore the function $v(x,t)$ satisfies homogeneous boundary conditions (5.121).

First, consider the situation when parameters β_1 and β_2 are not zero simultaneously.

Let us take $P_{1,2}(x)$ as polynomials of the first order and search for the function $w(x,t)$ in the form

$$w(x,t) = (\gamma_1 + \delta_1 x)g_1(t) + (\gamma_2 + \delta_2 x)g_2(t). \tag{5.126}$$

Substituting this into boundary conditions (5.118), we obtain two equations which may be written as

$$\begin{cases} \alpha_1[\delta_1 g_1(t) + \delta_2 g_2(t)] + \beta_1 [\gamma_1 g_1(t) + \gamma_2 g_2(t)] = g_1(t), \\ \alpha_2[\delta_1 g_1(t) + \delta_2 g_2(t)] + \beta_2 [(\gamma_1 + \delta_1 l)g_1(t) + (\gamma_2 + \delta_2 l)g_2(t)] = g_2(t), \end{cases}$$

or as

$$\begin{cases} (\alpha_1\delta_1 + \beta_1\gamma_1 - 1)g_1(t) + (\alpha_1\delta_2 + \beta_1\gamma_2)g_2(t) = 0, \\ (\alpha_2\delta_1 + \beta_2\gamma_1 + \beta_2\delta_1 l)g_1(t) + (\alpha_2\delta_2 + \beta_2\gamma_2 + \beta_2\delta_2 l - 1)g_2(t) = 0. \end{cases}$$

To be true, the coefficients of $g_1(t)$ and $g_2(t)$ should be zero, in which case the following system of equations is valid for arbitrary t:

$$\begin{cases} \alpha_1\delta_1 + \beta_1\gamma_1 - 1 = 0, \\ \alpha_1\delta_2 + \beta_1\gamma_2 = 0, \\ \alpha_2\delta_1 + \beta_2\gamma_1 + \beta_2\delta_1 l = 0, \\ \alpha_2\delta_2 + \beta_2\gamma_2 + \beta_2\delta_2 l - 1 = 0. \end{cases}$$

From this system of equations we obtain coefficients γ_1, δ_1, γ_2 and δ_2 as

$$\gamma_1 = \frac{\alpha_2 + \beta_2 l}{\beta_1\beta_2 l + \beta_1\alpha_2 - \beta_2\alpha_1}, \quad \delta_1 = \frac{-\beta_2}{\beta_1\beta_2 l + \beta_1\alpha_2 - \beta_2\alpha_1}, \tag{5.127}$$

$$\gamma_2 = \frac{-\alpha_1}{\beta_1\beta_2 l + \beta_1\alpha_2 - \beta_2\alpha_1}, \quad \delta_2 = \frac{\beta_1}{\beta_1\beta_2 l + \beta_1\alpha_2 - \beta_2\alpha_1}. \tag{5.128}$$

Therefore, the choice of function $w(x,t)$ in the form (5.118) is consistent with the boundary conditions (5.127).

Reading Exercise: Obtain the results (5.127) and (5.128).

Now, consider the case when $\beta_1 = \beta_2 = 0$, i.e. $u_x|_{x=0} = g_1(t)$, $u_x|_{x=l} = g_2(t)$.

In this situation, the system of Equations (5.126) is inconsistent, which is why the polynomials $P_{1,2}(x)$ in the expression for $w(x,t)$ must be of higher order. Let us take the second order polynomials:

$$P_1(x) = \gamma_1 + \delta_1 x + \xi_1 x^2, \quad P_2(x) = \gamma_2 + \delta_2 x + \xi_2 x^2. \tag{5.129}$$

Equations (5.118) yield a system of simultaneous equations which, to be valid for arbitrary t, leads to $\gamma_1 = \gamma_2 = \delta_2 = 0$ and the expression for $w(x,t)$ given by

$$w(x,t) = \left[x - \frac{x^2}{2l}\right]g_1(t) + \frac{x^2}{2l}g_2(t). \tag{5.130}$$

Reading Exercise: Check that when $\beta_1 = \beta_2 = 0$ such a choice of $w(x,t)$ satisfies the boundary conditions (5.118).

Combining all described types of boundary conditions, we obtain nine different auxiliary functions, which are given in Appendix C part 2.

The following two examples are applications of the wave equation with nonhomogeneous boundary conditions.

Example 5.9 The left end of a string is moving according to

$$g(t) = A\sin\omega t,$$

and the right end, $x = l$, is secured. Initially, the string is at rest. Describe oscillations when there are no external forces and the resistance of the surrounding medium is zero.

Solution. The boundary value problem is

$$u_{tt} = a^2 u_{xx}, \quad 0 < x < l, \quad t > 0,$$

$$u(x,0) = 0, \quad u_t(x,0) = 0,$$

$$u(0,t) = A\sin\omega t, \quad u(l,t) = 0.$$

Let us search for the solution as the sum

$$u(x,t) = v(x,t) + w(x,t),$$

where we use an auxiliary function

$$w(x,t) = \left(1 - \frac{x}{l}\right) A\sin\omega t,$$

which satisfies the same as $u(x,t)$, boundary conditions, i.e., $w(0,t) = A\sin\omega t$ and $w(l,t) = 0$. This obvious choice of $w(x,t)$ can also be obtained from Equations (5.127), (5.128) which give $\gamma_1 = 1$, $\delta_1 = -1/l$, $\gamma_2 = 0$, $\delta_2 = 1/l$. For $v(x,t)$ we arrive to the boundary value problem for the nonhomogeneous wave equation with external force (5.122):

$$\tilde{f}(x,t) = A\omega^2 \left(1 - \frac{x}{l}\right)\sin\omega t,$$

initial conditions (5.123), (5.124):

$$\tilde{\varphi}(x,t) = 0, \quad \tilde{\psi}(x,t) = -A\omega\left(1 - \frac{x}{l}\right)$$

and homogeneous boundary conditions.

The eigenvalues and eigenfunctions are

$$\lambda_n = \left(\frac{n\pi}{l}\right)^2, \quad X_n(x) = \sin\frac{n\pi x}{l}, \quad \|X_n\|^2 = \frac{l}{2}, \quad n = 1,2,3,\dots.$$

From Equations (5.80), (5.81), and (5.93), we have

$$a_n = 0, \quad b_n = \frac{1}{\omega_n \|X_n\|^2}\int_0^l \psi^*(x)\sin\frac{n\pi x}{l}xdx = -\frac{2A\omega l}{an^2\pi^2},$$

$$f_n(t) = \frac{2A\omega^2}{n\pi}\sin\omega t.$$

We can find $T_n(t)$ by using Equation (5.96):

$$T_n(t) = \frac{2A\omega^2 l}{a\pi^2 n^2} \cdot \frac{\omega\sin\omega_n t - \omega_n\sin\omega t}{\omega^2 - \omega_n^2}.$$

This expression applies, when, for any n, the frequency of the external force $\tilde{f}(x,t)$ is not equal to any of the natural frequencies $\omega_n = \frac{n\pi a}{l}$ of the string (i.e., $\omega \neq \omega_n$). In this case, the solution is

$$u(x,t) = w(x,t) + \sum_{n=1}^{\infty} [T_n(t) + b_n\sin\omega_n t]\sin\frac{n\pi x}{l}$$

$$= \left(1 - \frac{x}{l}\right) A\sin\omega t + \frac{2A\omega l}{a\pi^2}\sum_{n=1}^{\infty}\frac{1}{n^2}\frac{\omega_n(\omega_n\sin\omega_n t - \omega\sin\omega t)}{\omega^2 - \omega_n^2}\sin\frac{n\pi x}{l}.$$

Figure 5.15 shows the spatial-time-dependent solution $u(x,t)$ for Example 5.9 for the case when $a^2 = 1$, $l = 100$, $A = 4$, $\omega = 0.3$. With these parameters the period of the main harmonic is $2l/a = 200$.

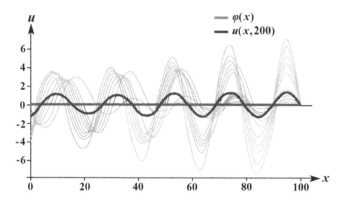

FIGURE 5.15
Graph of solutions, $u(x, t)$ at various times for Example 5.9.

Example 5.10 Find the longitudinal vibrations of a rod, $0 \leq x \leq l$, with the left end fixed. To the right end at $x = l$ the force

$$F(t) = Bt \quad (A = \text{const})$$

is applied starting at time $t = 0$. The initial deflection and speed are zero. Neglect friction.

Solution. We should solve the equation

$$u_{tt} = a^2 u_{xx}, \quad 0 < x < l, \quad t > 0$$

with the boundary and initial conditions

$$u(x, 0) = 0, \quad u_t(x, 0) = 0,$$

$$u(0, t) = 0, \quad u_x(l, t) = Bt/EA.$$

Assuming

$$u(x, t) = v(x, t) + w(x, t)$$

it is clear that we may choose the function $w(x, t)$ to have the form

$$w(x, t) = \frac{B}{EA} xt$$

such that $w(x, t)$ satisfies both boundary conditions (this choice also can be formally obtained as explained above). For $v(x, t)$, we obtain the boundary value problem

$$\nu_{tt} = a^2 \nu_{xx},$$

$$v(x, t)|_{t=0} = 0, \quad \nu_t(x, t)|_{t=0} = -Bx/EA,$$

$$v(0, t) = 0, \quad \nu_x(l, t) = 0.$$

Eigenvalues and eigenfunctions of this problem are

$$\lambda_n = \frac{(2n-1)\pi}{2l}, \quad X_n(x) = \sin \frac{(2n-1)\pi x}{2l}, \quad n = 1, 2, 3, \ldots.$$

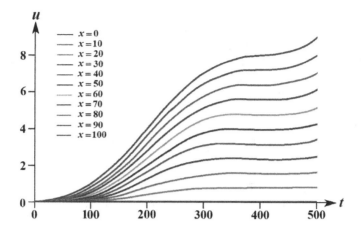

FIGURE 5.16
Time-traces of solutions $u(x,t)$ at various values of x for Example 5.10.

From Equations (5.80) and (5.81) we have

$$a_n = 0, \quad b_n = (-1)^n \frac{16Bl^2}{EAa\,(2n-1)^3\pi^3}$$

in which case the solution for $v(x,t)$ is

$$v(x,t) = \frac{16Al^2}{ESa\pi^3} \sum_{n=1}^{\infty} \frac{(-1)^n}{(2n-1)^3} \sin \frac{(2n-1)a\pi t}{2l} \sin \frac{(2n-1)\pi x}{2l}.$$

Thus, the solution of this problem is

$$u(x,t) = \frac{B}{EA}xt + \frac{16Bl^2}{EAa\pi^3} \sum_{n=1}^{\infty} \frac{(-1)^n}{(2n-1)^3} \sin \frac{(2n-1)a\pi t}{2l} \sin \frac{(2n-1)\pi x}{2l}.$$

Figure 5.16 shows the solution $u(x,t)$ as a function of t ($0 \le t \le 500$) for several values of coordinate x for Example 5.10. This solution was obtained in the case when $a^2 = 1$, $l = 100$, $E = 1$, $A = 1$, $B = 0.0002$.

5.10   The Consistency Conditions and Generalized Solutions

Let us briefly discuss the consistency between initial and boundary conditions and situations when functions representing these conditions are not perfectly smooth. Assuming that the solution of a boundary value problem, $u(x,t)$, and its first derivatives are continuous on the closed interval $[0,l]$, the initial condition $u(x,t)|_{t=0} = \varphi(x)$ and $u_t(x,t)|_{t=0} = \psi(x)$ at $x = 0$ and $x = l$ lead to the relations

$$\alpha_1\varphi_x + \beta_1\varphi|_{x=0} = g_1(0), \quad \alpha_2\varphi_x + \beta_2\varphi|_{x=l} = g_2(0), \tag{5.131}$$

and

$$\alpha_1\psi_x + \beta_1\psi|_{x=0} = \frac{\partial g_1}{\partial t}(0), \quad \alpha_2\psi_x + \beta_2\psi|_{x=l} = \frac{\partial g_2}{\partial t}(0). \tag{5.132}$$

If not all of these *consistency relations* are satisfied, or the functions $f(x,t)$, $\varphi(x)$, $\psi(x)$, $g_1(x,t)$, $g_2(x,t)$ and their derivatives are not continuous on the entire interval $[0,l]$, we still can solve a problem by the methods described above. In such cases, as well as for real, physical problems in which these functions always have some limited precision, such obtained solutions are *generalized solutions*.

A simple example is the problem of string oscillations in a gravitational field. The external force (nonhomogeneous term) is not zero at the ends of the string and if the ends are fixed, the boundary conditions are not consistent with the equation. But the situation is not that bad because the formal (generalized) solution converges uniformly at any point in the interval.

Also notice that in Example 5.4 from Section 5.6 the boundary condition $u_x(0,t) = 0$ and initial condition $u(x,0) = F_0\left(l+1/h-x\right)/EA$ are inconsistent. This inconsistency may lead to the non-uniform convergence of the series for derivatives.

Reading Exercise: Solve the problem of a string oscillating in gravity field

$$\frac{\partial^2 u}{\partial t^2} = a^2 \frac{\partial^2 u}{\partial x^2} - g$$

if the ends are fixed and initially the string was displaced at some point from the equilibrium position and released with zero speed. Discuss the properties of the analytical solution and its behavior at the end of the interval. Suggest inconsistent initial and boundary conditions and analyze the obtained (generalized) solution.

5.11   Energy in the Harmonics

Next, we consider the energy associated with the motion of the string. The kinetic energy density (energy per unit length) of the string is

$$\frac{1}{2}\rho\left(\frac{\partial u}{\partial t}\right)^2$$

and the total kinetic energy then is

$$E_{kin}(t) = \frac{1}{2}\int_0^l \rho\left(\frac{\partial u}{\partial t}\right)^2 dx. \tag{5.133}$$

Next, we determine an expression for the potential energy. If a portion of the string of initial length dx is stretched to a length ds when displaced, the increase in length is

$$ds - dx = \left[\sqrt{1 + \left(\frac{\partial u}{\partial x}\right)^2} - 1\right] dx.$$

In the approximation of "small vibrations" $(\partial u/\partial x \ll 1)$, this becomes

$$\frac{1}{2}\left(\frac{\partial u}{\partial x}\right)^2 dx.$$

Since this stretching takes place against a force of tension T, the potential energy gain on the interval dx, which is the work done against tension, is

$$\frac{1}{2}T\left(\frac{\partial u}{\partial x}\right)^2 dx.$$

Thus, the potential energy of the string is

$$E_{pot}(t) = \frac{1}{2}\int_0^l T\left(\frac{\partial u}{\partial x}\right)^2 dx. \tag{5.134}$$

The total energy is therefore

$$E_{tot} = E_{kin} + E_{pot} = \frac{\rho}{2}\int_0^l\left[\left(\frac{\partial u}{\partial t}\right)^2 + a^2\left(\frac{\partial u}{\partial x}\right)^2\right] dx, \tag{5.135}$$

where we have used the fact that $T = \rho a^2$.

It is easy to see (we leave the proof as the *Reading Exercise*), that for homogeneous boundary conditions of Dirichlet or Neumann type (e.g., the ends of a string are rigidly fixed or free) the derivatives of functions $X_n(x)$ are orthogonal (like the functions $X_n(x)$ themselves):

$$\int_0^l \frac{dX_n}{dx}\frac{dX_m}{dx} dx = 0. \tag{5.136}$$

The physical significance of this equation is that in this case the individual terms in the Fourier series solution are independent and energy in a harmonic cannot be exchanged with the energies associated with the other terms.

The nth harmonic is

$$u_n(x,t) = \left\{T_n(t) + \left[a_n y_n^{(1)}(t) + b_n y_n^{(2)}(t)\right]\right\} X_n(x) \tag{5.137}$$

(for homogeneous boundary conditions the auxiliary function $w(x,t) \equiv 0$).
Its kinetic energy is

$$E_{kin}^{(n)}(t) = \frac{\rho}{2}\int_0^l\left(\frac{\partial u_n}{\partial t}\right)^2 dx = \frac{\rho}{2}\left[\frac{dT_n}{dt} + a_n\frac{dy_n^{(1)}}{dt} + b_n\frac{dy_n^{(2)}}{dt}\right]^2\int_0^l X_n^2(x)dx \tag{5.138}$$

and its potential energy is

$$E_{pot}^{(n)}(t) = \frac{T}{2}\int_0^l\left(\frac{\partial u_n}{\partial x}\right)^2 dx = \frac{T}{2}\left[T_n(t) + a_n y_n^{(1)}(t) + b_n y_n^{(2)}(t)\right]^2\int_0^l\left(\frac{dX_n}{dx}\right)^2 dx. \tag{5.139}$$

It is evident that there are no terms representing interactions of the harmonics in either the kinetic or potential energy expressions. The total energy of the nth harmonic is the sum of these two energies.

Consider, as an example, free vibrations that occur in a medium without damping,

$$\frac{\partial^2 u}{\partial t^2} = a^2\frac{\partial^2 u}{\partial t^2}. \tag{5.140}$$

In this case

$$T_n(t) = 0, \quad y_n^{(1)}(t) = \cos a\sqrt{\lambda_n}t, \quad y_n^{(2)}(t) = \sin a\sqrt{\lambda_n}t, \tag{5.141}$$

and the kinetic energy of the nth harmonic is

$$E_{kin}^{(n)}(t) = \frac{\rho a^2 \lambda_n}{2}\left[-a_n \sin a\sqrt{\lambda_n}t + b_n \cos a\sqrt{\lambda_n}t\right]^2 \|X_n\|^2; \tag{5.142}$$

the potential energy of nth harmonic is

$$E_{pot}^{(n)}(t) = \frac{T}{2}\left[a_n \cos a\sqrt{\lambda_n}t + b_n \sin a\sqrt{\lambda_n}t\right]^2 \left\|\frac{dX_n}{dx}\right\|^2. \tag{5.143}$$

For boundary conditions of Dirichlet or Neumann type we have

$$\|X_n\|^2 = \int_0^l X_n^2(x)dx = \frac{l}{2}, \quad \left\|\frac{dX_n}{dx}\right\|^2 = \int_0^l \left(\frac{dX_n}{dx}\right)^2 dx = \lambda_n\|X_n\|^2 = \frac{\lambda_n l}{2}. \tag{5.144}$$

The potential and kinetic energies of a harmonic taken together give a constant, the total energy, which is (taking into account $T = \rho a^2$)

$$E_{tot}^{(n)}(t) = E_{kin}^{(n)}(t) + E_{pot}^{(n)}(t) = \frac{\rho a^2 \lambda_n l}{4}\left(a_n^2 + b_n^2\right). \tag{5.145}$$

In each harmonic, the energy oscillates between kinetic and potential forms as the string itself oscillates. The periods of the oscillations of the string are

$$\tau_n = \frac{2\pi}{a\sqrt{\lambda_n}} \tag{5.146}$$

and those of the energy are half that value.

Reading Exercises: We leave it to the reader to work out a few more results:

1. If both ends of the string are rigidly fixed or both ends are free, then $\lambda_n = \frac{\pi^2 n^2}{l^2}$ and the total energy is

$$E_{tot}^{(n)}(t) = \frac{\rho a^2 \pi^2 n^2}{4l}\left(a_n^2 + b_n^2\right). \tag{5.147}$$

2. If one end of the string is rigidly fixed and the other is free, then $\lambda_n = \frac{\pi^2(2n+1)^2}{4l^2}$ and the total energy is

$$E_{tot}^{(n)}(t) = \frac{\rho a^2 \pi^2(2n+1)^2}{16l}\left(a_n^2 + b_n^2\right). \tag{5.148}$$

3. As an example of another physical system, consider energy oscillations in the RLC circuit. Magnetic and electric energies are

$$E_L = \frac{Li^2}{2} \quad \text{and} \quad E_C = \frac{CV^2}{2}.$$

With $i = \partial q/\partial t$ this gives (for a wire from 0 to l)

$$E_L = \frac{1}{2}\int_0^l L\left(\frac{\partial q}{\partial t}\right)^2 dx \quad \text{and} \quad E_C = \frac{1}{2}\int_0^l R^2 C\left(\frac{\partial q}{\partial t}\right)^2 dx. \tag{5.149}$$

Reading Exercises: Using the material described in Section 5.3.1, obtain the expressions for these energies for boundary conditions of Dirichlet and Neumann type.

Let us use Examples 5.3 and 5.5 from Sections 5.6 and 5.7 to find the energy of a string under various conditions. In both examples, as the *Reading Exercises*, find analytical expressions for kinetic, potential and total energies of the string as functions of time.

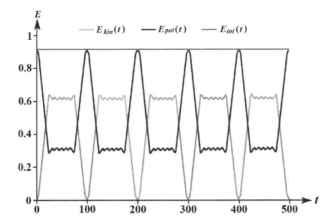

FIGURE 5.17
Energies of the string for Example 5.3.

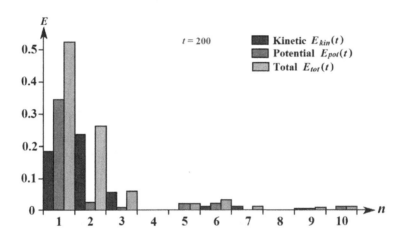

FIGURE 5.18
Distribution of kinetic, potential and total energies in harmonics at $t = 200$ for Example 5.3.

In Example 5.3 (fixed ends, no external forces), the string was displaced from equilibrium by shifting the point $x = x_0$ by a distance h at time $t = 0$ and then released with zero initial speed. Figure 5.17 shows the energies on the string; the following values of the parameters were chosen: $a^2 = 1$, $\kappa = 0$, $l = 100$, $h = 6$, and $x_0 = 25$.

Figures 5.18 demonstrate energy in harmonics at time $t = 200$.

We may solve the same problem when the force of friction is not zero. Here we again use the same values of the parameters and with the friction coefficient, $\kappa = 0.001$ (see Figure 5.19).

In Example 5.5 from Section 5.7, we considered a string with rigidly fixed ends and no friction. Starting at time $t = 0$ a harmonic force with linear density $F(x,t) = F_0 \sin \omega t$ acts on the string. In Figure 5.20 we present the energy graphs for the following values of the parameters: $a^2 = 1$, $\kappa = 0$, $l = 100$, $\rho = 1$, $F_0 = 0.025$, $\omega = 0.3$. Obviously, the energy is not conserved.

Figure 5.21 represents a resonance case. As the *Reading Exercises* describe and explain the difference between the cases presented in Figures 5.20 and 5.21.

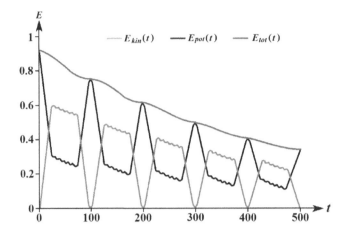

FIGURE 5.19
Energies of the string for Example 5.3 with dissipation included.

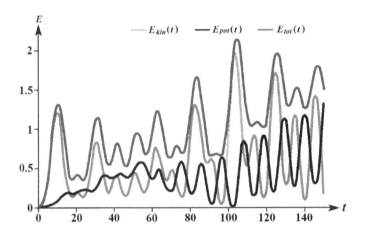

FIGURE 5.20
Energies of the string for Example 5.5.

5.12 Dispersion of Waves

5.12.1 Cauchy Problem in an Infinite Region

Let us return to hyperbolic equations in an infinite region. Is it possible to extend the method of separation of variables to the case of an infinite region? In the case of the wave equation,

$$\frac{\partial^2 u}{\partial t^2}(x,t) = a^2 \frac{\partial^2 u}{\partial x^2}(x,t), \tag{5.150}$$

there is no need in that, because the D'Alembert's formula (5.30) provides the full solution of the Cauchy problem. How can we tackle more general equations? As the basic example, let us consider the following equation with constant coefficients,

$$\frac{\partial^2 u}{\partial t^2} + 2\kappa \frac{\partial u}{\partial t} - a^2 \frac{\partial^2 u}{\partial x^2} + \gamma u = 0, \quad \kappa \geq 0,\ \gamma \geq 0. \tag{5.151}$$

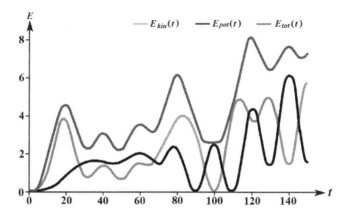

FIGURE 5.21

Energies of the string for the same values of the parameters and with a driving frequency $\omega = \omega_5 = 5\pi/100$.

In Section 5.6 the initial-boundary value problem for that equation was solved in a finite spatial region. Now we define that equation in an infinite region,

$$-\infty < x < \infty, \quad t \geq 0. \tag{5.152}$$

We impose initial conditions

$$u(x,t)|_{t=0} = \varphi(x), \quad u_t(x,t)|_{t=0} = \psi(x) \tag{5.153}$$

and the boundary conditions

$$\lim_{x \to \pm\infty} u(x,t) = 0. \tag{5.154}$$

Recall that in the case of a finite region, we presented the solution of the problem as an infinite sum of particular solutions of a special kind, $u_n(x,t) = X_n(x)T_n(t)$. In the case of an infinite region, a similar role is played by particular solutions

$$u_k(x,t) = \hat{u}(k,t)e^{ikx}, \tag{5.155}$$

which are spatially periodic with the period $2\pi/k$, where k is an arbitrary real number called *wavenumber*. Substituting (5.155) into Equation (5.151), we obtain the following ordinary differential equation for function $\hat{u}(k,t)$:

$$\frac{d^2\hat{u}}{dt^2} + 2\kappa\frac{d\hat{u}}{dt} + (\gamma + a^2k^2)\hat{u} = 0. \tag{5.156}$$

The general solution of Equation (5.156) is

$$\hat{u}(k,t) = a_+(k)e^{-i\omega_+(k)t} + a_-(k)e^{-i\omega_-(k)t}, \tag{5.157}$$

where $a_\pm(k)$ are arbitrary numbers and

$$\omega_\pm(k) = -i\kappa \pm \sqrt{\gamma - \kappa^2 + a^2k^2} \tag{5.158}$$

are the roots of the *dispersion relation*

$$\omega^2(k) + 2i\kappa\omega(k) - \gamma - a^2k^2 = 0. \tag{5.159}$$

If $\gamma - \kappa^2 + a^2 k^2 > 0$, solution (5.157) can be written as

$$\hat{u}(k,t) = a_+(k)e^{i(kx-\sqrt{\gamma-\kappa^2+a^2k^2}\,t)-\kappa t} + a_-(k)e^{i(kx+\sqrt{\gamma-\kappa^2+a^2k^2}\,t)-\kappa t}. \qquad (5.160)$$

The first term in the right-hand side of (5.160) corresponds to a wave moving to the right with *phase velocity*

$$v_{ph}(k) = \frac{\sqrt{\gamma - \kappa^2 + a^2 k^2}}{k},$$

while the second term describes a wave moving to the left with the same velocity; both waves decay with the *decay rate* κ. If $\gamma - \kappa^2 + a^2 k^2 < 0$, both terms monotonically decay with time:

$$\hat{u}(k,t) = a_+(k)e^{ikx-(\kappa-\sqrt{\kappa^2-\gamma-a^2k^2})t} + a_-(k)e^{ikx-(\kappa+\sqrt{\kappa^2-\gamma-a^2k^2})t}.$$

Note that functions (5.157) satisfy Equation (5.151) but do not satisfy conditions (5.153)-(5.154).

Equation (5.151) is linear; therefore any linear combinations of solutions (5.155) with different k are also solutions of that equation (this is the *superposition principle*). Because k is changed continuously, the most general expression for the linear superposition of spatially periodic waves (5.155) is the *integral*

$$u(x,t) = \int_{-\infty}^{\infty} \hat{u}(k,t)e^{ikx}\frac{dk}{2\pi} \qquad (5.161)$$

(the factor $1/2\pi$ is introduced for convenience). Thus, the general solution of Equation (5.151) can be written as

$$u(x,t) = \int_{-\infty}^{\infty} \left[a_+(k)e^{i(kx-\omega_+(k)t)} + a_-(k)e^{i(kx-\omega_-(k)t)}\right]\frac{dk}{2\pi}, \qquad (5.162)$$

where $a_+(k)$ and $a_-(k)$ are arbitrary functions of k.

Formula (5.161) shows that $\hat{u}(k,t)$ is just the *Fourier transform* of the solution $u(x,t)$ (see Appendix A). If $\hat{u}(k,t)$ is absolutely integrable, i.e., $\int_{-\infty}^{\infty} |u(k,t)|\frac{dk}{2\pi}$ is finite, then $u(x,t)$ satisfies condition (5.154) (see Appendix A).

Let us determine the coefficients $a_\pm(k)$ that allow us to satisfy the initial conditions (5.153). Apply the Fourier transform to both sides of (5.153):

$$\hat{u}(k,0) = \hat{\varphi}(k), \quad \hat{u}_t(k,0) = \hat{\psi}(k), \qquad (5.163)$$

where

$$\hat{\varphi}(k) = \int_{-\infty}^{\infty} \varphi(x)e^{-ikx}\frac{dk}{2\pi}, \quad \hat{\psi}(k) = \int_{-\infty}^{\infty} \psi(x)e^{-ikx}\frac{dk}{2\pi}. \qquad (5.164)$$

Substituting (5.157) into (5.163), we find

$$a_+(k) + a_-(k) = \hat{\varphi}(k), \quad -i\omega_+(k)a_+(k) - i\omega_-(k)a_-(k) = \hat{\psi},$$

hence

$$a_+(k) = \frac{-\omega_-(k)\hat{\varphi}(k) + i\hat{\psi}(k)}{\omega_+(k) - \omega_-(k)}, \quad a_-(k) = \frac{\omega_+(k)\hat{\varphi}(k) - i\hat{\psi}(k)}{\omega_+(k) - \omega_-(k)}. \qquad (5.165)$$

Formulas (5.162), (5.164)-(5.165) describe the solution of problem (5.151)-(5.154). Note that the solution $u(x,t)$ is real, if $a_\pm(k) = a_\pm^*(-k)$.

We emphasize that wave equation (5.150) provides a unique example where the phase velocity $v_{ph}(k) = a$ does not depend on k, hence all the waves with different k move (to the left and to the right) with the same velocity. Therefore, their superposition (5.162) consists of two waves moving to the left and to the right without changing their shapes. In any other cases, different constituents of (5.162) move with different velocities and/or decay with different rates; therefore their superposition changes its shape with time. That phenomenon is called *dispersion* of waves.

5.12.2 Propagation of a Wave Train

To better understand the evolution of dispersive waves, let us consider some simple examples.

Example 5.11 First, consider a wave moving rightwards with *a definite wavenumber* k_0

$$u(x, t; k_0) = a_+ e^{i(k_0 x - \omega_+(k_0)t)}, \tag{5.166}$$

which is perfectly periodic in space and time,

$$u(x + 2\pi/k_0, t; k_0) = u(x, t; k_0), \quad u(x, t + 2\pi/\omega_+ t; k_0) = u(x, t).$$

Note that this solution is complex; the real solution can be constructed as

$$\begin{aligned}
u(x, t) &= u(x, t; k_0) + u^*(x, t; k_0) \\
&= 2|a_+| \cos(k_0 - \operatorname{Re}\omega_+(k_0)t + \arg a_+)e^{\operatorname{Im}\omega_+(k_0)t}, \tag{5.167}
\end{aligned}$$

where

$$\arg a_+ = \tan^{-1}(\operatorname{Im} a_+/\operatorname{Re} a_+).$$

The Fourier transform of (5.167) is

$$\begin{aligned}
\hat{u}(k, t; k_0) &= \int_{-\infty}^{\infty} u(x, t)e^{-ikx}dx \\
&= 2\pi a_+ \delta(k - k_0)e^{-i\omega_+(k_0)t} + 2\pi a_+^* \delta(k + k_0)e^{i\omega_+^*(k_0)t}. \tag{5.168}
\end{aligned}$$

If $\omega_+(k_0)$ is real

$$u(x, t) = 2|a_+| \cos(k_0 x - i\omega_+(k_0)t + \arg a_+) \tag{5.169}$$

describes a *monochromatic traveling wave*, periodic in time and space, which moves with velocity $v_{ph}(k_0) = \omega_+(k_0)/k_0$ without changing its shape. Otherwise, it simultaneously grows or decreases exponentially in time, depending on the sign of $\operatorname{Im}\omega_+(k_0)$.

Example 5.12 Let us consider a particular solution which is the superposition of two monochromatic waves (5.169) with *different wave numbers* and equal amplitudes,

$$\begin{aligned}
u(x, t) &= 2a \cos(k_1 x - i\omega_+(k_1)t) + 2a \cos(k_2 x - i\omega_+(k_2)t) \\
&= 4a \cos\left(\frac{k_1 - k_2}{2}x - \frac{\omega_+(k_1) - \omega_+(k_2)}{2}t\right) \\
&\quad \times \cos\left(\frac{k_1 + k_2}{2}x - \frac{\omega_+(k_1) + \omega_+(k_2)}{2}t\right). \tag{5.170}
\end{aligned}$$

Let k_1 and k_2 be close: $k_1 = k_0 + \Delta k$, $k_2 = k_0 - \Delta k$, $0 < \Delta k \ll k_0$; then the solution can be written as

$$u(x,t) = 4a \cos\left[\Delta k\left(x - \frac{\omega_+(k_0 + \Delta k) - \omega_+(k_0 - \Delta k)}{2\Delta k}t\right)\right]$$

$$\times \cos\left[k_0\left(x - \frac{\omega_+(k_0 + \Delta k) + \omega_+(k_0 - \Delta k)}{2k_0}t\right)\right].$$

This *modulated wave* is a product of a rapidly oscillating *carrier wave* with the spatial period $l = 2\pi/k_0$, which moves with the velocity

$$\frac{\omega_+(k_0 + \Delta k) + \omega_+(k_0 - \Delta k)}{2k_0} \approx \frac{\omega_+(k_0)}{k_0} = v_{ph}(k_0),$$

and the long-periodic *envelope function* with the spatial period $L = 2\pi/\Delta k \gg l$, which moves with the velocity

$$\frac{\omega_+(k_0 + \Delta k) - \omega_+(k_0 - \Delta k)}{2\Delta k} \approx \frac{d\omega_+}{dk}(k_0).$$

The latter expression,

$$v_{gr}(k_0) = \frac{d\omega_+}{dk}(k_0),$$

is called *group velocity*. Note that

$$L \cdot \Delta k = 2\pi = O(1)$$

when Δk is small.

Of course, solutions (5.167) and (5.170) are not physically feasible: the wave can actually occupy only a finite part of the space of a certain length Δx. As mentioned in Appendix A, the corresponding characteristic interval Δk is connected with Δx by the relation $\Delta x \Delta k \sim 1$ which in quantum mechanics is called the *uncertainty relation*.

Let us consider a *nearly periodic wave train* moving rightwards with the Fourier transform concentrated in a narrow interval of the width $\Delta k \ll 1$ around the mean value k_0. For sake of simplicity, assume that $\omega_0(k)$ is real (e.g., $\kappa = 0$ in Equation (5.151)). The real solution can be written as

$$u(x,t) = \int_{-\infty}^{\infty} a_+(k)e^{i[kx - \omega_+(k)t]}\frac{dk}{2\pi} + \text{c.c.}, \tag{5.171}$$

where c.c. means the expression which is a complex conjugate to the written one.

Define $q \equiv k - k_0$, $a_+(k_0 + q) \equiv \hat{f}(q)$ and present solution (5.171) as

$$u(x,0) = e^{i(k_0 x - \omega_+(k_0)t)}\int_{-\infty}^{\infty} \hat{f}(q)e^{iqx - i[\omega_+(k_0 + q) - \omega_+(k_0)]t}\frac{dq}{2\pi} + \text{c.c.}$$

Let us expand $\omega_+(k_0 + q)$ into the Taylor series:

$$\omega_+(k_0 + q) - \omega_+(k_0) \approx \omega'_+(k_0)q + \frac{1}{2}\omega''_+(k_0)q^2 + \ldots,$$

hence

$$u(x,t) = e^{i[k_0 x - \omega_+(k_0)t]}I(x,t) + \text{c.c.},$$

$$I(x,t) \approx \int_{-\infty}^{\infty} \hat{f}(q) e^{iq(x - \omega'_+(k_0)t)} e^{-i\omega''_+(k_0)q^2 t} \frac{dq}{2\pi}. \tag{5.172}$$

Because the main contribution into the integral is given by the interval of $|q| \ll \Delta k$, one can omit the term with $\omega''_+(k_0)$, if $t \ll [\omega''_+]^{-1} (\Delta k)^{-2}$. In that case,

$$u(x,t) \approx e^{i[k_0 x - \omega_+(k_0)t]} \int_{-\infty}^{\infty} \hat{f}(q) e^{iq(x - \omega'_+(k_0)t)} \frac{dq}{2\pi} + \text{c.c.}$$

$$= e^{ik_0[x - (\omega_+(k_0)/k_0)t]} f(x - \omega'_+(k_0)t) + \text{c.c.}$$

Thus, the solution is a product of a rapidly oscillating *carrier wave* moving with the phase velocity $v_{ph}(k_0) = \omega_+(k_0)/k_0$ and the slowly changing (with the characteristic spatial scale $1/\Delta k \gg 1$) *envelope* moving with the *group velocity* $v_{gr}(k_0) = \omega'_+(k_0)$. For dispersive waves $(\omega_+(k_0) \neq ak)$ these two velocities do not coincide.

At large t, the term with $\omega''_+(k_0)$ in (5.172) cannot be ignored. Let us calculate $I(x,t)$ in the point $x = \omega'_+(k_0)t$ moving with the group velocity of the wave:

$$I = \int_{-\infty}^{\infty} \hat{f}(q) e^{-i\omega''_+(k_0)q^2 t} \frac{dq}{2\pi}.$$

In order to calculate the integral, let us carry out formally the change of variables,

$$q = z\sqrt{\frac{2}{i\omega''_+(k_0)t}}$$

(the justification of this transformation can be done using the complex analysis). Thus,

$$I = \sqrt{\frac{2}{i\omega''_+(k_0)t}} \int_{-\infty}^{\infty} \hat{f}\left(z\sqrt{\frac{2}{i\omega''_+(k_0)t}}\right) e^{-z^2} 2\pi.$$

Because only the region $z = O(1)$ gives an essential contribution into the integral, we can replace $\hat{f}\left(z\sqrt{\frac{2}{i\omega''_+(k_0)t}}\right)$ with $f(0)$. Using the relation

$$\int_{-\infty}^{\infty} e^{-z^2} dz = \sqrt{\pi},$$

we get

$$I \approx \frac{f(0)}{\sqrt{2\pi i\omega''_+(k_0)t}}.$$

We come to the conclusion that in the case of the dispersion $(\omega''_+(k_0) \neq 0)$, the intensity of the wave decreases with time. We shall return to this phenomenon in Chapter 10 when considering the properties of the Schrödinger equation.

5.13 Wave Propagation on an Inclined Bottom: Tsunami Effect

In the previous sections of this chapter we considered equations with constant coefficients. It may happen however that the wave velocity depends on the coordinate. For instance, the velocity a of a long gravity wave in a liquid layer is determined by the relation $a^2 = gh$, where g is the gravity acceleration and h is the layer depth. When the depth x is not constant, $h = h(x)$, $x \geq 0$, the wave velocity $a(x) = \sqrt{gh(x)}$ is not constant as well. Using equations of fluid dynamics, one can derive the following modification of the wave equation,

$$\frac{\partial^2 u}{\partial t^2} = \frac{\partial}{\partial x}\left[a^2(x)\frac{\partial u}{\partial x}\right], \ \ a^2(x) = gh(x), \tag{5.173}$$

where $u(x,t)$ is the deviation of the liquid surface.

Let us consider a solution of (5.173) corresponding to a wave with frequency ω,

$$u(x,t) = v(x)e^{i\omega t} + \text{c.c..} \tag{5.174}$$

For $v(x)$ we obtain the ordinary differential equation (check this result as a *Reading Exercise*):

$$\frac{d^2 v}{dx^2} + \frac{h'(x)}{h(x)}\frac{dv}{dx} + \frac{\omega^2}{gh(x)}v = 0. \tag{5.175}$$

Let us consider a gravity wave near the coast. Assume that the bottom slope is constant, i.e., the depth is changed linearly,

$$h(x) = \alpha x, \ x > 0.$$

Then we obtain,

$$\frac{d^2 v}{dx^2} + \frac{1}{x}\frac{dv}{dx} + \frac{\omega^2}{g\alpha x}v = 0, \ x > 0. \tag{5.176}$$

Let us apply the transformation

$$s = \sqrt{\frac{x}{2}}, \ v(x) \equiv w(s).$$

Substituting

$$x = 2s^2, \ \frac{d}{dx} = \frac{ds}{dx}\frac{d}{ds} = \frac{1}{2\sqrt{2x}}\frac{d}{ds} = \frac{1}{4s}\frac{d}{ds},$$

in (5.176) we obtain the Bessel equation:

$$\frac{d^2 w}{ds^2} + \frac{1}{s}\frac{dw}{ds} + q^2 w = 0, \ q^2 = \frac{8\omega^2}{g\alpha}, \ s > 0. \tag{5.177}$$

The appropriate bounded solution of (5.177) is

$$w(s) = cJ_0(qs),$$

where c is a complex constant, i.e.,

$$u(x,t) = J_0\left(\frac{2\omega}{\sqrt{g\alpha}}\sqrt{x}\right)\left(ce^{i\omega t} + c^*e^{-i\omega t}\right), \ x > 0. \tag{5.178}$$

One can see that the wave amplitude grows significantly towards the coast, i.e., with the decrease of x. That phenomenon, called *tsunami effect*, can be explained qualitatively in the following way. The wavelength of the wave, $\lambda(x) = 2\pi a(x)/\omega$, decreases when the wave moves towards the coast, i.e., the wave is compressed in the horizontal direction. The conservation of the water volume leads to the growth of the wave in the vertical direction.

Actually, everybody has observed that phenomenon when being on a beach: near the coast, the waves are higher than far in the sea. In the case of long waves created far in the ocean by an earthquake, that effect can lead to a catastrophic tsunami.

Problems

Write an analytical solution, representing the motion of a string (a rod, etc.). Solve these problems analytically which means the following: formulate the equation and initial and boundary conditions, obtain the eigenvalues and eigenfunctions, write the formulas for coefficients of the series expansion and the expression for the solution of the problem.

The reader can also obtain the graphs of eigenfunctions and illustrate the spatial-time-dependent solution $u(x,t)$ with Maple, Mathematica or software from books [7, 8].

When you choose the parameters of the problem and coefficients of the functions (initial and boundary conditions, and external forces) do not forget that the amplitudes of oscillations should remain small. All the parameters and the variables (time, coordinates, deflection $u(x,t)$), are considered to be dimensionless.

Infinite strings

In problems 1 through 4 an infinite stretched string is excited by the initial deflection $u(x,0) = \varphi(x)$ with no initial velocities.

Find the vibrations of the string.

1. $\varphi(x) = \dfrac{A}{1 + Bx^2}$.

2. $\varphi(x) = Ae^{-Bx^2}$.

3. $\varphi(x) = \begin{cases} A(l^2 - x^2), & x \in [-l, l], \\ 0, & x \notin [-l, l]. \end{cases}$

4. $\varphi(x) = \begin{cases} A\sin\frac{\pi x}{l}, & x \in [-l, l], \\ 0, & x \notin [-l, l]. \end{cases}$

In problems 5 through 8 an infinite stretched string is initially at rest. Assume at time $t = 0$ the initial distribution of velocities is given by $u_t(x,0) = \psi(x)$.

Find the vibrations of the string.

5. $\psi(x) = \dfrac{A}{1 + x^2}$.

6. $\psi(x) = Axe^{-Bx^2}$.

7. $\psi(x) = \begin{cases} v_0 = \text{const}, & x \in [-l, l], \\ 0, & x \notin [-l, l]. \end{cases}$

8. $\psi(x) = \begin{cases} x, & x \in [-l, l], \\ 0, & x \notin [-l, l]. \end{cases}$

9. Let an infinite string be at rest prior to $t = 0$. At time $t = 0$ it is excited by a sharp blow from a hammer that transmits an impulse I at point $x = x_0$ to the string.

    Find the vibrations of the string.

10. Consider an infinite thin wire with resistance R, capacitance C, inductance L and leakage G distributed along its length. Find the electrical current oscillations in a circuit if $GL = CR$. For the following initial voltage and current in the wire

    $$V(x,0) = f_1(x) = e^{-x^2}, \quad i(x,0) = f_2(x) = e^{-x^2}.$$

    find electrical voltage oscillations in the wire.

Loaded infinite string

Find the solutions of the following initial value problems.

11. $u_{tt} - u_{xx} = \cos x$, $u(x,0) = \cos x$, $u_t(x,0) = 1$.

12. $u_{tt} - u_{xx} = 2\sin t$, $u(x,0) = \cos x$, $u_t(x,0) = 1$.

13. $u_{tt} - u_{xx} = x + \cos t$, $u(x,0) = \sin x - 1$, $u_t(x,0) = \cos x$.

14. $u_{tt} - u_{xx} = -\tanh^3(x/\sqrt{2})\cos t$, $u(x,0) = \tanh(x/\sqrt{2})$, $u_t(x,0) = 0$.

15. $u_{tt} - u_{xx} = H(x) - H(x-1)$, where $H(x)$ is the Heaviside function;
    $u(x,0) = u_t(x,0) = 0$.

Transverse oscillations in finite strings

Problems 16 through 39 refer to uniform finite strings with the ends at $x = 0$ and $x = l$. In problems 16 through 20 the initial shape of a string with fixed ends is $u(x,0) = \varphi(x)$, the initial speed is $u_t(x,0) = \psi(x)$. External forces and dissipation are absent. Find the vibrations of the string.

16. $\varphi(x) = Ax\left(1 - \dfrac{x}{l}\right)$, $\psi(x) = 0$.

17. $\varphi(x) = A\sin\dfrac{\pi x}{l}$, $\psi(x) = 0$.

18. $\varphi(x) = 0$, $\psi(x) = v_0 = \text{const.}$

19. $\varphi(x) = 0$, $\psi(x) = \dfrac{B}{l}x(l - x)$.

20. $\varphi(x) = Ax\left(1 - \dfrac{x}{l}\right)$, $\psi(x) = \dfrac{B}{l}x(l - x)$.

21. A string with fixed ends is displaced at point $x = x_0$ by a small distance h from equilibrium and released at $t = 0$ without initial speed. No external forces or dissipation act. Describe the string oscillations. Find the location of x_0 so that the following overtones are absent: a) 3rd; b) 5th; c) 7th.

22. A string with fixed ends is displaced at point $x = x_0$ by a small distance h from the x axis and released at $t = 0$ without initial speed. Find the vibrations of the string if it vibrates in the constant gravitational field g, and the resistance of a medium is proportional to speed.

In problems 23 through 27 the end at $x = 0$ of a string is fixed while the end at $x = l$ is attached to a mass less ring that can slide along a frictionless rod perpendicular to the x axis such that the tangent line to the string is always horizontal. The initial shape and speed of the string are $u(x,0) = \varphi(x)$ and $u_t(x,0) = \psi(x)$. No external forces act except that the surrounding medium has a resistance with coefficient κ. Find string oscillations.

23. $\varphi(x) = A\dfrac{x}{l}$, $\psi(x) = 0$.

24. $\varphi(x) = A\sin\dfrac{\pi x}{2l}$, $\psi(x) = 0$.

25. $\varphi(x) = 0$, $\psi(x) = v_0 = \text{const.}$

26. $\varphi(x) = 0$, $\psi(x) = Ax\left(1 - \dfrac{x}{l}\right)$.

27. $\varphi(x) = Ax\left(1 - \dfrac{x}{l}\right), \quad \psi(x) = v_0 \cos \dfrac{\pi x}{l}.$

28. The ends of a string are rigidly fixed. The string is excited by the sharp blow of a hammer, supplying an impulse I at point x_0. No external forces act, but the surrounding medium supplies a resistance with coefficient κ. Find the string oscillations.

In problems 29 through 32 the ends of a string are rigidly fixed. An external force $F(t)$ begins to acts at point x_0 at time $t = 0$. Find the string oscillations in a medium with resistance coefficient κ. The zero initial conditions are zero.

29. $F(t) = F_0 \sin \omega t.$

30. $F(t) = F_0 \cos \omega t.$

31. $F(t) = F_0 e^{-At} \sin \omega t.$

32. $F(t) = F_0 e^{-At} \cos \omega t.$

In problems 33 through 36 the ends of a string are rigidly fixed. The initial conditions are zero with no resistance. Starting at time $t = 0$ a uniformly distributed force with linear density $f(x, t)$ acts on the string. Find oscillations of the string.

33. $f(x) = Ax(x/l - 1).$

34. $f(x, t) = Axe^{-Bt}.$

35. $f(x, t) = Ax(x/l - 1) \sin \omega t.$

36. $f(x, t) = Ax(x/l - 1) \cos \omega t.$

In problems 37 through 39 the left end, $x = 0$, of a string is driven according to $u(0, t) = g_1(t)$, the right end, $x = l$, is fixed. Initially the string is at rest.
Find oscillations of the string when there are no external forces and the resistance of a medium is zero. Find the solution in the case of resonance.

37. $g_1(t) = A \sin \omega t.$

38. $g_1(t) = A(1 - \cos \omega t).$

39. $g_1(t) = A(\sin \omega t + \cos \omega t).$

Longitudinal oscillations in rods

In problems 40 through 42 starting at $t = 0$ the end of a rod at $x = 0$ is moving horizontally according to $u(0, t) = g(t)$ and an external force $F = F(t)$ is applied to the end at $x = l$ along the axis. Assume zero initial conditions and an embedding medium which has a resistance proportional to speed. Describe the oscillations $u(x, t)$ of the rod.
Hint: The equation and the boundary conditions are:

$$u_{tt} + 2\kappa u_t - a^2 u_{xx} = 0,$$
$$u(x, 0) = 0, \quad u_t(x, 0) = 0,$$
$$u(0, t) = g(t), \quad u_x(l, t) = F(t)/EA.$$

40. $F(t) = F_0(1 - \cos \omega t), \quad g(t) = g_0.$

41. $F(t) = F_0, \quad g(t) = g_0 \sin \omega t.$

42. $F(t) = F_0 e^{-At} \sin \omega t, \quad g(t) = g_0 \sin \omega t.$

Electrical oscillations in circuits

In problems 43 through 47 a conductor of length l is perfectly insulated $(G = 0)$ and R, L and C are known. Current at $t = 0$ is absent and the voltage is $V(x,t)|_{t=0} = V(x)$. The end at $x = 0$ is insulated and the end at $x = l$ is grounded. Find the current $i(x,t)$ in the wire.

43. $V(x) = V_0 \sin \dfrac{\pi x}{l}$.

44. $V(x) = V_0 \left(1 - \cos \dfrac{\pi x}{l}\right)$.

45. $V(x) = V_0 x(l - x)$.

46. $V(x) = \dfrac{V_0}{(x - l/2)^2 + A}$.

47. $V(x) = V_0 \exp\left[-A(x - l/2)^2\right]$.

In problems 48 through 50 initially the current and voltage in an insulated $(G = 0)$ conductor are zero. The left end at $x = 0$ is insulated; the right end at $x = l$ is attached to a source of *emf* $E(t)$ at $t = 0$. The parameters R, L, G and C are known. Find the voltage $V(x,t)$ in the wire.

48. $E(t) = E_0 \sin \omega t$.

49. $E(t) = E_0(1 - \cos \omega t)$.

50. $E(t) = E_0(\sin \omega t + \cos \omega t)$.

Dispersive waves

51. The evolution of the complex wavefunction $u(x,t)$ of a quantum particle is governed by the (non-dimensional) Schrödinger equation

$$iu_t = -\frac{1}{2}u_{xx}, \quad -\infty < x < \infty, \quad t > 0.$$

(a) find the dispersion relation, phase and group velocities of waves;
(b) find the exact solution of the equation with initial condition $u(x,0) = \delta(x)$;
(c) find the exact solution of the equation with initial condition

$$u(x,0) = \exp(-x^2 + ik_0 x).$$

52. The evolution of small-amplitude long water waves is governed by the linear Korteweg-de Vries equation

$$u_t + u_{xxx} = 0, \quad -\infty < x < \infty, \quad t > 0.$$

(a) find the dispersion relation, phase and group velocities of waves;
(b) find the exact solution of the equation with initial condition $u(x,0) = \delta(x)$.

Hint: use the Airy function described in Appendix B.

6

One-Dimensional Parabolic Equations

In this chapter we consider a general class of equations known as parabolic equations, and their solutions. We start in Section 6.1 with two physical examples: heat conduction and diffusion. General properties of parabolic equations and their solutions are discussed in subsequent sections.

6.1 Heat Conduction and Diffusion: Boundary Value Problems

6.1.1 Heat Conduction

Heat may be defined as the flow of energy through a body due to a difference in temperature. This is a kinetic process at the molecular level and it involves energy transfer due to molecular collisions. We introduce the equations which model heat transfer in solids where there is no macroscopic mass transfer.

Heat flow through a solid due to a temperature change is assumed to obey the linear heat flow equation established by Fourier:

$$\vec{q} = -\kappa \nabla T. \tag{6.1}$$

Here $\vec{q}$ is the heat flux (or current density) which is the heat (or energy) that flows through a unit cross-sectional surface area of the body per unit time. The quantity ∇T is the temperature gradient (the difference in temperature along a line parallel to the flux) and the coefficient κ is called the thermal conductivity.

Functions $\vec{q}$ and T are also connected by the *continuity equation*:

$$\rho c \frac{\partial T}{\partial t} + \operatorname{div} \vec{q} = 0, \tag{6.2}$$

where ρ is the mass density of medium and c is its heat capacity.

Equation (6.2) may be stated as:

The amount of heat (or energy) obtained by a unit volume of a body during some unit time interval equals the negative of the divergence of the current density.

Using the relation

$$\operatorname{div} \vec{q} = \operatorname{div} \ [-\kappa \ \nabla T] = -\kappa \ \nabla^2 T,$$

Equations (6.1) and (6.2) lead to the *heat conduction equation* given by

$$\frac{\partial T}{\partial t} = \chi \nabla^2 T. \tag{6.3}$$

The coefficient $\chi = \kappa/c\rho$ is called *the thermal diffusivity*.

In a steady state – for instance with the help of some external heat source – the temperature in the solid becomes time independent and the heat equation reduces to *Laplace's equation* given by

$$\nabla^2 T = 0. \tag{6.4}$$

If Q is the rate at which heat is added (or removed) per unit time and unit volume, the heat conduction equation becomes

$$\frac{\partial T}{\partial t} = \chi \nabla^2 T + \frac{Q}{\rho c} \,. \tag{6.5}$$

6.1.2   Diffusion Equation

Diffusion is a mixing process which occurs when one substance is introduced into a second substance. The introduced quantity, by means of molecular or atomic transport, spreads from locations where its concentration is higher to locations where the concentration is lower. Examples include the diffusion of perfume molecules into a room when the bottle is opened and the diffusion of neutrons in a nuclear reactor. Given sufficient time, diffusion will lead to an equalizing of the concentration of the introduced substance. We may imagine situations, however, where equilibrium is not reached. This could occur, for example, by continually adding more of the introduced substance at one location and/or removing it at another location.

For low concentration gradients, diffusion obeys *Fick's law* in which the current density of each component of a mixture is proportional to the concentration gradient of this component:

$$\vec{I} = -D\nabla c. \tag{6.6}$$

Here I is the current density, and D is the diffusion coefficient. The case $I = const.$ corresponds to steady-state diffusion where a substance is introduced and removed at the same rate.

Let us derive an equation that describes changes of concentration of the introduced substance. The continuity equation

$$\frac{\partial c}{\partial t} + \mathrm{div}\ \vec{I} = 0 \tag{6.7}$$

is a statement of conservation of mass:

>   *Any increase in the amount of molecules in some volume must equal the amount of molecules entering through the surface enclosing this volume.*

Substituting $\vec{I}$ from Equation (6.6) we obtain the diffusion equation

$$\frac{\partial c}{\partial t} = D\nabla^2 c. \tag{6.8}$$

If the introduced substance is being created (or destroyed) – for example in chemical reactions – we may describe this action by some function f of time and coordinates on the right side of the continuity equation. The diffusion equation then takes the form

$$\frac{\partial c}{\partial t} = D\nabla^2 c + f. \tag{6.9}$$

If we now compare Equations (6.8) and (6.9) to Equations (6.3) and (6.5) we see they are identical *parabolic equations*: (6.3) and (6.8) are homogeneous and (6.5) and (6.9) are nonhomogeneous.

6.1.3   One-dimensional Parabolic Equations and Initial and Boundary Conditions

We start by considering *the boundary value problem* on the bounded interval $[0, l]$ for the *one-dimensional* heat conduction equation (we use "heat terminology" just for convenience) in its most general form

$$\frac{\partial u}{\partial t} = a^2 \frac{\partial^2 u}{\partial x^2} + \xi \frac{\partial u}{\partial x} - \gamma u + f(x, t) . \tag{6.10}$$

The term linear in u in the heat equation can appear ([7]) if one considers the possibility of heat exchange with the environment in the case when the lateral surface of the body is not insulated. In the diffusion equation such a term appears if the diffusing substance is unstable in the sense that particles may disappear (such as an unstable gas or a gas being absorbed) or multiply (as with neutron diffusion). If the rates of these processes at each point in space are proportional to the concentration, the process is described by

$$\frac{\partial c}{\partial t} = D \nabla^2 c + \beta c, \tag{6.11}$$

where $D > 0$, β is the coefficient of disintegration ($\beta < 0$) or multiplication ($\beta > 0$).

The term with u_x can appear, for instance, if one considers the possibility of a diffusing substance participating in the motion of the material in which it is diffusing (which can be a fluid or gas). Suppose the fluid and diffusing substance flow along the x-axis with velocity v. Selecting the element $(x, \, x + \Delta x)$ and considering the amount of substance which flows through cross-sections at x and $x + \Delta x$ due to diffusion as well as fluid motion we arrive to an equation that includes the first derivative of c with respect to x in addition to the second derivative:

$$\frac{\partial c}{\partial t} = D \frac{\partial^2 c}{\partial x^2} - v \frac{\partial c}{\partial x}. \tag{6.12}$$

Knowledge of the initial temperature distribution means that, at time $t = 0$, the *initial condition* is

$$u(x, 0) = \varphi(x), \quad 0 < x < l, \tag{6.13}$$

where $\varphi(x)$ is a given function.

Boundary conditions can be specified in several ways.

1. *Dirichlet's condition*

   The temperature at the end $x = a$ (here $a = 0$ or l) of the bar changes by a specified law given by

   $$u(a, t) = g(t), \tag{6.14}$$

   where $g(t)$ is a known function of time t. In particular, g is constant if the end of the bar is maintained at a steady temperature.

2. *Neumann's condition*

   The heat current is given at the end $x = a$ of the bar in which case

   $$q = -\kappa \frac{\partial u}{\partial x}.$$

   This may be written as

   $$\left. \frac{\partial u}{\partial x} \right|_{x=a} = g(t), \tag{6.15}$$

where $g(t)$ is a known function.

If one of the ends of the bar is insulated, then the boundary condition at this end takes the form

$$\frac{\partial u}{\partial x}\bigg|_{x=a} = 0. \tag{6.16}$$

3. *Mixed condition*

In this case the ends of the bar are exchanging heat with the environment which has a temperature, θ. Actual heat exchange in real physical situations is very complicated but we may simplify the problem by assuming that it obeys *Newton's law*, $q = H(u - \theta)$, where H is the coefficient of heat exchange.

Thus, in the case of free heat exchange, the boundary conditions at the $x = a$ end of the rod have the form

$$\frac{\partial u}{\partial x} \pm hu\bigg|_{x=a} = g(t) \tag{6.17}$$

with given function $g(t)$ and $h = \mathrm{const}$.

It may also be the case that the external environments at the ends of the bar are different. In this case boundary conditions at the ends become

$$
\begin{aligned}
\kappa S \Delta t \frac{\partial u}{\partial x}\bigg|_{x=0} &= H_1 \left[u|_{x=0} - \theta_1 \right], \\
-\kappa S \Delta t \frac{\partial u}{\partial x}\bigg|_{x=l} &= H_2 \left[u|_{x=l} - \theta_2 \right],
\end{aligned}
\tag{6.18}
$$

where the temperatures of the environment at the left and right ends, θ_1 and θ_2, are considered to be known functions of time. In the simplest case they are constants.

In general, the boundary conditions are

$$
\begin{aligned}
P_1[u] &= \alpha_1 \frac{\partial u}{\partial x} + \beta_1 u\bigg|_{x=0} = g_1(t), \\
P_2[u] &\equiv \alpha_2 \frac{\partial u}{\partial x} + \beta_2 u\bigg|_{x=l} = g_2(t).
\end{aligned}
\tag{6.19}
$$

At the end $x = 0$ when $\alpha_1 = 0$ we have Dirichlet's condition, when $\beta_1 = 0$ we have Neumann's condition, and when both constants α_1 and β_1 are not zero we have the mixed condition (clearly the constants α_1 and β_1 cannot be zero simultaneously). The same holds at the end $x = l$. As we have discussed in Chapter 4, physical limitations most often lead to the restrictions $\alpha_1/\beta_1 < 0$ and $\alpha_2/\beta_2 > 0$ for the signs of the coefficients in boundary condition (6.19).

Now let us show how Equation (6.10) with the help of a proper substitution can be reduced to the equation without the term with $\partial u/\partial x$.

Substituting

$$u(x,t) = e^{\mu x} v(x,t),$$

where $\mu = -\xi/\left(2a^2\right)$ into Equation (6.10) yields the equation

$$\frac{\partial v}{\partial t} = a^2 \frac{\partial^2 v}{\partial x^2} - \tilde{\gamma} v + \tilde{f}(x,t),$$

where

$$\tilde{\gamma} = \gamma + \frac{\xi^2}{4a^2}, \quad \tilde{f}(x,t) = e^{-\mu x} f(x,t).$$

The initial condition for the function $v(x,t)$ has the form

$$v(x,0) = \tilde{\varphi}(x)$$

with $\tilde{\varphi}(x) = e^{-\mu x} \varphi(x)$. Boundary conditions for the function $v(x,t)$ are as follows:

$$P_1[\nu] \equiv \alpha_1 \frac{\partial v}{\partial x} + \tilde{\beta}_1 v \bigg|_{x=0} = g_1(t), \quad P_2[\nu] \equiv \alpha_2 \frac{\partial v}{\partial x} + \tilde{\beta}_2 v \bigg|_{x=l} = \tilde{g}_2(t),$$

where

$$\tilde{\beta}_1 = \beta_1 + \mu \cdot \alpha_1, \quad \tilde{\beta}_2 = \beta_2 + \mu \cdot \alpha_2, \quad \tilde{g}_2(t) = e^{-\mu l} \cdot g_2(t)$$

(check these results as *Homework*). Keeping in mind this result, in future calculations we will consider a parabolic equation without the term with $\partial u/\partial x$.

We have focused here on the heat conduction equation because the associated terminology is more concrete and intuitively fruitful than that for diffusion. But because the diffusion and heat conduction equations have identical forms, the solutions to diffusion problems can be obtained by a trivial replacement of D and c by χ and T. The boundary condition in Equation (6.14) corresponds to the concentration maintained at the ends, condition (6.15) corresponds to an impenetrable end, and condition (6.17) corresponds to a semi-permeable end (when diffusion through this end is similar to that described by Newton's law for heat exchange). An analogue from chemistry is the case of a reaction on the boundary of a body when the speed of reaction, i.e. the speed of creation or absorption of one of the chemical components, is proportional to the concentration of this component.

Let us briefly notice the *uniqueness of the solution* of the heat conduction equation under the conditions in Equations (6.12), (6.13) and (6.19) and the continuous dependence of this solution on the right-hand terms of the boundary and initial conditions. This material and the important *principle of the maximum* we discuss in Section 6.7.

6.2 The Fourier Method for Homogeneous Equations

Let us first find the solution of the *homogeneous equation*

$$\frac{\partial u}{\partial t} = a^2 \frac{\partial^2 u}{\partial x^2} - \gamma u, \quad 0 < x < l, \ t > 0, \tag{6.20}$$

which satisfies the initial condition

$$u(x,t)|_{t=0} = \varphi(x) \tag{6.21}$$

and has *homogeneous boundary conditions*

$$P_1[u] = \alpha_1 \frac{\partial u}{\partial x} + \beta_1 u \bigg|_{x=0} = 0, \quad P_2[u] = \alpha_2 \frac{\partial u}{\partial x} + \beta_2 u \bigg|_{x=l} = 0. \tag{6.22}$$

Notice that the linear parabolic equation (6.20) has no solutions in the form of propagating waves, $f(x \pm at)$, like hyperbolic equations have.

The Fourier method of separation of variables supposes that a solution of Equation (6.20) can be found as a product of two functions, one depending only on x, the second depending only on t:

$$u(x,t) = X(x)\, T(t). \tag{6.23}$$

Substituting Equation (6.23) into Equation (6.20) we obtain

$$X(x)\left[T'(t) + \gamma T(t)\right] - a^2 X''(x)T(t) = 0.$$

The variables can be separated and denoting a *separation constant as* $-\lambda$, we obtain

$$\frac{T'(t) + \gamma T(t)}{a^2 T(t)} \equiv \frac{X''(x)}{X(x)} = -\lambda. \tag{6.24}$$

Thus Equation (6.24) gives two ordinary linear homogeneous differential equations, a first order equation for function $T(t)$:

$$T'(t) + \left(a^2 \lambda + \gamma\right) T(t) = 0 \tag{6.25}$$

and a second order equation for function $X(x)$:

$$X''(x) + \lambda\, X(x) = 0. \tag{6.26}$$

To find the allowed values of λ we apply the boundary conditions. Homogenous boundary condition (6.22) imposed on $u(x,t)$, yields homogeneous boundary conditions on the function $X(x)$ given by

$$\alpha_1 X' + \beta_1 X\big|_{x=0} = 0, \qquad \alpha_2 X' + \beta_2 X\big|_{x=l} = 0. \tag{6.27}$$

Thus, we obtain the Sturm-Liouville boundary problem for eigenvalues, λ, and the corresponding eigenfunctions, $X(x)$. As we know from Chapter 4, there exist *infinite* sets of the real non-negative discrete spectrum of eigenvalues $\{\lambda_n\}$ and corresponding set of eigenfunctions $\{X_n(x)\}$ (clearly $\lambda = 0$ is also possible if $\beta_1 = \beta_2 = 0$).

As we obtained in Chapter 4, the eigenvalues of the Sturm-Liouville problem stated in Equations (6.26) and (6.27) are

$$\lambda_n = \left(\frac{\mu_n}{l}\right)^2, \tag{6.28}$$

where μ_n is nth non-negative root of the equation

$$\tan \mu = \frac{(\alpha_1 \beta_2 - \alpha_2 \beta_1)l\mu}{\mu^2 \alpha_1 \alpha_2 + l^2 \beta_1 \beta_2}. \tag{6.29}$$

The corresponding eigenfunctions can be written as

$$X_n(x) = \frac{1}{\sqrt{\alpha_1^2 \lambda_n + \beta_1^2}} \left[\alpha_1 \sqrt{\lambda_n} \cos \sqrt{\lambda_n}x - \beta_1 \sin \sqrt{\lambda_n}x\right]. \tag{6.30}$$

Now consider Equation (6.25). It is a linear first order differential equation and the general solution with $\lambda = \lambda_n$ is

$$T_n(t) = C_n e^{-(a^2 \lambda_n + \gamma)t}, \tag{6.31}$$

where C_n is an arbitrary constant. Non-negative values of λ_n are required so that the solution cannot grow to infinity with time. Now we have that each function

$$u_n(x,t) = T_n(t)X_n(x) = C_n e^{-(a^2 \lambda_n + \gamma)t} X_n(x)$$

is a solution of Equation (6.20) satisfying boundary conditions (6.22). To satisfy the initial conditions (6.21), we compose the series

$$u(x,t) = \sum_{n=1}^{\infty} C_n e^{-(a^2 \lambda_n + \gamma)t} X_n(x). \tag{6.32}$$

If this series converges uniformly as well as the series obtained by differentiating twice by x and once by t, the sum gives a solution to Equation (6.20) and satisfies the boundary conditions (6.22). The initial condition in Equation (6.21) gives

$$u|_{t=0} = \varphi(x) = \sum_{k=1}^{\infty} C_k X_k(x), \tag{6.33}$$

where we have expanded the function $\varphi(x)$ in a series of the eigenfunctions of the boundary value problem given by Equations (6.26) and (6.27).

Assuming uniform convergence of the series (6.32) we can find the coefficients C_n. Multiplying both sides of Equation (6.33) by $X_n(x)$, integrating from 0 to l and imposing the orthogonality condition of the functions $X_n(x)$, we obtain

$$C_n = \frac{1}{\|X_n\|^2} \int_0^l \varphi(x) X_n(x) dx. \tag{6.34}$$

If the series (6.32) and the series obtained from it by differentiating by t and twice differentiating by x are uniformly convergent then by substituting these values of coefficients, C_n, into the series (6.32) we obtain the solution of the problem stated in Equations (6.20) through (6.22).

Equation (6.32) gives a solution for *free heat exchange* (heat exchange without sources of heat within the body). It can be considered as the decomposition of an unknown function, $u(x,t)$, into a Fourier series over an orthogonal set of functions $\{X_n(x)\}$.

Let us substitute (6.34) into (6.32). Changing the order of summation and integration (that is allowed due the uniform convergence), we obtain:

$$u(x,t) = \int_0^l g(x,\xi,t)\varphi(\xi)d\xi, \tag{6.35}$$

where

$$g(x,\xi,t) = \sum_{n=1}^{\infty} \frac{X_n(x)X_n(\xi)}{\|X_n\|^2} e^{-(a^2 \lambda_n + \gamma)t}. \tag{6.36}$$

Note that $u(x,0) = \varphi(x)$, therefore, $g(x,\xi,0)$ in (6.35) should be equal to $\delta(x-\xi)$. That means that

$$\sum_{n=1}^{\infty} \frac{X_n(x)X_n(\xi)}{\|X_n\|^2} = \delta(x - \xi). \tag{6.37}$$

Actually, this completeness relation is correct for an arbitrary complete set of eigenfunctions of a Sturm-Liouville problem (see Chapter 4).

Example 6.1 Let zero temperature be maintained on both the ends, $x = 0$ and $x = l$, of a uniform isotropic bar of length l with a heat-insulated lateral surface. Initially the temperature distribution inside the bar is given by

$$u(x,0) = \varphi(x) = \begin{cases} \dfrac{x}{l} u_0 & \text{for } 0 < x \le \dfrac{l}{2}, \\[2mm] \dfrac{l-x}{l} u_0 & \text{for } \dfrac{l}{2} < x < l, \end{cases}$$

where $u_0 = $ const . There are no sources of heat inside the bar. Find the temperature distribution for the interior of the bar for time $t > 0$.

Solution. The problem is described by the equation

$$\frac{\partial u}{\partial t} = a^2 \frac{\partial^2 u}{\partial x^2}, \quad 0 < x < l.$$

with initial and boundary conditions (in this Example they are consistent)

$$u(x,0) = \varphi(x), \quad u(0,t) = u(l,t) = 0.$$

The boundary conditions of the problem are Dirichlet homogeneous boundary conditions, therefore eigenvalues and eigenfunctions of problem are

$$\lambda_n = \left(\frac{n\pi}{l}\right)^2, \quad X_n(x) = \sin\frac{n\pi x}{l}, \quad ||X_n||^2 = \frac{l}{2}, \quad n = 1,2,3,....$$

Equation (6.34) gives:

$$C_n = \frac{2}{l}\int_0^l \varphi(x)\sin\frac{n\pi x}{l}dx = \frac{4u_0}{n^2\pi^2}\sin\frac{n\pi}{2}$$

$$= \begin{cases} 0, & n = 2k, \\ \dfrac{4u_0}{(2k-1)^2\pi^2}(-1)^k, & n = 2k-1. \end{cases} \quad k = 1,2,3,\dots$$

Hence the distribution of temperature inside the bar for some moment is described by the series

$$u(x,t) = \frac{4u_0}{\pi^2}\sum_{k=1}^{\infty}\frac{(-1)^k}{(2k-1)^2}e^{-\frac{a^2(2k-1)^2\pi^2}{l^2}t}\sin\frac{(2k-1)\pi x}{l}.$$

Figure 6.1 shows the spatial-time-dependent solution $u(x,t)$. This solution was obtained for the case when $l = 10$, $u_0 = 5$ and $a^2 = 0.25$. All parameters are dimensionless. The dark gray line represents the initial temperature, black line is temperature at time $t = 100$. The gray lines in between show the temperature evolution within the period of time from 0 until 100, step $\Delta t = 2$. The series for the solution converges rather rapidly because of the exponential factor.

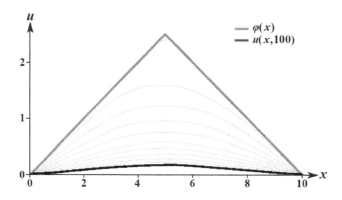

FIGURE 6.1
Solution $u(x,t)$ for Example 6.1.

Clearly the approach to equilibrium (the time for temperature fluctuations to decay) is governed by the factor $\exp[-(a^2\lambda_n + \gamma)t]$.

As a *Reading Exercise*, choose values of parameters in real physical units and estimate how much time it takes to reach equilibrium for different materials and for samples of different size. Make sure that the answers are physically reasonable.

Example 6.2 Consider the case when the ends, $x = 0$ and $x = l$, of a bar are thermally insulated from the environment. The lateral surface is also insulated. In this case the derivatives of temperature with respect to x on the ends of the bar equal zero. Initially the temperature is distributed as in the previous example:

$$\varphi(x) = \begin{cases} \dfrac{x}{l}u_0 & \text{for } 0 < x \le \dfrac{l}{2}, \\[2ex] \dfrac{l-x}{l}u_0 & \text{for } \dfrac{l}{2} < x < l, \end{cases}$$

where $u_0 = \text{const}$. Sources of heat are absent. Find the temperature distribution inside the bar for $t > 0$.

Solution. We are to solve the equation

$$\frac{\partial u}{\partial t} = a^2 \frac{\partial^2 u}{\partial x^2}, \quad 0 < x < l$$

with initial and boundary conditions

$$u(x,0) = \varphi(x), \quad \frac{\partial u}{\partial x}(0,t) = \frac{\partial u}{\partial x}(l,t) = 0.$$

Notice that the initial and boundary conditions in this case are not consistent (they are contradictory); in such situations we can only obtain a *generalized solution*. The boundary conditions of the problem are Neumann homogeneous boundary conditions. The eigenvalues and eigenfunctions are (see Chapter 4):

$$\lambda_n = \left(\frac{n\pi}{l}\right)^2, \quad X_n(x) = \cos\frac{n\pi x}{l}, \quad \|X_n\|^2 = \begin{cases} l, & n = 0, \\ l/2, & n > 0, \end{cases} \quad n = 0,1,2,\ldots.$$

Having applied Equation (6.34), we obtain

$$C_0 = \frac{u_0}{l^2}\left[\int_0^{l/2} x\,dx + \int_{l/2}^l (l-x)\,dx\right] = \frac{u_0}{4},$$

$$C_n = \frac{2u_0}{l^2}\int_0^{l/2} x\cos\frac{n\pi x}{l}\,dx + \frac{2u_0}{l^2}\int_{l/2}^l (l-x)\cos\frac{n\pi x}{l}\,dx$$

$$= \begin{cases} 0, & n = 2k-1, \\[2ex] \dfrac{u_0}{k^2\pi^2}\left[(-1)^k - 1\right], & n = 2k \end{cases} \quad k = 1,2,3,\ldots .$$

Thus the only nonzero C_n are those for which $n = 2k$ with $k = 2m+1$, $m = 0,1,2,\ldots$, i.e. $n = 4m + 2$; so we have

$$C_{4m+2} = -\frac{2u_0}{(2m+1)^2\pi^2}.$$

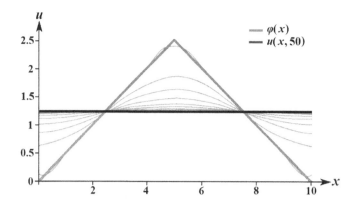

FIGURE 6.2
Solution, $u(x, t)$, for Example 6.2.

Finally, the temperature distribution inside the bar for some moment is expressed by the series

$$u(x, t) = \frac{u_0}{4} - \frac{2u_0}{\pi^2} \sum_{m=0}^{\infty} \frac{1}{(2m+1)^2} e^{-\frac{(4m+2)^2 a^2 \pi^2}{l^2} t} \cos \frac{(4m+2)\pi x}{l}.$$

At $x = l/4$ and $x = 3l/4$ all cosine terms equal zero, hence at these points $u = u_0/4$ for any $t \geq 0$. It is also clear that

$$\int_0^l u(x, t) dx = \frac{1}{4} u_0 l.$$

Notice that this area (the definite integral) is proportional to the amount of energy (heat) in the bar. The insulated ends of the bar correspond to a graph of $u(x, t)$ which has horizontal tangents at $x = 0$ and $x = l$. As $t \to \infty$ the first term of the series dominates. From this and from physical considerations we conclude that $u \to u_0/4$ as $t \to \infty$.

Figure 6.2 shows the space-time dependent solution $u(x, t)$. This solution was obtained for the case when $l = 10$, $u_0 = 5$ and $a^2 = 0.25$. All parameters are dimensionless. The dark gray line represents the initial temperature; the black line is temperature at time $t = 50$. The gray lines in between show the temperature evolution within the period of time from 0 until 50, step $\Delta t = 2$.

Example 6.3 Consider the situation where heat flux is governed by Newton's law of cooling and a constant temperature environment occurs at each end of a uniform isotropic bar of length l with insulated lateral surface. The initial temperature of the bar is equal: $u_0 = $ const. Internal sources of heat in the bar are absent. Find the temperature distribution inside the bar for $t > 0$.

Solution. We need to solve the equation

$$\frac{\partial u}{\partial t} = a^2 \frac{\partial^2 u}{\partial x^2}, \quad 0 < x < l, \quad t > 0$$

for the initial condition

$$u(x, 0) = u_0$$

and boundary conditions

$$\frac{\partial u}{\partial x}(0, t) - hu(0, t) = 0, \quad \frac{\partial u}{\partial x}(l, t) + hu(l, t) = 0.$$

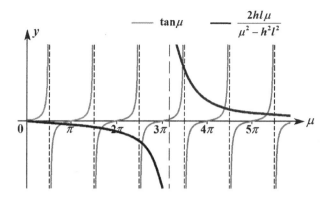

FIGURE 6.3
Graphical solution of the eigenvalue equation for Example 6.3.

Obviously, as in the previous example, we can only obtain a generalized solution.

The boundary conditions of the problem are mixed homogeneous boundary conditions, so eigenvalues are

$$\lambda_n = \left(\frac{\mu_n}{l}\right)^2, \quad n = 1, 2, 3, \ldots,$$

where μ_n is the nth root of the equation $\tan\mu = \frac{2hl\mu}{\mu^2 - h^2 l^2}$. Figure 6.3 shows curves of the two functions, $y = \tan\mu$ and $y = \frac{2hl\mu}{\mu^2 - h^2 l^2}$, plotted on the same set of axes. The eigenvalues are the squares of the values of μ at the intersection points of these curves divided by length l.

Each eigenvalue corresponds to an eigenfunction

$$X_n(x) = \frac{1}{\sqrt{\lambda_n + h^2}} \left[\sqrt{\lambda_n} \cos\sqrt{\lambda_n}x + h \sin\sqrt{\lambda_n}x\right]$$

with the norm $\|X_n\|^2 = \frac{l}{2} + \frac{h}{\lambda_n + h^2}$.

Figure 6.4 shows the first four eigenfunctions of the given boundary value problem.

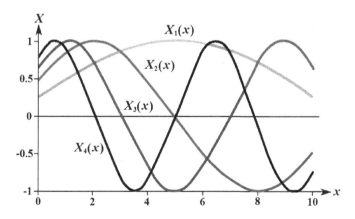

FIGURE 6.4
Eigenfunctions $X_1(x)$ through $X_4(x)$ for Example 6.3.

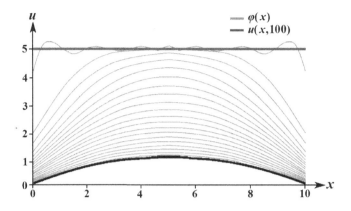

FIGURE 6.5
Solution $u(x,t)$ for Example 6.3.

Applying Equation (6.34) we obtain

$$C_n = \frac{1}{\|X_n\|^2} \int_0^l \frac{u_0}{\sqrt{\lambda_n + h^2}} \left[\sqrt{\lambda_n} \cos \sqrt{\lambda_n} x + h \sin \sqrt{\lambda_n} x \right] \, dx$$

$$= \frac{2u_0\sqrt{\lambda_n + h^2}}{l(\lambda_n + h^2) + 2h} \left[\sin \sqrt{\lambda_n} l - \frac{h}{\sqrt{\lambda_n}} \left(\cos \sqrt{\lambda_n} l - 1 \right) \right].$$

Hence the temperature distribution inside the bar for some moment of time is expressed by the series

$$u(x,t) = \sum_{n=1}^{\infty} C_n e^{-a^2 \lambda_n t} \frac{1}{\sqrt{\lambda_n + h^2}} \left[\sqrt{\lambda_n} \cos \sqrt{\lambda_n} x + h \sin \sqrt{\lambda_n} x \right].$$

Figure 6.5 shows the spatial time-dependent solution $u(x,t)$ for Example 6.3. This solution was obtained for the case when $l = 10$, $u_0 = 5$ and $a^2 = 0.25$. All parameters are dimensionless. The dark gray line represents the initial temperature; the black line is temperature at time $t = 100$, step $\Delta t = 2$.

In this example the boundary and initial conditions do not match each other; as a result at $t = 0$ the temperature, $u(x,0)$, given by the solution in the form of an eigenfunction expansion does not converge uniformly. The convergence is poor at points close to the ends of the rod and the solution appears to have unphysical oscillations of temperature initially. Increasing the number of terms in the series smoothes these oscillations at all points except the ends – this is the Gibbs phenomenon which is discussed in Appendix A. These oscillations do not occur physically and, in fact, they disappear from the solution for any finite time $t > 0$ in which case $u(x,t)$ converges rapidly to a physically reasonable result.

Next, let us give an example of the solution of a BVP with *Laplace transform*.

Example 6.4 The heat propagation in a semi-infinite rod is governed by the heat equation,

$$u_t = a^2 u_{xx}, \quad x > 0, \ t > 0.$$

Initially, the temperature of the rod was u_1. At $t = 0$, the temperature at the left end suddenly becomes equal to $u_2 \neq u_1$. Thus, the initial condition is $u(x,0) = u_1$, the boundary conditions are $u(0,t) = u_2$, $|u(\infty,t)| < \infty$. Find $u(x,t)$ for $t > 0$.

Applying the Laplace transform (Appendix A) to the equation we obtain

$$p\hat{u}(x,p) = a^2 \hat{u}_{xx}(x,p).$$

This ordinary differential equation with boundary conditions

$$\hat{u}(0,p) = u_2/p, |\hat{u}(\infty,p)| < \infty$$

has the solution

$$u(x,p) = \frac{u_1}{p} + \frac{u_2 - u_1}{p} e^{-x\sqrt{p}/a}.$$

The inverse Laplace transform gives (see the Table of LT in Appendix A).

$$u(x,t) = u_1 + (u_2 - u_1)\operatorname{erfc}\left(\frac{x}{2a\sqrt{t}}\right).$$

6.3 Nonhomogeneous Equations

First, let us consider an example of a nonhomogeneous equation with a simple right side:

$$u_t - a^2 u_{xx} = \cos\frac{\pi x}{l}, \quad t > 0, \tag{6.38}$$

$$u(x,0) = 0, \quad u_t(x,0) = 0, \tag{6.39}$$

$$u_x(0,t) = u_x(l,t) = 0, \quad t > 0. \tag{6.40}$$

Similarly to the example considered in Chapter 5, Section 5.8, we apply the Laplace transform to Equation (6.38) while taking into account the initial conditions (6.39):

$$p\hat{u}(x,p) - a^2 \hat{u}_{xx}(x,p) = \frac{1}{p}\cos\frac{\pi x}{l},$$

or

$$\hat{u}_{xx}(x,p) - \frac{p}{a^2}\hat{u}(x,p) = -\frac{1}{pa^2}\cos\frac{\pi x}{l}. \tag{6.41}$$

The solution of (6.41) satisfying the boundary conditions

$$\hat{u}(0,p) = \hat{u}(l,p) = 0, \tag{6.42}$$

is

$$\hat{u}(x,p) = \frac{1}{p[p + (\pi a/l)^2]}\sin\frac{\pi x}{l} = \left(\frac{l}{\pi a}\right)^2 \left[\frac{1}{p} - \frac{p}{p + (\pi a/l)^2}\right]\cos\frac{\pi x}{l}.$$

Taking the inverse Laplace transform (see the table in Appendix A), we obtain

$$u(x,t) = \left(\frac{l}{\pi a}\right)^2 \left\{1 - \exp\left[-\left(\frac{\pi a}{l}\right)^2 t\right]\right\}\cos\frac{\pi x}{l}.$$

Now consider the general form of the *nonhomogeneous* linear equation

$$\frac{\partial u}{\partial t} = a^2\frac{\partial^2 u}{\partial x^2} - \gamma u + f(x,t), \tag{6.43}$$

where $f(x,t)$ is a known function, with initial condition

$$u(x,t)|_{t=0} = \varphi(x) \tag{6.44}$$

and *homogeneous boundary conditions*

$$P_1[u] \equiv \alpha_1 u_x + \beta_1 u|_{x=0} = 0, \quad P_2[u] \equiv \alpha_2 u_x + \beta_2 u|_{x=l} = 0. \tag{6.45}$$

To start, let us express the function $u(x,t)$ as the sum of two functions:

$$u(x,t) = u_1(x,t) + u_2(x,t),$$

where $u_1(x,t)$ satisfies the *homogeneous equation* with the given boundary conditions and the initial condition:

$$\frac{\partial u_1}{\partial t} = a^2 \frac{\partial^2 u_1}{\partial x^2} - \gamma u_1,$$

$$u_1(x,t)|_{t=0} = \varphi(x),$$

$$P_1[u_1] \equiv \alpha_1 u_{1_x} + \beta_1 u_1|_{x=0} = 0, \quad P_2[u_1] \equiv \alpha_2 u_{1_x} + \beta_2 u_1|_{x=l} = 0.$$

The function $u_2(x,t)$ satisfies the *nonhomogeneous* equation with *zero boundary and initial conditions*:

$$\frac{\partial u_2}{\partial t} = a^2 \frac{\partial^2 u_2}{\partial x^2} - \gamma u_2 + f(x,y), \tag{6.46}$$

$$u_2(x,t)|_{t=0} = 0, \tag{6.47}$$

$$P_1[u_2] \equiv \alpha_1 u_{2_x} + \beta_1 u_2|_{x=0} = 0, \quad P_2[u_2] \equiv \alpha_2 u_{2_x} + \beta_2 u_2|_{x=l} = 0. \tag{6.48}$$

The methods for finding $u_1(x,t)$ have been discussed in the previous section; therefore here we concentrate our attention on finding the solutions $u_2(x,t)$. As for the case of free heat exchange inside the bar let us expand function $u_2(x,t)$ as a series

$$u_2(x,t) = \sum_{n=1}^{\infty} T_n(t) X_n(x), \tag{6.49}$$

where $X_n(x)$ are eigenfunctions of the corresponding homogeneous boundary value problem and $T_n(t)$ are unknown functions of t.

Boundary conditions in Equation (6.48) for $u_2(x,t)$ are valid for any choice of functions $T_n(t)$ (when the series converge uniformly) because they are valid for the functions $X_n(x)$. Substituting the series (6.49) into Equation (6.46) we obtain

$$\sum_{n=1}^{\infty} \left[T_n'(t) + \left(a^2 \lambda_n + \gamma \right) T_n(t) \right] X_n(x) = f(x,t). \tag{6.50}$$

Using the completeness property we can expand the function $f(x,t)$, as function of x, into a Fourier series of the functions $X_n(x)$ on the interval $(0,l)$ such that

$$f(x,t) = \sum_{n=1}^{\infty} f_n(t) X_n(x). \tag{6.51}$$

Using the orthogonality property of the functions $X_n(x)$ we find that

$$f_n(t) = \frac{1}{\|X_n\|^2} \int_0^l f(x,t) X_n(x) dx. \tag{6.52}$$

Comparing the two expansions in Equations (6.50) and (6.51) for the same function $f(x,t)$ we obtain a differential equation for the functions $T_n(t)$:

$$T_n'(t) + \left(a^2\lambda_n + \gamma\right)T_n(t) = f_n(t). \tag{6.53}$$

In order that $u_2(x,t)$ given by Equation (6.49) satisfies the initial condition (6.47) it is necessary that the functions $T_n(t)$ obey the condition

$$T_n(0) = 0. \tag{6.54}$$

The solution to the ordinary differential equation of the first order, Equation (6.53), with initial condition (6.54) can be represented in the integral form

$$T_n(t) = \int_0^t f_n(\tau)e^{-(a^2\lambda_n+\gamma)(t-\tau)}d\tau, \tag{6.55}$$

or

$$T_n(t) = \int_0^t f_n(\tau)Y_n(t-\tau)d\tau, \quad \text{where} \quad Y_n(t-\tau) = e^{-(a^2\lambda_n+\gamma)(t-\tau)}.$$

Thus, the solution of the nonhomogeneous heat conduction problem for a bar with boundary conditions equal to zero has the form

$$u(x,t) = u_1(x,t) + u_2(x,t) = \sum_{n=1}^{\infty}\left[T_n(t) + C_ne^{-(a^2\lambda_n+\gamma)t}\right]X_n(x), \tag{6.56}$$

where functions $T_n(t)$ are defined by Equation (6.55) and coefficients C_n have been found earlier when we considered the homogeneous heat equation.

Example 6.5 A point-like heat source with power $Q = \text{const}$ is located at x_0 $(0 < x_0 < l)$ in a uniform isotropic bar with insulated lateral surfaces. The initial temperature of the bar is zero. Temperatures at the ends of the bar are maintained at zero. Find the temperature inside the bar for $t > 0$.

Solution. The boundary value problem modeling heat propagation for this example is

$$\frac{\partial u}{\partial t} = a^2\frac{\partial^2 u}{\partial x^2} + \frac{Q}{c\rho}\delta(x-x_0),$$
$$u(x,0) = 0, \quad u(0,t) = u(l,t) = 0,$$

where $\delta(x-x_0)$ is the delta function.

The boundary conditions are Dirichlet homogeneous boundary conditions, so the eigenvalues and eigenfunctions of problem are

$$\lambda_n = \left(\frac{n\pi}{l}\right)^2, \quad X_n(x) = \sin\frac{n\pi x}{l}, \quad \|X_n\|^2 = \frac{l}{2} \quad (n = 1, 2, 3, ...).$$

In the case of homogeneous initial conditions, $\varphi(x) = 0$, we have $C_n = 0$ and the solution $u(x,t)$ is defined by the series

$$u(x,t) = \sum_{n=1}^{\infty} T_n(t)\sin\frac{n\pi x}{l},$$

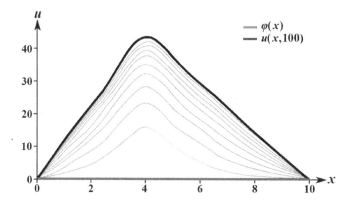

FIGURE 6.6
Solution $u(x,t)$ for Example 6.5.

where f_n and T_n are defined by Equations (6.52) and (6.56):

$$f_n(t) = \frac{2}{l} \int_0^l \frac{Q}{c\rho} \delta\left(x - x_0\right) \sin \frac{n\pi x}{l} dx = \frac{2Q}{lc\rho} \sin \frac{n\pi x_0}{l} dx,$$

$$T_n(t) = \frac{2Ql}{c\rho a^2 n^2 \pi^2} \left(1 - e^{-n^2 a^2 \pi^2 t/l^2}\right) \sin \frac{n\pi x_0}{l}.$$

Substituting the expression for $T_n(t)$ into the general formulas we obtain the solution of the problem:

$$u(x,t) = \frac{2Ql}{c\rho a^2 \pi^2} \sum_{n=1}^\infty \frac{1}{n^2} \left(1 - e^{-n^2 a^2 \pi^2 t/l^2}\right) \sin \frac{n\pi x_0}{l} \sin \frac{n\pi x}{l}.$$

Figure 6.6 shows the spatial-time-dependent solution $u(x,t)$ for Example 4. This solution was obtained for the case when $l = 10$, $Q/c\rho = 5$, $x_0 = 4$ and $a^2 = 0.25$. All parameters are dimensionless. The dark gray line represents the initial temperature; black line is temperature at time $t = 100$. The gray lines in between show the temperature evolution within the period of time from 0 until 100, step $\Delta t = 2$.

6.4   Green's Function and Duhamel's Principle

In the present section, we consider the relation between the solution of the homogeneous equation with nonzero initial condition, $u_1(x,t)$, and that of the nonhomogeneous equation with zero initial condition, $u_2(x,t)$.

In Section 6.2. we have found that $u_1(x,t)$ can be presented in the form

$$u_1(x,t) = \int_0^l g(x,\xi,t)\varphi(\xi)d\xi, \tag{6.57}$$

where

$$g(x,\xi,t) = \sum_{n=1}^\infty \frac{X_n(x)X_n(\xi)}{\|X_n\|^2} e^{-(a^2\lambda_n + \gamma)t}. \tag{6.58}$$

Let us obtain a similar formula for $u_2(x,t)$. Substituting (6.52) and (6.55) into (6.49), we find (a uniform convergence is assumed):

$$u_2(x,t) = \int_0^t d\tau \int_0^l d\xi \sum_{n=1}^\infty \frac{X_n(x)X_n(\xi)}{\|X_n\|^2} e^{-(a^2\lambda_n+\gamma)(t-\tau)} f(\xi,\tau). \tag{6.59}$$

Define $f(\xi,\tau) = 0$, if $\tau < 0$. Then expression (6.59) can be rewritten as

$$u_2(x,t) = \int_{-\infty}^\infty d\tau \int_0^l d\xi G(x,\xi,t-\tau) f(\xi,\tau), \tag{6.60}$$

where *Green's function*

$$G(x,\xi,t-\tau) = H(t-\tau)g(x,\xi,t-\tau) \tag{6.61}$$

($H(t-\tau)$ is the Heaviside step function). Thus, we see that the same function g appears in expressions (6.57) and (6.60), but (6.60) contains an additional integration in time. That circumstance allows us to find the relation between the solution of the nonhomogeneous problem (6.46)-(6.48) with zero initial condition and that of a homogeneous problem with an appropriate initial condition. That relation is determined below.

Let us define the following homogeneous problem for a function $v(x,t;\tau)$, which depends on the parameter τ:

$$v_t(x,t;\tau) = a^2 v_{xx}(x,t;\tau) - \gamma v(x,t;\tau); \quad 0 < x < l, \ t > \tau; \tag{6.62}$$

$$v(x,\tau;\tau) = f(x,\tau), \ 0 \le x \le l; \ P_1[v] = 0, \ P_2[v] = 0. \tag{6.63}$$

By the change of variables $s = t - \tau$, $\tilde{v}(x,s;\tau) = v(x,t;\tau)$, we obtain the standard problem

$$\tilde{v}_s = a^2 \tilde{v}_{xx}; \quad 0 < x < l, \ s > 0;$$

$$\tilde{v}(x,0;\tau) = f(x,\tau)), \quad 0 \le x \le l; \ P_1[\tilde{v}] = 0, \ P_2[\tilde{v}] = 0.$$

Its solution, according to (6.35), is

$$\tilde{v}(x,s;\tau) = \int_0^l g(x,\xi,s) f(\xi,\tau) d\xi, \quad s > 0,$$

hence

$$v(x,t;\tau) = \int_0^l g(x,\xi,t-\tau) f(\xi,\tau) d\xi, \quad t > \tau.$$

Let us show that the solution of the nonhomogeneous problem (6.46)-(6.48) can be found as

$$u_2(x,t) = \int_0^t v(x,t;\tau) d\tau. \tag{6.64}$$

Indeed, $u_2(x,t)$ satisfies the initial condition (6.42) and the boundary conditions (6.48). Substituting (6.64) into (6.46), we find:

$$u_{2,t}(x,t) = v(x,t;t) + \int_0^t v_t(x,t;\tau) d\tau,$$

$$u_{2,xx}(x,t) = \int_0^t v_{xx}(x,t;\tau) d\tau,$$

hence

$$u_{2,t}(x,t) - a^2 u_{2,xx}(x,t) + \gamma u_2$$

$$= v(x,t;t) + \int_0^t \left[v_t(x,t;\tau) - a^2 v_{xx}(x,t;\tau) + \gamma v(x,t;\tau) \right] d\tau = f(x,t).$$

Relation (6.64) is called *Duhamel's principle*.

Note that the dimension of the space does not influence the proof presented above. Hence, Duhamel's principle is valid also for parabolic problems in higher dimensions considered in Chapter 9.

Example 6.6 Using Duhamel's principle, find the solution of the following BVP for a nonhomogeneous heat equation:

$$u_t = a^2 u_{xx} + t \sin 2x, \quad 0 < x < \pi, \ t > 0;$$

$$u(x,0) = 0, \quad 0 \le x \le \pi;$$

$$u(0,t) = u(\pi,t) = 0, \quad t \ge 0.$$

First, let us formulate problem (6.62)-(6.63) for function $v(x,t;\tau)$:

$$v_t(x,t;\tau) = a^2 v_{xx}(x,t;\tau), \quad 0 < x < \pi, \ t > \tau;$$

$$v(x,\tau;\tau) = \tau \sin 2x, \quad 0 \le x \le \pi;$$

$$v(0,t;\tau) = v(\pi,t;\tau) = 0, \quad t \ge \tau.$$

The latter problem can be solved by separation of variables (see Section 6.2). Because $X(x) = \sin 2x$ is an eigenfunction satisfying (6.26), (6.27) for $\lambda = 4$, only one term in expansion (6.32) is different from zero. The corresponding function $T(t)$, which satisfies equation

$$T_t(t;\tau) + 4a^2 T(t;\tau) = 0, \quad t > \tau$$

with initial condition

$$T(\tau;\tau) = \tau,$$

is

$$T(t;\tau) = \tau e^{-4a^2(t-\tau)},$$

thus

$$v(x,t;\tau) = \tau e^{-4a^2(t-\tau)} \sin 2x.$$

Substituting $v(x,t;\tau)$ into formula (6.64), we find

$$u(x,t) = \int_0^t \tau e^{-4a^2(t-\tau)} \sin(2x) d\tau = e^{-4a^2 t} \sin(2x) \int_0^t \tau e^{4a^2 \tau} d\tau$$

$$= \frac{1}{(4a^2)^2} \left(4a^2 t - 1 + e^{-4a^2 t} \right) \sin(2x).$$

Note also that a similar relation exists for hyperbolic problems. For instance, let us consider the nonhomogeneous wave equation with homogeneous initial conditions

$$u_{tt}(x,t) = a^2 u_{xx}(x,t) + f(x,t), \quad 0 < x < l, \ t > 0;$$

$$u(x,0) = 0, \quad u_t(x,0) = 0, \quad 0 \le x \le l; \quad u(0,t) = u(l,t) = 0, \quad t \ge 0.$$

Define function $v(x, t; \tau)$, $t \geq \tau \geq 0$, which solves the problem

$$v_{tt}(x, t; \tau) = a^2 v_{xx}(x, t; \tau), \quad 0 < x < l, \; t > \tau;$$

$$v(x, \tau; \tau) = 0; \quad v_t(x, \tau; \tau) = f(x, \tau), \quad 0 \leq x \leq l.$$

$$v(x, \tau; \tau) = 0; \quad v_t(x, \tau; \tau) = f(x, \tau), \quad 0 \leq x \leq l.$$

$$v(0, t; \tau) = v(l, t; \tau), \quad \tau \geq t.$$

One can show that

$$u(x, t) = \int_0^t v(x, t; \tau) d\tau.$$

Example 6.7 Using Duhamel's principle, find the solution of the following BVP for a nonhomogeneous wave equation:

$$u_{tt} = a^2 u_{xx} + \sin 3x, \quad 0 < x < \pi, \; t > 0;$$

$$u(x, 0) = 0, \quad 0 \leq x \leq \pi;$$

$$u(0, t) = u(\pi, t) = 0, \quad t \geq 0.$$

Duhamel's principle allows us to reformulate the problem using a homogeneous wave equation:

$$v_{tt}(x, t; \tau) = a^2 v_{xx}(x, t; \tau), \quad 0 < x < \pi, \; t > \tau;$$

$$v(x, \tau; \tau) = 0; \quad v_t(x, \tau; \tau) = \sin 3x, \quad 0 \leq x \leq \pi;$$

$$v(0, t; \tau) = v(\pi, t; \tau) = 0, \quad t \geq \tau.$$

That problem can be solved by the method of separation of variables (see Section 5.6). Because function $X(x) = \sin 3x$ is an eigenfunction of problem (5.57)-(5.58) with boundary conditions $X(0) = X(\pi) = 0$, corresponding to $\lambda = 9$, only one term is nonzero in expansion (5.77). The corresponding function $T(t)$ satisfies equation

$$T_{tt}(t; \tau) + 9a^2 T(t; \tau) = 0$$

with initial condition $T_t(\tau; \tau) = 1$, hence

$$T(t; \tau) = \frac{1}{3a} \sin(3at)$$

and

$$v(x, t; \tau) = \frac{1}{3a} \sin(3at) \sin(3x).$$

Thus,

$$u(x, t) = \int_0^t v(x, t; \tau) d\tau = \frac{1}{9a^2} [1 - \cos(3at)] \sin(3x).$$

6.5 The Fourier Method for Nonhomogeneous Equations with Nonhomogeneous Boundary Conditions

Now consider the general boundary problem for heat conduction, Equation (6.43), given by

$$\frac{\partial u}{\partial t} = a^2 \frac{\partial^2 u}{\partial x^2} - \gamma u + f(x,t)$$

with nonhomogeneous initial (Equation (6.44)) and boundary conditions

$$u(x,t)|_{t=0} = \varphi(x),$$

$$P_1[u] \equiv \alpha_1 u_x + \beta_1 u|_{x=0} = g_1(t), \quad P_2[u] \equiv \alpha_2 u_x + \beta_2 u|_{x=l} = g_2(t). \tag{6.65}$$

We cannot apply the Fourier method directly to obtain a solution of the problem because the boundary conditions are nonhomogeneous. However, the problem can easily be reduced to a problem with boundary conditions equal to zero in the following way.

Let us search for the solution of the problem in the form

$$u(x,t) = v(x,t) + w(x,t),$$

where $v(x,t)$ is a new unknown function, and the function $w(x,t)$ is chosen so that it satisfies the given nonhomogeneous boundary conditions

$$P_1[w] \equiv \alpha_1 w_x + \beta_1 w|_{x=0} = g_1(t), \quad P_2[w] \equiv \alpha_2 w_x + \beta_2 w|_{x=l} = g_2(t).$$

For the function $v(x,t)$ we obtain the following boundary value problem:

$$\frac{\partial v}{\partial t} = a^2 \frac{\partial^2 v}{\partial x^2} - \gamma v + \tilde{f}(x,t),$$

$$v(x,t)|_{t=0} = \tilde{\varphi}(x),$$

$$P_1[v] \equiv \alpha_1 v_x + \beta_1 v|_{x=0} = 0, \quad P_2[v] \equiv \alpha_2 v_x + \beta_2 v|_{x=l} = 0,$$

where

$$\tilde{f}(x,t) = f(x,t) - \frac{\partial w}{\partial t} + a^2 \frac{\partial^2 w}{\partial x^2} - \gamma w,$$

$$\tilde{\varphi}(x) = \varphi(x) - w(x,0).$$

The solution of such a problem with homogeneous boundary conditions has been considered in the previous section.

The *auxiliary function* $w(x,t)$ is ambiguously defined. The simplest way to proceed is to use polynomials and construct it in the form

$$w(x,t) = P_1(x)g_1(t) + P_2(x)g_2(t),$$

where $P_1(x)$ and $P_2(x)$ are polynomials of the 1st or 2nd order. Coefficients of these polynomials will be chosen so that the function $w(x,t)$ satisfies the given boundary conditions. We have the following possibilities.

Case I. If β_1 and β_2 in Equation (6.65) are not zero simultaneously we may seek the function $w(x,t)$ in the form

$$w(x,t) = (\gamma_1 + \delta_1 x)g_1(t) + (\gamma_2 + \delta_2 x)g_2(t).$$

Substituting this into boundary conditions (6.65), and taking into account that the derived system of equations must be valid for arbitrary t, we obtain coefficients γ_1, δ_1, γ_2 and δ_2 as

$$\gamma_1 = \frac{\alpha_2 + \beta_2 l_x}{\beta_1 \beta_2 l_x + \beta_1 \alpha_2 - \beta_2 \alpha_1}, \quad \delta_1 = \frac{-\beta_2}{\beta_1 \beta_2 l_x + \beta_1 \alpha_2 - \beta_2 \alpha_1},$$

$$\gamma_2 = \frac{-\alpha_1}{\beta_1 \beta_2 l_x + \beta_1 \alpha_2 - \beta_2 \alpha_1}, \quad \delta_2 = \frac{\beta_1}{\beta_1 \beta_2 l_x + \beta_1 \alpha_2 - \beta_2 \alpha_1}.$$

Reading Exercise: We leave it to the reader to obtain the results above.

Case II. If $\beta_1 = \beta_2 = 0$, that is $\begin{cases} u_x(0,t) = g_1(t) \\ u_x(l,t) = g_2(t) \end{cases}$ the auxiliary function has the form

$$w(x,t) = \left[x - \frac{x^2}{2l} \right] \cdot g_1(t) + \frac{x^2}{2l} \cdot g_2(t).$$

Reading Exercise: Verify the statement above.

It is easily checked that, defined in such a way, the auxiliary functions $w(x,t)$ satisfy the boundary conditions in Equation (6.65).

Reading Exercise: Prove the statement above.

Combining the different kinds of boundary conditions listed above, we obtain nine different auxiliary functions which are listed in Appendix C part 2.

In the following examples we consider problems which involve the nonhomogeneous heat conduction equation with nonhomogeneous boundary conditions.

Example 6.8 Let the pressure and temperature of air in a cylinder be equal to the atmospheric pressure. One end of the cylinder at $x = 0$ is opened at the instant $t = 0$, and the other, at $x = l$, remains closed. The concentration of some gas in the external environment is constant ($u_0 = \text{const}$). Find the concentration of gas in the cylinder for $t > 0$ if at the instant $t = 0$ the gas begins to diffuse into the cylinder through the opened end.

Solution. This problem can be represented by the equation

$$\frac{\partial u}{\partial t} = a^2 \frac{\partial^2 u}{\partial x^2}, \quad a^2 = D,$$

under conditions

$$u(x,0) = 0, \quad u(0,t) = u_0, \quad DS\frac{\partial u}{\partial x}(l,t) = 0,$$

where D is the diffusion coefficient.

Clearly the eigenvalues and eigenfunctions of the problem are (see Chapter 4):

$$\lambda_n = \left[\frac{(2n-1)\pi}{2l} \right]^2, \quad X_n(x) = \sin\frac{(2n-1)\pi x}{2l}, \quad ||X_n||^2 = \frac{l}{2}, \quad n = 1,2,3,\ldots.$$

and the solution will be of the form

$$u(x,t) = v(x,t) + w(x,t).$$

In general, for a specific problem an auxiliary function is easily obtained from general formulas for $w(x,t)$ found in Appendix C part 2. We may often guess what the function

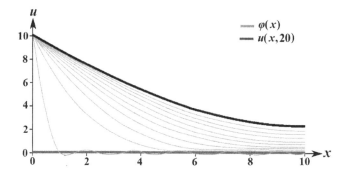

FIGURE 6.7
Solution $u(x, t)$ for Example 6.8.

must look like based on the physical observation that we are searching for a terminal or steady state solution. In the present case

$$w(x, t) = u_0 .$$

We also have

$$\tilde{f}(x, t) = 0, \quad \tilde{\varphi}(x) = -u_0.$$

Applying formula (6.34), we obtain

$$C_n = -\frac{2u_0}{l} \int_0^l \sin \frac{(2n - 1)\pi x}{2l} dx = -\frac{4u_0}{(2n - 1)\pi}.$$

Substituting the expression for C_n into the general formula, we obtain the final solution:

$$u(x, t) = w(x, t) + \sum_{n=1}^{\infty} C_n e^{-\frac{a^2 (2n-1)^2 \pi^2}{4l^2} t} \sin \frac{(2n - 1)\pi x}{2l}$$

$$= u_0 - \frac{4u_0}{\pi} \sum_{n-1}^{\infty} \frac{1}{2n - 1} e^{-\frac{a^2 (2n-1)^2 \pi^2}{4l^2} t} \sin \frac{(2n - 1)\pi x}{2l}.$$

Figure 6.7 shows the spatial-time-dependent solution $u(x, t)$. This solution was obtained for the case when $l = 10$, $u_0 = 10$ and $a^2 = D = 1$. All parameters are dimensionless. The dark gray line represents the initial temperature and the black line is temperature at time $t = 20$, step $dt = 2$.

Example 6.9 Find the temperature change in a homogeneous isotropic bar of length l $(0 \leq x \leq l)$ with a heat-insulated lateral surface during free heat exchange if the initial temperature is given by

$$u(x, 0) = \varphi(x) = u_0 \frac{x^2}{l^2}.$$

The left end of the bar at $x = 0$ is insulated and at the right end temperature is held constant:

$$u(l, t) = u_0, \quad \text{where} \quad u_0 = \text{const} > 0.$$

Solution. The problem is described by the equation

$$\frac{\partial u}{\partial t} = a^2 \frac{\partial^2 u}{\partial x^2}$$

with the conditions

$$u(x,0) = \varphi(x) = u_0 x^2/l^2,$$

$$\frac{\partial u}{\partial x}(0,t) = 0, \quad u(l,t) = u_0.$$

The solution of the problem will be of the form

$$u(x,t) = v(x,t) + w(x,t).$$

The auxiliary function can be easily obtained from the general formulas above (do this as a *Reading Exercise*; the answers are found in Appendix C part 2) and we find

$$w(x,t) = u_0.$$

This function corresponds to the steady-state regime as $t \to \infty$. The eigenvalues and eigenfunctions are easy to obtain (you can also find them in Appendix C part 1) and we leave it to the reader as a *Reading Exercise* to check that they are:

$$\lambda_n = \left[\frac{(2n-1)\pi}{2l}\right]^2, \quad X_n(x) = \cos\frac{(2n-1)\pi x}{2l}, \quad ||X_n||^2 = \frac{l}{2}, \quad (n = 1, 2, 3, \ldots).$$

Following the same logic as in previous problems, for the function $v(x,t)$ we obtain the conditions

$$\tilde{f}(x,t) = 0,$$

$$\tilde{\varphi}(x) = u_0\frac{x^2}{l^2} - u_0 = u_0\left(\frac{x^2}{l^2} - 1\right).$$

We apply Equation (6.34) to obtain

$$C_n = \frac{2}{l}\int_0^l \tilde{\varphi}(x)\cos\frac{(2n-1)\pi x}{2l}dx$$

$$= \frac{2}{l}\int_0^l u_0\left(\frac{x^2}{l^2} - 1\right)\cos\frac{(2n-1)\pi x}{2l}dx = -\frac{32u_0}{(2n-1)^3\pi^3}(-1)^n.$$

With these coefficients we obtain the solution of the problem:

$$u(x,t) = u_0 - \frac{32u_0}{\pi^3}\sum_{n=1}^{\infty}\frac{(-1)^n}{(2n-1)}e^{-\frac{(2n-1)^2 a^2 \pi^2}{l^2}t}\cos\frac{(2n-1)\pi x}{2l}.$$

Figure 6.8 shows the spatial-time-dependent solution $u(x,t)$ for Example 6.9. This solution was obtained for the case when $l = 10$, $u_0 = 5$ and $a^2 = 0.25$. All parameters are dimensionless. The dark gray line represents the initial temperature and the black line is temperature at time $t = 250$, step $\Delta t = 2$.

Example 6.10 Let the pressure and temperature of air in a cylinder length l ($0 \le x \le l$) be equal to the atmospheric pressure. One end of the cylinder at $x = 0$ is opened at the instant $t = 0$, and the other, at $x = l$, remains closed. The concentration of some gas in the external environment is constant ($u_0 = $ const). Find the concentration of gas in the cylinder for $t > 0$ if at the instant $t = 0$ the gas begins to diffuse into the cylinder through the opened end.

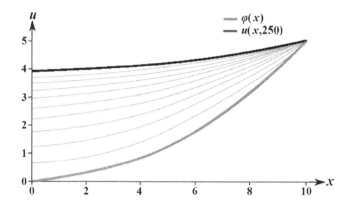

FIGURE 6.8
Solution $u(x,t)$ for Example 6.9.

Solution. This problem can be represented by the equation

$$\frac{\partial u}{\partial t} = a^2 \frac{\partial^2 u}{\partial x^2}, \quad a^2 = D,$$

under conditions

$$u(x,0) = 0, \quad u(0,t) = u_0, \quad DS\frac{\partial u}{\partial x}(l,t) = 0,$$

where D is the diffusion coefficient.

Clearly the eigenvalues and eigenfunctions of the problem are (see Appendix C part 1):

$$\lambda_n = \left[\frac{(2n-1)\pi}{2l}\right]^2, \quad X_n(x) = \sin\frac{(2n-1)\pi x}{2l}, \quad \|X_n\|^2 = \frac{l}{2}, \quad n = 1,2,3,\dots.$$

and the solution will be of the form

$$u(x,t) = v(x,t) + w(x,t).$$

In general, for a specific problem an auxiliary function is easily obtained from general formulas for $w(x,t)$ found in Appendix C part 2. We may often guess what the function must look like based on the physical observation that we are searching for a terminal or steady state solution. In the present case

$$w(x,t) = u_0 .$$

We also have

$$\tilde{f}(x,t) = 0, \quad \tilde{\varphi}(x) = -u_0.$$

Applying formula (6.34), we obtain

$$C_n = -\frac{2u_0}{l} \int_0^l \sin\frac{(2n-1)\pi x}{2l} dx = -\frac{4u_0}{(2n-1)\pi}.$$

Substituting the expression for C_n into the general formula, we obtain the final solution:

$$u(x,t) = w(x,t) + \sum_{n=1}^{\infty} C_n e^{-\frac{a^2(2n-1)^2\pi^2}{4l^2}t} \sin\frac{(2n-1)\pi x}{2l}$$

$$= u_0 - \frac{4u_0}{\pi} \sum_{n=1}^{\infty} \frac{1}{2n-1} e^{-\frac{a^2(2n-1)^2\pi^2}{4l^2}t} \sin\frac{(2n-1)\pi x}{2l}.$$

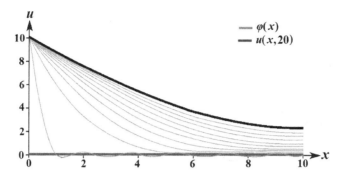

FIGURE 6.9
Solution $u(x,t)$ for Example 6.10.

Figure 6.9 shows the spatial-time-dependent solution $u(x,t)$ for Example 6.10 for $l = 10$, $u_0 = 10$ and $a^2 = D = 1$.

Example 6.11 The initial temperature of a homogeneous isotropic bar of length l ($0 \leq x \leq l$) is

$$u(x,0) = u_0 = \text{const.}$$

There exists a steady heat flux from the environment into the ends of the bar which is given by

$$\frac{\partial u}{\partial x}(0,t) = Q_1 = -\frac{q_1}{\kappa S}, \quad \frac{\partial u}{\partial x}(l,t) = Q_2 = \frac{q_2}{\kappa S}.$$

Convective heat transfer occurs with the environment through the lateral surface. Find the temperature $u(x,t)$ of the bar for $t > 0$.

Solution. The problem is described by the equation

$$\frac{\partial u}{\partial t} = a^2 \frac{\partial^2 u}{\partial x^2} - \gamma u, \quad a^2 = \frac{\kappa}{c\rho}, \quad \gamma = \frac{h}{c\rho}$$

with initial and boundary conditions

$$u(x,0) = \varphi(x) = u_0,$$

$$\frac{\partial u}{\partial x}(0,t) = Q_1, \quad \frac{\partial u}{\partial x}(l,t) = Q_2.$$

The boundary conditions of the problem are Neumann boundary conditions and the eigenvalues and eigenfunctions are, respectively,

$$\lambda_n = \left(\frac{n\pi}{l}\right)^2, \quad X_n(x) = \cos\frac{n\pi x}{l}, \quad \|X_n\|^2 = \begin{cases} l, & n = 0, \\ l/2, & n > 0. \end{cases}$$

The auxiliary function is (see Appendix C part 2):

$$w(x,t) = \left(x - \frac{x^2}{2l}\right)Q_1 + \frac{x^2}{2l}Q_2 = xQ_1 + \frac{x^2}{2l}(Q_2 - Q_1).$$

We leave it to the reader as a *Reading Exercise* to obtain this auxiliary function and the eigenvalues and eigenfunctions.

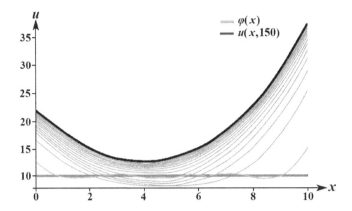

FIGURE 6.10
Solution $u(x,t)$ for Example 6.11.

The solution is

$$u(x,t) = w(x,t) + \sum_{n=0}^{\infty} \left[T_n(t) + C_n \cdot e^{-(a^2\lambda_n+\gamma)t} \right] \cos \frac{n\pi x}{l}.$$

To obtain the coefficients we find first

$$\tilde{f}(x,t) = \frac{a^2}{l}(Q_2 - Q_1) - \gamma \left[\left(x - \frac{x^2}{2l} \right) Q_1 + \frac{x^2}{2l}Q_2 \right],$$

$$\tilde{\varphi}(x) = u_0 - \left(x - \frac{x^2}{2l} \right) Q_1 - \frac{x^2}{2l}Q_2.$$

Applying Equations (6.34), (6.52) and (6.55), we obtain

$$C_0 = \frac{1}{l} \int_0^l \left[u_0 - \left(x - \frac{x^2}{2l} \right) Q_1 - \frac{x^2}{2l}Q_2 \right] dx = u_0 - \frac{l}{6}(2Q_1 + Q_2),$$

$$C_n = \frac{2}{l} \int_0^l \left[u_0 - Q_1 x + (Q_1 - Q_2)\frac{x^2}{2l} \right] \cos\frac{n\pi x}{l} dx = \frac{2l}{n^2\pi^2} \left[Q_1 - (-1)^n Q_2 \right],$$

$$f_0(t) = \frac{1}{l} \int_0^l f^*(x,t)\, dx = \frac{a^2(Q_2 - Q_1)}{l} - \frac{\gamma l}{6}(2Q_1 + Q_2),$$

$$T_0(t) = \int_0^t f_0(\tau) \cdot e^{-\gamma(t-\tau)}\, d\tau = \frac{1}{\gamma} \cdot \left[\frac{a^2(Q_2 - Q_1)}{l} - \frac{\gamma l}{6}(2Q_1 + Q_2) \right] \cdot \left(1 - e^{-\gamma t} \right).$$

For $n > 0$,

$$f_n(t) = \frac{2}{l} \int_0^l f^*(x,t) \cos\frac{n\pi x}{l}\, dx = \frac{2l\gamma}{n^2\pi^2} \left[Q_1 - (-1)^n Q_2 \right],$$

$$T_n(t) = \int_0^t f_n(\tau) \cdot e^{-(a^2\lambda_n+\gamma)(t-\tau)}\, d\tau$$

$$= \frac{2l\gamma}{n^2\pi^2 (a^2\lambda_n + \gamma)} \cdot \left[Q_1 - (-1)^n Q_2 \right] \cdot \left(1 - e^{-(a^2\lambda_n+\gamma)t} \right).$$

Figure 6.10 shows the spatial-time-dependent solution $u(x,t)$. This solution was obtained for the case when $l = 10$, $\gamma = 0.02$, $u_0 = 10$, $Q_1 = -5$, $Q_2 = 10$ and $a^2 = 0.25$. All parameters are dimensionless. The dark gray line represents the initial temperature and the black line is temperature at time $t = 150$, step $\Delta t = 2$.

Example 6.12 Find the temperature distribution inside a thin homogeneous isotropic bar of length l ($0 \leq x \leq l$) with insulated lateral surface if the initial temperature is zero. The temperature is maintained at zero on the right end of the bar ($x = l$), and on the left it changes as governed by

$$u(0, t) = u_0 \cos \omega t,$$

where u_0, ω are known constants. There are no sources or absorbers of heat inside the bar.

Solution. The temperature is given by a solution of the equation

$$\frac{\partial u}{\partial t} = a^2 \frac{\partial^2 u}{\partial x^2}$$

with conditions

$$u(x, 0) = 0,$$
$$u(0, t) = u_0 \cos \omega t, \quad u(l, t) = 0.$$

For these Dirichlet boundary conditions the eigenvalues and eigenfunctions of the problem are

$$\lambda_n = \left(\frac{n\pi}{l}\right)^2, \quad X_n(x) = \sin \frac{n\pi x}{l}, \quad \|X_n\|^2 = \frac{l}{2} \quad (n = 1, 2, 3, \ldots).$$

The solution will be of the form

$$u(x, t) = v(x, t) + w(x, t).$$

The auxiliary function follows from the general case (also see Appendix C part 2):

$$w(x, t) = \left(1 - \frac{x}{l}\right) u_0 \cos \omega t.$$

Then

$$\tilde{f}(x, t) = u_0 \omega \left(1 - \frac{x}{l}\right) \sin \omega t \quad \text{and} \quad \tilde{\varphi}(x) = -u_0 \left(1 - \frac{x}{l}\right).$$

Applying Equations (6.34), (6.52) and (6.55), we obtain

$$C_n = \frac{2}{l} \int_0^l \tilde{\varphi}(x) \sin \frac{n\pi x}{l} dx = -\frac{2u_0}{n\pi},$$

$$f_n(t) = \frac{2}{l} \int_0^l \tilde{f}(x) \sin \frac{n\pi x}{l} dx = \frac{2u_0 \omega}{n\pi} \sin \omega t,$$

$$T_n(t) = \int_0^t f_n(\tau) e^{-a^2 \lambda_n (t-\tau)} d\tau = \frac{2u_0 \omega}{n\pi} \int_0^t \sin \omega\tau \, e^{-a^2 \lambda_n (t-\tau)} d\tau$$

$$= \frac{2u_0 \omega}{n\pi (a^4 \lambda_n^2 + \omega^2)} \left[a^2 \lambda_n \sin \omega t - \omega \cos \omega t + \omega e^{-a^2 \lambda_n t} \right].$$

Substituting the expressions for C_n and $T_n(t)$ into the general formulas, we obtain the solution:

$$u(x, t) = \left(1 - \frac{x}{l}\right) u_0 \cos \omega t + \sum_{n=1}^{\infty} \left[T_n(t) + C_n e^{-a^2 \lambda_n t} \right] \sin \frac{n\pi x}{l}$$

$$= \left(1 - \frac{x}{l}\right) u_0 \cos \omega t + \frac{2u_0}{\pi} \sum_{n=1}^{\infty} \frac{1}{n(a^4 \lambda_n^2 + \omega^2)}$$

$$\times \left[a^2 \lambda_n \omega \sin \omega t - \omega^2 \cos \omega t + a^2 \lambda_n e^{-a^2 \lambda_n t} \right] \sin \frac{n\pi x}{l}.$$

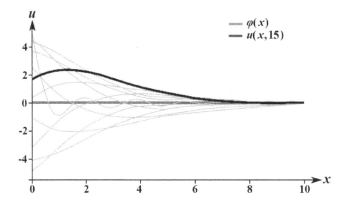

FIGURE 6.11
Solution $u(x, t)$ for Example 6.12.

Figure 6.11 shows the spatial-time-dependent solution $u(x, t)$. This solution was obtained for the case when $l = 10$, $\omega = 0.5$, $u_0 = 5$ and $a^2 = 4$. All parameters are dimensionless. The dark gray line represents the initial temperature and the black line is temperature at time $t = 15$, step $\Delta t = 2$.

Reading Exercise: Check that for $t \to \infty$ the solution obtained above is a sum of purely periodic harmonics and can be presented in the form

$$u(x, t) = \left(1 - \frac{x}{l}\right) u_0 \cos \omega t + \frac{2u_0}{\pi} \sum_{n=1}^{\infty} \frac{1}{n} \sin \delta_n \cdot \sin(\omega t - \delta_n) \cdot \sin \frac{n\pi x}{l},$$

where the phase shifts are given by $\delta_n = \tan^{-1}(\omega/a^2 \lambda_n)$.

6.6 Large Time Behavior of Solutions

Next, we consider frequently encountered physical situations where knowledge of initial conditions is not important. It is seen from the previous examples that the influence of the initial conditions decreases with time for cases where heat propagates through a body. If the moment of interest is long enough after the initial time, the temperature of a bar, for example, is for all purposes defined by the boundary conditions since the effects of the initial conditions have had time to decay. In this case we may suppose that after a long enough time the initial condition vanishes. This situation also frequently applies when *boundary conditions change periodically*, for example in the previous Example 6.12. In these cases we may assume that after a large interval of time the temperature of a body varies periodically with the same frequency as the boundary condition. After this time the initial temperature can always be assumed to be equal to zero (even if it is not). In problems like Example 6.12 we can specify this time – as can be seen from the solution temperature does not depend on the initial condition when $t \gg (l/a)^2$.

Another situation when initial conditions are not important occurs when the equation contains periodically changing terms so that the solution varies with the same frequency after a long enough time. The example below presents such a situation.

Example 6.13 Consider the motion of fluid between two parallel plates, located at $x = 0$ and $x = H$, under a periodically changing pressure gradient parallel to the y-axis. Clearly this is a one-dimensional problem; the function for which we are searching is the y-component of the fluid speed, $u(x, t)$. Since we are searching for a steady-state regime we assume the solution does not depend on the initial condition and formally we set $u(x, 0) = 0$.

Solution. This problem is described by the equation that follows from the Navier-Stokes equation for fluid motion given by

$$\frac{\partial u}{\partial t} = a^2 \frac{\partial^2 u}{\partial x^2} + b \cos \omega t, \tag{6.66}$$

where $a^2 \equiv \nu$ is the coefficient of kinematic viscosity. The boundary conditions

$$u(0, t) = u(H, t) = 0$$

correspond to zero velocity at the plates and the initial condition is

$$u(x, 0) = 0.$$

The eigenvalues and eigenfunctions of the problem we have seen a number of times before and are

$$\lambda_n = \left(\frac{n\pi}{H}\right)^2, \quad X_n(x) = \sin \frac{n\pi x}{H}, \quad ||X_n||^2 = \frac{H}{2}, \quad n = 1, 2, 3, \ldots.$$

The coefficient, $C_n = 0$, since $\varphi(x) = 0$. Applying Equations (6.52) and (6.55) we obtain

$$f_n(t) = \frac{2}{H} \int_0^H b \cos \omega t \cdot \sin \frac{n\pi x}{H} dx = \frac{2b \cos \omega t}{n\pi} [1 - (-1)^n],$$

$$T_n(t) = \frac{2b [1 - (-1)^n]}{n\pi} \int_0^t \cos \omega \tau \, e^{-\frac{a^2 n^2 \pi^2}{H^2}(t-\tau)} d\tau = \frac{2bH^2 [1 - (-1)^n]}{n\pi [a^4 n^4 \pi^4 + \omega^2 H^4]}$$

$$\times \left[a^2 n^2 \pi^2 \cos \omega t + \omega H^2 \sin \omega t - a^2 n^2 \pi^2 e^{-\frac{a^2 n^2 \pi^2}{H^2} t} \right].$$

Obviously from above we see that $T_{2k}(t) = 0$. Also

$$a^2 n^2 \pi^2 \cos \omega t + \omega H^2 \sin \omega t = \sqrt{a^4 n^4 \pi^4 + \omega^2 H^4} \sin(\omega t + \theta_n),$$

where $\theta_n = \arctan \frac{a^2 n^2 \pi^2}{\omega H^2}$. Hence, the solution of the problem has the form

$$u(x, t) = \frac{4bH^2}{\pi} \sum_{k=1}^{\infty} \frac{1}{2k-1} \left\{ \frac{\sin(\omega t + \theta_{2k-1})}{\sqrt{a^4 (2k-1)^4 \pi^4 + \omega^2 H^4}} \right.$$

$$\left. - \frac{a^2 (2k-1)^2 \pi^2}{[a^4 (2k-1)^4 \pi^4 + \omega^2 H^4]} e^{-\frac{a^2 (2k-1)^2 \pi^2}{H^2} t} \right\} \sin \frac{(2k-1)\pi x}{H}.$$

Figure 6.12 shows the spatial-time-dependent solution $u(x, t)$. This solution was obtained for the case when $H = 10$, $\omega = 0.5$, $b = 5$ and $a^2 = \nu = 1$. All parameters are dimensionless. The dark gray line represents the initial temperature and the black line is velocity at time $t = 30$, step $\Delta t = 2$.

Clearly as $t \to \infty$ (more accurately, when $t \gg H^2/a^2$) we have

$$u(x, t) \to \frac{4bH^2}{\pi} \sum_{k=1}^{\infty} \frac{\sin(\omega t + \theta_{2k-1})}{(2k-1)\sqrt{a^4 (2k-1)^4 \pi^4 + \omega^2 H^4}} \sin \frac{(2k-1)\pi x}{H}.$$

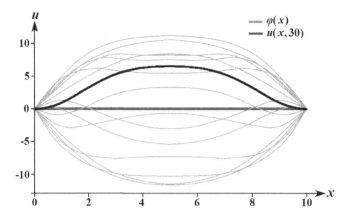

FIGURE 6.12
Solution $u(x,t)$ for Example 6.13.

We can also see that $u(0,t) = u(H,t) = 0$ as it should be. Also $u(x,t)$ takes its maximum value at $x = H/2$.

Notice that the initial conditions and the solution formally disagree but in this periodic problem the role of the initial condition becomes negligible for times $t \gg H^2/a^2$.

In this and similar problems it is easy to obtain an asymptotic solution in a closed form (not as the series). From the physical point of view it follows that such a solution is periodic with frequency ω because in systems with dissipation (described by parabolic equations) the internal oscillations decay exponentially with time (which is exactly the opposite of the case of hyperbolic equations). This allows us to search for a solution in the form

$$u(x,t) = X(x)\exp(i\omega t). \tag{6.67}$$

Notice that we cannot look for a time dependence for the solution in the form of an external force $\cos\omega t$ only - the reason is that the derivative in time mix the *sin* and *cos* functions and we have to take into account both of them. Another way is to use an exponential form (6.67).

Substituting Equation (6.67) into Equation (6.66) we obtain the ordinary differential equation

$$X'' - \frac{i\omega}{a^2}X = -\frac{b}{a^2}. \tag{6.68}$$

First, we solve the homogeneous equation,

$$X'' - \frac{i\omega}{a^2}X = 0. \tag{6.69}$$

The characteristic equation for this linear equation has two roots:

$$\frac{\sqrt{i\omega}}{a} = \pm\frac{\sqrt{\omega/2}}{a}(1+i). \tag{6.70}$$

Thus, a general solution to the homogeneous Equation (6.69) is

$$X(x) = C_1\exp[q(1+i)x] + C_2\exp[-q(1+i)x], \quad q = \frac{1}{a}\sqrt{\frac{\omega}{2}}. \tag{6.71}$$

A particular solution of the nonhomogeneous Equation (6.68) is $(-ib/\omega)$, in which case we may write a solution of this equation which satisfies zero boundary conditions as

$$X(x) = -\frac{ib}{\omega}\left(1 - \frac{\cos\alpha\left(x - \frac{H}{2}\right)}{\cos\alpha\frac{H}{2}}\right), \quad \alpha = q\left(1 - i\right). \tag{6.72}$$

From this we have

$$u(x,t) = \operatorname{Re}X\cos\omega t - \operatorname{Im}X\sin\omega t. \tag{6.73}$$

$X(x)$ is complex and the system response will be, as follows from the previous formula, shifted in phase relative to the external influence.

To obtain a final form we use the identity

$$\cos z = \cos\left(x + iy\right) = \cos x\cos iy + \sin x\sin iy = \cos x\cosh y + i\sin x\sinh y \tag{6.74}$$

to yield the result for $\operatorname{Re}X$ and $\operatorname{Im}X$ that, with Equation (6.73), gives the final answer for the steady state solution:

$$\operatorname{Re}X = -\frac{b}{\omega}\frac{\sin qx\sinh q\left(x - H\right) - \sin q\left(x - H\right)\sinh qx}{\cos qH + \cosh qH},$$

$$\operatorname{Im}X = -\frac{b}{\omega}\left(1 - \frac{\cos qx\cosh q\left(x - H\right) + \cos q\left(x - H\right)\cosh qx}{\cos qH + \cosh qH}\right). \tag{6.75}$$

6.7 Maximum Principle

Let us consider the homogeneous heat equation

$$\frac{\partial u}{\partial t} = a^2\frac{\partial^2 u}{\partial x^2}, \quad 0 < x < l, \ 0 < t < T \tag{6.76}$$

with the initial condition

$$u(x,t)|_{t=0} = \varphi(x), \quad 0 \le x \le l \tag{6.77}$$

and Dirichlet boundary conditions

$$P_1[u] = u|_{x=0} = g_1(t), \quad P_2[u] = u|_{x=l} = g_2(t), \quad 0 \le t \le T. \tag{6.78}$$

The initial-boundary value problem (6.76)-(6.78) describes the temperature field in a rod heated or cooled at the ends, when there are no sources of heating or cooling inside the rod.

Let M_1 be the maximum initial temperature, $M_1 = \max_x\varphi(x)$, M_2 be the maximum temperature ever imposed on the rod's ends,

$$M_2 = \max\left[\max_t g_1(t), \max_t g_2(t)\right],$$

and M is the largest number among M_1 and M_2, $M = \max(M_1, M_2)$. It is intuitively clear that the temperature inside the rod can never exceed M. Indeed, when the temperature does not exceed M in the beginning, the ends of the rod are never heated to the temperature higher than M, and there is no internal heating, there is no reason for the temperature inside the rod to exceed M somewhere. Thus, the maximum of function $u(x,t)$ in the whole region $0 \le x \le l, 0 \le t \le T$ is reached either at the initial time instant $t = 0$, or at the ends $x = 0$ or $x = l$, i.e., on the boundary γ shown in bold in Figure 6.13. The maximum cannot be reached in an internal point $x = x_0$ at a certain time instant $t = t_0$, neither for

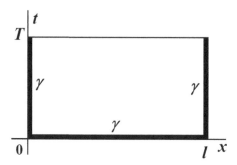

FIGURE 6.13
Boundary γ is shown in bold.

$0 < t_0 < T$, nor at $t = T$. This property of the solution is called *the maximum principle*. Though this property looks rather obvious, its mathematical proof is a bit tricky, and we present it below.

First, let us prove the maximum principle for a function $v(x,t)$ satisfying the inequality $v_t - a^2 v_{xx} = f(x,t) < 0$, i.e., in the case of internal cooling. Assume that the maximum of v is reached in a certain point (x_0, t_0), where $0 < x_0 < l$, $0 < t_0 < T$. According to the properties of a maximum, $v_t = 0$ and $v_{xx} \leq 0$ in that point, hence, $v_t - a^2 v_{xx} \geq 0$, i.e., we get a contradiction. If we assume that the maximum of $v(x,t)$ in the whole region $0 \leq x \leq l$, $0 \leq t \leq T$ is reached at $t = T$ in a certain point x_0, $0 < x_0 < l$, then it should be $v_t \geq 0$, otherwise u will take higher values at $t < T$. Still, $v_{xx} \leq 0$ at $x = x_0$, hence we get again a contradiction: $v_t - a^2 v_{xx} \geq 0$. Thus, we have proved the maximum principle for function $v(x,t)$.

Let us prove now the maximum principle for function $u(x,t)$ satisfying (6.76)-(6.78). Let M be the maximum of u on the boundary γ indicated in Figure 6.13. Define function $v(x,t) = u(x,t) + \varepsilon x^2$, where $\varepsilon > 0$. Then $v_t - a^2 v_{xx} = -2a^2 \varepsilon < 0$. Applying the maximum principle to function $v(x,t)$, we obtain the following sequence of inequalities:

$$u(x,t) = v(x,t) - \varepsilon x^2 \leq v(x,t) \leq v(x,t)|_\gamma \leq M + \varepsilon l^2, \quad 0 \leq x \leq l, \ 0 \leq t \leq T.$$

In the limit $\varepsilon \to 0$, we find that $u(x,t) \leq M$ for $0 \leq x \leq l$, $0 \leq t \leq T$, i.e., $u(x,t)$ satisfies the maximum principle.

A similar property is correct for the *minimum* of function u: it is reached at γ. That is just the maximum principle for function $-u(x,t)$, which is the solution of the heat equation with function $-\varphi(x)$ in the initial condition and functions $-g_1(t)$, $-g_2(t)$ in boundary conditions.

Note that the maximum/minimum principle guarantees the well-posedness of problem (6.76)-(6.78). Indeed, assume that we have two solutions u_1 and u_2 of the homogeneous heat equation for slightly different initial and boundary conditions, and apply the maximum and minimum principles to function $w = u_1 - u_2$ which is also a solution of the homogeneous heat equation. We find that the absolute value of the difference between solutions in the whole definition region of the problem cannot exceed its maximum on γ. Thus, the solution depends continuously on the initial and boundary conditions. If both functions u_1 and u_2 satisfy the same initial and boundary conditions, then their difference $w = u_1 - u_2$ is equal to zero on γ. Because of the maximum/minimum principle, $w = 0$ everywhere, i.e., the solution of (6.76)-(6.78) is unique. The conditions of the existence of a classical solution were discussed in Section 6.2.

6.8 The Heat Equation in an Infinite Region

Let us consider the Cauchy problem for the homogeneous heat equation in the infinite region:

$$\frac{\partial u_1}{\partial t} - a^2 \frac{\partial u_1^2}{\partial x^2} = 0, \quad \infty < x < \infty, \ t > 0, \tag{6.79}$$

$$u_1(x,t)|_{t=0} = \varphi(x), \quad \infty < x < \infty. \tag{6.80}$$

It is assumed that at infinity

$$\lim_{x \to \pm\infty} u_1(x,t) = 0, \quad t \geq 0. \tag{6.81}$$

We can solve the initial value problem (6.79)-(6.81) by means of the Fourier transform, as was done in Section 5.11.1 in the case of a hyperbolic equation. It would be more instructive, however, to obtain the same result taking a limit $l \to \infty$ in the problem in a finite region. That is done below.

First, let us consider the Cauchy problem in a symmetric finite region $-l/2 < x < l/2$ with homogeneous Dirichlet boundary conditions

$$u_1\left(-\frac{l}{2}, t\right) = u_1\left(\frac{l}{2}, t\right) = 0, \quad t > 0.$$

The solution can be written in the form

$$u_1(x,t;l) = \int_0^l g(x,\xi,t;l)\varphi(\xi)d\xi, \tag{6.82}$$

$$g(x,\xi,t;l) = \sum_{n=1}^{\infty} \frac{X_n(x;l)X_n(\xi;l)}{\|X_n\|^2} e^{-a^2\lambda_n t}, \tag{6.83}$$

where

$$X_n(x;l) = \sin\frac{n\pi}{l}\left(x + \frac{l}{2}\right), \quad \|X_n\|^2 = \frac{l}{2}, \quad \lambda_n = \left(\frac{n\pi}{l}\right)^2, \quad n = 1,2,3,\ldots$$

(see (6.57), (6.58)). Note that for even $n = 2m$, $m = 1, 2, \ldots$,

$$\sin\frac{2m\pi}{l}\left(x + \frac{l}{2}\right) = \sin\left(\frac{2m\pi}{l}x + m\pi\right) = (-1)^m \sin\frac{2m\pi x}{l},$$

and for odd $n = 2m + 1$, $m = 0, 1, \ldots$,

$$\sin\left[\frac{(2m+1)\pi}{l}\left(x + \frac{l}{2}\right)\right] = \sin\left[\frac{(2m+1)\pi}{l}x + \left(m + \frac{1}{2}\right)\pi\right] = (-1)^m \cos\frac{(2m+1)\pi x}{l},$$

thus, the kernel (6.83)

$$g(x,\xi,t;l) = g_s(x,\xi,t;l) + g_c(x,\xi,t;l),$$

where

$$g_s(x,\xi,t;l) = \frac{2}{l}\sum_{m=1}^{\infty}\exp\left[-\left(\frac{2m\pi a}{l}\right)^2 t\right]\sin\frac{2m\pi x}{l}\sin\frac{2m\pi\xi}{l}, \tag{6.84}$$

$$g_c(x, \xi, t; l) = \frac{2}{l} \sum_{m=0}^{\infty} \exp\left[-\left[\frac{(2m+1)\pi a}{l}\right]^2 t\right] \cos\frac{(2m+1)\pi x}{l} \cos\frac{(2m+1)\pi \xi}{l}. \tag{6.85}$$

In order to find the limit of (6.84) at $l \to \infty$, let us define

$$\eta_m = \frac{2m\pi}{l}, \ \Delta\eta = \frac{2\pi}{l},$$

then

$$g_s(x, \xi, t; l) = \frac{1}{\pi} \sum_{m=1}^{\infty} e^{-a^2\eta_m^2 t} \sin\eta_m x \sin\eta_m\xi\Delta\eta.$$

In the limit $\Delta\eta \to 0$ we obtain

$$g_s(x, \xi, t) = \lim_{l\to\infty} g_s(x, \xi, t; l) = \frac{1}{\pi} \int_0^\infty e^{-a^2\eta^2 t} \sin\eta x \sin\eta\xi d\eta.$$

Similarly, we can obtain

$$g_c(x, \xi, t) = \lim_{l\to\infty} g_c(x, \xi, t; l) = \frac{1}{\pi} \int_0^\infty e^{-a^2\eta^2 t} \cos\eta x \cos\eta\xi d\eta.$$

Thus,

$$g(x, \xi, t) = \lim_{l\to\infty} g(x, \xi, t; l) = g_s(x, \xi, t) + g_c(x, \xi, t)$$

$$= \frac{1}{\pi} \int_0^\infty e^{-a^2\eta^2 t} \cos\eta(x - \xi)d\eta. \tag{6.86}$$

To calculate this integral, let us present it as

$$g(x, \xi, t) = \frac{1}{2\pi} \int_{-\infty}^\infty e^{-a^2\eta^2 t} \cos\eta(x-\xi)d\eta = \frac{1}{2\pi}\text{Re} \int_{-\infty}^\infty e^{-a^2\eta^2 t + i\eta(x-\xi)} d\eta$$

$$= \frac{1}{2\pi} \int_{-\infty}^\infty \exp\left(-a^2 t\left[\eta^2 - \frac{i(x-\xi)}{a^2 t}\eta\right]\right) d\eta$$

$$-\frac{1}{2\pi} \exp\left[-\frac{(x-\xi)^2}{4a^2 t}\right] \text{Re} \int_{-\infty}^\infty \exp\left(-a^2 t\left[\eta - \frac{i(x-\xi)}{2a^2 t}\right]^2\right).$$

Define

$$\zeta = a\sqrt{t}\left[\eta - \frac{i(x-\xi)}{2a^2 t}\right],$$

then

$$g(x, \xi, t) = \frac{1}{2\pi a\sqrt{t}} e^{-(x-\xi)^2/4a^2 t} \int_{-\infty}^\infty e^{-\zeta^2} d\zeta = \frac{e^{-(x-\xi)^2/4a^2 t}}{2a\sqrt{\pi t}}. \tag{6.87}$$

Finally, the solution of the problem (6.79)-(6.80) is

$$u_1(x, t) = \int_{-\infty}^\infty g(x, \xi, t)\varphi(\xi)d\xi,$$

where $g(x, \xi, t)$ is determined by formula (6.87).

Similarly to (6.60), (6.61), the solution of the nonhomogeneous heat equation with zero initial conditions,

$$\frac{\partial u_2}{\partial t} - a^2\frac{\partial u_1^2}{\partial x^2} = f(x, t), \quad \infty < x < \infty, \ t > 0; \tag{6.88}$$

$$u_2(x,t)|_{t=0} = 0, \quad \infty < x < \infty; \qquad \lim_{x \to \pm\infty} u_2(x,t) = 0, \quad t \geq 0, \tag{6.89}$$

can be found as

$$u_2(x,t) = \int_{-\infty}^{\infty} d\tau \int_{-\infty}^{\infty} d\xi \, G(x,\xi,t-\tau) f(\xi,\tau), \tag{6.90}$$

where Green's function

$$G(x,\xi,t-\tau) = H(t-\tau)g(x,\xi,t) = \frac{H(t-\tau)}{2a\sqrt{\pi(t-\tau)}} \exp\left[-\frac{(x-\xi)^2}{4a^2(t-\tau)}\right]. \tag{6.91}$$

The solution of the nonhomogeneous heat equation with non-zero initial condition can be found as $u(x,t) = u_1(x,t) + u_2(x,t)$.

Example 6.14 Using the expression for Green's function (6.91), solve the following BVP for a nonhomogeneous heat equation:

$$u_t - a^2 u_{xx} = \delta(x), \quad -\infty < x < \infty, \quad t > 0,$$

$$u(x,0) = 0, \quad -\infty < x < \infty; \qquad \lim_{x \to \pm\infty} u(x,t) = 0, \quad t \geq 0.$$

Substituting $f(\xi,\tau) = \delta(\xi)H(\tau)$ into Equation (6.90), we find:

$$u(x,t) = \int_{-\infty}^{\infty} d\tau \frac{H(t-\tau)H(\tau)}{2a\sqrt{\pi(t-\tau)}} \int_{-\infty}^{\infty} d\xi \exp\left[-\frac{(x-\xi)^2}{4a^2(t-\tau)}\right]\delta(\xi)$$

$$= \frac{1}{2a\sqrt{\pi}} \int_0^t \frac{1}{\sqrt{t-\tau}} \exp\left[-\frac{x^2}{4a^2(t-\tau)}\right].$$

To calculate that integral, it is convenient first to apply the following change of the integration variable:

$$y = \frac{1}{t-\tau}; \quad \tau = t - \frac{1}{y}.$$

Then

$$u(x,t) = \frac{1}{2a\sqrt{\pi}} \int_{1/t}^{\infty} y^{-3/2} \exp\left[-\frac{x^2}{4a^2}y\right].$$

Using tables of integrals, or Mathematica one finds:

$$u(x,t) = \frac{\sqrt{t}}{a\sqrt{\pi}} \exp\left(-\frac{x^2}{4a^2t}\right) - \frac{|x|}{4a^2}\text{erfc}\frac{|x|}{2a\sqrt{t}}.$$

Problems

Problems 1 through 30 involve the temperature distribution inside a homogeneous isotropic rod (or bar) of length l. Solve these problems analytically which means the following: formulate the equation and initial and boundary conditions, obtain the eigenvalues and eigenfunctions, write the formulas for coefficients of the series expansion and the expression for the solution of the problem. You can obtain the pictures of several eigenfunctions and screenshots of the solution and of the auxiliary functions with Maple, Mathematica or software from books [7, 8].

In problems 1 through 9 we consider rods which are thermally insulated over their lateral surfaces. In the initial time, $t = 0$, the temperature distribution is given by $u(x,0) = \varphi(x)$, $0 < x < l$. There are no heat sources or absorbers inside the rod.

1. The ends of the rod are kept at zero temperature. The initial temperature of the rod is given as

$$\varphi_1(x) = 1, \quad \varphi_2(x) = x, \quad \varphi_3(x) = x(l - x).$$

2. The left end of the rod is kept at zero temperature and the right end is thermally insulated from the environment. The initial temperature of the rod is

$$\varphi_1(x) = x^2, \quad \varphi_2(x) = x, \quad \varphi_3(x) = x\left(l - \frac{x}{2}\right).$$

3. The left end of the rod is thermally insulated and the right end is kept at zero temperature. The initial temperature of the rod is

$$\varphi_1(x) = x, \quad \varphi_2(x) = 1, \quad \varphi_3(x) = l^2 - x^2.$$

4. Both ends of the rod are thermally insulated. The initial temperature of the rod is

$$\varphi_1(x) = x, \quad \varphi_2(x) = l^2 - x^2, \quad \varphi_3(x) = x^2\left(1 - \frac{2x}{3}\right).$$

5. The left end of the rod is kept at zero temperature and the right end is subject to convective heat transfer with the environment. The initial temperature of the rod is

$$\varphi_1(x) = 1, \quad \varphi_2(x) = x, \quad \varphi_3(x) = x\left(l - \frac{x}{2}\right).$$

6. The left end of the rod is subject to convective heat transfer with the environment, which has zero temperature, and the right end is kept at zero temperature. The initial temperature of the rod is

$$\varphi_1(x) = 1, \quad \varphi_2(x) = l - x, \quad \varphi_3(x) = \frac{x}{3}(l - 2x) + \frac{1}{3}.$$

7. The left end of the rod is thermally insulated and the right end is subject to convective heat transfer with the environment which has constant temperature of zero. The initial temperature of the rod is

$$\varphi_1(x) = 1, \quad \varphi_2(x) = x, \quad \varphi_3(x) = \left(1 - \frac{x^3}{3}\right).$$

8. The left end of the rod is subject to convective heat transfer with the environment (whose temperature is zero) and the right end is thermally insulated. The initial temperature of the rod is

$$\varphi_1(x) = 1, \quad \varphi_2(x) = x, \quad \varphi_3(x) = 1 - \frac{(x - 1)^3}{3}.$$

9. Both ends of the rod are subject to convective heat transfer with the environment, which has a temperature of zero. The initial temperature of the rod is

$$\varphi_1(x) = 1, \quad \varphi_2(x) = x, \quad \varphi_3(x) = \left(1 - \frac{x^3}{3}\right).$$

In problems 10 through 18 we consider rods whose lateral surfaces are subject to heat transfer according to Newton's law. In problems 10 through 14 the environment has constant temperature, which of course can be taken as zero. The initial temperature of the rod is given as $u(x, 0) = \varphi(x)$. Again there are no heat sources or absorbers inside the rod.

10. The ends of the rod are kept at a constant temperature, the left end has temperature $u(0,t) = u_1$ and the right end has temperature $u(l,t) = u_2$. The initial temperature of the rod is

$$\varphi_1(x) = x^2, \quad \varphi_2(x) = x, \quad \varphi_3(x) = x\left(l - \frac{x}{2}\right).$$

11. The left end of the rod is kept at a constant temperature $u(0,t) = u_1$, and a constant heat flow is supplied to the right end of the rod. The initial temperature of the rod is

$$\varphi_1(x) = x^2, \quad \varphi_2(x) = x, \quad \varphi_3(x) = x\left(l - \frac{x}{2}\right).$$

12. A constant heat flow is supplied to the left end of the rod from outside and the right end of the rod is kept at a constant temperature $u(l,t) = u_2$. The initial temperature of the rod is

$$\varphi_1(x) = x, \quad \varphi_2(x) = 1, \quad \varphi_3(x) = l^2 - x^2.$$

13. Constant heat flows are supplied to both ends of the rod. The initial temperature of the rod is

$$\varphi_1(x) = x, \quad \varphi_2(x) = l^2 - x^2, \quad \varphi_3(x) = x^2\left(1 - \frac{2x}{3}\right).$$

14. The left end of the rod is kept at a constant temperature $u(0,t) = u_1$ and the right end is subject to convective heat transfer with the environment, which has a constant temperature of u_0. The initial temperature of the rod is

$$\varphi_1(x) = 1, \quad \varphi_1(x) = 1, \quad \varphi_3(x) = x\left(l - \frac{x}{2}\right).$$

15. The left end of the rod is subject to convective heat transfer with an environment which has a constant temperature of u_0, the right end is kept at the constant temperature $u(l,t) = u_2$. The initial temperature of the rod is

$$\varphi_1(x) = 1, \quad \varphi_2(x) = l - x, \quad \varphi_3(x) = \frac{x}{3}(l - 2x) + \frac{1}{3}.$$

16. A constant heat flow is supplied to the left end of the rod from outside and the right end of the rod is subject to convective heat transfer with an environment of constant temperature, u_0. The initial temperature of the rod is

$$\varphi_1(x) = 1, \quad \varphi_2(x) = x, \quad \varphi_3(x) = \left(1 - \frac{x^3}{3}\right).$$

17. The left end of the rod is subject to convective heat transfer with an environment of constant temperature u_0, and a constant heat flow is supplied to the right end of the rod. The initial temperature of the rod is

$$\varphi_1(x) = 1, \quad \varphi_2(x) = x, \quad \varphi_3(x) = 1 - \frac{(x-1)^3}{3}.$$

18. Both ends of the rod are subject to convective heat transfer with an environment of constant temperature u_0. The initial temperature of the rod is

$$\varphi_1(x) = 1, \quad \varphi_2(x) = x, \quad \varphi_3(x) = \left(1 - \frac{x^3}{3}\right).$$

Problems 19 through 21 consider rods whose lateral surfaces are subjected to heat transfer according to Newton's law and the environment has a constant temperature θ. One internal source of heat acts at the point x_0 inside the rod and the power of this source is Q.

19. The ends of the rod are kept at constant temperatures – the left end has a temperature $u(0,t) = u_1$ and the right end has a temperature $u(l,t) = u_2$. The initial temperature of the rod is

$$\varphi_1(x) = x^2, \quad \varphi_2(x) = x, \quad \varphi_3(x) = x\left(l - \frac{x}{2}\right).$$

20. Constant heat flows, q_1 and q_2, are supplied to both ends of the rod from outside. The initial temperature of the rod is

$$\varphi_1(x) = x, \quad \varphi_2(x) = l^2 - x^2, \quad \varphi_3(x) = x^2\left(1 - \frac{2x}{3}\right).$$

21. At both ends of the rod a convective heat transfer occurs with the environment which has a constant temperature of θ. The initial temperature of the rod is

$$\varphi_1(x) = 1, \quad \varphi_2(x) = x, \quad \varphi_3(x) = \left(1 - \frac{x^3}{3}\right).$$

In problems 22 through 24 we consider rods whose lateral surfaces are subjected to heat transfer according to Newton's law and the environment has a constant temperature θ. Internal heat sources and absorbers are active in the rod and their intensity (per unit mass of the rod) is given by $f(x,t)$. The initial temperature of the rod is zero, $u(x,0) = 0$.

22. The ends of the rod are kept at constant temperatures – the left end has temperature $u(0,t) = u_1$ and the right end has temperature $u(l,t) = u_2$. The intensities of heat sources and absorbers are

$$\begin{array}{ll} f_1(x,t) = A\sin\omega t, & f_2(x,t) = Ae^{-\alpha t}\sin\omega t, \\ f_3(x,t) = A\cos\omega t, & f_4(x,t) = Ae^{-\alpha t}\cos\omega t, \\ f_5(x,t) = A\sin\omega t\cos\omega t, & f_6(x,t) = Ae^{-\alpha t}(\sin\omega t + \cos\omega t). \end{array}$$

23. The constant heat flows, q_1 and q_2, are supplied to both ends of the rod from outside. The intensities of heat sources and absorbers are

$$\begin{array}{ll} f_1(x,t) = A\sin\omega t, & f_2(x,t) = Ae^{-\alpha t}\sin\omega t, \\ f_3(x,t) = A\cos\omega t, & f_4(x,t) = Ae^{-\alpha t}\cos\omega t, \\ f_5(x,t) = A\sin\omega t\cos\omega t, & f_6(x,t) = Ae^{-\alpha t}(\sin\omega t + \cos\omega t). \end{array}$$

24. Both ends of the rod are subjected to convective heat transfer with the environment at constant temperature θ. The intensities of heat sources and absorbers are

$$\begin{array}{ll} f_1(x,t) = A\sin\omega t, & f_2(x,t) = Ae^{-\alpha t}\sin\omega t, \\ f_3(x,t) = A\cos\omega t, & f_4(x,t) = Ae^{-\alpha t}\cos\omega t, \\ f_5(x,t) = A\sin\omega t\cos\omega t, & f_6(x,t) = Ae^{-\alpha t}(\sin\omega t + \cos\omega t). \end{array}$$

In problems 25 through 30 we consider rods with thermally insulated lateral surfaces. The initial temperature of the rod is zero $u(x,0) = 0$. Generation (or absorption) of heat by internal sources is absent. Find the temperature change inside the rod for the following cases:

25. The left end of the rod is kept at the constant temperature $u(0,t) = u_1$ and the temperature of the right end changes according to $g_2(t) = A\cos\omega t$.

26. The temperature of the left end changes as $g_1(t) = A\cos\omega t$ and the right end of the rod is kept at the constant temperature $u(l,t) = u_2$.

27. The left end of the rod is kept at the constant temperature $u(0,t) = u_1$, and the heat flow $g_2(t) = A\sin\omega t$ is supplied to the right end of the rod from outside.

28. The heat flow $g_1(t) = A\cos\omega t$ is supplied to the left end of the rod from outside while the right end of the rod is kept at the constant temperature $u(l,t) = u_2$.

29. The left end of the rod is kept at the constant temperature $u(0,t) = u_1$ and the right end is subjected to a convective heat transfer with the environment which has a temperature that varies as $u_{md}(t) = A\sin\omega t$.

30. The left end of the rod is subjected to a convective heat transfer with the environment which has a temperature that varies as $u_{md}(t) = A\cos\omega t$ and the right end is kept at the constant temperature $u(l,t) = u_2$.

In problems 31 through 32 find Green's functions for the following problems:

31. $u_t - a^2 u_{xx} = F(x,t)$, $\quad u(0,t) = u_x(L,t) = 0$, $t \geq 0$; $\quad u(x,0) = 0$.

32. $u_t - a^2 u_{xx} = F(x,t)$, $\quad 0 < x < \infty$, $\quad u(0,t) = 0$, $t \geq 0$; $\quad u(x,0) = 0$, $x \geq 0$.

33. Show that Duhamel's principle is valid for the heat equation in an infinite region.

34. Show that the maximum principle holds for equation $u_t + vu_x = a^2 u_{xx}$, where $v = const$.

35. The following boundary value problem is given:

    $u_t = u_{xx}$ (heat equation), $0 < x < \pi$, $0 < t < T$; $u(0,t) = u(\pi,t) = 0$, $0 \leq t \leq T$; $u(x,0) = \sin^2 x$, $0 \leq x \leq \pi$.

    Without solving that problem, show that $0 \leq u(x,t) \leq e^{-t}\sin x$ in the region $0 \leq x \leq \pi$, $0 \leq t \leq T$.

7

Elliptic Equations

7.1   Elliptic Differential Equations and Related Physical Problems

When studying different stationary (time-independent) processes, very often we meet the *Laplace equation*

$$\nabla^2 u = 0. \tag{7.1}$$

The nonhomogeneous equation

$$\nabla^2 u = -f \tag{7.2}$$

with a given function f of the coordinates is called the *Poisson equation.*

Laplace and Poisson partial differential equations are of *Elliptic type.*

The Laplace operator (Laplacian) in Cartesian coordinates is

$$\nabla^2 = \frac{\partial^2}{\partial x^2} + \frac{\partial^2}{\partial y^2} + \frac{\partial^2}{\partial z^2}. \tag{7.3}$$

In spherical coordinates, r, θ, φ, related with Cartesian coordinates as (see Figure 7.1)

$$x = r \sin\theta \cos\varphi, \quad y = r \sin\theta \sin\varphi, \quad z = r \cos\theta,$$

the Laplacian is

$$\nabla^2 = \frac{1}{r^2} \frac{\partial}{\partial r} \left(r^2 \frac{\partial}{\partial r} \right) + \frac{1}{r^2 \sin\theta} \frac{\partial}{\partial \theta} \left(\sin\theta \frac{\partial}{\partial \theta} \right) + \frac{1}{r^2 \sin^2\theta} \frac{\partial^2}{\partial \varphi^2} \equiv \nabla_r^2 + \frac{1}{r^2} \nabla_{\theta\varphi}^2, \tag{7.4}$$

where $\nabla_r^2 = \frac{1}{r^2} \frac{\partial}{\partial r} \left(r^2 \frac{\partial}{\partial r} \right) = \frac{\partial^2}{\partial r^2} + \frac{2}{r} \frac{\partial}{\partial r}$.

In cylindrical coordinates r, φ, z, related with Cartesian coordinates as

$$x = r \cos\varphi, \quad y = r \sin\varphi, \quad z = z,$$

the Laplacian is

$$\nabla^2 = \frac{1}{r} \frac{\partial}{\partial r} \left(r \frac{\partial}{\partial r} \right) + \frac{1}{r^2} \frac{\partial^2}{\partial \varphi^2} + \frac{\partial^2}{\partial z^2}. \tag{7.5}$$

For the particular case of cylindrical coordinates, a polar coordinate system r, φ with no dependence on z, we have

$$x = r \cos\varphi, \quad y = r \sin\varphi$$

and the Laplacian is

$$\nabla^2 = \frac{1}{r} \frac{\partial}{\partial r} \left(r \frac{\partial}{\partial r} \right) + \frac{1}{r^2} \frac{\partial^2}{\partial \varphi^2}. \tag{7.6}$$

Consider several physical problems.

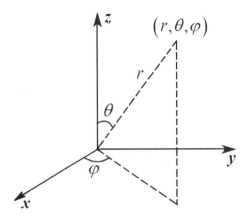

FIGURE 7.1
Spherical coordinates.

1. If a temperature distribution created by external heating does not change with time, $\partial T/\partial t = 0$, the homogeneous heat equation (7.2), Chapter 6, reduces to the Laplace equation,

$$\nabla^2 T = 0. \tag{7.7}$$

   If a medium contains a heat source or heat absorber, Q, and the temperature is time independent, the heat conduction equation (7.7), Chapter 6, becomes

$$\nabla^2 T = -\frac{Q}{\rho c \chi} \tag{7.8}$$

   which is a particular example of the Poisson equation (7.2).

2. The diffusion equation (6.9), Chapter 6, for stationary diffusion, $\partial c/\partial t = 0$, becomes

$$\nabla^2 c = -f/D, \tag{7.9}$$

   where c is the concentration of a material diffusing through a medium, D is the diffusion coefficient and f is a source or sink of the diffusing material. This is also the Poisson equation (or Laplace equation when $f = 0$).

3. The electrostatic potential due to a point charge q is

$$\varphi = \frac{q}{r}, \tag{7.10}$$

   where r is the distance from the charge to the point where the electric field is measured.

For continuous charge distribution with charge density ρ, the potential φ is related to ρ as:

$$\nabla^2 \varphi = -4\pi\rho. \tag{7.11}$$

Equation (7.11) is the Poisson equation for an electrostatic potential. In regions that do not contain electric charges, for instance at points outside of a charged body, $\rho = 0$ and the potential which the body creates obeys the Laplace equation

$$\nabla^2 \varphi = 0. \tag{7.12}$$

Reading exercise: Prove that the scalar potential (7.10) satisfies the Laplace equation for any $r > 0$.

7.2   Harmonic Functions

Let us consider the two-dimensional Laplace equation,

$$\frac{\partial^2 u}{\partial x^2} + \frac{\partial^2 u}{\partial y^2} = 0 \tag{7.13}$$

in the entire x, y-plane.

In Section 3.2.2, we have obtained the general solution of elliptic equation (3.38) with constant coefficients. The Laplace equation is its particular case with $a = c = 1$, $b = 0$. Formula (3.42) gives the general solution of the Laplace equation:

$$u(x, y) = F_+(y + ix) + F_-(y - ix), \tag{7.14}$$

where functions F_+ and F_- are continuously differentiable twice. Define the complex variables

$$z = x + iy, \quad z^* = x - iy, \tag{7.15}$$

then

$$u(x, y) = F_+(iz^*) + F_-(-iz). \tag{7.16}$$

Define $G_1(z) \equiv F_-(-iz)$ and $G_2(z^*) \equiv F_+(iz^*)$, then

$$u(x, y) = G_1(z) + G_2(z^*). \tag{7.17}$$

Functions G_1 and G_2 should be twice differentiable. It is proved in the theory of functions of complex variables that a differentiable function of a complex variable z is actually differentiable infinitely many times; such a function is called an *analytical function*. Thus, we come to the conclusion that the arbitrary solution of the Laplace equation is a sum of two functions, an analytical function of z and a function which is a complex conjugate to an analytical function of z.

Solution (7.17) is real if $G_2(z^*) = (G_1(z))^*$, i.e., $u(x, y) = G_1(z) + (G_1(z))^*$. A particular real solution of the Laplace equation is called the *harmonic function*. For any non-constant analytical function $G(z)$, we can obtain two harmonic functions taking its real part,

$$u_1(x, y) = \text{Re}\, G(z) = \frac{1}{2}\left[G(z) + (G(z))^*\right],$$

and its imaginary part,

$$u_2(x, y) = \text{Im}\, G(z) = \frac{1}{2i}\left[G(z) - (G(z))^*\right].$$

For instance, the analytical function $G(z) = e^z$ generates two harmonic functions, $u_1(x, y) = \text{Re}\, G(z) = e^z \cos y$ and $u_2(x, y) = \text{Im}\, G(z) = e^z \sin y$.

If $G(z)$ is a polynomial of z, then functions $u_1(x, y)$ and $u_2(x, y)$ are *harmonic polynomials*. For instance, the analytical functions

$$G(z) = z^2 = (x + iy)^2 = x^2 - y^2 + 2ixy$$

give two harmonic polynomials,

$$u_1(x, y) = \text{Re}\, G(z) = x^2 - y^2 \quad \text{and} \quad u_2(x, y) = \text{Im}\, G(z) = 2xy.$$

A real constant function is both analytical and harmonic.

According to the superposition principle, any linear combination of harmonic functions is also a harmonic function.

7.3 Boundary Conditions

As in the case of hyperbolic and parabolic equations, a unique solution of an elliptic equation can be obtained when some additional conditions are imposed. Note that because the problems described by elliptic equations do not contain time, they do not need initial conditions. Physically it is also clear that the Laplace and Poisson equations by themselves are not sufficient to determine, for example, the temperature in all points of a body, or the electric potential outside the conductor.

7.3.1 Example of an Ill-posed Problem

In order to understand the difference between elliptic boundary value problems and appropriate problems for other types of PDEs, let us consider the Laplace equation in an infinite stripe:

$$u_{xx} + u_{yy} = 0, \quad -\infty < x < \infty, \quad 0 < y < y_m. \tag{7.18}$$

In the case of a hyperbolic equation,

$$u_{yy} - u_{xx} = 0, \quad -\infty < x < \infty, \quad 0 < y < y_m. \tag{7.19}$$

(variable y plays role of time) we have seen that conditions

$$u(x,y)|_{y=0} = \varphi(x), \quad u_y(x,y)|_{y=0} = \psi(x), \quad -\infty < x < \infty \tag{7.20}$$

with sufficiently smooth functions $\varphi(x)$ and $\psi(x)$ determine a unique classical solution which depends continuously on these functions, i.e., the *Cauchy problem* (7.19), (7.20) for hyperbolic equation is *well-posed*.

In the case of an elliptic equation, the situation is different. Though solution of problem (7.18), (7.20) exists, and it is unique, it does not depend continuously on the boundary functions $\varphi(x)$ and $\psi(x)$. That means that arbitrary small errors in measured boundary functions lead to significant errors in the solution. Hence, the problem is *ill-posed*.

As an example, let us find the solutions $u_n(x,y)$ of problem (7.18), (7.20) for the following sequence of boundary conditions:

$$\varphi_n(x) = 0, \quad \psi_n(x) = \frac{1}{n}\sin nx, \quad -\infty < x < \infty, \quad n = 1, 2, \ldots \tag{7.21}$$

Obviously,

$$\varphi(x) = \lim_{n \to \infty} \varphi_n(x) = 0, \quad \psi(x) = \lim_{n \to \infty} \psi_n(x) = 0, \quad -\infty < x < \infty,$$

and the solution corresponding to the boundary conditions $\varphi(x) = \psi(x) = 0$ is just $u(x,y) = 0$. To find $u_n(x,y)$, we can apply the method of separation of variables. One can guess that the solution of (7.18), (7.20), (7.21) has the structure

$$u_n(x,y) = Y_n(y)\sin nx. \tag{7.22}$$

Substituting (7.22) into (7.18), (7.20), we obtain the ordinary differential equation,

$$Y_n''(y) - n^2 Y_n(y) = 0, \quad 0 < y < y_m \tag{7.23}$$

with initial conditions

$$Y_n(0) = 0, \quad Y_n'(y) = \frac{1}{n}. \tag{7.24}$$

Solving (7.23), (7.24), we find

$$Y_n(y) = \frac{1}{n^2} \sinh ny,$$

hence, we obtain a unique classical solution,

$$u_n(x, y) = \frac{1}{n^2} \sinh ny \sin nx, \quad 0 \le y \le y_m.$$

For any $y > 0$, the limit $\lim_{n \to \infty} u_n(x, y)$ does not exist, because the expression $\sinh ny/n^2$ tends to infinity, instead of tending to zero. Thus, time small-scale oscillations in the boundary conditions create a catastrophic change of the solution. That means that the problem is ill-posed, and it cannot be applied to any real physical problem.

What boundary conditions for the Laplace and Poisson equations would create well-posed boundary value problems?

7.3.2 Well-posed Boundary Value Problems

In the present subsection, we give several examples of well-posed elliptic boundary value problems. From physical reasoning it is clear that, for instance, if the temperature distribution on the surface of a body is known, the solution of such a boundary value problem which consists of the Laplace or Poisson equation together with a boundary condition should exist and be unique.

Boundary conditions can be set in several ways and in fact in our discussion above of various physical problems we have already met several kinds of boundary conditions. Below we categorize three primary kinds of boundary conditions which correspond to the three different heat regimes on the surface (here we use temperature terminology but the arguments apply equally to other physical systems).

Consider some volume V bounded by a surface S. The boundary value problem for a stationary distribution of temperature, $u(x, y, z)$, inside the body is stated in the following way:
Find the function $u(x, y, z)$ inside the volume V satisfying the equation

$$\nabla^2 u = -f(x, y, z)$$

and satisfying one of the following kinds of boundary conditions:

1. $u = f_1$ on S (boundary value problem of the 1st kind)
2. $\frac{\partial u}{\partial n} = f_2$ on S (boundary value problem of the 2nd kind)
3. $\frac{\partial u}{\partial n} + h(u - f_3) = 0$ on S ($h > 0$) (boundary value problem of the 3rd kind),

where f_1, f_2, f_3, h are known functions and $\partial u/\partial n$ is the derivative in the direction of the outward normal to the surface, S.

We may use the above formulation in two distinct ways. If we want to find the temperature, *inside* the volume V for the bounded region we have what is referred as an *interior boundary value problem*. If instead we need to find the temperature outside of a heater, or the electrostatic potential for an unbounded region outside the charged volume V, we have an *exterior boundary value problem*.

The physical sense of each of these boundary conditions is clear. The first boundary value problem when the *surface temperature is prescribed* is called *Dirichlet's problem*. The second boundary value problem when the *flux across the surface is prescribed* is called *Neumann's problem*. The third boundary value problem is called *the mixed problem*. This boundary condition corresponds to the well known Newton's law of cooling, which governs the heat flux from the surface into the ambient medium.

Obviously, a stationary temperature distribution is possible only if the net heat flow across the boundary is equal to zero. From here it follows that for the *interior Neumann's problem* the function f_2 should obey the additional requirement:

$$\iint_S f_2 dS = 0. \tag{7.25}$$

In a similar way a boundary value problem can be formulated for the two-dimensional case when an area is bounded by a closed contour L. In this case the requirement that the net heat flow through the boundary for the interior problem is equal to zero becomes

$$\oint_L f_2 dl = 0. \tag{7.26}$$

Let us demonstrate the solution of a boundary value problem for a one-dimensional case. When the function $u(x)$ depends only on one variable, the Laplace equation becomes an ordinary differential equation and the solution is trivial.

Example 7.1 Solve the one-dimensional Laplace equation in Cartesian coordinates, $d^2u/dx^2 = 0$, and apply Dirichlet boundary conditions to find a simple solution.

Solution. Integrating the equation gives $u = ax + b$. Dirichlet's problem with boundary conditions $u(x = 0) = u_1$ and $u(x = l) = u_2$ gives the solution as $u(x) = (u_2 - u_1)x/l + u_1$.

Reading exercise: In Cartesian coordinates obtain a solution of the one-dimensional Neumann's problem for the Laplace equation.

Reading exercise: Solve Dirichlet's problem in the case of axial symmetry with boundary conditions $u(r = a) = u_1$, $u(r = b) = u_2$. The result will give a solution to the problem of a stationary distribution of heat between two cylinders with common axis when cylinders' surfaces are kept at constant temperatures. The same solution also gives the electric potential between two equipotential cylindrical surfaces. (The solution is a harmonic function between the surfaces with the axis $r = 0$ excluded.)

Hint: Use the Laplace operator in cylindrical coordinates when there is no dependence on φ and z.

Reading exercise: Solve Dirichlet's problem in the case of spherical symmetry with boundary conditions $u(r = a) = u_1$, $u(r = b) = u_2$. The result will give a solution to the problem of a static distribution of heat between two spheres with a common center when the surfaces are kept at constant temperatures. It also gives the electric potential between two equipotential spherical surfaces. As in the previous case, the solution is a harmonic function between the surfaces and the center at $r = 0$ is excluded.

Hint: Use the Laplace operator in spherical coordinates for the case of no dependence on φ and θ.

7.3.3 Maximum Principle and its Consequences

In this and the following sections, except 7.10, 7.11 and 7.15, we consider two-dimensional problems.

Let us prove the well-posedness of the Dirichlet problem. For that goal, we will use the maximum principle for the Laplace equation.

Theorem 1 *If the function u is harmonic in some domain D bounded by the curve S, it reaches its maximum (minimum) value at the boundary.*

The physical reasoning for that principle is clear: there is no reason for maximum stationary temperature to be inside the bulk if there are no internal sources of heat. The proof, which is similar to the proof of the maximum principle for the heat equation (see Chapter 6), is presented below.

First, let us consider a function $v(x, y)$ satisfying the equation

$$\nabla^2 v = -f > 0, \tag{7.27}$$

which corresponds to the internal cooling. If its maximum is reached in an internal point (x_0, y_0), then $v_{xx}(x_0, y_0) \leq 0$, $v_{yy}(x_0, y_0) \leq 0$, thus $\nabla^2 v \leq 0$, which contradicts to (7.27). Thus, the maximum principle is valid for $v(x, y)$.

For the harmonic function u, which satisfies the Laplace equation $\nabla^2 u = 0$, define $v(x, y) = u(x, y) + \varepsilon(x^2 + y^2)$, where $\varepsilon > 0$. Assume that $M = \max_S u(x, y)$ and define $R = \max_D(x^2 + y^2)$. Because $\nabla^2 v = 4\varepsilon > 0$, the maximum of v is reached on the boundary S, hence in any point (x, y),

$$v(x, y) \leq \max_S v(x, y) \leq M + \varepsilon R.$$

Then in any point

$$u(x, y) = v(x, y) - \varepsilon(x^2 + y^2) \leq v(x, y) \leq \max_S v(x, y) \leq M + \varepsilon R.$$

Taking the limit $\varepsilon \to 0$, we get

$$u(x, y) \leq M = \max_S u(x, y),$$

i.e., the maximum of the harmonic function u is reached at the boundary.

The proof for minimum is similar.

The maximum/minimum principle has important consequences.

1. The solution of the Dirichlet problem depends continuously on the boundary conditions.

   Indeed, if

   $$\nabla^2 u_1 = 0 \text{ in } D, \quad u_1 = f_1 \text{ on } S$$

   and

   $$\nabla^2 u_2 = 0 \text{ in } D, \quad u_2 = f_2 \text{ on } S,$$

   then the difference $w = u_1 - u_2$ satisfies the problem

   $$\nabla^2 w = 0 \text{ in } D, \quad w = f_1 - f_2 \text{ on } S.$$

   Applying the maximum/minimum principle to w, we find

   $$|u_1 - u_2| = |w| \leq \max_S |f_1 - f_2|.$$

2. The solution of the Dirichlet problem is unique.

   If u_1 and u_2 are solutions of the same Dirichlet problem, then for $w = u_1 - u_2$,

   $$\nabla^2 w = 0 \text{ in } D, \quad w = 0 \text{ on } S.$$

   Applying the maximum/minimum principle to w, we find that $|u_1 - u_2| = 0$ everywhere, i.e., $u_1 = u_2$.

   In the consequent sections, we present examples showing that the classical solution of the Dirichlet problem exists for a continuous boundary function. In that case, the Dirichlet problem is well-posed.

7.4   Laplace Equation in Polar Coordinates

In Sections 7.4–7.9 we consider two-dimensional problems which have a symmetry allowing the use of polar coordinates. Solutions to corresponding problems contain simple trigonometric functions of the polar angle as well as power and logarithmic functions of the radius.

In polar coordinates, (r, φ), the Laplacian has the form

$$\nabla^2 = \frac{\partial^2}{\partial r^2} + \frac{1}{r}\frac{\partial}{\partial r} + \frac{1}{r^2}\frac{\partial^2}{\partial \varphi^2} \equiv \nabla_r^2 + \frac{1}{r^2}\nabla_\varphi^2. \tag{7.28}$$

We begin by solving the Laplace equation $\nabla^2 u = 0$ using the Fourier method of separation of variables. First, we represent $u(r, \varphi)$ in the form

$$u(r, \varphi) = R(r)\Phi(\varphi). \tag{7.29}$$

Substituting Equation (7.29) into $\nabla^2 u = 0$ and separating the variables we have

$$\frac{r^2 \nabla_r R}{R} = -\frac{\nabla_\varphi^2 \Phi}{\Phi} \equiv \lambda.$$

Because the first term does not depend on the angular variable φ and the second does not depend on r, each term must equal a constant which we denoted as λ. From here we obtain two separate equations for $R(r)$ and $\Phi(\varphi)$:

$$\Phi'' + \lambda\Phi = 0, \tag{7.30}$$

$$r^2 R'' + rR' - \lambda R = 0. \tag{7.31}$$

Clearly, the periodic solution of Equation (7.31),

$$\Phi(\varphi + 2\pi) = \Phi(\varphi), \tag{7.32}$$

exists only for positive integer values of λ(this problem was considered in detail in Chapter 4 (Example 4.4)) The periodicity condition leads to a discrete spectrum of eigenvalues:

$$\lambda_n = n^2, \quad n = 0, 1, 2, \ldots;$$

thus, the eigenfunctions are:

$$\Phi = \Phi_n(\varphi) = \begin{cases} \cos n\varphi, \\ \sin n\varphi. \end{cases}$$

Negative values of n correspond to the same eigenfunctions and therefore need not be included in the list of eigenvalues.

The equation for $R(r)$

$$r^2 R'' + r R' - n^2 R = 0 \tag{7.33}$$

is known as the *Euler equation*. The general solution to this equation is

$$
\begin{aligned}
R = R_n(r) &= C_1 r^n + C_2 r^{-n}, & n \neq 0, \\
R_0(r) &= C_1 + C_2 \ln r, & n = 0.
\end{aligned}
\tag{7.34}
$$

Combining the above results for $\Phi_n(\varphi)$ and $R_n(r)$ we obtain the following particular and general solutions of the Laplace equation:

a) Under the condition that the solution be finite at $r = 0$ we have

$$u_n(r, \varphi) = r^n \begin{Bmatrix} \cos n\varphi \\ \sin n\varphi \end{Bmatrix}, \quad n = 0, 1, \ldots$$

We can write a general solution for the Laplace problem for an *interior boundary value problem*, $0 \leq r \leq l$, via the expansion with these particular solutions $u_n(r, \varphi)$ as

$$u(r, \varphi) = \sum_{n=0}^{\infty} r^n \left(A_n \cos n\varphi + B_n \sin n\varphi \right).$$

The term with $n = 0$ is more conveniently written as $A_0/2$, thus we have

$$u(r, \varphi) = \frac{A_0}{2} + \sum_{n=1}^{\infty} r^n \left(A_n \cos n\varphi + B_n \sin n\varphi \right). \tag{7.35}$$

b) For the case that the solution is finite at $r \to \infty$ we have

$$u_n(r, \varphi) = \frac{1}{r^n} \begin{Bmatrix} \cos n\varphi \\ \sin n\varphi \end{Bmatrix}, \quad n = 0, 1, \ldots$$

These functions may be used as solutions to the Laplace problem for regions outside of a circle. The general solution of the Laplace equation for such an *exterior boundary value problem*, $r \geq l$, limited (i.e. bounded) at infinity, can be written as

$$u(r, \varphi) = \frac{A_0}{2} + \sum_{n=1}^{\infty} \frac{1}{r^n} \left(A_n \cos n\varphi + B_n \sin n\varphi \right). \tag{7.36}$$

c) We also have a third set of solutions,

$$u_n(r, \varphi) = 1, \quad \ln r, \quad r^n \begin{Bmatrix} \cos n\varphi \\ \sin n\varphi \end{Bmatrix}, \quad \frac{1}{r^n} \begin{Bmatrix} \cos n\varphi \\ \sin n\varphi \end{Bmatrix}, \quad n = 1, 2, \ldots$$

for the cases where the solution is unbounded as $r \to 0$, as well as $r \to \infty$. This set of functions is used to solve the Laplace equation for regions which form a circular ring, $l_1 \leq r \leq l_2$.

7.5 Laplace Equation and Interior BVP for Circular Domain

In this section we consider the first of the three cases presented in the previous section – solve the boundary value problem for a disk:

$$\nabla^2 u = 0 \quad \text{in} \quad 0 \le r < l, \tag{7.37}$$

with boundary condition

$$u(r, \varphi)|_{r=l} = f(\varphi). \tag{7.38}$$

Applying (7.38) to formula (7.35) we obtain

$$\frac{A_0}{2} + \sum_{n=1}^{\infty} l^n \left(A_n \cos n\varphi + B_n \sin n\varphi \right) = f(\varphi). \tag{7.39}$$

From this we see that $l^n A_n$ and $l^n B_n$ are the Fourier coefficients of expansion of the function $f(\varphi)$ in the system (or basis) of trigonometric functions $\{\cos n\varphi, \ \sin n\varphi\}$. We may evaluate the coefficients using the formulas

$$A_n l^n = \frac{1}{\pi} \int_0^{2\pi} f(\varphi) \cos n\varphi d\varphi, \quad B_n l^n = \frac{1}{\pi} \int_0^{2\pi} f(\varphi) \sin n\varphi d\varphi, \quad n = 0, 1, 2, \dots. \tag{7.40}$$

Thus, the solution of the *interior Dirichlet problem* for the Laplace equation is

$$u(r, \varphi) = \frac{A_0}{2} + \sum_{n=1}^{\infty} \left(\frac{r}{l} \right)^n \left[A_n \cos n\varphi + B_n \sin n\varphi \right]. \tag{7.41}$$

Example 7.2 Find the temperature distribution inside a circle if the boundary is kept at the temperature $T_0 = C_1 + C_2 \cos \varphi + C_3 \sin 2\varphi$.

Solution. It is obvious that for this particular case the series given by Equation (7.39) reduces to three nonzero terms:

$$A_0 = 2C_1, \quad lA_1 = C_2, \quad l^2 B_2 = C_3.$$

In this case the solution given by Equation (7.41) is

$$T = C_1 + C_2 \frac{r}{l} \cos \varphi + C_3 \left(\frac{r}{l} \right)^2 \sin 2\varphi.$$

Similarly, we can obtain solutions of the *second and third boundary value problems* for the Laplace equation for a disk. We leave to the reader as *Reading Exercises* to check that the resulting formulas presented below are correct for the following two cases of Neumann and mixed boundary conditions.

The *Neumann problem* for the Laplace equation with boundary condition

$$\frac{\partial u}{\partial r} \bigg|_{r=l} = f(\varphi) \tag{7.42}$$

has the solution

$$u(r, \varphi) = \sum_{n=1}^{\infty} \frac{r^n}{n l^{n-1}} \left[A_n \cos n\varphi + B_n \sin n\varphi \right] + C, \tag{7.43}$$

where C is an arbitrary constant. We remind the reader that a solution of the interior Neumann problem can exist only under the condition

$$\int_{C_l} f dl = \int_0^{2\pi} f(\varphi) d\varphi = 0. \tag{7.44}$$

We discussed the meaning of that condition in Section 7.3.2: if heating and cooling of the body through its boundary are not balanced, the temperature of the body cannot be time-independent. It grows if integral (7.44) is positive and decreased otherwise, hence the problem is described by the heat equation rather than the Laplace equation.

Note that the average temperature of the body

$$\frac{1}{\pi l^2} \int_0^{2\pi} d\varphi \int_0^l r u(r, \varphi) dr = C$$

is not determined by BVP (7.37), (7.42). Indeed, the boundary condition (7.42) does not provide any reference temperature for the body: a spatially homogeneous change of the body's temperature does not violate that boundary condition. Thus, the solution is unique up to an arbitrary constant.

The *mixed problem* for the Laplace equation with boundary condition

$$\left. \frac{\partial u}{\partial r} + h u \right|_{r=l} = f(\varphi), \quad h = \text{const}, \tag{7.45}$$

has the solution

$$u(r, \varphi) = \frac{A_0}{2h} + \sum_{n=1}^{\infty} \frac{r^n}{(n + lh)l^{n-1}} [A_n \cos n\varphi + B_n \sin n\varphi]. \tag{7.46}$$

Clearly, in the case of homogeneous boundary conditions ($f(\varphi) = 0$) all three problems have only trivial solutions (i.e. equal to zero, or for the Neumann problem, any constant).

Coefficients in the expansions in Equations (7.41), (7.43) and (7.46) are determined using Equation (7.40). Let us briefly discuss the convergence of the series in these expansions. If the function $f(\varphi)$ defining the boundary condition can be integrated absolutely, its Fourier coefficients are bounded and, as can be seen from the structure of these series, they converge in any interior point of the circle ($r < l$). The smoother the function $f(\varphi)$ is, the faster these series converge. The series can be differentiated term by term any number of times and the sums satisfy the Laplace equation, i.e. they are harmonic functions. The same can be said for the problems discussed in the following sections.

Example 7.3 Let $u(r, \varphi)|_{r=l} = \sin(\varphi/2)$ at $0 \leq \varphi < 2\pi$. Find the temperature at several points $P(r, \varphi)$ (here r is in units of l) of the circle: $P_1(0, \varphi)$, $P_2(0.2, \pi/18)$, $P_3(0.3, \pi/18)$. Evaluate the precision of the calculations.

Solution. The coefficients in (7.39) are

$$A_0 = \frac{4}{\pi}, \quad A_n = \frac{4}{\pi(1 - 4n^2)}$$

(for $n = 1, 2, \ldots$), and $B_n = 0$. Thus, we have the expansion

$$u(r, \varphi) = \frac{2}{\pi} + \sum_{n=1}^{\infty} \left(\frac{r}{l}\right)^n \frac{4}{\pi(1 - 4n^2)} \cos n\varphi.$$

At point $P_1(0, 0)$, the temperature is $u(0, 0) = 2/\pi$ and the error is zero because $r = 0$. At point $P_2(0.2, \pi/18)$, keeping six terms in the partial sum and rounding off the result with accuracy, $\varepsilon_1 = 10^{-4}$, we obtain $u(0.2, \pi/18) = 0.5496$. Because the terms in the series above monotonically decrease, the error when we keep six terms is less than or equal to the absolute value of the 7th term in the series,

$$\varepsilon \leq \left| \left(\frac{2}{10}\right)^7 \frac{4}{\pi (1 - 4 \cdot 7^2)} \cdot \frac{1}{1 - \frac{2}{10}} \right| \approx 0.5 \cdot 10^{-5}.$$

The total error is $\varepsilon + \varepsilon_1 \approx 10^{-4}$.

Similarly, $u(0.3, \pi/18) = 0.5031$, $\varepsilon \approx 10^{-4}$ and $\varepsilon + \varepsilon_1 \approx 2 \cdot 10^{-4}$.

Example 7.4 Let an infinite homogenous cylinder with a circular surface of radius a be kept at a constant temperature

$$u(r, \varphi)|_{r=l} = \begin{cases} T_0, & 0 \leq \varphi < \pi, \\ -T_0, & \pi \leq \varphi < 2\pi \end{cases}$$

for any z. After a long period of time the temperature inside the cylinder will become constant, i.e. the system reaches equilibrium. Find the temperature distribution inside the cylinder when this occurs.

Solution. This is the Dirichlet's interior problem for a circle. The solution to this problem is given by the series (7.41); the coefficients (7.40) are

$$A_n = 0, \quad B_{2k} = 0, \quad B_{2k+1} = \frac{4T_0}{\pi (2k + 1)}, \quad k = 0, 1, 2 \ldots.$$

The coefficients decrease as $1/n$, rather than $1/n^2$ in Example 7.3, because of discontinuity in the boundary condition.

The results of numerical summation of the series in Equation (7.41) with N terms are presented in Figure 7.2 using the dimensionless variables, length in units of the radius l and temperature in units of T_0. Because of the symmetry of the problem we have

$$u(r, \pi - \varphi) = -u(r, \pi + \varphi)$$

and we can search for the solution in the half-domain ($\varphi < \pi$),

Figure 7.2 shows the temperature distribution on the boundary of the cylinder for different values of N. As can be seen from the graphs, closer to the points $\varphi \neq 0$, π we need more terms to keep the same precision. It is clear that at the discontinuity points $\varphi = 0$ and $\varphi = \pi$ the series gives a value of zero for the temperature.

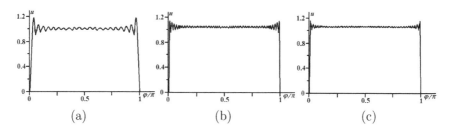

(a)   (b)   (c)

FIGURE 7.2

Temperature distribution on the circle boundary ($r = 1$) obtained with partial sums of the series in Equation (7.41) for the number of terms; (a) $N = 20$, (b) $N = 50$, and (c) $N = 100$.

7.6 Laplace Equation and Exterior BVP for Circular Domain

In this section we consider the second of the three cases presented in Section 7.4. This problem is formulated as

$$\nabla^2 u = 0 \quad \text{for} \quad r > l.$$

With the boundary condition $u|_{r=l} = f(\varphi)$ we directly obtain the solution of the *Dirichlet problem* as (7.36):

$$u(r, \varphi) = \frac{A_0}{2} + \sum_{n=1}^{\infty} \left(\frac{l}{r}\right)^n [A_n \cos n\varphi + B_n \sin n\varphi]. \tag{7.47}$$

The *Neumann problem* with boundary condition

$$\left.\frac{\partial u}{\partial r}\right|_{r=l} = f(\varphi) \tag{7.48}$$

has the solution

$$u(r, \varphi) = -\sum_{n=1}^{\infty} \frac{1}{n} \frac{l^{n+1}}{r^n} [A_n \cos n\varphi + B_n \sin n\varphi] + C, \tag{7.49}$$

where C is an arbitrary constant. We remind the reader again that the exterior Neumann problem for a plane has a solution only under the condition

$$\int_{C_l} f dl = \int_0^{2\pi} f(\varphi) d\varphi = 0 \tag{7.50}$$

and its solution has an arbitrary additive constant.

The *mixed problem*, with a boundary condition

$$\left.\frac{\partial u}{\partial r} - hu\right|_{r=l} = f(\varphi) \tag{7.51}$$

has the solution

$$u(r, \varphi) = -\frac{A_0}{2h} - \sum_{n=1}^{\infty} \frac{l^{n+1}}{(n+lh)r^n} [A_n \cos n\varphi + B_n \sin n\varphi]. \tag{7.52}$$

Notice that different signs for h in Equation (7.51) as compared to Equation (7.45) are because of different directions of the vector normal to the boundary. For the interior problem this vector is directed outward; for the exterior problem it is inward.

Coefficients A_n and B_n in the series (7.47), (7.49) and (7.52) are the Fourier coefficients of function $f(\varphi)$ and are calculated with Equations (7.40). We leave to the reader as useful *Reading Exercises* to prove the results in Equations (7.47), (7.49) and (7.52).

7.7 Poisson Equation: General Notes and a Simple Case

Let us briefly discuss how to find a solution of a Poisson equation

$$\nabla^2 u = -f \tag{7.53}$$

with the Dirichlet boundary condition:

$$u(r,\varphi)|_{r=l} = g(\varphi). \tag{7.54}$$

Either the interior or exterior problem can be considered. Note that, contrary to the solution of the Laplace equation, the solution of the Poisson equation is not zero even for homogeneous boundary conditions $g(\varphi) \equiv 0$.

The way to solve the problem (7.53), (7.54) is to solve the nonhomogeneous equation (Poisson equation) without taking into consideration the boundary condition, and then to add a solution of the Laplace equation in a way that the sum satisfies the boundary condition. In other words, the function u is presented as the sum of two functions, $u = u_p + u_0$, where u_p is a particular solution of the Poisson equation

$$\nabla^2 u_p = -f \tag{7.55}$$

and u_0 is a solution of the Laplace equation

$$\nabla^2 u_0 = 0. \tag{7.56}$$

The function u should satisfy the necessary boundary condition from which follows the boundary condition for u_0:

$$u_0(r,\varphi)|_{r=l} = g(\varphi) - u_p. \tag{7.57}$$

The boundary value problem defined by Equations (7.56) and (7.57) was considered in Sections 7.5 and 7.6, thus now we turn to the solution of Equation (7.55). The question is how to find a particular solution of the Poisson equation which is finite in the center (for the interior problem) or at infinity (for the exterior problem) irrespective of the boundary condition at $r = l$. It should be emphasized that changing the type of boundary condition for the Poisson equation, (7.53), we need only to change the boundary condition for the Laplace Equation (7.56).

Let us consider a particular case: very often the inhomogeneous term $f(r,\varphi)$ has the form

$$f(r,\varphi) = r^m \cos n\varphi. \tag{7.58}$$

(or perhaps $r^m \sin n\varphi$). Here m is an arbitrary real number, n is an integer since the function $f(r,\varphi)$ should be periodic in φ, thus $f(r,\varphi+2\pi) = f(r,\varphi)$. In this case a particular solution of the Poisson equation can be obtained using the method of undetermined coefficients. Note that $m > -2$ corresponds to an interior problem, while $m < -2$ to an exterior one. Indeed, the function $f(r,\varphi)$ can be infinite at $r = 0$ for the interior problem; we should only ensure that the integral

$$\int_S |f(r,\varphi)|\, dS$$

remains finite (for example, in electrostatics it means that the full charge inside the domain is finite). With $dS = rdrd\varphi$ in polar coordinates, it is obvious that at $m > -2$ corresponds to a finite value of

$$\int_0^l f(r,\varphi)rdr.$$

On the other hand, at $m < -2$ both $f(r \to \infty, \varphi)$ and the integral $\int_l^\infty f(r,\varphi)rdr$ remain finite – this situation corresponds to an exterior problem. The value $m = -2$ can be used only for a boundary problem involving solutions inside an annulus (ring).

In polar coordinates Equation (7.53) is

$$\frac{\partial^2 u}{\partial r^2} + \frac{1}{r}\frac{\partial u}{\partial r} + \frac{1}{r^2}\frac{\partial^2 u}{\partial \varphi^2} = -f(r,\varphi)$$

and because of

$$\nabla^2 r^{m+2} \cos n\varphi = \left[(m+2)^2 - n^2\right] r^m \cos n\varphi,$$

the particular solution of equation

$$\nabla^2 u = -r^m \cos n\varphi$$

is

$$u_p = -\frac{r^{m+2} \cos n\varphi}{(m+2)^2 - n^2}. \tag{7.59}$$

A difficulty occurs if $m + 2 = \pm n$ in which case we cannot apply Equation (7.59). In this case we may seek the solution in the form

$$u_p = R(r) \cos n\varphi$$

and obtain for $R(r)$ the equation

$$R'' + \frac{1}{r}R' - \frac{n^2}{r^2}R = -r^{\pm n - 2},$$

where the derivatives with respect to r are denoted by primes. The particular solution of this equation is

$$\mp \frac{r^{\pm n} \ln r}{2n} \quad \text{at} \quad n \neq 0 \tag{7.60}$$

and

$$-\frac{\ln^2 r}{2} \quad \text{at} \quad n = 0. \tag{7.61}$$

Recall that $m > -2$ corresponds to an interior problem, whereas n stays in the argument of $\cos n\varphi$, i.e. there is no need to consider negative values of n. The result is the solution in Equation (7.60) with the upper sign used for the interior problem, $r \leq l$, and the lower sign for the exterior problem, $r \geq l$. The solution in Equation (7.61) with $n = 0$ is the particular solution for a boundary value problem inside an annulus.

Example 7.5 Solve the boundary value problem for a disk given by

$$\nabla^2 u = -Axy, \quad r \leq l, \quad u|_{r=l} = 0. \tag{7.62}$$

Solution. The function, f, on the right side of the Poisson equation is $f = Axy = Ar^2 \sin 2\varphi / 2$, thus we have the situation described by Equation (7.58) with $m = n = 2$. Using the result of Equation (7.59) we obtain a particular solution of Equation (7.62) as

$$u_p(r, \varphi) = -\frac{A}{2 \cdot 12} r^4 \sin 2\varphi \tag{7.63}$$

Taking the solution given by Equation (7.63) into account, the boundary value problem of Equations (7.56) and (7.57) takes the following form:

$$\nabla^2 u_0 = 0,$$

$$r = l: \quad u_0 = -u_p = \frac{A}{24} l^4 \sin 2\varphi.$$

Since the boundary condition contains only one Fourier harmonic, we can conclude that $u_0(r, \varphi) = Cr^2 \sin 2\varphi$ with $C = -Al^2/24$. Thus, the function $u = u_p + u_0$ that satisfies the given boundary condition is

$$u(r, \varphi) = \frac{A}{24} r^2 \left(l^2 - r^2\right) \sin 2\varphi.$$

7.8 Poisson Integral

It is possible to present a solution of Dirichlet's problem for the Laplace equation as an integral formula. Let us do this first for the interior problem for a circle. Substituting the formulas for Fourier coefficients in Equation (7.40) into Equation (7.41) and switching the order of summation and integration, we obtain

$$u(r, \varphi) = \frac{1}{\pi} \int_0^{2\pi} f(\phi) \left\{ \frac{1}{2} + \sum_{n=1}^{\infty} \left(\frac{r}{l} \right)^n [\cos n\phi \cos n\varphi + \sin n\phi \sin n\varphi] \right\} d\phi$$

$$= \frac{1}{\pi} \int_0^{2\pi} f(\phi) \left\{ \frac{1}{2} + \sum_{n=1}^{\infty} \left(\frac{r}{l} \right)^n \cos n(\phi - \varphi) \right\} d\phi. \tag{7.64}$$

Since $t \equiv r/l < 1$, the expression in the parentheses can be transformed as follows:

$$Z \equiv \frac{1}{2} + \sum_{n=1}^{\infty} t^n \cos n(\phi - \varphi) = \frac{1}{2} + \frac{1}{2} \sum_{n=1}^{\infty} t^n \left[e^{in(\varphi - \phi)} + e^{-in(\varphi - \phi)} \right]$$

$$= \frac{1}{2} \left\{ 1 + \sum_{n=1}^{\infty} \left[\left(t e^{i(\varphi - \phi)} \right)^n + \left(t e^{-i(\varphi - \phi)} \right)^n \right] \right\}.$$

Using

$$\sum_{n=0}^{\infty} x^n = \frac{1}{1-x}, \quad \sum_{n=1}^{\infty} x^n = \sum_{n=0}^{\infty} x^n - 1 = \frac{x}{1-x},$$

we have

$$Z = \frac{1}{2} \left[1 + \frac{t e^{i(\varphi - \phi)}}{1 - t e^{i(\varphi - \phi)}} + \frac{t e^{-i(\varphi - \phi)}}{1 - t e^{-i(\varphi - \phi)}} \right] = \frac{1}{2} \frac{1 - t^2}{1 - 2t \cos(\varphi - \phi) + t^2}.$$

Therefore Equation (7.64) becomes

$$u(r, \varphi) = \frac{1}{2\pi} \int_0^{2\pi} f(\phi) \frac{l^2 - r^2}{r^2 - 2lr \cos(\varphi - \phi) + l^2} d\phi. \tag{7.65}$$

This formula gives the solution to the first boundary value problem inside a circle and is called the Poisson integral. The expression

$$u(r, \varphi, l, \phi) = \frac{l^2 - r^2}{r^2 - 2lr \cos(\varphi - \phi) + l^2} \tag{7.66}$$

is called the Poisson kernel.

Expression (7.65) is not applicable at $r = l$, but its limit as $r \to l$ for any fixed value of φ is equal to $f(\varphi)$ because the series we used to obtain Equation (7.65) is a continuous function in the closed region $r \leq l$. Thus, the function defined by the formula

$$u(r, \varphi) = \begin{cases} \dfrac{1}{2\pi} \displaystyle\int_0^{2\pi} f(\phi) \dfrac{l^2 - r^2}{r^2 - 2lr \cos(\varphi - \phi) + l^2} d\phi, & \text{if } r < l, \\ f(\varphi), & \text{if } r = l \end{cases}$$

is a harmonic function satisfying the Laplace equation $\nabla^2 u = 0$ for $r < l$ and continuous in the closed region $r \leq l$.

Similarly, we obtain the solution to the exterior boundary value problem for a circle as

$$u(r, \varphi) = \begin{cases} \dfrac{1}{2\pi} \displaystyle\int_0^{2\pi} f(\phi) \dfrac{r^2 - l^2}{r^2 - 2lr \cos(\varphi - \phi) + l^2} d\phi, & \text{if } r > l, \\ f(\varphi), & \text{if } r = l. \end{cases} \tag{7.67}$$

Poisson integrals cannot be evaluated analytically for an arbitrary function $f(\varphi)$; however they are often very useful in certain applications. In particular, they can be more useful for numerical calculations than the infinite series solution.

Example 7.6 Consider a stationary membrane's deflection from the equilibrium position. For the stationary case the membrane surface is described by the function $u = u(x, y)$ which satisfies the equation

$$\frac{\partial^2 u}{\partial x^2} + \frac{\partial^2 u}{\partial y^2} = 0.$$

If the membrane contour projection onto the xy-plane is a circle of radius l, we can consider this problem as an interior Dirichlet's problem for a circle.

Let the equation of the contour be given by function $u = f(\varphi)$, where f is a z-coordinate of the contour at angle φ. As an example, consider a film fixed on a firm frame that has a circular projection onto the xy-plane with radius l and center at point O. The equation of the film contour in polar coordinates is $u = C \cos 2\varphi$ $(0 \le \varphi \le 2\pi)$, $r = l$. Find the shape, $u(r, \varphi)$, of the film.

Solution. The solution is given by the Poisson integral:

$$u(r, \varphi) = \frac{1}{2\pi} \int_0^{2\pi} C \cos 2\phi \cdot \frac{l^2 - r^2}{r^2 - 2lr \cos(\phi - \varphi) + l^2} d\phi.$$

To evaluate this integral we use the substitution $\phi - \varphi = \zeta$. The limits of integration will not change because the integrand is periodic with period 2π (an integral in the limits from $-\varphi$ to $2\pi - \varphi$ is equal to the same integral in the limits from 0 to 2π). Thus, we have

$$\begin{aligned} u(r, \varphi) &= \frac{C(l^2 - r^2)}{2\pi} \int_0^{2\pi} \frac{\cos(2\zeta + 2\varphi)}{r^2 - 2lr \cos \zeta + l^2} d\zeta \\ &= \frac{C(l^2 - r^2)}{2\pi} \left[\cos 2\varphi \int_0^{2\pi} \frac{\cos 2\zeta \, d\zeta}{r^2 - 2lr \cos \zeta + l^2} - \sin 2\varphi \int_0^{2\pi} \frac{\sin 2\zeta \, d\zeta}{r^2 - 2lr \cos \zeta + l^2} \right]. \end{aligned}$$

The second of these integrals is equal to zero as the integral of the odd function on the interval $(0, 2\pi)$. So, for this case

$$u(r, \varphi) = \frac{C(l^2 - r^2)}{2\pi} \cos 2\varphi \int_{-\pi}^{\pi} \frac{\cos 2\zeta}{r^2 - 2lr \cos \zeta + l^2} d\zeta.$$

With the substitution $\tan(\zeta/2) = v$ finally we obtain

$$u(r, \varphi) = \frac{C(l^2 - r^2)}{2\pi} \cos 2\varphi \frac{2\pi r^2}{l^2(l^2 - r^2)} = \frac{Cr^2}{l^2} \cos 2\varphi.$$

The function $C(r/l)^2 \cos 2\varphi$ is harmonic and takes the values $C \cos 2\varphi$ on the contour of the circle. It has a shape of a saddle, shown in Figure 7.3.

Reading Exercise. Solve this problem using the method of Section 7.6.

FIGURE 7.3
Shape of the film in Example 7.6.

7.9   Application of Bessel Functions For the Solution of Poisson Equations in a Circle

In this section we will show how to solve Poisson equations for the general form of functions $f(r, \varphi)$, not necessarily $r^{\pm n} \sin n\varphi$, like in Section 7.7. Such solutions can be obtained with Bessel functions.

Let us consider the following *interior* boundary value problem for the Poisson equation:

$$\nabla^2 u = \frac{\partial^2 u}{\partial r^2} + \frac{1}{r}\frac{\partial u}{\partial r} + \frac{1}{r^2}\frac{\partial^2 u}{\partial \varphi^2} = -f(r, \varphi), \quad 0 \le r < l, \quad 0 \le \varphi < 2\pi, \qquad (7.68)$$

$$\alpha \frac{\partial u(r, \varphi)}{\partial r} + \beta u(r, \varphi) \bigg|_{r=l} = g(\varphi), \qquad (7.69)$$

$$u(r, \varphi) = u(r, \varphi + 2\pi), \quad |\alpha| + |\beta| \ne 0.$$

To separate variables when a boundary condition is nonhomogeneous, we should split function $u(r, \varphi)$ into two functions:

$$u(r, \varphi) = v(r, \varphi) + w(r, \varphi), \qquad (7.70)$$

where the introduced auxiliary function $w(r, \varphi)$ must satisfy the nonhomogeneous boundary condition (7.69) and leaves the boundary condition for the function $v(r, \varphi)$ homogeneous.

The function $w(r, \varphi)$ satisfying the boundary condition (7.69) can be chosen in different ways; the only restriction is that it should be continuous and finite. Let us seek it in the form

$$w(r, \varphi) = (c_0 + c_1 r + c_2 r^2) \cdot g(\varphi).$$

Because

$$\frac{1}{r}\frac{\partial w(r, \varphi)}{\partial r} = \left(\frac{c_1}{r} + 2c_2\right) g(\varphi),$$

we have to put $c_1 = 0$ and as the result

$$w(r, \varphi) = (c_0 + c_2 r^2) g(\varphi). \qquad (7.71)$$

Case 1. For $\alpha = 0$, $\beta = 1$ we have:

a) Boundary condition $u(r, \varphi)|_{r=l} = g(\varphi)$ with auxiliary function

$$w(r, \varphi) = \frac{r^2}{l^2} g(\varphi). \qquad (7.72)$$

b) Boundary condition $u(r, \varphi)|_{r=l} = g_0 = \text{const}$ with auxiliary function

$$w(r, \varphi) = g_0. \tag{7.73}$$

Case 2. For $\alpha = 1$, $\beta = 0$ we have the boundary condition $\frac{\partial u}{\partial r}(r, \varphi)\big|_{r=l} = g(\varphi)$ with auxiliary function

$$w(r, \varphi) = \frac{r^2}{2l}g(\varphi) + C, \tag{7.74}$$

where C is an arbitrary constant.

Case 3. For $\alpha = 1$, $\beta = h > 0$ we have the boundary condition $\frac{\partial u}{\partial r}(r, \varphi) + hu(r, \varphi)\big|_{r=l} = g(\varphi)$ with auxiliary function

$$w(r, \varphi) = \frac{r^2}{l(2 + hl)}g(\varphi). \tag{7.75}$$

It is easy to verify by direct substitution that the above expressions for $w(r, \varphi)$ satisfy boundary condition (7.69).

Thus, the simple expressions for auxiliary functions $w(r, \varphi)$ given by one of the Equations (7.72) through (7.75) allow us to reduce nonhomogeneous conditions given in Equation (7.69) to homogeneous.

Let us consider now the solution of Poisson's equation with *homogeneous* boundary condition. For function $v(r, \varphi)$ we have the BVP

$$\nabla^2 v = \frac{\partial^2 v}{\partial r^2} + \frac{1}{r}\frac{\partial v}{\partial r} + \frac{1}{r^2}\frac{\partial^2 v}{\partial \varphi^2} = -\tilde{f}(r, \varphi). \tag{7.76}$$

$$\alpha\frac{\partial v(l, \varphi)}{\partial r} + \beta v(l, \varphi)\bigg| = 0 \tag{7.77}$$

with function $\tilde{f}(r, \varphi)$

$$\tilde{f}(r, \varphi) = f(r, \varphi) + \frac{\partial^2 w}{\partial r^2} + \frac{1}{r}\frac{\partial w}{\partial r} + \frac{1}{r^2}\frac{\partial^2 w}{\partial \varphi^2}. \tag{7.78}$$

Check the result (7.78) as a *Reading Exercise*.

Let us show that the solution to the problem of Equations (7.76) and (7.77) can be obtained by expansion of function $v(r, \varphi)$ in a series by eigenfunctions of the Sturm-Liouville problem for the Laplace operator as

$$v(r, \varphi) = \sum_{n=0}^{\infty}\sum_{m=0}^{\infty}\left[A_{nm}V_{nm}^{(1)}(r, \varphi) + B_{nm}V_{nm}^{(2)}(r, \varphi)\right], \tag{7.79}$$

where $V_{nm}^{(1)}(r, \varphi)$, $V_{nm}^{(2)}(r, \varphi)$ are the eigenfunctions of the Laplacian satisfying the corresponding *homogeneous* boundary value problem. As shown in Appendix D part 1, the eigenfunctions of the Laplacian in polar coordinates for Dirichlet, Neumann and mixed *interior* problems can be expressed in terms of the Bessel functions

$$V_{nm}^{(1)} = J_n\left(\frac{\mu_m^{(n)}}{l}r\right)\cos n\varphi, \quad V_{nm}^{(2)} = J_n\left(\frac{\mu_m^{(n)}}{l}r\right)\sin n\varphi. \tag{7.80}$$

Different types of boundary conditions lead to different eigenvalues $\mu_m^{(n)}$.

Because of orthogonality, the coefficients in Equation (7.79) can be obtained via the (unknown) function $v(r, \varphi)$ by multiplying (7.79) by $dS = rdrd\varphi$ and integrating over the disc's area:

$$A_{nm} = \frac{1}{||V_{nm}^{(1)}||^2} \int_0^{2\pi} \int_0^l v(r, \varphi) V_{nm}^{(1)}(r, \varphi) rdrd\varphi,$$

$$B_{nm} = \frac{1}{||V_{nm}^{(2)}||^2} \int_0^{2\pi} \int_0^l v(r, \varphi) V_{nm}^{(2)}(r, \varphi) rdrd\varphi. \tag{7.81}$$

To find a final form for these coefficients we first multiply Equation (7.76) by $V_{nm}^{(1)}(r, \varphi)$ and $V_{nm}^{(2)}(r, \varphi)$, and integrate over the circular domain:

$$\int_0^{2\pi} \int_0^l \nabla^2 v \, V_{nm}^{(1)}(r, \varphi) rdrd\varphi = - \int_0^{2\pi} \int_0^l \tilde{f}(r, \varphi) \, V_{nm}^{(1)}(r, \varphi) rdrd\varphi, \tag{7.82}$$

$$\int_0^{2\pi} \int_0^l \nabla^2 v \, V_{nm}^{(2)}(r, \varphi) rdrd\varphi = - \int_0^{2\pi} \int_0^l \tilde{f}(r, \varphi) \, V_{nm}^{(2)}(r, \varphi) rdrd\varphi. \tag{7.83}$$

Now substitute Equation (7.79) into Equations (7.82) and (7.83). Because $V_{nm}^1(r, \varphi)$ and $V_{nm}^2(r, \varphi)$ are the eigenfunctions of the Laplacian we have

$$\nabla^2 V_{nm}^{(1,2)}(r, \varphi) = -\lambda_{nm} V_{nm}^{(1,2)}(r, \varphi), \tag{7.84}$$

where the eigenvalues $\lambda_{nm} = \left(\mu_m^{(n)}/l \right)^2$. The left sides of Equations (7.82) and (7.83) become:

$$-\sum_{i=0}^{\infty} \sum_{j=0}^{\infty} \lambda_{ij} \int_0^{2\pi} \int_0^l \left[A_{ij} V_{ij}^{(1)}(r, \varphi) + B_{ij} V_{ij}^{(2)}(r, \varphi) \right] V_{nm}^{(p)}(r, \varphi) rdrd\varphi.$$

Due to the orthogonality relation for the functions $V_{nm}^{(p)}(r, \varphi) (p = 1, 2)$, the only term in the sums that differs from zero is

$$-\lambda_{nm} A_{nm} \left\| V_{nm}^{(p)} \right\|^2 \delta_{p1} \quad \text{or} \quad -\lambda_{nm} B_{nm} \left\| V_{nm}^{(p)} \right\|^2 \delta_{p2}.$$

Comparing with the right sides in Equations (7.82) and (7.83) we obtain

$$\lambda_{nm} A_{nm} = f_{nm}^{(1)}, \quad \lambda_{nm} B_{nm} = f_{nm}^{(2)}, \quad n, m = 0, 1, 2, \ldots,$$

where

$$f_{nm}^{(p)} = \frac{1}{||V_{nm}^{(p)}||^2} \int_0^{2\pi} \int_0^a \tilde{f}(r, \varphi) V_{nm}^{(p)}(r, \varphi) rdrd\varphi. \tag{7.85}$$

From this the coefficients A_{nm}, B_{nm} can be obtained.

In the case of boundary conditions of the 1st or 3rd type (Dirichlet's condition or mixed condition) the eigenvalues $\lambda_{nm} \neq 0$ for all n, $m = 0, 1, 2, \ldots$ in this case the solution is defined uniquely and has the form

$$v(r, \varphi) = \sum_{n=0}^{\infty} \sum_{m=0}^{\infty} \frac{1}{\lambda_{nm}} \left[f_{nm}^{(1)} V_{nm}^{(1)}(r, \varphi) + f_{nm}^{(2)} V_{nm}^{(2)}(r, \varphi) \right]. \tag{7.86}$$

In the case of boundary conditions of the 2nd type (Neumann's condition) the eigenvalue $\lambda_{00} = 0$ $\left(V_{00}^{(1)} = 1, V_{00}^{(2)} = 0 \right)$ and all other eigenvalues are nonzero. There are two options.

If

$$f_{00}^{(1)} = \int_0^{2\pi} \int_0^l \tilde{f}(r, \varphi) \, r dr d\varphi = 0,$$

then the coefficient A_{00} is undefined. The other coefficients are defined uniquely. A solution to the given problem exists but is determined only up to an arbitrary additive constant. The solution in this case is

$$v(r, \varphi) = \sum_{n=0}^{\infty} \sum_{m=0}^{\infty} \frac{1}{\lambda_{nm}} \left[f_{nm}^{(1)} V_{nm}^{(1)}(r, \varphi) + f_{nm}^{(2)} V_{nm}^{(2)}(r, \varphi) \right] + \text{const.} \qquad (7.87)$$

If

$$\int_0^{2\pi} \int_0^l \tilde{f}(r, \varphi) \, r dr d\varphi \neq 0,$$

then the solution to the given problem does not exist.

Thus, the *general solution* of the Poisson problem in a circular domain with nonzero boundary condition has the form

$$u(r, \varphi) = w(r, \varphi) + v(r, \varphi)$$

$$= w(r, \varphi) + \sum_{n=0}^{\infty} \sum_{m=0}^{\infty} \left[A_{nm} V_{nm}^{(1)}(r, \varphi) + B_{nm} V_{nm}^{(2)}(r, \varphi) \right], \qquad (7.88)$$

where

$$A_{nm} = \frac{f_{nm}^{(1)}}{\lambda_{nm}} = \frac{1}{\lambda_{nm} ||V_{nm}^{(1)}||^2} \int_0^{2\pi} \int_0^l \tilde{f}(r, \varphi) \, V_{nm}^{(1)}(r, \varphi) r \, dr d\varphi, \qquad (7.89)$$

$$B_{nm} = \frac{f_{nm}^{(2)}}{\lambda_{nm}} = \frac{1}{\lambda_{nm} ||V_{nm}^{(2)}||^2} \int_0^{2\pi} \int_0^l \tilde{f}(r, \varphi) \, V_{nm}^{(2)}(r, \varphi) r \, dr d\varphi \qquad (7.90)$$

with $V_{nm}^{(p)}(r, \varphi)$ defined by Equation (7.80) and the auxiliary function $w(r, \varphi)$ (7.71). Notice, that the norms are different for different types of the BVP. The expressions for them are given in Appendix D part 1.

Example 7.7 Find a stationary temperature distribution in a thin circular plate of radius l if a boundary of the plate is kept at zero temperature and the plate contains a distributed source of heat with

$$Q(r, \varphi) = r^2 \cos 2\varphi.$$

Solution. The BVP is formulated as the Poisson equation

$$\frac{\partial^2 u}{\partial r^2} + \frac{1}{r} \frac{\partial u}{\partial r} + \frac{1}{r^2} \frac{\partial^2 u}{\partial \varphi^2} = -r^2 \cos 2\varphi \quad (0 \leq r < l, \, 0 \leq \phi < 2\pi)$$

with zero boundary condition

$$u(l, \varphi, t) = 0.$$

This is the Dirichlet BVP and the eigenvalues $\mu_m^{(n)}$ are positive roots of equation $J_n(\mu r/l) = 0$. From there, $\lambda_{nm} = \left(\mu_m^{(n)}/l \right)^2$ and $\left\| V_{nm}^{(1,2)} \right\|^2 = \sigma_n \pi \frac{l^2}{2} \left[J_n' \left(\mu_m^{(n)} \right) \right]^2$, $\sigma_n = \begin{cases} 2, & \text{if } n = 0, \\ 1, & \text{if } n \neq 0, \end{cases}$ (see Appendix D part 1).

The boundary condition is zero, so the solution $u(r, \varphi)$ is defined by the series (7.79)

$$u(r, \varphi) = \sum_{n=0}^{\infty} \sum_{m=0}^{\infty} \left[A_{nm} V_{nm}^{(1)}(r, \varphi) + B_{nm} V_{nm}^{(2)}(r, \varphi) \right],$$

where $A_{nm} = \frac{f_{nm}^{(1)}}{\lambda_{nm}} = \frac{l^2 f_{nm}^{(1)}}{\left(\mu_m^{(n)} \right)^2}$, $B_{nm} = \frac{f_{nm}^{(2)}}{\lambda_{nm}} = \frac{l^2 f_{nm}^{(2)}}{\left(\mu_m^{(n)} \right)^2}$.

Next find $f_{nm}^{(1)}$, $f_{nm}^{(2)}$ using formulas (7.85). Integrals in (7.85) contain

$$\int_0^{2\pi} \cos 2\varphi \cos n\varphi \, d\varphi = \begin{cases} \pi, & \text{if } n = 2 \\ 0, & \text{if } n \neq k \end{cases}, \quad \int_0^{2\pi} \cos 2\varphi \sin n\varphi \, d\varphi = 0,$$

thus $f_{nm}^{(1)} = 0$ for $n \neq 2$ and $f_{nm}^{(2)} = 0$ for all values of n.

Figure 7.5 shows the graph of function $J_2(\mu)$ and the respective table lists the roots of equation $J_2(\mu) = 0$.

To find $f_{2m}^{(1)}$ let us use the recurrence formula for Bessel functions

$$\int x^n J_{n-1}(x) dx = x^n J_n(x),$$

which gives

$$\int_0^l r^3 J_2 \left(\frac{\mu_m^{(2)}}{l} r \right) dr = \frac{l^4}{\mu_m^{(2)}} J_3 \left(\mu_m^{(2)} \right).$$

Thus

$$f_{2m}^{(1)} = \frac{1}{\left\| V_{2m}^{(1)} \right\|^2} \int_0^{2\pi} \int_0^l r^2 \cos 2\varphi \cdot \cos 2\varphi \cdot J_2 \left(\frac{\mu_m^{(2)}}{l} r \right) r \, dr \, d\varphi$$

$$= \frac{2}{\pi l^2 \left[J_2' \left(\mu_m^{(2)} \right) \right]^2} \cdot \pi \cdot \frac{l^4}{\mu_m^{(2)}} J_3 \left(\mu_m^{(2)} \right) = \frac{2l^2}{\mu_m^{(2)} \cdot \left[J_2' \left(\mu_m^{(2)} \right) \right]^2} \cdot J_3 \left(\mu_m^{(2)} \right).$$

Therefore, the solution of the problem $u(r, \varphi)$ is the series (see Figure 7.4):

$$u(r, \varphi) = l^2 \cos 2\varphi \sum_{m=0}^{\infty} \frac{f_{2m}^{(1)}}{\left(\mu_m^{(2)} \right)^2} \cdot J_2 \left(\frac{\mu_m^{(2)}}{l} r \right).$$

The coefficients in $f_{2m}^{(1)}$ can be evaluated, for instance by using Maple, Mathematica or software from books [7, 8]).

7.10　Three-dimensional Laplace Equation for a Cylinder

Up to now we did not discuss three-dimensional problems. In this section we will show that they can be solved in a way similar to two-dimensional ones.

Let us separate the variables in the *three-dimensional* Laplace equation

$$\nabla^2 u = 0 \tag{7.91}$$

inside a circular bounded cylinder, $r \leq a$, $0 \leq \varphi < 2\pi$, $0 \leq z \leq l$.

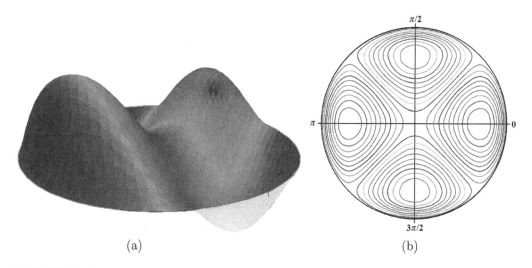

(a)   (b)

FIGURE 7.4
Surface plot (a) and lines of equal temperature (b) of the solution $u(r, \varphi)$ for Example 7.7.

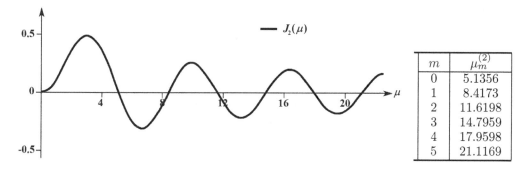

m	$\mu_m^{(2)}$
0	5.1356
1	8.4173
2	11.6198
3	14.7959
4	17.9598
5	21.1169

FIGURE 7.5
Graph of function $J_2(\mu)$ (a short notation for $J_2(\mu r/l)$ and table for the roots of equation $J_2(\mu) = 0$.

Let us represent the unknown function u in the following form:

$$u(r, \varphi, z) = V(r, \varphi) Z(z). \tag{7.92}$$

Substituting (7.92) into Equation (7.91), after the separation of variables, we get

$$\frac{1}{Vr} \frac{\partial}{\partial r} \left(r \frac{\partial V}{\partial r} \right) + \frac{1}{Vr^2} \frac{\partial^2 V}{\partial \varphi^2} = -\frac{1}{Z} \frac{\partial^2 Z}{\partial z^2} = -\lambda, \tag{7.93}$$

where $\lambda > 0$ is the separation constant which will be determined from the conditions of existence of a non-trivial solution of the problem. As a result, we obtain the equation for the function $V(r, \varphi)$

$$\frac{1}{r} \frac{\partial}{\partial r} \left(r \frac{\partial V}{\partial r} \right) + \frac{1}{r^2} \frac{\partial^2 V}{\partial \varphi^2} + \lambda V = 0 \tag{7.94}$$

and the equation for the function $Z(z)$

$$Z'' - \lambda Z = 0 \tag{7.95}$$

with a solution which can be written in the form

$$Z(z) = d_1 \cosh \sqrt{\lambda} z + d_2 \sinh \sqrt{\lambda} z. \tag{7.96}$$

A separation of variables in Equation (7.94)

$$V(r, \varphi) = R(r)\Phi(\varphi) \tag{7.97}$$

results in

$$\frac{1}{r}\frac{\partial}{\partial r}\left(r\frac{\partial R}{\partial r}\right) + \left(\lambda - \frac{\nu}{r^2}\right)R = 0 \tag{7.98}$$

and

$$\Phi'' + \nu\Phi = 0, \tag{7.99}$$

where ν is a separation constant. From periodicity condition, $\Phi(\varphi) = \Phi(\varphi + 2\pi)$, we have $\nu = n^2$, where $n = 0, 1, 2, \ldots$, and two sets of eigenfunctions $\Phi_n(\varphi)$:

$$\Phi_n(\varphi) = \sin n\varphi \quad \text{and} \quad \Phi_n(\varphi) = \cos n\varphi. \tag{7.100}$$

Equation (7.98) for the function $R(r)$ is the Bessel equation which is bounded at $r = 0$; solutions are the Bessel functions

$$R(r) = J_n\left(\sqrt{\lambda} r\right). \tag{7.101}$$

This result explains our choice of the sign, $\lambda > 0$ - only in this case the separation of variables leads to the Bessel function $R(r)$. If $\lambda < 0$, the solutions of Equation (7.98) for $R(r)$ give the modified Bessel functions, $I_n\left(\sqrt{-\lambda} r\right)$. Also, when $\lambda < 0$, the solutions of Equation (7.95) for function $Z(z)$ are periodic functions, $\sin\left(\sqrt{-\lambda} z\right)$ and $\cos\left(\sqrt{-\lambda} z\right)$, with the eigenvalues $\lambda = \lambda_m$ determined by the boundary conditions at $z = 0$ and $z = l$. The physics of the problems governs what sign of λ has to be chosen: in one case we expect oscillatory behavior of function $u(x, y, z)$ in z, in the other the exponential behavior. In books [7, 8] the reader can find a solution of Equation (7.93) for both signs of λ.

Consider the Dirichlet boundary value problem with zero boundary condition at the lateral surface

$$u|_{r=a} = 0 \tag{7.102}$$

and boundary conditions at the bottom and top surfaces

$$u|_{z=0} = g(r, \varphi), \qquad u|_{z=l} = F(r, \varphi), \tag{7.103}$$

where $g(r, \varphi)$ and $F(r, \varphi)$ are given functions.

Assume that an expected solution is not periodic in z, thus $\lambda > 0$. The boundary condition at the lateral surface (7.102) results in $R(a) = 0$ which gives

$$J_n\left(\mu_m^{(n)}\right) = 0, \tag{7.104}$$

where $\mu_m^{(n)} = \sqrt{\lambda} a$, $m = 0, 1, 2 \ldots$ numerates the roots of this equation. Therefore, Equation (7.94) gives the eigenvalues $\lambda_{nm} = \left(\mu_m^{(n)}/a\right)^2$ for Dirichlet BVP; the corresponding eigenfunctions are

$$V_{nm}^{(1)} = J_n\left(\frac{\mu_m^{(n)}}{a}r\right)\cos n\varphi, \quad V_{nm}^{(2)} = J_n\left(\frac{\mu_m^{(n)}}{a}r\right)\sin n\varphi. \tag{7.105}$$

The norms in the case of Dirichlet BVP are

$$\left\|V_{nm}^{(1)}\right\|^2 = \pi\sigma_n \frac{a^2}{2}\left|J_n'\left(\mu_m^{(n)}\right)\right|^2 \quad \text{and} \quad \left\|V_{nm}^{(2)}\right\|^2 = \pi\frac{a^2}{2}\left|J_n'\left(\mu_m^{(n)}\right)\right|^2, \tag{7.106}$$

where $\sigma_n = 2$ for $n = 0$ and $\sigma_n = 1$ for $n \neq 0$.

Using the above results, the solution of the first BVP for Equations (7.91) with zero boundary condition at the lateral surface can be represented as the series

$$u(r,\varphi,z) = \sum_{n=0}^{\infty}\sum_{m=0}^{\infty}\left\{\left[a_{1nm}V_{nm}^{(1)}(r,\varphi) + b_{1nm}V_{nm}^{(2)}(r,\varphi)\right]\cosh\left(\sqrt{\lambda_{nm}}z\right)\right.$$

$$\left. + \left[a_{2nm}V_{nm}^{(1)}(r,\varphi) + b_{2nm}V_{nm}^{(2)}(r,\varphi)\right]\sinh\left(\sqrt{\lambda_{nm}}z\right)\right\}. \tag{7.107}$$

From the boundary condition at $z = 0$ we have

$$\sum_{n=0}^{\infty}\sum_{m=0}^{\infty}\left[a_{1nm}V_{nm}^{(1)}(r,\varphi) + b_{1nm}V_{nm}^{(2)}(r,\varphi)\right] = g(r,\varphi),$$

where the coefficients a_{1nm} and b_{1nm} may be determined by expanding the function $g(r,\varphi)$ in a Fourier series in the basis functions $V_{nm}^{(1)}(r,\varphi)$ and $V_{nm}^{(2)}(r,\varphi)$:

$$a_{1nm} = \frac{1}{\left\|V_{nm}^{(1)}\right\|^2}\int_0^a\int_0^{2\pi} g(r,\varphi)V_{nm}^{(1)}(r,\varphi)rdrd\varphi,$$

$$b_{1nm} = \frac{1}{\left\|V_{nm}^{(2)}\right\|^2}\int_0^a\int_0^{2\pi} g(r,\varphi)V_{nm}^{(2)}(r,\varphi)rdrd\varphi. \tag{7.108}$$

Analogously, we find the coefficients a_{2nm} and b_{2nm} using the boundary condition at $z = l$:

$$a_{2nm} = \frac{1}{\left\|V_{nm}^{(1)}\right\|^2}\int_0^a\int_0^{2\pi} F(r,\varphi)V_{nm}^{(1)}(r,\varphi)rdrd\varphi,$$

$$b_{2nm} = \frac{1}{\left\|V_{nm}^{(2)}\right\|^2}\int_0^a\int_0^{2\pi} F(r,\varphi)V_{nm}^{(2)}(r,\varphi)rdrd\varphi. \tag{7.109}$$

In a similar way the *three-dimensional* Laplace and Poisson equations can be solved in a cylindrical domain for other types of boundary conditions on the lateral surface. The only difference is that Equations (7.104) and (7.106) should be replaced by the proper ones for the corresponding types of the boundary conditions.

Example 7.8 Find an expression for the potential of the electrostatic field inside a cylinder $r \leq a$, $0 \leq z \leq l$, the upper end and outside surfaces of which are grounded, and the lower end is held at potential $A\sin 2\varphi$.

Solution. The problem is formulated as

$$\nabla^2 u = 0 \; u|_{r=a} = 0 \; u|_{z=0} = A\sin 2\varphi, \quad u|_{z=l} = 0.$$

Coefficients $a_{2mn} = b_{2mn} = 0$. Clearly all $a_{1nm} = 0$ because of orthogonality of function $\sin 2\varphi$ and functions $\cos n\varphi$ on $[0, 2\pi]$. Among coefficients b_{1nm}, only coefficients $b_{12m} \neq 0$ are not zero:

$$b_{12m} = \frac{A}{\left\| V_{2m}^{(2)} \right\|^2} \int_0^a \int_0^{2\pi} J_2\left(\frac{\mu_m^{(2)}}{a} r \right) \sin 2\varphi \sin 2\varphi r dr d\varphi = \frac{A\pi}{\left\| V_{2m}^{(2)} \right\|^2} \int_0^a J_2\left(\frac{\mu_m^{(2)}}{a} r \right) r dr,$$

and with (7.2.1) we have

$b_{12m} = \frac{2A}{a^2 \left| J_2'\left(\mu_m^{(2)} \right) \right|^2} \int_0^a J_2\left(\frac{\mu_m^{(2)}}{a} r \right) r dr$ – this integral can be calculated numerically.

Thus,

$$u(r, \varphi, z) = \sin 2\varphi \sum_{m=0}^{\infty} b_{12m} J_2\left(\frac{\mu_m^{(2)}}{a} r \right) \cosh\left(\mu_m^{(2)} z/a \right).$$

7.11 Three-dimensional Laplace Equation for a Ball

The next important example is the three-dimensional Laplace equation

$$\nabla^2 u = 0$$

in a ball of radius a. Using the *spherical coordinates*, we obtain the following equation for function $u = u(r, \theta, \varphi)$,

$$\nabla^2 u = \frac{1}{r^2} \frac{\partial}{\partial r}\left(r^2 \frac{\partial u}{\partial r} \right) + \frac{1}{r^2 \sin \theta} \frac{\partial}{\partial \theta}\left(\sin \theta \frac{\partial u}{\partial \theta} \right) + \frac{1}{r^2 \sin^2 \theta} \frac{\partial^2 u}{\partial \varphi^2} = 0, \qquad (7.110)$$

$r < a, 0 < \theta < \pi, 0 \leq \varphi \leq 2\pi$. We shall discuss only the *Dirichlet boundary value problem*,

$$u(a, \theta, \varphi) = F(\theta, \varphi), \quad 0 \leq \theta \leq \pi, \quad 0 \leq \varphi \leq 2\pi. \qquad (7.111)$$

7.11.1 Axisymmetric Case

Let us start with the case where the function in the boundary condition does not depend on the azimuthal angle φ,

$$u(a, \theta, \varphi) = F(\theta), \quad 0 \leq \theta \leq \pi, \qquad (7.112)$$

hence the solution also does not depend on φ,

$$u = u(r, \theta). \qquad (7.113)$$

Equation (7.110) is reduced to

$$\nabla^2 u = \frac{1}{r^2} \frac{\partial}{\partial r}\left(r^2 \frac{\partial u}{\partial r} \right) + \frac{1}{r^2 \sin \theta} \frac{\partial}{\partial \theta}\left(\sin \theta \frac{\partial u}{\partial \theta} \right) = 0. \qquad (7.114)$$

As usual, we apply the method of separation of variables and find particular solutions in the form

$$u(r, \theta) = R(r)\Theta(\theta). \qquad (7.115)$$

Substituting (7.115) into (7.114), we get

$$\left(r^2 R'(r) \right)' \Theta(\theta) + \frac{R(r)}{\sin \theta} \left[\sin \theta \Theta'(\theta) \right]' = 0, \qquad (7.116)$$

hence

$$\frac{\left(r^2 R'(r)\right)'}{R(r)} = -\frac{[\sin\theta\Theta'(\theta)]'}{\sin\theta\Theta(\theta)} = \lambda, \tag{7.117}$$

where λ is constant.

It is convenient to introduce the variable $x = \cos\theta$,

$$\frac{d}{dx} = -\frac{1}{\sin\theta}\frac{d}{d\theta},$$

which allows us to rewrite the equation for $\Theta(\theta) \equiv X(x)$ as

$$\frac{d}{dx}\left[\left(1-x^2\right)\frac{dX}{dx}\right] + \lambda X = 0, \quad 1 \le x \le 1. \tag{7.118}$$

Thus, function $X(x)$ is determined by *the Legendre equation*.

The properties of solutions of that equation are described in Appendix B. It is shown that the bounded solutions of that equation, *the Legendre polynomials*, $X_n(x) \equiv P_n(x)$, exist only for $\lambda_n = n(n+1)$, where $n = 0, 1, 2, \ldots$

The equation for $R_n(r)$

$$\left[r^2 R'_n(r)\right]' - n(n+1)R_n = 0 \tag{7.119}$$

has two particular solutions,

$$r^n \quad \text{and} \quad r^{-(n+1)}.$$

The solution $r^{-(n+1)}$ has to be dropped because it is unbounded when $r \to 0$.

Finally, the bounded solution of Equation (7.114) can be presented in the form

$$u(r,\theta) = \sum_{n=0}^{\infty} A_n \left(\frac{r}{a}\right)^n P_n(\cos\theta). \tag{7.120}$$

The boundary condition (7.112) prescribes

$$u(a,\theta) = \sum_{n=0}^{\infty} A_n P_n(\cos\theta) = F(\theta). \tag{7.121}$$

Using the orthogonality property of the Legendre polynomials,

$$\int_{-1}^{1} P_n(x)P_{n'}(x)dx = \frac{2}{2n+1}\delta_{nn'}, \tag{7.122}$$

we find that

$$A_n = \frac{2n+1}{2}\int_{0}^{\pi} F(\theta)P_n(\cos\theta)\sin\theta d\theta. \tag{7.123}$$

7.11.2 Non-axisymmetric Case

Let is consider now the general case of (7.110)-(7.111). First, we search particular solutions in the form

$$u(r,\theta,\varphi) = R(r)\Theta(\theta)\Phi(\varphi). \tag{7.124}$$

Similarly to (7.117), we find

$$\frac{\left(r^2 R'(r)\right)'}{R(r)} = -\frac{[\sin\theta\Theta'(\theta)]'}{\sin\theta\Theta(\theta)} - \frac{1}{\sin^2\theta}\frac{\Phi''(\varphi)}{\Phi(\varphi)} = \lambda. \tag{7.125}$$

The second equality in (7.125) leads to relation

$$\lambda \sin^2 \theta + \sin \theta \frac{[\sin \theta \Theta'(\theta)]'}{\Theta(\theta)} = -\frac{\Phi''(\varphi)}{\Phi(\varphi)} = \mu,$$

where μ is a constant.

Equation

$$\Phi'' + \mu \Phi = 0$$

is solved with periodic boundary condition $\Phi(\varphi + 2\pi) = \Phi(\varphi)$, hence

$$\mu_m = m^2, \quad m = 0, 1, 2, \ldots$$

The eigenfunctions are

$$\Phi_0 = a_0, \Phi_m = a_m \cos m\varphi + b_m \sin m\varphi, \quad m \neq 0,$$

where a_0, a_m and b_m are arbitrary constants.

The equation for $\Theta(\theta)$ becomes

$$\sin \theta \left[\sin \theta \Theta'(\theta)\right]' + \left(\lambda \sin \theta - m^2\right) \Theta(\theta) = 0.$$

By means of the change of variables

$$x = \cos \theta, \quad X(x) = \Theta(\theta),$$

it is transformed to

$$\frac{d}{dx} \left[\left(1 - x^2\right) \frac{dX}{dx}\right] + \left(\lambda - \frac{m^2}{1 - x^2}\right) X = 0. \tag{7.126}$$

Its bounded solutions are *associate Legendre polynomials*:

$$X_{mn}(x) = P_n^m(x), \quad \lambda_n = n(n+1); \quad n = 0, 1, 2, \ldots, \quad 0 \leq m \leq n$$

(see Appendix B). Thus

$$\Theta_{mn}(\theta) = P_n^m(\cos \theta).$$

For $R_n(r)$ we obtain the same Equation (7.119) for any m, hence the bounded solution is $R_n(r) = r^n$.

Finally, we obtain

$$u(r, \theta, \varphi) = \sum_{n=0}^{\infty} \left(\frac{r}{a}\right)^n \sum_{m=0}^{n} P_n^m(\cos \theta)(A_{mn} \cos m\varphi + B_{mn} \sin m\varphi).$$

The coefficients A_{mn} and B_{mn} are found from the boundary condition (7.111),

$$u(a, \theta, \varphi) = \sum_{n=0}^{\infty} \sum_{m=0}^{n} P_n^m(\cos \theta) \left[A_{mn} \cos m\varphi + B_{mn} \sin m\varphi\right].$$

We leave the computation of coefficients A_{mn} and B_{mn} to the reader. When carrying out the computations, one has to use the known orthogonality properties of trigonometric functions, orthogonality properties of associate Legendre polynomials (with the weight $\sin \theta$), and the relation (see Appendix B).

$$\int_{-1}^{1} \left[P_n^m(x)\right]^2 dx = \int_0^{\pi} \left[P_n^m(\cos \theta)\right]^2 \sin \theta d\theta = \frac{2}{2n+1} \frac{(n+m)!}{(n-m)!} \tag{7.127}$$

7.12 BVP for Laplace Equation in a Rectangular Domain

Boundary value problems for the Laplace equation in a rectangular domain can be solved with the method of separation of variables. We begin with the *Dirichlet problem* defined by

$$\nabla^2 u = 0 \quad (0 < x < l_x, \quad 0 < y < l_y),$$
$$u(x,y)|_{x=0} = g_1(y), \quad u(x,y)|_{x=l_x} = g_2(y), \tag{7.128}$$

$$u(x,y)|_{y=0} = g_3(x), \quad u(x,y)|_{y=l_y} = g_4(x). \tag{7.129}$$

Let us split the problem in Equations (7.128) through (7.129) into two parts, each of which has *homogeneous (zero) boundary conditions in one variable*. To proceed we introduce

$$u(x,y) = u_1(x,y) + u_2(x,y), \tag{7.130}$$

where $u_1(x,y)$ and $u_2(x,y)$ are the solutions to the following problems on a rectangular boundary:

$$\nabla^2 u_1 = 0, \tag{7.131}$$

$$u_1(x,y)|_{x=0} = u_1(x,y)|_{x=l_x} = 0, \tag{7.132}$$

$$u_1(x,y)|_{y=0} = g_3(x), \quad u_1(x,y)|_{y=l_y} = g_4(x), \tag{7.133}$$

and

$$\nabla^2 u_2 = 0, \tag{7.134}$$

$$u_2(x,y)|_{y=0} = u_2(x,y)|_{y=l_y} = 0, \tag{7.135}$$

$$u_2(x,y)|_{x=0} = g_1(y), \quad u_2(x,y)|_{x=l_x} = g_2(y). \tag{7.136}$$

First, we consider the problem for the function $u_1(x,y)$ and search for the solution in the form

$$u_1(x,y) = X(x)Y(y). \tag{7.137}$$

Substituting Equation (7.137) into the Laplace equation and separating the variables yields

$$\frac{X''(x)}{X(x)} \equiv -\frac{Y''(y)}{Y(y)} = -\lambda, \tag{7.138}$$

where we take $\lambda > 0$ for the further solution.

From here we obtain equations for $X(x)$ and $Y(y)$. With the homogeneous boundary conditions in Equation (7.132) we obtain the one-dimensional Sturm-Liouville problem for $X(x)$ given by

$$X'' + \lambda X = 0, \quad 0 < x < l_x,$$
$$X(0) = X(l_x) = 0.$$

The solution to this problem is

$$X_n = \sin\sqrt{\lambda_{xn}}x, \quad \lambda_{xn} = \left(\frac{\pi n}{l_x}\right)^2, \quad n = 1,2,3,\dots. \tag{7.139}$$

With these eigenvalues, λ_{xn}, we obtain an equation for $Y(y)$ from Equation (7.138):

$$Y'' - \lambda_{xn}Y = 0, \quad 0 < y < l_y. \tag{7.140}$$

A general solution to this equation can be written as

$$Y_n = C_n^{(1)} \exp\left(\sqrt{\lambda_{xn}} y\right) + C_n^{(2)} \exp\left(-\sqrt{\lambda_{xn}} y\right). \qquad (7.141)$$

Such a form of solution does not fit well the purposes of the further analysis. It is more suitable to take a fundamental system of solution $\{Y_n^{(1)}, Y_n^{(2)}\}$ of Equation (7.140) in the way that function $Y_n^{(1)}$ and $Y_n^{(2)}$ satisfy the homogeneous boundary condition, one at $y = 0$ and another at $y = l_y$:

$$Y_n^{(1)}(0) = 0, \quad Y_n^{(2)}(l_y) = 0.$$

It is convenient to choose the following conditions at two other boundaries:

$$Y_n^{(1)}(l_y) = 1, \quad Y_n^{(2)}(0) = 1.$$

As a result the proper fundamental solutions of Equations (7.140) are:

$$Y_n^{(1)} = \frac{\sinh \sqrt{\lambda_{xn}} y}{\sinh \sqrt{\lambda_{xn}} l_y} \quad \text{and} \quad Y_n^{(2)} = \frac{\sinh \sqrt{\lambda_{xn}}(l_y - y)}{\sinh \sqrt{\lambda_{xn}} l_y}. \qquad (7.142)$$

It is easily verified that they both satisfy Equation (7.140) and are linearly independent; thus they can serve as a fundamental set of particular solutions for this equation.

Using the above relations we may write a general solution of the Laplace equation satisfying the homogeneous boundary conditions at the boundaries $x = 0$ and $x = l_x$ in Equation (7.132), as a series in the functions $Y_n^{(1)}(y)$ and $Y_n^{(2)}(y)$:

$$u_1 = \sum_{n=1}^{\infty} \left[A_n Y_n^{(1)}(y) + B_n Y_n^{(2)}(y) \right] \sin \sqrt{\lambda_{xn}} x. \qquad (7.143)$$

The coefficients of this series are determined from the boundary conditions (7.133):

$$u_1(x,y)\big|_{y=0} = \sum_{n=1}^{\infty} B_n \sin \sqrt{\lambda_{xn}} x = g_3(x),$$

$$u_1(x,y)\big|_{y=l_y} = \sum_{n=1}^{\infty} A_n \sin \sqrt{\lambda_{xn}} x = g_4(x). \qquad (7.144)$$

We see from here that B_n and A_n are Fourier coefficients of functions $g_3(x)$ and $g_4(x)$ in the system of eigenfunctions $\left\{ \sin \sqrt{\lambda_{xn}} x \right\}_1^{\infty}$:

$$B_n = \frac{2}{l_x} \int_0^{l_x} g_3(\xi) \sin \sqrt{\lambda_{xn}} \xi \, d\xi,$$

$$A_n = \frac{2}{l_x} \int_0^{l_x} g_4(\xi) \sin \sqrt{\lambda_{xn}} \xi \, d\xi. \qquad (7.145)$$

This completes the solution of the problem given in Equations (7.131) through (7.133).

Obviously, the solution of the similar problem given in Equations (7.144) through (7.145) can be obtained from Equations (7.143) and (7.145) by replacing y for x, l_y for l_x and $g_3(x)$, $g_4(x)$ for $g_1(y)$ and $g_2(y)$. Carrying out this procedure yields

$$u_2 = \sum_{n=1}^{\infty} \left[C_n X_n^{(1)}(x) + D_n X_n^{(2)}(x) \right] \sin \sqrt{\lambda_{yn}} y, \qquad (7.146)$$

where

$$X_n^{(1)}(x) = \frac{\sinh \sqrt{\lambda_{yn}} x}{\sinh \sqrt{\lambda_{yn}} l_x}, \quad X_n^{(2)}(x) = \frac{\sinh \sqrt{\lambda_{yn}} (l_x - x)}{\sinh \sqrt{\lambda_{yn}} l_x}, \quad \lambda_{yn} = \left(\frac{\pi n}{l_y}\right)^2, \quad (7.147)$$

and

$$C_n = \frac{2}{l_y} \int_0^{l_y} g_2(\xi) \sin \sqrt{\lambda_{yn}} \xi d\xi, \quad D_n = \frac{2}{l_y} \int_0^{l_y} g_1(\xi) \sin \sqrt{\lambda_{yn}} \xi d\xi. \quad (7.148)$$

Finally, the solution to the problem (7.128) and (7.129) has the form

$$u(x, y) = u_1(x, y) + u_2(x, y), \quad (7.149)$$

where functions $u_1(x, y)$ and $u_2(x, y)$ are defined by formulas (7.143) and (7.146), respectively.

In the same way can be solved a BVP for the Laplace equation in a rectangular domain with *other types of boundary conditions*. The only difference is that the other fundamental solutions should be used. Fundamental systems of solutions for different types of boundary conditions are collected in Appendix E part 1.

Example 7.9 Find a steady state temperature distribution inside a rectangular material which has boundaries maintained under the following conditions:

$$T(x, y)|_{x=0} = T_0 + (T_3 - T_0) \frac{y}{l_y}, \quad T(x, y)|_{x=l_x} = T_1 + (T_2 - T_1) \frac{y}{l_y},$$

and

$$T(x, y)|_{y=0} = T_0 + (T_1 - T_0) \frac{x}{l_x}, \quad T(x, y)|_{y=l_y} = T_3 + (T_2 - T_3) \frac{x}{l_x},$$

i.e. at the corners of the rectangle the temperatures are T_0, T_1, T_2, T_3, and on the boundaries the temperatures are linear functions.

Solution. Introduce the function $u = T - T_0$ so that we measure the temperature relative to T_0. Then $g_1(y) = (T_3 - T_1)\frac{y}{l_y}$, etc. and evaluating the integrals in Equations (7.145) and (7.148), we obtain

$$A_n = \frac{2}{\pi n}[T_3 - T_0 - (-1)^n T_2], \quad B_n = -2\frac{(-1)^n}{\pi n}(T_1 - T_0),$$

$$C_n = \frac{2}{\pi n}[T_1 - T_0 - (-1)^n T_2], \quad D_n = -2\frac{(-1)^n}{\pi n}(T_3 - T_0).$$

These coefficients decay only as $1/n$; thus the series in Equations (7.145) and (7.147) converge rather slowly.

7.13 The Poisson Equation with Homogeneous Boundary Conditions

Consider a boundary value problem for the Poisson equation

$$\frac{\partial^2 u}{\partial x^2} + \frac{\partial^2 u}{\partial y^2} = -f(x, y) \quad (0 < x < l_x, 0 < y < l_y) \quad (7.150)$$

with homogeneous boundary conditions

$$P_1[u] \equiv \alpha_1 \frac{\partial u}{\partial x} + \beta_1 u \bigg|_{x=0} = 0, \quad P_2[u] \equiv \alpha_2 \frac{\partial u}{\partial x} + \beta_2 u \bigg|_{x=l_x} = 0,$$

$$P_3[u] \equiv \alpha_3 \frac{\partial u}{\partial x} + \beta_3 u \bigg|_{y=0} = 0, \quad P_4[u] \equiv \alpha_4 \frac{\partial u}{\partial x} + \beta_4 u \bigg|_{y=l_y} = 0. \tag{7.151}$$

The solution to the problem (7.150), (7.151) can be expanded in a series by eigenfunctions of the Sturm-Liouville problem for the Laplace operator over a rectangular domain

$$u(x,y) = \sum_{n=1}^{\infty} \sum_{m=1}^{\infty} C_{nm} V_{nm}(x,y), \tag{7.152}$$

where $V_{nm}(x,y)$ are eigenfunctions of the respective Laplace boundary value problem and coefficients are

$$C_{nm} = \frac{1}{\|V_{nm}\|^2} \int_0^{l_x} \int_0^{l_y} u(x,y) V_{nm}(x,y) dx dy. \tag{7.153}$$

Let us multiply the Equation (7.150) by $V_{nm}(x,y)$ and integrate over the rectangle $[0, l_x; 0, l_y]$:

$$\int_0^{l_x} \int_0^{l_y} \left[\frac{\partial^2 u}{\partial x^2} + \frac{\partial^2 u}{\partial y^2} \right] \cdot V_{nm}(x,y) dx dy = - \int_0^{l_x} \int_0^{l_y} f(x,y) V_{nm}(x,y) dx dy. \tag{7.154}$$

Now substitute Equation (7.152) into Equation (7.154). Because $V_{nm}(x,y)$ are the eigenfunctions of the Laplacian we have

$$\nabla^2 V_{nm}(x,y) = -\lambda_{nm} V_{nm}(x,y), \tag{7.155}$$

where the eigenvalues λ_{nm} correspond to the boundary condition (7.151). The left sides of Equations (7.154) become:

$$-\sum_{l-1}^{\infty} \sum_{j-1}^{\infty} \lambda_{ij} \int_0^{l_x} \int_0^{l_y} C_{ij} \cdot V_{ij}(x,y) V_{nm}(x,y) dx dy.$$

Due to the orthogonality relation for the functions $V_{nm}(x,y)$, the only term in the sums that differs from zero is

$$-\lambda_{nm} C_{nm} \|V_{nm}\|^2.$$

Comparing with the right side in Equation (7.154) we obtain

$$\lambda_{nm} C_{nm} = f_{nm}, \quad n, m = 1, 2, 3, \ldots, \tag{7.156}$$

where

$$f_{nm} = \frac{1}{\|V_{nm}\|^2} \int_0^{l_x} \int_0^{l_y} f(x,y) V_{nm}(x,y) dx dy. \tag{7.157}$$

From (7.156) and (7.157) the coefficients C_{nm} can be obtained.

In the case of boundary conditions of the 1st or 3rd type (*Dirichlet condition* or *mixed condition*) eigenvalues $\lambda_{nm} \neq 0$ for all n, $m = 1,2,3, \ldots$, thus $C_{nm} = f_{nm}/\lambda_{nm}$ and the solution (7.152) is defined uniquely:

$$u(x,y) = \sum_{n=1}^{\infty} \sum_{m=1}^{\infty} \frac{f_{nm}}{\lambda_{nm}} V_{nm}(x,y). \tag{7.158}$$

In the case of boundary conditions of the 2nd type (*Neumann conditions*) eigenvalue $\lambda_{00} = 0$ ($V_{00} = 1$) and all other eigenvalues are nonzero. Then there are two options.
If

$$f_{00} = \int_0^{l_x} \int_0^{l_y} f(x,y)dxdy = 0,$$

then coefficient C_{00} is uncertain, and the other coefficients are defined uniquely. The solution to the given problem exists but is determined only up to an arbitrary additive constant. The solution is

$$u(x,y) = \sum_{n=1}^{\infty} \sum_{m=1}^{\infty} \frac{f_{nm}}{\lambda_{nm}} V_{nm}(x,y) + \text{const.} \tag{7.159}$$

If

$$f_{00} = \int_0^{l_x} \int_0^{l_y} f(x,y)dxdy \neq 0,$$

then the *solution to the given problem does not exist*.

The solution of the Poisson equations with nonhomogeneous boundary conditions needs to introduce auxiliary functions to switch to homogeneous boundary conditions. It can be done similarly to the cases we already considered several times, but for a rectangular domain a determination of an auxiliary function needs more technical steps. That is why this problem is placed into Appendix C part 2.

7.14 Green's Function for Poisson Equations

7.14.1 Homogeneous Boundary Conditions

In the present subsection, we apply Green's function approach for finding the solution of the Poisson equation

$$\frac{\partial^2 u}{\partial x^2} + \frac{\partial^2 u}{\partial y^2} = -f(x,y) \tag{7.160}$$

defined in a certain region D, with some homogeneous boundary conditions imposed on the boundary L of that region. Our goal is to find the kernel $G(x,y;\xi,\eta)$ of the integral transformation

$$u(x,y) = -\iint_D G(x,y;\xi,\eta)f(\xi,\eta)d\xi d\eta \tag{7.161}$$

that transforms the right-hand side of Equation (7.160) (heat source or charge density) into the solution of the boundary value problem (temperature or potential field). Recall that formerly we obtained such integral transformations for the wave equation (Section 5.5.2) and the heat equation (Section 6.4). For that goal, it is necessary to solve the Poisson equation with the source localized in the definite point in the region D, i.e.,

$$\frac{\partial^2 G(x,y;\xi,\eta)}{\partial x^2} + \frac{\partial^2 G(x,y;\xi,\eta)}{\partial y^2} = \delta(x-\xi)\delta(y-\eta) \tag{7.162}$$

with the same homogeneous boundary conditions at the boundary L. Due to the superposition principle, for an arbitrary right-hand side, $f(x,y)$, the solution will be determined by integral formula (7.161).

Let us consider some basic examples.

Example 7.10 First, let us find the solution of the Poisson equation (7.162) in the whole plane $-\infty < x < \infty$, $-\infty < y < \infty$.

Of course, any harmonic function can be added to that solution. However, we are interested in the particular solution which corresponds to the action of the localized source rather than external heating. Therefore, we impose the condition of zero flux on infinity:

$$\frac{\partial G(x,y;\xi,\eta)}{\partial x} = \frac{\partial G(x,y;\xi,\eta)}{\partial y} = 0 \quad \text{as } x^2 + y^2 \to \infty. \tag{7.163}$$

Because the right-hand side of Equation (7.127) depends only on differences

$$X = x - \xi, \quad Y = y - \eta,$$

it is natural to expect that the solution of the problem (7.162), (7.163) also depends only on those differences,

$$G(x,y;\xi,\eta) = \Gamma(X,Y), \tag{7.164}$$

hence

$$\frac{\partial^2 \Gamma(X,Y)}{\partial X^2} + \frac{\partial^2 \Gamma(X,Y)}{\partial Y^2} = \delta(X)\delta(Y), \quad -\infty < X < \infty, \ -\infty < Y < \infty. \tag{7.165}$$

Moreover, because the point heat source is localized at the origin $X = Y = 0$, the problem is rotationally invariant; hence we expect that the temperature field depends only on the radial coordinate:

$$\Gamma = \Gamma(R), \quad R = \sqrt{X^2 + Y^2}. \tag{7.166}$$

Except for the point $R = 0$, function $\Gamma(R)$ satisfies the Laplace equation,

$$\frac{1}{R}\frac{d}{dR}\left(R\frac{d\Gamma}{dR}\right) = 0,$$

hence

$$\Gamma(R) = c_1 \ln R + c_2.$$

The constant c_2 is arbitrary, and it is not related to the point source. Later on, we choose $c_2 = 0$. The constant c_1 has to be found using the full Poisson equation (7.165).

Let us integrate both sides of (7.165) over the disk D of radius R_0. Recall the formula of vector analysis,

$$\iint_D \nabla \cdot \mathbf{v}\,dxdy = \oint_L \mathbf{v} \cdot \mathbf{n}\,ds, \tag{7.167}$$

where $\mathbf{v}$ is a vector field, L is the boundary of D, $\mathbf{n}$ is the outward normal vector to L, and the line integral over L is taken counterclockwise. Using this formula, we obtain in the left-hand side of the equation:

$$\iint_D \Delta\Gamma\,dxdy = \iint_D \nabla \cdot \nabla\Gamma\,dxdy = \oint_L \frac{d\Gamma}{dR}ds = \frac{c_1}{R_0} \cdot 2\pi R_0 = 2\pi c_1,$$

while the right-hand side is equal to 1. Hence, $c_1 = 1/2\pi$.

Solution

$$\Gamma(x-\xi, y-\eta) = \frac{1}{2\pi}\ln R = \frac{1}{4\pi}\ln[(x-\xi)^2 + (y-\eta)^2] \tag{7.168}$$

is called *the fundamental solution* of the Poisson equation.

We can use the obtained solution in order to transform the boundary value problem (7.162) for the *Poisson equation* to a certain boundary value problem for the *Laplace equation*. For instance, let us consider Equation (7.162) in D with a homogeneous Dirichlet boundary condition,

$$G(x, y; \xi, \eta) = 0, \quad (x, y) \in L. \tag{7.169}$$

Let us present the solution in the form

$$G(x, y; \xi, \eta) = \Gamma(x - \xi, y - \eta) + v(x, y; \xi, \eta). \tag{7.170}$$

Then function $v(x, y; \xi, \eta)$ satisfies the Laplace equation

$$\frac{\partial^2 v(x, y; \xi, \eta)}{\partial x^2} + \frac{\partial^2 v(x, y; \xi, \eta)}{\partial y^2} = 0, \quad (x, y) \in D \tag{7.171}$$

with the nonhomogeneous Dirichlet boundary condition

$$v(x, y; \xi, \eta) = -\Gamma(x - \xi, y - \eta), \quad (x, y) \in L. \tag{7.172}$$

Example 7.11 Let us find Green's function for the Poisson equation in the half-plane $y > 0$ with the homogeneous Dirichlet boundary condition,

$$G(x, 0; \xi, \eta) = 0.$$

Using transformation (7.170), we obtain the Laplace equation (7.171) in the region $y > 0$ with the nonhomogeneous Dirichlet boundary condition,

$$v(x, 0, \xi, \eta) = -\Gamma(x - \xi, -\eta) = -\frac{1}{4\pi} \ln[(x - \xi)^2 + \eta^2], \quad -\infty < x < \infty. \tag{7.173}$$

Thus, the addition $v(x, y; \xi, \eta)$ is produced in the region $y > 0$ by the boundary temperature distribution (7.173) at $y = 0$.

The natural way to create such a temperature distribution is to extend the problem to the region of negative y and put a *negative* point heat source ("*an image*") in the point $x = \xi$, $y = -\eta$. Indeed, let us consider the problem

$$\frac{\partial^2 v(x, y; \xi, \eta)}{\partial x^2} + \frac{\partial^2 v(x, y; \xi, \eta)}{\partial y^2} = -\delta(x - \xi)\delta(y + \eta) \tag{7.174}$$

in the whole plane. Using the results of Example 7.10, we find that

$$v(x, y; \xi, \eta) = -\Gamma(x - \xi, y + \eta) = -\frac{1}{4\pi} \ln[(x - \xi)^2 + (y + \eta)^2]. \tag{7.175}$$

Solution (7.175) satisfies condition (7.173) at $y = 0$.

Substituting (7.175) into (7.170), we obtain Green's function for the Dirichlet problem in a half-plane,

$$G(x, y; \xi, \eta) = \Gamma(x - \xi, y - \eta) - \Gamma(x - \xi, y + \eta) = \frac{1}{4\pi} \ln \frac{(x - \xi)^2 + (y - \eta)^2}{(x - \xi)^2 + (y + \eta)^2},$$

$$-\infty < x < \infty, \quad y > 0; \quad -\infty < \xi < \infty, \quad \eta > 0.$$

Example 7.12 Let us find now Green's function for the Poisson equation in the half-plane $y > 0$ with the homogeneous Neumann boundary condition,

$$\frac{\partial G}{\partial y}(x, 0; \xi, \eta) = 0. \tag{7.176}$$

It is easy to guess that in this case we have to put the image *of the same sign* in the point $(\xi, -\eta)$, i.e., $v(x, y; \xi, \eta)$ should satisfy the equation

$$\frac{\partial^2 v(x, y; \xi, \eta)}{\partial x^2} + \frac{\partial^2 v(x, y; \xi, \eta)}{\partial y^2} = \delta(x - \xi)\delta(y + \eta),$$

thus

$$v(x, y; \xi, \eta) = \Gamma(x - \xi, y + \eta).$$

Indeed, Green's function

$$G(x, y; \xi, \eta) = \Gamma(x - \xi, y - \eta) + \Gamma(x - \xi, y + \eta)$$

$$= \frac{1}{4\pi} \ln \left\{ \left[(x - \xi)^2 + (y - \eta)^2 \right] \cdot \left[(x - \xi)^2 + (y + \eta)^2 \right] \right\}$$

is an even function of y, therefore it satisfies boundary condition (7.176).

Example 7.13 Consider now Equation (7.162) in a disk $x^2 + y^2 < 1$ with the homogeneous Dirichlet boundary condition,

$$G(x, y; \xi, \eta) = 0 \quad \text{as} \quad x^2 + y^2 = 1,$$

According to formula (7.172), we have to find the solution of the Laplace equation that creates the distribution

$$v(x, y; \xi, \eta) = -\Gamma(x - \xi, y - \eta) = -\frac{1}{4\pi} \ln[(x - \xi)^2 + (y - \eta)^2] \tag{7.177}$$

on the circle $x^2 + y^2 = 1$. Following the approach applied above, we search that solution in the form

$$\tilde{v}(x, y; \xi, \eta) = C - \frac{1}{4\pi} \ln[(x - \tilde{\xi})^2 + (y - \tilde{\eta})^2], \tag{7.178}$$

where $(\tilde{\xi}, \tilde{\eta})$ are the coordinates of the "image". Using the symmetry arguments, we can suggest that the image is located on the straight line that passes the points $(0, 0)$ and (ξ, η), i.e., it is in a certain point $(\tilde{\xi}, \tilde{\eta}) = c(\xi, \eta)$, where c is a constant to be found.

Equating (7.177) and (7.178) on the circle $x^2 + y^2 = 1$,

$$C - \frac{1}{4\pi} \ln[(x - c\xi)^2 + (y - c\eta)^2] = -\frac{1}{4\pi} \ln[(x - \xi)^2 + (y - \eta)^2]$$

we obtain

$$e^{-4\pi C}[(x - c\xi)^2 + (y - c\eta)^2] = [(x - \xi)^2 + (y - \eta)^2]. \tag{7.179}$$

Equation (7.179) is satisfied on the whole circle $x^2 + y^2 = 1$, if

$$e^{-4\pi C}c = 1, \quad e^{-4\pi C}[1 + c^2(\xi^2 + \eta^2)] = 1 + \xi^2 + \eta^2. \tag{7.180}$$

Solving (7.180), we find

$$c = \frac{1}{\xi^2 + \eta^2}, \quad C = -\frac{1}{4\pi} \ln(\xi^2 + \eta^2).$$

Thus,

$$\tilde{v}(x, y; \xi, \eta) = -\frac{1}{4\pi} \ln(\xi^2 + \eta^2) - \frac{1}{4\pi} \ln[(x - \tilde{\xi})^2 + (y - \tilde{\eta})^2]$$

and

$$G(x, y; \xi, \eta) = \frac{1}{4\pi} \ln \frac{(x - \xi)^2 + (y - \eta)^2}{(\xi^2 + \eta^2)[(x - \tilde{\xi})^2 + (y - \tilde{\eta})^2]}, \tag{7.181}$$

where

$$(\tilde{\xi}, \tilde{\eta}) = \frac{(\xi, \eta)}{\xi^2 + \eta^2}. \tag{7.182}$$

In polar coordinates,

$$x = r \cos \theta, \quad y = r \sin \theta, \quad \xi = \rho \cos \varphi, \quad \eta = \rho \sin \varphi,$$

formulas (7.181), (7.182) can be written as

$$G(r, \theta; \rho, \varphi) = \frac{1}{4\pi} \ln \frac{r^2 + \rho^2 - 2r\rho \cos(\theta - \varphi)}{r^2 \rho^2 + 1 - 2r\rho \cos(\theta - \varphi)}; \quad 0 < r < 1, \; 0 < \varphi < 1. \tag{7.183}$$

7.14.2 Nonhomogeneous Boundary Conditions

Green's function can be used also for solving the nonhomogeneous boundary value problem,

$$u_{xx} + u_{yy} = -f(x, y), \quad (x, y) \in D, \tag{7.184}$$

$$u(x, y) = \varphi(x, y), \quad (x, y) \in L, \tag{7.185}$$

where L is the boundary of region D.

In order to obtain the generalization of formula (7.161) in the case of non-homogeneous boundary conditions, we have first to derive some relations.

Let us integrate the obvious identity

$$\nabla \cdot (f \nabla g - g \nabla f) = f \nabla^2 g - g \nabla^2 f$$

over the region D. Using formula (7.167), we obtain *Green's identity*

$$\int_L (f\mathbf{n} \cdot \nabla g - g\mathbf{n} \cdot \nabla f) \, ds = \iint_D (f \nabla^2 g - g \nabla^2 f) \, dx dy. \tag{7.186}$$

Let us take now $f = G(x, y; \xi_1, \eta_1)$ and $g = G(x, y; \xi_2, \eta_2)$, where

$$\nabla^2 G(x, y; \xi_i, \eta_i) = \delta(x - \xi_i)\delta(y - \eta_i), \quad (x, y) \in D, \tag{7.187}$$

$$G(x, y; \xi_i, \eta_i) = 0, \quad (x, y) \in L; \quad i = 1, 2; \tag{7.188}$$

here (ξ_1, η_1) and (ξ_2, η_2) are two different points in D. Substituting f and g into Green's identity, we find that Green's function is symmetric,

$$G(\xi_1, \eta_1; \xi_2, \eta_2) = G(\xi_2, \eta_2; \xi_1, \eta_1). \tag{7.189}$$

Let us take now $f = u(x, y)$ and $g = G(x, y; \xi, \eta)$. Substituting into the Green's identity, we obtain,

$$\int_L (u\mathbf{n} \cdot \nabla G - G\mathbf{n} \cdot \nabla u) \, ds = \iint_D (u \nabla^2 G - G \nabla^2 u) dx dy.$$

Taking into account (7.187), (7.188), we find that

$$u(\xi, \eta) = \iint_D G(x, y; \xi, \eta)\nabla^2 u(x, y)dx dy + \int_L u(x, y) \left(\mathbf{n}(x, y) \cdot \nabla_{\mathbf{x}} G(x, y; \xi, \eta)\right) ds,$$

where $\nabla_x = (\partial/\partial x, \partial/\partial y)$. Let us interchange the notations of (x, y) and (ξ, η), and take into account the symmetry of Green's function (7.189):

$$u(x,y) = \iint_D G(x,y;\xi,\eta)\Delta u(\xi,\eta)d\xi d\eta + \int_L u(\xi,\eta)\left(\mathbf{n}(\xi,\eta)\cdot\nabla_\xi G(x,y;\xi,\eta)\right)d\sigma,$$

where $\nabla_\xi = (\partial/\partial\xi, \partial/\partial\eta)$, and integration over $d\sigma$ is performed along the region's boundary in the plane (ξ, η). Using (7.184), (7.185), we obtain *Green's representation formula*

$$u(x,y) = -\iint_D G(x,y;\xi,\eta)f(\xi,\eta)d\xi d\eta$$
$$+ \int_L \varphi(\xi,\eta)\left(\mathbf{n}(\xi,\eta)\cdot\nabla_\xi G(x,y;\xi,\eta)\right)d\sigma, \qquad (7.190)$$

which provides the contributions of both the source and the boundary condition into the solution.

Note that the Poisson integral formula (7.62) obtained in Section 7.8 is just a consequence of formula (7.190) with Green's function given by (7.183).

Example 7.14 Solve the following boundary value problem:

$$\nabla^2 u = -f(x,y), \quad -\infty < x < \infty, \quad y > 0;$$

$$u(x,y) = \varphi(x), \quad -\infty < x < \infty.$$

Formerly, we have found Green's function for the Poisson equation in the half-plane $y > 0$ (see Example 7.12):

$$G(x,y,\xi,\eta) = \frac{1}{4\pi}\ln\frac{(x-\xi)^2 + (y-\eta)^2}{(x-\xi)^2 + (y+\eta)^2}.$$

The outward normal vector is $(0, -1)$, therefore

$$\mathbf{n}(\xi,\eta)\cdot\nabla_\xi G(x,y,\xi,\eta)\big|_{\eta=0} = -\frac{\partial G(x,y,\xi,\eta)}{\partial\eta}\bigg|_{\eta=0}$$
$$= \frac{1}{2\pi}\left[\frac{y-\eta}{(x-\xi)^2+(y-\eta)^2} + \frac{y+\eta}{(x-\xi)^2+(y+\eta)^2}\right]\bigg|_{\eta=0}$$
$$= \frac{1}{\pi}\frac{y}{(x-\xi)^2+y^2}.$$

Thus,

$$u(x,y) = -\frac{1}{4\pi}\int_{-\infty}^{\infty}d\xi\int_0^\eta d\eta\ln\frac{(x-\xi)^2+(y-\eta)^2}{(x-\xi)^2+(y+\eta)^2}f(\xi,\eta) + \frac{y}{\pi}\int_{-\infty}^{\infty}d\xi\frac{\varphi(\xi)}{(x-\xi)^2+y^2}.$$

7.15 Some Other Important Equations

In the present section we consider some other important equations.

7.15.1 Helmholtz Equation

The Laplace equation considered above can be obtained as a *static limit* of dynamic equations which describe temporal evolution of physical fields, e.g., the wave equation

$$\frac{\partial^2 u}{\partial t^2} = a^2 \nabla^2 u, \tag{7.191}$$

or the heat equation

$$\frac{\partial u}{\partial t} = a^2 \nabla^2 u, \tag{7.192}$$

where $u = u(\mathbf{x}, t)$ is a function of two or three variables ($\mathbf{x}$ is (x, y) or (x, y, z)). Problems (7.191) and (7.192) will be considered in detail in Chapters 8 and 9. Here we note that instead of *time-independent* solutions of Equation (7.192) we can consider a *monochromatic wave*

$$u(\mathbf{x}, t) = U(\mathbf{x}) \cos(\omega t + C). \tag{7.193}$$

Solutions of this kind, which correspond to oscillations and waves with a *definite frequency*, appear in a natural way when the method of separation of variables is applied. Substituting (7.193) into (7.192), we find that $U(\mathbf{x})$ satisfies the *Helmholtz equation*

$$\nabla^2 U + k^2 U = 0, \tag{7.194}$$

where $k = \omega/a$ (the physical meaning of k is the wavenumber).

Note that the wave velocity a can depend on the coordinate $\mathbf{x}$, when the wave propagates in a heterogeneous medium. For instance, the velocity of light in a medium depends on the local refraction index. In that case, the Helmholtz equation for a monochromatic wave is

$$\nabla^2 U + k^2(\mathbf{x})U = 0. \tag{7.195}$$

For solving the Helmholtz equation, we can apply approaches similar to those used for the Laplace equation. Let us consider some examples.

Example 7.15 Find the general solution of the Helmholtz equation in a plane *in polar coordinates*,

$$\frac{1}{r} \frac{\partial}{\partial r} \left(r \frac{\partial U}{\partial r} \right) + \frac{1}{r^2} \frac{\partial^2 U}{\partial \varphi^2} + k^2 U = 0.$$

Using the separation of variables,

$$U(r, \varphi) = R(r)\Phi(\varphi),$$

we obtain two ordinary differential equations,

$$\Phi'' + \lambda\Phi = 0 \tag{7.196}$$

and

$$r^2 R'' + rR' + \left(k^2 - \lambda \right) R = 0. \tag{7.197}$$

Equation (7.196) along with periodic condition, $\Phi(\varphi + 2\pi) = \Phi(\varphi)$, gives the eigenvalues

$$\lambda_m = m^2, \quad m = 0, 1, 2, \ldots \tag{7.198}$$

and the eigenfunctions

$$\Phi_0 = A_0, \quad \Phi_m(\varphi) = A_m \cos m\varphi + B_m \sin m\varphi \text{ for } m \neq 0. \tag{7.199}$$

Equation (7.197) is the Bessel equation,

$$R'' + \frac{1}{r}R' + \left(k^2 - \frac{m^2}{r^2}\right)R = 0;$$

its solution is

$$R_m(r) = C_m J_m(kr) + D_m N_m(kr)$$

(see Appendix B). Thus, the general solution of the Helmholtz equation can be written as

$$U(r,\varphi) = \sum_{n=0}^{\infty}(A_m \cos m\varphi + B_m \sin m\varphi)(C_m J_m(kr) + D_m N_m(kr)). \qquad (7.200)$$

If the general solution is applied for a boundary value problem in a disk, so that the solution has to be regular in the point $r = 0$, thus $D_n = 0$ for all n. We shall use the obtained solution in Chapter 8 when considering oscillations of a membrane.

The solution of a three-dimensional Helmholtz equation in *cylindrical coordinates*,

$$\frac{1}{r}\frac{\partial}{\partial r}\left(r\frac{\partial U}{\partial r}\right) + \frac{1}{r^2}\frac{\partial^2 U}{\partial \varphi^2} + \frac{\partial^2 U}{\partial z^2} + k^2 U = 0,$$

can be obtained in a similar way using the separation of variables,

$$U(r,z,\varphi) = R(r)Z(z)\Phi(\varphi).$$

For $\Phi(\varphi)$, we obtain the same equation (7.196) and the same set of eigenvalues (7.198) and eigenfunctions (7.199). Also, we obtain equation

$$\frac{Z''}{Z} = \mu \qquad (7.201)$$

for function $Z(z)$ and equation

$$R'' + \frac{1}{r}R' + \left(k^2 + \mu - \frac{m^2}{r^2}\right)R = 0 \qquad (7.202)$$

for $R(r)$. We find that

$$Z(z) = E\exp(\sqrt{\mu}z) + F\exp(-\sqrt{\mu}z).$$

In an infinite space, μ is arbitrary; in the case of a boundary value problem, the set of allowed values of μ is discrete, $\mu = \mu_n$, $n = 0,1,2,\ldots$ (see Section 7.10). In that case, we get

$$R_{mn} = C_{mn}J_m(r\sqrt{\mu_n + k^2}) + D_{mn}N_m(r\sqrt{\mu_n + k^2})$$

and

$$U(r,z,\varphi) = \sum_{n=0}^{\infty}\sum_{m=0}^{\infty}(A_m \cos m\varphi + B_m \sin m\varphi)$$
$$\times \left[C_{mn}J_m\left(r\sqrt{\mu_n + k^2}\right) + D_{mn}N_m\left(r\sqrt{\mu_n + k^2}\right)\right]$$
$$\times [E_n \exp\left(\sqrt{\mu_n}z\right) + F_n \exp\left(-\sqrt{\mu_n}z\right)].$$

Example 7.16 Let us consider the Helmholtz equation in *spherical coordinates*,

$$\frac{1}{r}\frac{\partial}{\partial r}\left(r\frac{\partial U}{\partial r}\right) + \frac{1}{r^2\sin\theta}\frac{\partial}{\partial \theta}\left(\sin\theta\frac{\partial U}{\partial \theta}\right) + \frac{1}{r^2}\frac{\partial^2 U}{\partial \varphi^2} + k^2 U = 0.$$

Similar to the case of the Laplace equation (see Section 7.11), the general solution can be obtained by separation of variables:

$$U(r, \theta, \varphi) = R(r)\Theta(\theta)\Phi(\varphi).$$

Like in the case of the Laplace equation, we find that functions $\Phi_m(\varphi)$ are linear combinations of $\cos m\varphi$ and $\sin m\varphi$, and functions

$$\Theta_{ml}(\theta) = P_l^m(\cos\theta).$$

The functions $\Phi_m(\varphi)$ and $\Theta_{ml}(\theta)$ can be combined into *spherical harmonics*

$$Y_l^m(\theta, \varphi) = N_{ml}P_l^m(\cos\theta)e^{im\varphi}; \quad l = 0, 1, 2, \ldots; \quad m = -l, -l+1, \ldots, l,$$

where the coefficients N_{ml} are determined by the normalization condition (see Appendix B).

$$\int_0^{2\pi} d\varphi \int_0^\pi d\theta \sin\theta Y_{l_1}^{m_1}(\theta, \varphi)^* Y_{l_2}^{m_2}(\theta, \varphi) = \delta_{l_1 l_2}\delta_{m_1 m_2}.$$

The parameter k^2 appears only in the equation for the radial function $R(r)$,

$$r^2 R'' + 2rR' + \left[k^2 r^2 - l(l+1)\right]R = 0. \tag{7.203}$$

Changing the variable,

$$R(r) = \frac{Z(r)}{(kr)^{1/2}},$$

we obtain the Bessel equation,

$$r^2 Z'' + 2rZ' + \left[k^2 r^2 - \left(l + \frac{1}{2}\right)^2\right]Z = 0. \tag{7.204}$$

Its solutions are $J_{l+1/2}(kr)$ and $N_{l+1/2}(kr)$.

Using *spherical Bessel and Neumann functions* (see Appendix B),

$$j_l(x) = \sqrt{\frac{\pi}{2x}}J_{l+1/2}(x), \quad n_l(x) = \sqrt{\frac{\pi}{2x}}N_{l+1/2}(x) \tag{7.205}$$

the general solution of the Helmholtz equation is

$$U(r, \theta, \varphi) = \sum_{l=0}^{\infty} \sum_{m=-l}^{l} \left[A_{lm}j_l(kr) + B_{lm}n_l(kr)\right] Y_l^m(\theta, \varphi). \tag{7.206}$$

A solution is regular in the point $r = 0$ if all $B_{lm} = 0$.

Note that the wave velocity a can depend on the coordinate $\mathbf{x}$, when the wave propagates in a heterogeneous medium. For instance, the velocity of light in a medium depends on the local refraction index. In that case, the Helmholtz equation for a monochromatic wave is

$$\nabla^2 U + k^2(\mathbf{x})U = 0. \tag{7.207}$$

7.15.2   Schrödinger Equation

Let us consider a three-dimensional Helmholtz equation

$$u_{xx} + u_{yy} + u_{zz} + k^2 u = 0,$$

which describes the propagation of a monochromatic electromagnetic wave. Assume that the wave has a form of a beam which propagates mostly in the direction of the z-axis. Substituting

$$u(x, y, z) = U(x, y, z)e^{ikz}, \tag{7.208}$$

we obtain the *reduced wave equation* for the *envelope function* $U(x, y, z)$,

$$U_{xx} + U_{yy} + U_{zz} + 2ikU_z = 0. \tag{7.209}$$

The physical field is the real part of (7.208). If the characteristic spatial scale of the beam is large compared to the wavelength $\lambda = 2\pi/k$, then $|U_{zz}| \ll |kU_z|$, hence the term U_{zz} can be neglected. We arrive at the *paraxial wave equation*,

$$U_{xx} + U_{yy} + 2ikU_z = 0. \tag{7.210}$$

Equation (7.210) can be also called the *Schrödinger equation*, because it is equivalent, up to rescaling of variables, to the equation governing the propagation of a free quantum particle,

$$i\hbar\frac{\partial \Psi}{\partial t} = -\frac{\hbar^2}{2m}\nabla^2\Psi. \tag{7.211}$$

Here Ψ is the wave function of the particle, $\hbar = h/2\pi$ is the reduced Planck constant, and m is the mass of the particle.

Note that in the case of the light beam propagation, the longitudinal coordinate z plays the role of time.

Example 7.17 As an example of the application of the paraxial wave/Schrödinger equation, let us consider the diffraction of a *Gaussian beam*.

Let us rewrite Equation (7.210) using cylindrical coordinates,

$$U_{rr} + \frac{1}{r}U_r + \frac{1}{r^2}U_{\psi\psi} + 2ikU_z = 0$$

and consider an axially symmetric beam ($U = U(r, z)$) governed by equation

$$U_{rr} + \frac{1}{r}U_r + 2ikU_z = 0. \tag{7.212}$$

Let us search for the particular solution of this equation in the form

$$U(r, z) = A(z)\exp\left[\frac{ikr^2}{2q(z)}\right], \tag{7.213}$$

where $A(z)$ and $q(z)$ are complex functions that depend only on z. Substituting the ansatz (7.213) into (7.212), we find:

$$2ik\left(\frac{A}{q} + \frac{dA}{dz}\right) + \frac{k^2 r^2 A}{q^2}\left(\frac{dq}{dz} - 1\right) = 0. \tag{7.214}$$

Relation (7.214) is satisfied for any r, if

$$\frac{dq}{dz} = 1, \quad \frac{dA}{dz} = -\frac{A}{q}. \tag{7.215}$$

Let us impose the initial condition

$$U(r, 0) = C \exp\left[-\left(\frac{r}{r_0}\right)^2\right],$$

i.e., at $z = 0$, the beam is characterized by a Gaussian distribution with the characteristic radius r_0. Then

$$A(0) = C, \quad q(0) = -ikr_0^2/2.$$

Solving (7.215) in the direction of the beam propagation, $z > 0$, we find that

$$q(z) = -ikr_0^2/2 + z, \quad A(z) = \frac{C}{1 + 2iz/kr_0^2},$$

i.e.,

$$U(r, z) = \frac{C}{1 + 2iz/kr_0^2} \exp\left[-\frac{r^2}{r_0^2 (1 + 2iz/kr_0^2)}\right].$$

Let us calculate the wave intensity

$$I(r, z) = |U(r, z)|^2 = \frac{C^2}{1 + 4z^2/k^2 r_0^4} \exp\left[-\frac{2r^2}{1 + 4z^2/k^2 r_0^4}\right].$$

One can see that the characteristic radius of the beam $\bar{r}(z)$ grows with z as

$$\bar{r}(z) = r_0 \sqrt{1 + 4z^2/k^2 r_0^4},$$

while the maximum intensity of the beam $I(0, z)$ decreases as

$$I(0, z) = \frac{C^2 r_0^2}{[\bar{r}(z)]^2}.$$

We have seen that the Schrödinger equation is obtained from the Helmholtz equation in the paraxial approximation. Vice versa, we obtain the Helmholtz equation from the time-dependent Schrödinger equation (7.211) when considering a particle with a definite energy E,

$$\Psi(\mathbf{x}, t) = U(\mathbf{x}) \exp\left(-i\frac{E}{\hbar} t\right). \tag{7.216}$$

Indeed, substituting (7.216) into (7.211), we obtain Equation (7.194) with

$$k^2 = 2mE/\hbar^2.$$

In quantum mechanics, equation

$$\nabla^2 U + k^2 U = 0$$

is called the time-independent Schrödinger equation. Thus, Example 16 of this section describes the motion of a free particle. The motion of a quantum particle with energy E in a certain external potential $V(\mathbf{x})$ is governed by Equation (7.207) with

$$k^2(\mathbf{x}) = 2m(E - V(\mathbf{x}))/\hbar^2.$$

If the potential is spherically symmetric, $V = V(r)$, equation

$$\nabla^2 U + k^2(r) U = 0$$

can be solved by separation of variables. Acting as in Example 16, we obtain the same functions of angular variables $Y_l^m(\theta, \varphi)$. Only the equation for the radial function $R(r)$ is modified,

$$r^2 R'' + 2r R' + \left[k^2(r) r^2 - l(l + 1)\right] R = 0. \tag{7.217}$$

Solutions of Equation (7.217) for typical potentials can be found in textbooks in quantum mechanics. For the one-electron atom the solution can be also found in books [7, 8].

Problems

Solve these problems analytically which means the following: formulate the equation and boundary conditions, obtain the eigenvalues and eigenfunctions, write the formulas for coefficients of the series expansion and the expression for the solution of the problem. You can obtain the pictures of several eigenfunctions and screenshots of the solution and of the auxiliary functions with Maple, Mathematica or software from [7, 8].

In problems 1 through 5 we consider rectangular plates $(0 \le x \le l_x, 0 \le y \le l_y)$ which are thermally insulated over their lateral surfaces. There are no heat sources or absorbers inside the plates. Find the steady-state temperature distribution in the plates.

1. The sides $x = 0$, $y = 0$ and $y = l_y$ have a fixed temperature of zero and the side $x = l_x$ follows the temperature distribution $u(l_x, y) = \sin^2(\pi y/l_y)$.

2. The sides $x = 0$, $x = l_x$ and $y = 0$ have a fixed temperature of zero and the side $y = l_y$ follows the temperature distribution $u(x, l_y) = \sin^2(\pi x/l_x)$.

3. The sides $x = 0$ and $y = 0$ have a fixed temperature of zero, the side $x = l_x$ is thermally insulated, and the side $y = l_y$ follows the temperature distribution

$$u(x, l_y) = \sin(5\pi x/l_x).$$

4. The sides $x = 0$ and $x = l_x$ have a fixed temperature of zero, and the sides $y = 0$ and $y = l_y$ follow the temperature distributions

$$u(x, 0) = \sin\frac{\pi x}{l_x} \quad \text{and} \quad u(x, l_y) = \sin\frac{3\pi x}{l_x}.$$

5. The sides $y = 0$ and $y = l_y$ have a fixed temperature of zero, and the constant heat flows

$$u_x(0, y) = u_x(l_x, y) = \sin(3\pi y/l_y)$$

are supplied to the sides $x = 0$ and $x = l_x$ of the plate from outside.

In problems 6 through 10 we consider a rectangular plate $(0 \le x \le l_x, 0 \le y \le l_y)$ which is thermally insulated over its lateral surfaces. One internal source of heat $Q = $ const. acts at the point (x_0, y_0) of the plate. Find the steady-state temperature distribution in the plate.

6. The edges $x = 0$, $y = 0$ and $y = l_y$ of the plate are kept at zero temperature and the edge $x = l_x$ is subjected to convective heat transfer with the environment which has a temperature of zero.

7. The edges $x = 0$ and $y = 0$ of the plate are kept at zero temperature, the edge $y = l_y$ is thermally insulated and the edge $x = l_x$ is subjected to convective heat transfer with the environment which has a temperature of zero.

8. The edges $x = l_x$ and $y = l_y$ of the plate are kept at zero temperature, the edge $x = 0$ is thermally insulated and the edge $y = 0$ is subjected to convective heat transfer with the environment which has a temperature of zero.

9. The edges $y = 0$ and $y = l_y$ are thermally insulated, the edge $x = 0$ is kept at zero temperature and the edge $x = l_x$ is subjected to convective heat transfer with the environment which has a temperature of zero.

10. The edges $x = 0$, $x = l_x$ and $y = l_y$ are thermally insulated and the edge $y = 0$ is subjected to convective heat transfer with the environment which has a temperature of zero.

In problems 11 through 15 we consider a heat-conducting rectangular plate ($0 \leq x \leq l_x$, $0 \leq y \leq l_y$) thermally insulated over its lateral surfaces. Let heat be generated throughout the plate; the intensity of internal sources (per unit mass of the plate) is $Q(x, y)$. Find the steady-state temperature distribution in the plate.

11. Part of the plate bound ($x = 0$ and $x = l_x$) is thermally insulated, and the other part is subjected to convective heat transfer with a medium. The temperature of the medium is $u_{md} = \text{const}$. The intensity of internal sources (per unit mass of the plate) is
$$Q(x, y) = A \cos \frac{\pi x}{l_x} \cos \frac{\pi y}{l_y}.$$

12. Part of the plate bound ($y = 0$ and $y = l_y$) is thermally insulated, and the other part is subjected to convective heat transfer with a medium. The temperature of the medium is $u_{md} = \text{const}$. The intensity of internal sources (per unit mass of the plate) is
$$Q(x, y) = A \cos \frac{\pi x}{l_x} \cos \frac{\pi y}{l_y}.$$

13. Sides $x = 0$ and $x = l_x$ of the plate are thermally insulated, and sides $y = 0$ and $y = l_y$ are held at fixed temperatures $u(x, 0) = 0$ and $u(x, l_y) = \cos(5\pi x/l_x)$. The intensity of internal sources (per unit mass of the plate) is
$$Q(x, y) = Ax \sin \frac{\pi y}{l_y}.$$

14. Sides $y = 0$ and $y = l_y$ of the plate are thermally insulated , and sides $x = 0$ and $x = l_x$ are held at fixed temperatures $u(0, y) = 0$ and $u(l_x, y) = \cos(3\pi y/l_y)$.
$$Q(x, y) = Ax \cos \frac{\pi y}{l_y}.$$

15. Sides $y = 0$ and $y = l_y$ of the plate are thermally insulated, side $x = 0$ is held at fixed temperature $u = u_1$ and side $x = l_x$ is subjected to convective heat transfer with a medium. The temperature of the medium is zero. The intensity of internal sources (per unit mass of the plate) is
$$Q(x, y) = Axy.$$

In problems 16 through 20 an infinitely long rectangular cylinder has its central axis along the z-axis and its cross-section is a rectangle with sides of length π. The sides of the cylinder are kept at an electric potential described by functions $u(x, y)|_\Gamma$ given below. Find the electric potential within the cylinder.

16. $u|_{x=0} = u|_{x=\pi} = y^2$,　$u|_{y=0} = x$,　$u|_{y=\pi} = 0$.

17. $u|_{x=0} = y$,　$u|_{x=\pi} = y^2$,　$u|_{y=0} = u|_{y=\pi} = 0$.

18. $u|_{x=0} = 0$,　$u|_{x=\pi} = y^2$,　$u|_{y=0} = 0$,　$u|_{y=\pi} = \cos x$.

19. $u|_{x=0} = u|_{x=\pi} = \cos 2y$,　$u|_{y=0} = u|_{y=\pi} = 0$.

20. $u|_{x=0} = \cos 3y, \quad u|_{x=\pi} = 0, \quad u|_{y=0} = x^2, \quad u|_{y=\pi} = 0.$

In problems 21 through 23 we consider a circular plate of radius l which is thermally insulated over its lateral surfaces. The circular periphery of the plate is kept at the temperature described by functions of the polar angle $u(l, \varphi) = g(\varphi)$, given below. Find the steady-state temperature distribution in the plate.

21. $g(\varphi) = \cos 3\varphi.$

22. $g(\varphi) = \cos \frac{\varphi}{2} + \frac{\varphi}{\pi}.$

23. $g(\varphi) = \cos \frac{\varphi}{2} + \sin \frac{\varphi}{2}.$

In problems 24 and 26 a thin homogeneous circular plate of radius l is electrically insulated over its lateral surfaces. The boundary of the plate is kept at an electric potential described by functions of polar angle $u(l, \varphi) = g(\varphi)$ given below. Find an electric potential in the plate.

24. $g(\varphi) = \sin 4\varphi.$

25. $g(\varphi) = \sin \frac{\varphi}{2} + \frac{\pi}{2}.$

26. $g(\varphi) = 2 \cos \varphi - 3 \sin \varphi.$

In problems 27 through 29 we consider a very long (infinite) cylinder of radius l. The constant heat flow $Q(\varphi)$ is supplied to the surface of the cylinder from outside. Find the steady-state temperature distribution in the cylinder.

27. $Q(\varphi) = 3 \sin \varphi + 2 \sin^3 \varphi.$

28. $Q(\varphi) = 4 \cos^3 \varphi + 2 \sin \varphi.$

29. $Q(\varphi) = 5 \sin \varphi - \cos \varphi.$

In problems 30 through 32 we consider a very long (infinite) cylinder of radius l. At the surface of the cylinder there is a heat exchange with the medium. The temperature of the medium is $u_{md}(\varphi)$. Find the steady-state temperature distribution in the cylinder.

30. $u_{md}(\varphi) = \sin \varphi + \cos 4\varphi.$

31. $u_{md}(\varphi) = 1 + 4 \cos^2 \varphi.$

32. $u_{md}(\varphi) = 2 \sin^2 \varphi + 1.$

In problems 33 through 35 we consider a circular plate of radius l which is thermally insulated over its lateral surfaces. One constant internal source of heat acts at the point (r_0, φ_0) of the plate. The value of this source is $Q = $ const. Find the steady-state temperature distribution in the plate.

33. The edge of the plate is kept at zero temperature.

34. The edge of the plate is subjected to convective heat transfer with the environment which has a temperature of zero.

35. The edge of the plate is subjected to convective heat transfer with the environment which has a temperature of $u_{md} = $ const.

In problems 36 through 38 we consider a circular plate of radius l which is thermally insulated over its lateral surfaces. The contour of the plate is maintained at zero temperature. A uniformly distributed source of heat with power $Q(r, \varphi)$ is acting in the plate. Find the steady-state temperature distribution in the plate.

36. $Q(r, \varphi) = r \cos 2\varphi$.

37. $Q(r, \varphi) = r^2 \sin \varphi$.

38. $Q(r, \varphi) = r^2 (\cos 3\varphi + \sin 3\varphi)$.

39. Find homogeneous harmonic polynomials of degree 3 and 4.

40. Solve the following Dirichlet problem:

$$\nabla^2 u = 0, \ \ 0 < r < a, \ \ 0 < \theta < \alpha, \ \ 0 \leq \varphi \leq 2\pi; \ \ 0 < \alpha < \pi;$$
$$u(a, \theta, \phi) = 1, \ \ 0 \leq \theta \leq \alpha, \ \ 0 \leq \varphi \leq 2\pi;$$
$$u(r, \alpha, \phi) = 0, \ \ 0 \leq r \leq a, \ \ 0 \leq \varphi \leq 2\pi.$$

41. Solve the following Dirichlet problem:

$$\nabla^2 u + k^2 u = 0, \ \ 0 < r < a, \ \ 0 < \theta < \pi, \ \ 0 \leq \varphi \leq 2\pi;$$
$$u = \cos^2 \theta, \ \ r = a, \ \ 0 < \theta < \pi, \ \ 0 \leq \varphi \leq 2\pi.$$

Hint: Use the transformation of the kind $f(x) = x^\alpha v(x)$ to transform the obtained ODE to the Bessel equation.

42. Derive Green's function of the Dirichlet problem for the Poisson equation in the region $0 < r < 1, 0 < \phi < \pi$.

8

Two-Dimensional Hyperbolic Equations

In this chapter we consider physical problems related to two-dimensional flexible surfaces called membranes. A membrane may be defined as a thin film which bends but, in the present analysis, does not stretch. The boundary of the membrane may be fixed or free or have forces applied to it. We will also consider cases where the membrane interacts with the material in which it is embedded and is thus subject to external forces such as driving forces or friction. Examples of membranes include drum heads, flags, trampolines, biological barriers such as cellular membranes. The surface of liquids may be treated as membranes when covered by surfactant that makes it rigid.

Our consideration of the membrane behavior will parallel our previous discussion of a vibrating string, but now we analyze the motion of a two-dimensional object oscillating in a third direction. First let us consider a membrane in equilibrium in the x-y plane limited by a smooth, closed boundary, l, under tension, T, which acts tangent to the surface of the membrane. In the following we will treat external forces acting on the membrane in a direction perpendicular to the x-y plane only, except at the boundary of the membrane. Under the action of such a force or in the case of an initial perturbation from equilibrium, points on the membrane move to a new position which we will describe by the distance from equilibrium, $u = u(x, y)$ at location (x, y). The distance of the membrane surface from equilibrium may also vary in time so that the displacement $u(x, y, t)$ is a function of time as well as location.

We consider only cases where the curvature of the membrane is small; hence we can neglect powers of u and derivatives (squared and higher orders): $u^2 \approx 0$, $u_x^2 \approx 0$, etc.

In Figure 8.1 a small section, σ, of the membrane whose equilibrium position is limited by the closed curve l is shown. When the membrane is displaced from the equilibrium position this section is deformed to the area σ', limited by the closed curve l' as shown in Figure 8.1. The new area σ' at some instant of time is given by

$$\sigma' = \iint_\sigma \sqrt{1 + u_x^2 + u_y^2} dx dy \approx \iint_\sigma dx dy = \sigma.$$

From this result we see that, for small oscillations with low curvature, we may neglect changes of area of the membrane. As in the case of small string vibrations, we assume that the tension in the membrane does not vary with x or y.

8.1 Derivation of the Equations of Motion

To derive an equation of motion for the membrane let us consider its fragment, the deformed area σ' limited by the curve l'. The tension acting on this area is evenly distributed on the contour, l' and is perpendicular to the contour and tangent to the surface of the deformed area. For a segment ds' of the curve l' the tension acting on the segment will be $T ds'$ (T is tension per unit length). Since motions of the membrane are constrained to be perpendicular

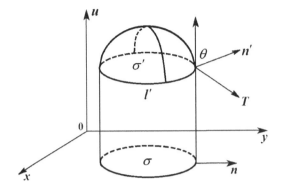

FIGURE 8.1
Small surface element of a membrane, σ displaced from equilibrium into stretched element σ'. The angle θ is between the force of tension, T, which is tangent to the curved surface element, and the direction of the displacement, u. The vectors n and n' are the normal vectors to the surfaces σ and σ', respectively.

to the x-y plane we consider the component of the tension in the direction u (perpendicular to the x-y plane) which is $T ds' \cos\theta$ where θ is the angle between T and the direction of the displacement u. For small oscillations of the membrane, $\cos\theta$ is approximately equal to $\frac{\partial u}{\partial n}$ where n is the normal perpendicular to the curve l, the boundary of the original equilibrium area σ. From this we have that the component of tension acting on element ds' of contour l' in the direction of displacement u is

$$T \frac{\partial u}{\partial n} ds'.$$

We now integrate over the contour l' to find the component of tension acting on area element σ' and perpendicular to the equilibrium surface as

$$T \int_{l'} \frac{\partial u}{\partial n} ds'.$$

For small oscillations of the membrane $ds \approx ds'$ (i.e. the boundary l does not deform much as the element σ is stretched). Using Green's formula we have, in rectangular coordinates,

$$T \int_{l} \frac{\partial u}{\partial n} ds = \iint_{\sigma} T \left(\frac{\partial^2 u}{\partial x^2} + \frac{\partial^2 u}{\partial y^2} \right) dxdy. \tag{8.1}$$

The above only includes forces due to the original tension on the membrane. If an additional external force per unit area $F(x, y, t)$ (which may vary in time) acts parallel to the direction $u(x, y, t)$, then the component in the u direction of this force acting on area σ' of the membrane is given by

$$\iint_{\sigma} F(x, y, t) dxdy. \tag{8.2}$$

The two forces in Equations (8.1) and (8.2) cause an acceleration of the area element σ'. If the mass per area of the membrane is given by the surface density, $\rho(x, y)$, the right-hand side of Newton's second law for the motion of this area element becomes

$$\iint_{\sigma} \rho(x, y) \frac{\partial^2 u}{\partial t^2} dxdy.$$

Setting the forces acting on this element equal to the mass times acceleration of the area element we have

$$\iint_{\sigma} \left[\rho(x,y)\frac{\partial^2 u}{\partial t^2} - T\left(\frac{\partial^2 u}{\partial x^2} + \frac{\partial^2 u}{\partial y^2}\right) + F(x,y,t) \right] dx dy = 0.$$

We began with an arbitrary surface element, σ, from which it follows that

$$\rho(x,y)\frac{\partial^2 u}{\partial t^2} - T\left(\frac{\partial^2 u}{\partial x^2} + \frac{\partial^2 u}{\partial y^2}\right) = F(x,y,t). \tag{8.3}$$

Equation (8.3) is the linear partial differential equation which describes *small, transverse, forced oscillations of a membrane*.

In the case of a membrane of uniform mass density ($\rho = $ const) we may write this equation as

$$\frac{\partial^2 u}{\partial t^2} - a^2\left(\frac{\partial^2 u}{\partial x^2} + \frac{\partial^2 u}{\partial y^2}\right) = f(x,y,t), \tag{8.4}$$

where $a = \sqrt{T/\rho}$, $f(x,y,t) = F(x,y,t)/\rho$. In cases where the external force is absent, i.e. $F(x,y,t) = 0$, then from Equation (8.4) we obtain the *homogeneous* equation for *free oscillations of a uniform membrane* given by

$$\frac{\partial^2 u}{\partial t^2} = a^2\left(\frac{\partial^2 u}{\partial x^2} + \frac{\partial^2 u}{\partial y^2}\right). \tag{8.5}$$

If, in addition to the internal tension, the membrane is subject to an external restoring force proportional to displacement, we may add a force $F = -\alpha u$ per unit of area of the membrane, where α is the elasticity coefficient of the ambient material. For such a membrane embedded in an elastic or spongy environment, Equation (8.4) becomes

$$\frac{\partial^2 u}{\partial t^2} - a^2\left(\frac{\partial^2 u}{\partial x^2} + \frac{\partial^2 u}{\partial y^2}\right) + \gamma u = f(x,y,t), \tag{8.6}$$

where $\gamma = \alpha/\rho$.

If the membrane is embedded in a material which produces a drag on the motion of the membrane such as the case for biological membranes, which are normally immersed in a liquid environment, a friction term must be added to Equation (8.4). Friction forces are generally proportional to velocity and we have $F = -ku_t$ as the force per unit area of membrane where k is the coefficient of friction. The equation of oscillation in this case includes the time derivative of displacement, $u_t(x,y,t)$ and we have

$$\frac{\partial^2 u}{\partial t^2} + 2\kappa\frac{\partial u}{\partial t} - a^2\left(\frac{\partial^2 u}{\partial x^2} + \frac{\partial^2 u}{\partial y^2}\right) = f(x,y,t), \tag{8.7}$$

where $2\kappa = k/\rho$.

All the equations from (8.4) through (8.7) are linear partial differential equations of hyperbolic type. In the following we solve the above equations for various cases and give examples. First, we consider the physical limitations presented by requirements at the boundaries of the membrane.

8.1.1   Boundary and Initial Conditions

The equations of motion (8.4), (8.5), (8.6) and (8.7) are not by themselves sufficient to entirely specify the motion of a membrane. Additional conditions need be specified: initial conditions and boundary conditions.

If the position and velocity of points on the membrane are known at some initial time, $t = 0$ and are given by the functions $\varphi(x, y)$ and $\psi(x, y)$, respectively, we have the *initial conditions*

$$u|_{t=0} = \varphi(x, y), \quad \left.\frac{\partial u}{\partial t}\right|_{t=0} = \psi(x, y). \tag{8.8}$$

As in the case of the vibrating string we may be given, along with initial conditions, information about the behavior of the membrane at its edges at all times, t, in which case we have *boundary conditions*. In the following we outline several variants of conditions on the boundary, L, of the membrane.

1. If the edge of the membrane is rigidly fixed then we have as the boundary condition
    $$u|_L = 0,$$
    which is referred to as a *fixed edge* boundary condition.

2. If the behavior over time of the displacement, $u(x, y, t)$, of the boundary is given by some function $g(t)$ then we have
    $$u|_L = g(t),$$
    which is called a *driven edge* boundary condition.

3. In the case of a boundary which is free (for example the edges of a flag under small oscillations) so that the displacement is only in a direction perpendicular to the x-y plane we have *free edge* boundary conditions given by
    $$\left.\frac{\partial u}{\partial n}\right|_L = 0.$$

4. The edge may also be subject to a force with linear density, f_1 in the x-y plane which affects the tension at the boundary. In this case we have the *stretched edge* boundary condition,
    $$\left.\left(-T\frac{\partial u}{\partial n} + f_1\right)\right|_L = 0. \tag{8.9}$$

5. If the force density, f_1, in the stretched edge condition, Equation (8.9), is a spring-like force (for example the boundary of a trampoline fixed to its support with springs) we may write $-ku$ for f_1 and we have
    $$\left.\left(\frac{\partial u}{\partial n} + hu\right)\right|_L = 0, \quad \text{where } h = k/T. \tag{8.10}$$

6. If the edges to which a membrane is elastically attached are moving in some prescribed way, the right sides of Equations (8.9) and (8.10) will contain some function of time, $g(t)$, describing the motion of the edges. In this case we have nonhomogeneous boundary conditions.

We may combine all of these conditions in a generic form given by

$$\left.\alpha\frac{\partial u}{\partial n} + \beta u\right|_L = g(t). \tag{8.11}$$

The Dirichlet boundary condition corresponds to $\alpha = 0$, $\beta = 1$ in which case we have the driven edge situation. The Neumann boundary condition corresponds to $\alpha = 1$, $\beta = 0$

and we have the stretched edge condition, or the free edge if $g(t) = 0$. Mixed boundary conditions, when both $\alpha \neq 0$ and $\beta \neq 0$, correspond to the two last cases; if $g(t) = 0$ we have homogeneous ($g(t) \neq 0$ nonhomogeneous) mixed boundary conditions. Clearly, the types of boundary conditions can vary along the boundary, and we will consider such a situation in the following section.

8.2 Oscillations of a Rectangular Membrane

In this section we consider the Fourier method for a rectangular membrane limited by the straight lines $x = 0$, $x = l_x$, $y = 0$ and $y = l_y$ (Figure 8.2).

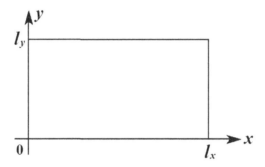

FIGURE 8.2
Rectangular membrane in its equilibrium position.

We begin with the most general case of a membrane subject to friction forces, a restoring force and external forcing, $f(x, y, t)$. From the previous discussion we see that the equation of motion for such a problem is given by

$$\frac{\partial^2 u}{\partial t^2} + 2\kappa \frac{\partial u}{\partial t} - a^2 \left(\frac{\partial^2 u}{\partial x^2} + \frac{\partial^2 u}{\partial y^2} \right) + \gamma u = f(x, y, t) \tag{8.12}$$

with generic boundary conditions given on the boundary of the rectangle as

$$P_1[u] \equiv \alpha_1 u_x + \beta_1 u|_{x=0} = g_1(y, t), \quad P_2[u] \equiv \alpha_2 u_x + \beta_2 u|_{x=l_x} = g_2(y, t),$$

$$P_3[u] \equiv \alpha_3 u_y + \beta_3 u|_{y=0} = g_3(x, t), \quad P_4[u] \equiv \alpha_4 u_y + \beta_4 u|_{y=l_y} = g_4(x, t), \tag{8.13}$$

where $g_1(y, t), \ldots, g_4(x, t)$ are the given functions of time and respective variable, and α_1, β_1, α_2, β_2, α_3, β_3, α_4 and β_4 are constants subject to the same restrictions from physical arguments which we saw in Chapter 4. We also consider initial conditions

$$u|_{t=0} = \varphi(x, y), \quad \left. \frac{\partial u}{\partial t} \right|_{t=0} = \psi(x, y), \tag{8.14}$$

where $\varphi(x, y)$ and $\psi(x, y)$ are given functions. The compatibility of initial and boundary conditions will be discussed in Subsection 8.2.3.

As in the case for the movement of a string we will use the Fourier method and separation of variables to solve this equation. In a manner exactly parallel to the solution of a vibrating string, but instead for an object described initially by two spatial dimensions, we will obtain solutions in the form of a series of eigenfunctions of the corresponding Sturm-Liouville problem.

8.2.1   The Fourier Method for Homogeneous Equations with Homogeneous Boundary Conditions

We start with the *homogeneous equation* (i.e. no external forcing)

$$\frac{\partial^2 u}{\partial t^2} + 2\kappa \frac{\partial u}{\partial t} - a^2 \left(\frac{\partial^2 u}{\partial x^2} + \frac{\partial^2 u}{\partial y^2} \right) + \gamma u = 0 \tag{8.15}$$

with *homogeneous boundary conditions*

$$
\begin{aligned}
P_1[u] &\equiv \alpha_1 u_x + \beta_1 u|_{x=0} = 0, \quad P_2[u] \equiv \alpha_2 u_x + \beta_2 u|_{x=l_x} = 0, \\
P_3[u] &\equiv \alpha_3 u_y + \beta_3 u|_{y=0} = 0, \quad P_4[u] \equiv \alpha_4 u_y + \beta_4 u|_{y=l_y} = 0,
\end{aligned}
\tag{8.16}
$$

and initial conditions in Equation (8.14) given by

$$u|_{t=0} = \varphi(x,y), \quad \left. \frac{\partial u}{\partial t} \right|_{t=0} = \psi(x,y).$$

As we did before, assume that solutions can be written as the product of two functions, one a function of time and the second a function of x and y:

$$u(x,y,t) = V(x,y)T(t). \tag{8.17}$$

Substituting Equation (8.17) into Equation (8.15), we get

$$V(x,y)T''(t) + 2\kappa V(x,y)T'(t) - a^2[V_{xx}(x,y) + V_{yy}(x,y)]T(t) + \gamma V(x,y)T(t) = 0$$

or, upon rearranging terms,

$$\frac{T''(t) + 2\kappa T'(t) + \gamma T(t)}{a^2 T(t)} = \frac{V_{xx}(x,y) + V_{yy}(x,y)}{V(x,y)},$$

where we have used the shorthand notation for the derivatives in x and y and primes denote derivatives with respect to time.

The left-hand side of the previous equality is a function of t only and the right-hand side only of x and y, which is only possible if both sides are equal to some constant value. Denoting this constant as $-\lambda$ we have

$$\frac{T''(t) + 2\kappa T'(t) + \gamma T(t)}{a^2 T(t)} \equiv \frac{V_{xx}(x,y) + V_{yy}(x,y)}{V(x,y)} = -\lambda.$$

For the function $T(t)$ we get the homogeneous linear differential equation of second order

$$T''(t) + 2\kappa T'(t) + (a^2\lambda + \gamma)T(t) = 0. \tag{8.18}$$

For the function $V(x,y)$ we have the equation

$$V_{xx}(x,y) + V_{yy}(x,y) + \lambda V(x,y) = 0 \tag{8.19}$$

with boundary conditions

$$
\begin{aligned}
P_1[V] &\equiv \alpha_1 V_x + \beta_1 V|_{x=0} = 0, \quad P_2[V] \equiv \alpha_2 V_x + \beta_2 V|_{x=l_x} = 0, \\
P_3[V] &\equiv \alpha_3 V_y + \beta_3 V|_{y=0} = 0, \quad P_4[V] \equiv \alpha_4 V_y + \beta_4 V|_{y=l_y} = 0.
\end{aligned}
$$

To solve Equation (8.19) for $V(x, y)$ we again make the assumption that the variables are independent and attempt to separate them using the substitution

$$V(x, y) = X(x)Y(y).$$

From here we obtain two separate BVP:

$$X''(x) + \lambda_x X(x) = 0 \tag{8.20}$$

with boundary conditions

$$\alpha_1 X'(0) + \beta_1 X(0) = 0, \quad \alpha_2 X'(l_x) + \beta_2 X(l_x) = 0,$$

and

$$Y''(y) + \lambda_y Y(y) = 0 \tag{8.21}$$

with boundary conditions

$$\alpha_3 Y'(0) + \beta_3 Y(0) = 0, \quad \alpha_4 Y'(l_y) + \beta_4 Y(l_y) = 0,$$

where λ_x and λ_y are constants from the division of variables linked by the correlation (see Appendix D, part 2) $\lambda_x + \lambda_y = \lambda$.

If λ_{xn} and $X_n(x)$ are eigenvalues and eigenfunctions of Equation (8.20), and λ_{ym} and $Y_m(y)$ are eigenvalues and eigenfunctions of Equation (8.21), then

$$\lambda_{nm} = \lambda_{xn} + \lambda_{ym} \tag{8.22}$$

and

$$V_{nm}(x, y) = X_n(x)Y_m(y) \tag{8.23}$$

are eigenvalues and eigenvectors, respectively, of the problem in Equation (8.19).

The functions $V_{nm}(x, y)$ are orthogonal and the square norms are given by

$$\|V_{nm}\|^2 = \|X_n\|^2 \|Y_m\|^2. \tag{8.24}$$

The system of eigenfunctions, V_{nm}, given in Equation (8.23) form a complete set of basis functions for a two-dimensional rectangular membrane. By this we mean that any smooth (i.e. twice differentiable) shape of the deformed rectangular membrane with the generic boundary conditions given above can be expanded in a converging series of the functions V_{nm}.

We now return to Equation (8.18) describing the time evolution of the membrane. This is an ordinary linear differential equation of 2nd order which we have seen previously for one-dimensional oscillations. It should clear in this case, however, that $T(t)$ now depends on two indexes corresponding to the eigenfunctions $X_n(x)$ and $Y_n(y)$. Specifically we may write $\lambda = \lambda_{nm}$ and denote $T(t)$ as $T_{nm}(t)$ which is a general solution of Equation (8.18):

$$T_{nm}(t) = a_{nm} y_{nm}^{(1)}(t) + b_{nm} y_{nm}^{(2)}(t), \tag{8.25}$$

where a_{nm} and b_{nm} are arbitrary constants. Similar to the case for the one-dimensional problem, we have

$$y_{nm}^{(1)}(t) = \begin{cases} e^{-\kappa t} \cos \omega_{nm} t, & \omega_{nm} = \sqrt{a^2 \lambda_{nm} + \gamma - \kappa^2}, & \kappa^2 < a^2 \lambda_{nm} + \gamma, \\ e^{-\kappa t} \cosh \omega_{nm} t, & \omega_{nm} = \sqrt{\kappa^2 - a^2 \lambda_{nm} - \gamma}, & \kappa^2 > a^2 \lambda_{nm} + \gamma, \\ e^{-\kappa t}, & & \kappa^2 = a^2 \lambda_{nm} + \gamma, \end{cases}$$

$$y_{nm}^{(2)}(t) = \begin{cases} e^{-\kappa t} \sin \omega_{nm} t, & \omega_{nm} = \sqrt{a^2 \lambda_{nm} + \gamma - \kappa^2}, & \kappa^2 < a^2 \lambda_{nm} + \gamma, \\ e^{-\kappa t} \sinh \omega_{nm} t, & \omega_{nm} = \sqrt{\kappa^2 - a^2 \lambda_{nm} - \gamma}, & \kappa^2 > a^2 \lambda_{nm} + \gamma, \\ t e^{-\kappa t}, & & \kappa^2 = a^2 \lambda_{nm} + \gamma. \end{cases} \tag{8.26}$$

Reading Exercise: Following the arguments used in the chapter for the one-dimensional case, verify the above formulas.

Thus, particular solutions for the free oscillations of a rectangular membrane may be written as

$$u_{nm}(x,y,t) = T_{nm}(t)V_{nm}(x,y) = \left[a_{nm}y_{nm}^{(1)}(t) + b_{nm}y_{nm}^{(2)}(t)\right]V_{nm}(x,y), \qquad (8.27)$$

which form a complete set of solutions to Equation (8.15) satisfying boundary conditions in Equation (8.16).

The general solution can be presented as a sum

$$u(x,y,t) = \sum_n \sum_m T_{nm}(t)V_{nm}(x,y). \qquad (8.28)$$

Substituting the initial conditions in the series (8.28) and, as always, assuming the uniform convergence of this series, which allows us to differentiate the series term-by-term, we have:

$$u|_{t=0} = \varphi(x,y) = \sum_n \sum_m a_{nm}V_{nm}(x,y), \qquad (8.29)$$

$$\frac{\partial u}{\partial t}\bigg|_{t=0} = \psi(x,y) = \sum_n \sum_m [\omega_{nm}b_{nm} - \kappa a_{nm}]V_{nm}(x,y) \qquad (8.30)$$

(to treat simultaneously all three cases for κ^2 we replace in (8.30) ω_{nm} by 1 when $\kappa^2 = a^2\lambda_{nm} + \gamma$).

Formulas (8.29) and (8.30) show that functions $\varphi(x,y)$ and $\psi(x,y)$ can be expanded in a complete set of functions, $V_{nm}(x,y)$, which form the solution of the Sturm-Liouville BVP of Equation (8.19).

Again, supposing the series in Equations (8.29) and (8.30) converge uniformly we may determine the coefficients a_{nm} and b_{nm} by using the orthogonality of the eigenfunctions $V_{nm}(x,y)$. Multiplying Equations (8.29) and (8.30) by $V_{nm}(x,y)$ and integrating over x from 0 to l_x and over y from 0 to l_y, yields the Fourier coefficients

$$a_{nm} = \frac{1}{||V_{nm}||^2}\int_0^{l_x}\int_0^{l_y}\varphi(x,y)V_{nm}(xy)dxdy, \qquad (8.31)$$

$$b_{nm} = \frac{1}{\omega_{nm}}\left[\frac{1}{||V_{nm}||^2}\int_0^{l_x}\int_0^{l_y}\psi(x,y)V_{nm}(xy)dxdy + \kappa a_{nm}\right]. \qquad (8.32)$$

The coefficients a_{nm} and b_{nm} substituted into the series (8.28) yield a complete solution to Equation (8.15) with boundary conditions (8.16), and initial conditions (8.14), under the assumption that the series in Equation (8.28) converges uniformly and can be differentiated term by term in x, y and t. Therefore, we may say that Equation (8.28) completely describes the *free oscillations* of a membrane. This solution thus has the form of a Fourier series on the orthogonal system of functions $\{V_{nm}(x,y)\}$, each function of which represents a mode characterized by two numbers, n and m.

Particular solutions $u_{nm}(x,y,t) = T_{nm}(t)V_{nm}(x,y)$ where the time and space components are separate are called *standing wave* solutions and are analogous to standing waves on a one-dimensional string. The profile of the standing wave is defined by the function $V_{nm}(x,y)$ with an amplitude which varies as a function of time $T_{nm}(t)$. Lines, along which $V_{nm}(x,y) = 0$ does not change with time, are called *node lines* of the standing wave. Loose sand placed on a vibrating membrane will collect along node lines because there is no motion at those locations. Locations where $V_{nm}(x, y)$ has a relative maximum or minimum at some instant of time are called antinodes of the standing wave. The general solution, $u(x, y, t)$, is an infinite sum of these standing waves as was the case for the vibrating string. This property of being able to construct arbitrary shapes from a sum of component waves (or modes) is referred to as the *superposition of standing waves* and is a general property of linear systems of all dimensions.

Consider a simple case of a rectangular membrane with sides clamped at the boundary. The vibrations are caused only by initial conditions; thus we want to solve the equation

$$\frac{\partial^2 u}{\partial t^2} = a^2 \left(\frac{\partial^2 u}{\partial x^2} + \frac{\partial^2 u}{\partial y^2} \right)$$

satisfying boundary conditions

$$u(0, y, t) = u(l_x, y, t) = u(x, 0, t) = u(x, l_y, t) = 0,$$

and initial conditions (8.14).

We leave to the reader to check as a *Reading Exercise* that, using the results from the generic case presented above, eigenvalues and eigenfunctions for this problem are

$$\lambda_{xn} = \left(\frac{n\pi}{l_x} \right)^2, \quad X_n(x) = \sin \frac{n\pi x}{l_x}, \quad \|X_n\|^2 = \frac{l_x}{2}, \quad n = 1, 2, 3, \ldots,$$

$$\lambda_{ym} = \left(\frac{m\pi}{l_y} \right)^2, \quad Y_m(y) = \sin \frac{m\pi y}{l_y}, \quad \|Y_m\|^2 = \frac{l_y}{2}, \quad m = 1, 2, 3, \ldots,$$

with

$$\lambda_{nm} = \lambda_{xn} + \lambda_{ym} = \pi^2 \left(\frac{n^2}{l_x^2} + \frac{m^2}{l_y^2} \right),$$

and we have that

$$V_{nm}(x, y) = X_n(x) Y_m(y) = \sin \frac{n\pi x}{l_x} \sin \frac{m\pi y}{l_y} \quad \text{with} \quad \|V_{nm}\|^2 = \frac{l_x l_y}{4}.$$

It is obvious that these functions form a complete set of orthogonal functions for oscillations of the rectangular membrane. The time evolution function can be written as

$$T_{nm}(t) = a_{nm} \cos \omega_{nm} t + b_{nm} \sin \omega_{nm} t,$$

where the frequencies are $\omega_{nm} = a\lambda_{nm} = a\pi \sqrt{\frac{n^2}{l_x^2} + \frac{m^2}{l_y^2}}$. Each pair of integers (n, m) corresponds to a particular characteristic mode (called a normal mode) of vibration of the membrane. An arbitrary membrane deflection may then be represented as a superposition of normal modes:

$$u(x, y, t) = \sum_{n=1}^{\infty} \sum_{m=1}^{\infty} T_{nm} V_{nm} = \sum_{n=1}^{\infty} \sum_{m=1}^{\infty} c_{nm} \sin \frac{n\pi x}{l_x} \sin \frac{m\pi y}{l_y} \cos(\omega_{nm} t + \delta_{nm}),$$

where we have introduced coefficients c_{nm} and phase shifts δ_{nm} via the relations $a_{nm} = c_{nm} \cos \delta_{nm}$, $b_{nm} = -c_{nm} \sin \delta_{nm}$.

If the membrane vibrates in one of its normal modes, then all points on the membrane participate in harmonic motion with frequency ω_{nm}. As an example consider the $(2,1)$ mode (i.e. $n = 2$, $m = 1$). The eigenfunction is

$$V_{21}(x, y) = X_2(x) Y_1(y) = \sin \frac{2\pi x}{l_x} \sin \frac{\pi y}{l_y}.$$

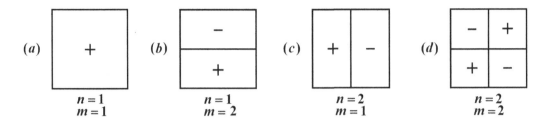

FIGURE 8.3
Modes of vibration $V_{nm}(x, y) = X_n(x)Y_m(y)$. Plus signs indicate motion out of the page, minus signs indicate simultaneous motion into the page.

The only nodal line is the straight line $x = l_x/2$. Similarly, the (1,2) mode ($n = 1$, $m = 2$) has the nodal line $y = l_y/2$ (see Figure 8.3). Nodal lines split the membrane into zones and all points of each zone move with the same phase, i.e. all up or all down (labeled with + and -) at some instant (although not necessarily with the same amplitude).

Generally speaking, each node vibrates with its own frequency, ω_{nm}. However, if l_y/l_x is a rational number, two or more modes could possesses the same frequency. As an example consider a square membrane where $l_x = l_y$, in which case $\omega_{12} = \omega_{21}$. This frequency is said to be two-fold degenerate, by which we mean there are two linearly independent eigenfunctions corresponding to the same eigenvalue.

Below we consider two examples of physical problems for free oscillations of a membrane with homogeneous boundary conditions.

Example 8.1 Find the transverse oscillations of a uniform rectangular membrane ($0 \leq x \leq l_x$, $0 \leq y \leq l_y$) having fixed edges and with an initial displacement of

$$u(x, y, 0) = Axy(l_x - x)(l_y - y),$$

assuming interactions with the surrounding medium can be neglected and the initial velocities of points on the membrane are zero.

Solution. This is an example of the case discussed above with the specified initial conditions; thus the problem reduces to solutions of the equation

$$\frac{\partial^2 u}{\partial t^2} - a^2 \left(\frac{\partial^2 u}{\partial x^2} + \frac{\partial^2 u}{\partial y^2} \right) = 0,$$

with initial and boundary conditions

$$u(x, y, 0) = Axy(l_x - x)(l_y - y), \quad \frac{\partial u}{\partial t}(x, y, 0) = 0,$$

$$u(0, y, t) = u(l_x, y, t) = u(x, 0, t) = u(x, l_y, t) = 0.$$

The general solution to this problem can be presented as a sum (8.28)

$$u(x, y, t) = \sum_{n=1}^{\infty} \sum_{m=1}^{\infty} T_{nm} V_{nm} = \sum_{n=1}^{\infty} \sum_{m=1}^{\infty} \left[a_{nm} y_{nm}^{(1)}(t) + b_{nm} y_{nm}^{(2)}(t) \right] V_{nm}(x, y).$$

As we obtained above, the eigenfunctions $V_{nm}(x, y)$ are

$$V_{nm}(x, y) = X_n(x)Y_m(y) = \sin \frac{n\pi x}{l_x} \sin \frac{m\pi y}{l_y}, \quad \|V_{nm}\|^2 = \frac{l_x l_y}{4}.$$

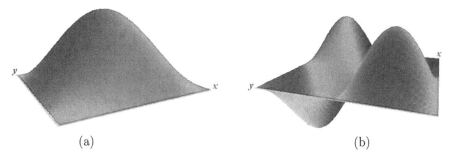

FIGURE 8.4
Eigenfunctions (a) $V_{11}(x, y)$ and (b) $V_{22}(x, y)$ for a membrane with fixed edges.

The three-dimensional view shown in Figure 8.4 depicts two eigenfunctions, $V_{11}(x, y)$ and $V_{22}(x, y)$, chosen as examples, for this problem.

In our case $\varphi(x, y) = Axy(l_x - x)(l_y - y)$ and $\psi(x, y) = 0$, so the expressions for the coefficients of the series are (8.31) and (8.32)

$$a_{nm} = \frac{1}{||V_{nm}||^2} \int_0^{l_x} \int_0^{l_y} \varphi(x, y) V_{nm}(xy) dx dy = \frac{4A}{l_x l_y} \int_0^{l_x} x \, (l_x - x) \sin \frac{n\pi x}{l_x} dx$$

$$\times \int_0^{l_y} y \, (l_y - y) \sin \frac{m\pi y}{l_y} dy = \begin{cases} \dfrac{64A \, l_x^2 l_y^2}{\pi^2 n^2 m^2}, & \text{if } n \text{ and } m - \text{ odd,} \\ 0, & \text{if } n \text{ or } m - \text{ even,} \end{cases}$$

$$b_{nm} = 0.$$

We leave to the reader as a *Reading Exercise* to check, using Equations (8.25) and (8.26), that the time evolution is given by

$$T_{nm} = a_{nm} \cos \omega_{nm} t = \begin{cases} \dfrac{64A \, l_x^2 l_y^2}{\pi^2 n^2 m^2} \cos \omega_{nm} t, & \text{if } n \text{ and } m - \text{ odd,} \\ 0, & \text{if } n \text{ or } m - \text{ even,} \end{cases}$$

where $\omega_{nm} = a\lambda_{nm} = a\pi\sqrt{(n/l_x)^2 + (m/l_y)^2}$.

Consequently, the displacements of the membrane as a function of time for this problem can be expressed by the series

$$u(x, y, t) = \frac{64Al_x^2 l_y^2}{\pi^2} \sum_{n=1}^{\infty} \sum_{m=1}^{\infty} \frac{\cos \omega_{(2n-1)(2m-1)} t}{(2n-1)^2 (2m-1)^2} \sin \frac{(2n-1)\pi x}{l_x} \sin \frac{(2m-1)\pi y}{l_y}.$$

Figure 8.5 shows two snapshots of the solution at the times $t = 2$ and $t = 10$. This solution was obtained for the case $a^2 = 1$, $l_x = 4$, $l_y = 6$, and $A = 0.01$.

Example 8.2 A uniform rectangular membrane $(0 \leq x \leq l_x, 0 \leq y \leq l_y)$ has edges $x = l_x$ and $y = l_y$ which are free and edges at $x = 0$ and $y = 0$ which are firmly fixed. Find the transverse oscillations of the membrane caused by an initial displacement $u(x, y, 0) = Axy$ assuming interactions with the surrounding medium can be neglected and the initial velocities of points on the membrane are zero.

Solution. This problem reduces to finding the solution of the equation

$$\frac{\partial^2 u}{\partial t^2} - a^2 \left(\frac{\partial^2 u}{\partial x^2} + \frac{\partial^2 u}{\partial y^2} \right) = 0,$$

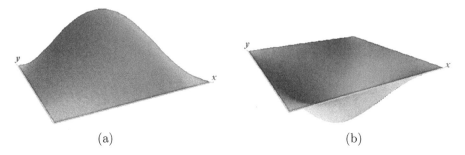

(a)   (b)

FIGURE 8.5
Graph of the membrane in Example 8.1 for different time instants: (a) $t = 7$, (b) $t = 10$.

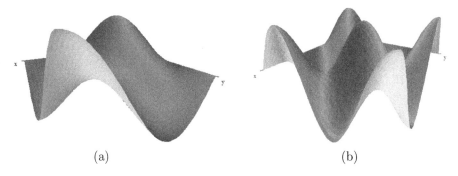

(a)   (b)

FIGURE 8.6
Sample eigenfunctions (a) $V_{22}(x, y)$ and (b) $V_{33}(x, y)$ for Example 8.2.

with initial and boundary conditions

$$u(x, y, 0) = Axy, \quad \frac{\partial u}{\partial t}(x, y, 0) = 0,$$

$$u(0, y, t) = \frac{\partial u}{\partial x}(l_x, y, t) = 0, \quad u(0, y, t) = \frac{\partial u}{\partial y}(x, l_y, t) = 0.$$

The general solution to this problem can be presented as a sum (8.28)

$$u(x, y, t) = \sum_{n=1}^{\infty} \sum_{m=1}^{\infty} T_{nm} V_{nm} = \sum_{n=1}^{\infty} \sum_{m=1}^{\infty} \left[a_{nm} y_{nm}^{(1)}(t) + b_{nm} y_{nm}^{(2)}(t) \right] V_{nm}(x, y).$$

Eigenfunctions $V_{nm}(x, y)$ of the problem are

$$V_{nm}(x, y) = X_n(x)Y_m(y) = \sin \frac{(2n-1)\pi x}{2l_x} \sin \frac{(2m-1)\pi y}{2l_y}, \quad \|V_{nm}\|^2 = \frac{l_x l_y}{4}.$$

The three-dimensional picture shown in Figure 8.6 depicts two eigenfunctions, $V_{11}(x, y)$ and $V_{22}(x, y)$ chosen as examples for this problem.

Using formulas (8.31) and (8.32) we obtain the coefficients

$$a_{nm} = (-1)^{n+m} \frac{64 A l_x l_y}{\pi^4 (2n-1)^2 (2m-1)^2} \quad \text{and} \quad b_{nm} = 0.$$

In this case displacements of the membrane as a function of time are expressed by the series

$$u(x, y, t) = \frac{64 A l_x l_y}{\pi^4} \sum_{n=1}^{\infty} \sum_{m=1}^{\infty} \frac{(-1)^{n+m} \cos \omega_{nm} t}{(2n-1)^2 (2m-1)^2} \sin \frac{(2n-1)\pi x}{2l_x} \sin \frac{(2m-1)\pi y}{2l_y},$$

where

$$\omega_{nm} = a\lambda_{nm} = a\pi\sqrt{\frac{(2n-1)^2}{4l_x^2} + \frac{(2m-1)^2}{4l_y^2}}.$$

8.2.2 The Fourier Method for Nonhomogeneous Equations with Homogeneous Boundary Conditions

Building on the previous sections we now consider the problem of solutions of the *nonhomogeneous* Equation (8.12) for a two-dimensional membrane:

$$\frac{\partial^2 u}{\partial t^2} + 2\kappa\frac{\partial u}{\partial t} - a^2\left(\frac{\partial^2 u}{\partial x^2} + \frac{\partial^2 u}{\partial y^2}\right) + \gamma u = f(x,y,t),$$

where $f(x,y,t)$ is a given function. First, we search for solutions which satisfy the *homogeneous boundary conditions* in Equation (8.16) given by

$$P_1[u] \equiv \alpha_1 u_x + \beta_1 u|_{x=0} = 0, \quad P_2[u] \equiv \alpha_2 u_x + \beta_2 u|_{x=l_x} = 0,$$

$$P_3[u] \equiv \alpha_3 u_y + \beta_3 u|_{y=0} = 0, \quad P_4[u] \equiv \alpha_4 u_y + \beta_4 u|_{y=l_y} = 0,$$

and *nonhomogeneous (non-zero) initial conditions* given in Equation (8.14)

$$u|_{t=0} = \varphi(x,y), \quad \frac{\partial u}{\partial t}\bigg|_{t=0} = \psi(x,y).$$

Because the equation of membrane oscillations is linear, the displacement, $u(x,y,t)$, may be written as the sum

$$u(x,y,t) = u_1(x,y,t) + u_2(x,y,t),$$

where $u_1(x,y,t)$ is the solution of the *homogeneous* equation with *homogeneous* boundaries and *nonhomogeneous* initial conditions:

$$\frac{\partial^2 u_1}{\partial t^2} + 2\kappa\frac{\partial u_1}{\partial t} - a^2\left(\frac{\partial^2 u_1}{\partial x^2} + \frac{\partial^2 u_1}{\partial y^2}\right) + \gamma u_1 = 0, \tag{8.33}$$

$$P_1[u_1]|_{x=0} = 0, \quad P_2[u_1]|_{x=l_x} = 0,$$
$$P_3[u_1]|_{y=0} = 0, \quad P_4[u_1]|_{y=l_y} = 0, \tag{8.34}$$

$$u_1|_{t=0} = \varphi(x,y), \quad \frac{\partial u_1}{\partial t}\bigg|_{t=0} = \psi(x,y). \tag{8.35}$$

The function $u_2(x,y,t)$ is the solution of the *nonhomogeneous* equation with *homogeneous* boundary conditions and initial conditions:

$$\frac{\partial^2 u_2}{\partial t^2} + 2\kappa\frac{\partial u_2}{\partial t} - a^2\left(\frac{\partial^2 u_2}{\partial x^2} + \frac{\partial^2 u_2}{\partial y^2}\right) + \gamma u_2 = f(x,y,t), \tag{8.36}$$

$$P_1[u_2]|_{x=0} = 0, \quad P_2[u_2]|_{x=l_x} = 0, \quad P_3[u_2]|_{y=0} = 0, \quad P_4[u_2]|_{y=l_y} = 0, \tag{8.37}$$

$$u_2|_{t=0} = 0, \quad \frac{\partial u_2}{\partial t}\bigg|_{t=0} = 0. \tag{8.38}$$

In other words the solution $u_1(x,y,t)$ is for *free oscillations,* i.e. such oscillations which occur only as a consequence of an initial perturbation and the solution $u_2(x,y,t)$ is for the case of *forced oscillations,* i.e. such oscillations which occur under the action of an external force $f(x, y, t)$ when initial perturbations are absent.

The problem of free oscillations was considered in the previous section for which case the solution $u_1(x, y, t)$ is known. To proceed we need only to find the solution $u_2(x, y, t)$ for forced oscillations. As in the case for free oscillations we may expand $u_2(x, y, t)$ in the series

$$u_2(x, y, t) = \sum_n \sum_m T_{nm}(t) V_{nm}(x, y), \qquad (8.39)$$

where $V_{nm}(x, y)$ are eigenfunctions of the corresponding homogeneous boundary problem and $T_{nm}(t)$ are, at this stage, unknown functions of t. Any choice of functions $T_{nm}(t)$ satisfies the homogeneous boundary conditions (8.37) for the function $u_2(x, y, t)$ because the functions $V_{nm}(x, y)$ satisfy these conditions.

To find the functions $T_{nm}(t)$ we proceed as follows. Substituting the series (8.39) into Equation (8.36) we have

$$\sum_n \sum_m \left[T''_{nm}(t) + 2\kappa T'_{nm}(t) + (a^2 \lambda_{nm} + \gamma) T_{nm}(t) \right] V_{nm}(x, y) = f(x, y, t). \qquad (8.40)$$

We may also expand the function $f(x, y, t)$ in a Fourier series using the basis functions $V_{nm}(x, y)$ on the rectangle $[0, l_x; 0, l_y]$:

$$f(x, y, t) = \sum_n \sum_m f_{nm}(t) V_{nm}(x, y), \qquad (8.41)$$

where the coefficients of expansion are given by

$$f_{nm}(t) = \frac{1}{||V_{nm}||^2} \int_0^{l_x} \int_0^{l_y} f(x, y, t) V_{nm}(x, y) dx dy. \qquad (8.42)$$

Comparing the expansions in Equations (8.40) and (8.41) for the same function $f(x, y, t)$, obtain a differential equation for the functions $T_{nm}(t)$:

$$T''_{nm}(t) + 2\kappa T'_{nm}(t) + (a^2 \lambda_{nm} + \gamma) T_{nm}(t) = f_{nm}(t). \qquad (8.43)$$

The solution $u_2(x, y, t)$, defined by the series in Equation (8.39) and satisfying initial conditions (8.38) requires that the functions $T_{nm}(t)$ in turn satisfy the conditions

$$T_{nm}(0) = 0, \quad T'_{nm}(0) = 0. \qquad (8.44)$$

The solution of the Cauchy problem defined by Equations (8.43) and (8.44) may be written as

$$T_{nm}(t) = \int_0^t f_{nm}(\tau) Y_{nm}(t - \tau) d\tau, \qquad (8.45)$$

where (Section 5.6)

$$Y_{nm}(t) = \begin{cases} \dfrac{1}{\omega_{nm}} e^{-\kappa t} \sin \omega_{nm} t, & \omega_{nm} = \sqrt{a^2 \lambda_{nm} + \gamma - \kappa^2}, \quad \kappa^2 < a^2 \lambda_{nm} + \gamma, \\[2ex] \dfrac{1}{\omega_{nm}} e^{-\kappa t} \sinh \omega_{nm} t, & \omega_{nm} = \sqrt{\kappa^2 - a^2 \lambda_{nm} - \gamma}, \quad \kappa^2 > a^2 \lambda_{nm} + \gamma, \\[2ex] t e^{-\kappa t}, & \kappa^2 = a^2 \lambda_{nm} + \gamma. \end{cases}$$

Reading Exercise: Verify the above formulas.

We can substitute the expression for $f_{nm}(t)$ given in Equation (8.42) to yield, finally,

$$T_{nm}(t) = \frac{1}{\|V_{nm}\|^2} \int_0^t d\tau \int_0^{l_x} \int_0^{l_y} f(x, y, \tau) V_{nm}(x, y) Y_{nm}(t - \tau) dx dy. \tag{8.46}$$

Substituting the above formulas for $T_{nm}(t)$ into the series (8.39) yields the solution of the boundary value problem defined by Equations (8.36) through (8.38) under the condition that the series (8.39) and the series obtained from Equation (8.39) by term-by-term differentiation (up to second order with respect to x, y, and t) converge uniformly. Thus, the solution of the original problem of forced oscillations is given by

$$u(x, y, t) = u_1(x, y, t) + u_2(x, y, t)$$
$$= \sum_n \sum_m \left\{ T_{nm}(t) + \left[a_{nm} y_{nm}^{(1)}(t) + b_{nm} y_{nm}^{(2)}(t) \right] \right\} V_{nm}(x, y), \tag{8.47}$$

where coefficients $T_{nm}(t)$ are defined by Equation (8.46) and a_{nm}, b_{nm} are defined in the previous section for free oscillations.

We now consider examples of solutions of physical problems involving a nonhomogeneous equation of oscillations with homogeneous boundary conditions.

Example 8.3 Consider transverse oscillations of a rectangular membrane $[0, l_x; 0, l_y]$ with fixed edges, subjected to a transverse driving force

$$F(t) = A \sin \omega t,$$

applied at the point (x_0, y_0), $0 < x_0 < l_x$, $0 < y_0 < l_y$. Assume that the reaction of the surrounding medium can be ignored.

Solution. The problem can be defined as

$$\frac{\partial^2 u}{\partial t^2} - a^2 \left(\frac{\partial^2 u}{\partial x^2} + \frac{\partial^2 u}{\partial y^2} \right) = \frac{A}{\rho} \delta(x - x_0) \delta(y - y_0) \sin \omega t$$

with initial conditions

$$u(x, y, 0) = 0, \quad \frac{\partial u}{\partial t}(x, y, 0) = 0,$$

and Dirichlet homogeneous boundary conditions

$$u(0, y, t) = u(l_x, y, t) = u(x, 0, t) = u(x, l_y, t) = 0.$$

Using the initial conditions, $\varphi(x, y) = \psi(x, y) = 0$, the solution $u(x, y, t)$ is determined by the series

$$u(x, y, t) = \sum_{n=1}^{\infty} \sum_{m=1}^{\infty} T_{nm}(t) \sin \frac{n\pi x}{l_x} \sin \frac{m\pi y}{l_y},$$

where

$$T_{nm}(t) = \frac{1}{\omega_{nm}} \int_0^t f_{nm}(\tau) \sin \omega_{nm}(t - \tau) d\tau,$$

$$f_{nm}(t) = \frac{4A}{\rho l_x l_y} \sin \omega t \sin \frac{n\pi x_0}{l_x} \sin \frac{m\pi y_0}{l_y},$$

and $\omega_{nm} = a\lambda_{nm} = a\pi \sqrt{\frac{n^2}{l_x^2} + \frac{m^2}{l_y^2}}$, $V_{nm}(x, y) = \sin \frac{n\pi x}{l_x} \sin \frac{m\pi y}{l_y}$, $\|V_{nm}\|^2 = \frac{l_x l_y}{4}$.

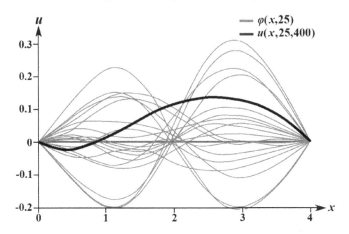

FIGURE 8.7
Solution profile $u(x, 2.5, t)$ for Example 8.3 at a driving frequency other than resonance.

Figure 8.7 shows the solution profile $u(x, 2.5, t)$ for Example 8.3. This solution was obtained for the case $a^2 = 1$, $l_x = 4$, $l_y = 6$, $\rho = 1$, $A = 0.5$, $x_0 = 1.5$, $y_0 = 2.5$, and $\omega = 1.5$ (the frequency of the external force).

If the frequency of the driving force is not equal to any of the natural frequencies of the membrane, i.e. $\omega \neq \omega_{nm}$, $n, m = 1, 2, 3 \ldots$, then

$$T_{nm}(t) = \frac{4A}{\rho l_x l_y (\omega_{nm}^2 - \omega^2)} \sin \frac{n\pi x_0}{l_x} \sin \frac{m\pi y_0}{l_y} \left[\sin \omega t - \frac{\omega}{\omega_{nm}} \sin \omega_{nm} t \right]$$

and

$$u(x, y, t) = \frac{4A}{\rho l_x l_y} \sum_{n=1}^{\infty} \sum_{m=1}^{\infty} \frac{1}{(\omega_{nm}^2 - \omega^2)} \left[\sin \omega t - \frac{\omega}{\omega_{nm}} \sin \omega_{nm} t \right]$$
$$\times \sin \frac{n\pi x_0}{l_x} \sin \frac{m\pi y_0}{l_y} \sin \frac{n\pi x}{l_x} \sin \frac{m\pi y}{l_y}.$$

In the case of *resonance*, where the frequency of the driving force *does* coincide with one of the normal mode frequencies of the membrane, $(\bar{n}, \bar{m})$, i.e. $\omega = \omega_{\bar{n}\bar{m}}$, we have

$$T_{\bar{n}\bar{m}}(t) = \frac{2A}{\rho l_x l_y \omega} \sin \frac{n_0 \pi x_0}{l_x} \sin \frac{m_0 \pi y_0}{l_y} \left[\sin \omega t - \omega t \frac{\omega}{\omega_{\bar{n}\bar{m}}} \cos \omega t \right]$$

and

$$u(x, y, t) = \frac{4A}{\rho l_x l_y} \sum_{n \neq \bar{n}} \sum_{m \neq \bar{m}} \frac{1}{(\omega_{nm}^2 - \omega^2)} \left[\sin \omega t - \frac{\omega}{\omega_{nm}} \sin \omega_{nm} t \right]$$
$$\times \sin \frac{n\pi x_0}{l_x} \sin \frac{m\pi y_0}{l_y} \sin \frac{n\pi x}{l_x} \sin \frac{m\pi y}{l_y}$$
$$+ \frac{2A}{\rho l_x l_y \omega} \left[\sin \omega t - \omega t \frac{\omega}{\omega_{\bar{n}\bar{m}}} \cos \omega t \right] \sin \frac{n_0 \pi x_0}{l_x} \sin \frac{m_0 \pi y_0}{l_y} \sin \frac{n_0 \pi x}{l_x} \sin \frac{m_0 \pi y}{l_y}.$$

Figure 8.8 shows the solution profile $u(x, 2.5, t)$ in the case of resonance where the frequency of the external force $\omega = \omega_{23} = \pi/\sqrt{2}$. Other parameters are the same as in Figure 8.7.

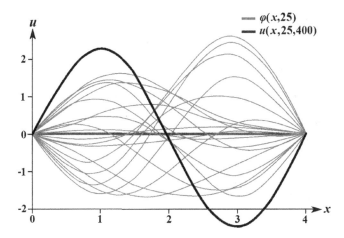

FIGURE 8.8
Solution profile $u(x, 2.5, t)$ for Example 8.3 at resonance.

8.2.3   The Fourier Method for Nonhomogeneous Equations with Nonhomogeneous Boundary Conditions

Consider now the boundary problem of forced oscillations of a membrane given by Equation (8.12) with *nonhomogeneous boundary and initial conditions* given by Equations (8.13) and (8.14):

$$\frac{\partial^2 u}{\partial t^2} + 2\kappa \frac{\partial u}{\partial t} - a^2 \left(\frac{\partial^2 u}{\partial x^2} + \frac{\partial^2 u}{\partial y^2} \right) + \gamma u = f(x, y, z),$$

$$P_1[u] \equiv \alpha_1 u_x + \beta_1 u|_{x=0} = g_1(y, t), \quad P_2[u] \equiv \alpha_2 u_x + \beta_2 u|_{x=l_x} = g_2(y, t),$$

$$P_3[u] \equiv \alpha_3 u_y + \beta_3 u|_{y=0} = g_3(x, t), \quad P_4[u] \equiv \alpha_4 u_y + \beta_4 u|_{y=l_y} = g_4(x, t),$$

$$u|_{t=0} = \varphi(x, y), \quad \left. \frac{\partial u}{\partial t} \right|_{t=0} = \psi(x, y).$$

We will consider situations when the boundary conditions along the edges of membrane are *consistent* at the corners of a membrane (which would be required in a physical occurrence), i.e. the following conforming conditions are valid:

$$P_3[g_1] \equiv \alpha_3 g_{1y} + \beta_3 g_1|_{y=0} = P_1[g_3] \equiv \alpha_1 g_{2y} + \beta_1 g_3|_{x=0},$$

$$P_4[g_1] \equiv \alpha_4 g_{1y} + \beta_4 g_1|_{y=l_y} = P_1[g_4] \equiv \alpha_1 g_{4y} + \beta_1 g_4|_{x=0},$$

$$P_3[g_2] \equiv \alpha_3 g_{2y} + \beta_3 g_2|_{y=0} = P_2[g_3] \equiv \alpha_2 g_{3y} + \beta_2 g_3|_{x=l_x},$$

$$P_4[g_2] \equiv \alpha_4 g_{2y} + \beta_4 g_2|_{y=l_y} = P_2[g_4] \equiv \alpha_2 g_{4y} + \beta_2 g_4|_{x=l_x}.$$

We know that it is not possible to use the Fourier method immediately since the boundary conditions are nonhomogeneous. However, this problem is easily reduced to the problem with zero boundary conditions. To proceed, let us search for solutions of the problem in the form

$$u(x, y, t) = v(x, y, t) + w(x, y, t), \tag{8.48}$$

where $v(x, y, t)$ is an unknown function and the function $w(x, y, t)$ satisfies the given non-homogeneous boundary conditions

$$P_1[w] \equiv \alpha_1 w_x + \beta_1 w|_{x=0} = g_1(y, t), \quad P_2[w] \equiv \alpha_2 w_x + \beta_2 w|_{x=l_x} = g_2(y, t),$$

$$P_3[w] \equiv \alpha_3 w_y + \beta_3 w|_{y=0} = g_3(x, t), \quad P_4[w] \equiv \alpha_4 w_y + \beta_4 w|_{y=l_y} = g_4(x, t) \tag{8.49}$$

and possesses the necessary number of continuous derivatives with respect to x, y and t.

For the function $v(x, y, t)$ we have following boundary value problem (check this result as Reading Exercise):

$$\frac{\partial^2 v}{\partial t^2} + 2\kappa \frac{\partial v}{\partial t} - a^2 \left(\frac{\partial^2 v}{\partial x^2} + \frac{\partial^2 v}{\partial y^2} \right) + \gamma v = \tilde{f}(x, y, t),$$

$$P_1[v]|_{x=0} = 0, \quad P_2[v]|_{x=l_x} = 0, \quad P_3[v]|_{y=0} = 0, \quad P_4[v]|_{y=l_y} = 0,$$

$$v|_{t=0} = \tilde{\varphi}(x, y), \quad \frac{\partial v}{\partial t}\bigg|_{t=0} = \tilde{\psi}(x, y),$$

where

$$\tilde{f}(x, y, t) = f(x, y, t) - \frac{\partial^2 w}{\partial t^2} - 2\kappa \frac{\partial w}{\partial t} + a^2 \left(\frac{\partial^2 w}{\partial x^2} + \frac{\partial^2 w}{\partial y^2} \right) - \gamma w, \tag{8.50}$$

$$\tilde{\varphi}(x, y) = \varphi(x, y) - w(x, y, 0),$$

$$\tilde{\psi}(x, y) = \psi(x, y) - w_t(x, y, 0). \tag{8.51}$$

Solutions of this problem were considered in the previous section.

We seek an auxiliary function $w(x, y, t)$ in a form

$$w(x, y, t) = g_1(y, t)\overline{X} + g_2(y, t)\overline{\overline{X}} + g_3(x, t)\overline{Y} + g_4(x, t)\overline{\overline{Y}}$$

$$+ A(t)\overline{X}\,\overline{Y} + B(t)\overline{X}\,\overline{\overline{Y}} + C(t)\overline{\overline{X}}\,\overline{Y} + D(t)\overline{\overline{X}}\,\overline{\overline{Y}}, \tag{8.52}$$

where $\overline{X}(x)$, $\overline{\overline{X}}(x)$ and $\overline{Y}(y)$, $\overline{\overline{Y}}(y)$ are polynomials of 1st or 2nd order. The coefficients of these polynomials should be adjusted to satisfy the boundary conditions. The detailed description of constructing function $w(x, y, t)$ is presented in Appendix C part 2.

Example 8.4 Consider oscillations of a homogeneous rectangular membrane ($0 \leq x \leq l_x$, $0 \leq y \leq l_y$), if the boundary conditions are given by

$$u(0, y, t) = u(l_x, y, t) = 0 \quad \text{and} \quad u(x, 0, t) = u(x, l_y, t) = h \sin \frac{\pi x}{l_x}$$

and initially the membrane has shape and velocity given, respectively, by

$$\varphi(x, y, 0) = h \sin \frac{\pi x}{l_x} \quad \text{and} \quad \psi(x, y, 0) = v_0 \sin \frac{\pi x}{l_x}.$$

Solution. The problem is described by equation

$$\frac{\partial^2 u}{\partial t^2} - a^2 \left(\frac{\partial^2 u}{\partial x^2} + \frac{\partial^2 u}{\partial y^2} \right) = 0$$

under the conditions

$$u(x, y, 0) = h \sin \frac{\pi x}{l_x}, \quad \frac{\partial u}{\partial t}(x, y, 0) = v_0 \sin \frac{\pi x}{l_x},$$

$$u(0, y, t) = u(l_x, y, t) = 0, \quad u(x, 0, t) = u(x, l_y, t) = h \sin \frac{\pi x}{l_x}.$$

We search for a solution of this problem as the sum

$$u(x, y, t) = v(x, y, t) + w(x, y, t),$$

where the auxiliary function $w(x, y, t)$ chosen to be

$$w(x, y, t) = h \sin \frac{\pi x}{l_x}$$

satisfies the boundary conditions of the problem and therefore obviously provides homogeneous boundary conditions for the function $v(x, y, t)$ (this result can be obtained from general formula (8.52) and the scheme presented in Appendix C part 2).

For the function $v(x, y, t)$ we have the boundary value problem for the nonhomogeneous equation of oscillation where

$$\tilde{f}(x, y, t) = -a^2 \frac{h\pi^2}{l_x^2} \sin \frac{\pi x}{l_x},$$

$$\tilde{\varphi}(x, y) = 0, \quad \tilde{\psi}(x, y) = v_0 \sin \frac{\pi x}{l_x}$$

and homogeneous boundary conditions. The solution to the problem is thus

$$u(x, y, t) = h \sin \frac{\pi x}{l_x} + \sum_{m=1}^{\infty} \left\{ \frac{4v_0}{(2m-1)\pi\omega_{1(2m-1)}} \sin \omega_{1(2m-1)} t \right.$$
$$\left. - \frac{4ha^2\pi}{(2m-1)\, l_x^2 \omega_{1(2m-1)}^2} \left[1 - \cos \omega_{1(2m-1)} t \right] \right\} \sin \frac{\pi x}{l_x} \sin \frac{(2m-1)\pi y}{l_y}.$$

8.3 Small Transverse Oscillations of a Circular Membrane

Suppose a membrane in its equilibrium position has the form of a circle with radius l, is located in the x-y plane and has its center as the origin of coordinates. As before for the case of rectangular membranes, we will consider transverse oscillations for which all points on the membrane move perpendicular to the x-y plane. In polar coordinates, (r, φ), the displacement of points of the membrane will be $u = u(r, \varphi, t)$. The domains of the independent variables are $0 \le r \le l$, $0 \le \varphi < 2\pi$, and $0 \le t < \infty$ respectively.

The Laplace operator in polar coordinates is given by

$$\nabla^2 u = \frac{\partial^2 u}{\partial r^2} + \frac{1}{r} \frac{\partial u}{\partial r} + \frac{1}{r^2} \frac{\partial^2 u}{\partial \varphi^2}$$

with the result that the equation of oscillations of a membrane in polar coordinates has the form

$$\frac{\partial^2 u}{\partial t^2} + 2\kappa \frac{\partial u}{\partial t} - a^2 \left(\frac{\partial^2 u}{\partial r^2} + \frac{1}{r} \frac{\partial u}{\partial r} + \frac{1}{r^2} \frac{\partial^2 u}{\partial \varphi^2} \right) + \gamma u = f(r, \varphi, t). \tag{8.53}$$

Boundary conditions in polar coordinates are particularly simple and in general from can be written as

$$\alpha \frac{\partial u}{\partial r} + \beta u \bigg|_{r=l} = g(\varphi, t), \tag{8.54}$$

where α, and β are constants that are not zero simultaneously, i.e. $|\alpha| + |\beta| \neq 0$. If $\alpha = 0$ we have the Dirichlet boundary condition, and if $\beta = 0$ we have the Neumann boundary condition. If $\alpha \neq 0$ and $\beta \neq 0$ then we have mixed boundary conditions. From physical arguments it will normally be the case that $\beta/\alpha > 0$.

The initial conditions are stated as

$$u|_{t=0} = \phi(r, \varphi), \quad \left.\frac{\partial u}{\partial t}\right|_{t=0} = \psi(r, \varphi). \tag{8.55}$$

Thus the deviation of points of membrane with coordinates (r, φ) at some arbitrary initial moment of time is $\phi(r, \varphi)$ and initial velocities of these points are given by the function $\psi(r, \varphi)$. It should be clear from physical arguments that the solution $u(r, \varphi, t)$ is to be single-valued, periodic in φ with period 2π and remains finite at all points of the membrane, including the center of membrane, $r = 0$.

8.3.1 The Fourier Method for Homogeneous Equations with Homogeneous Boundary Conditions

Here we consider the *homogeneous equation* of oscillations

$$\frac{\partial^2 u}{\partial t^2} + 2\kappa\frac{\partial u}{\partial t} - a^2\left(\frac{\partial^2 u}{\partial r^2} + \frac{1}{r}\frac{\partial u}{\partial r} + \frac{1}{r^2}\frac{\partial^2 u}{\partial \varphi^2}\right) + \gamma u = 0, \tag{8.56}$$

with *homogeneous* boundary conditions

$$\left.\alpha\frac{\partial u}{\partial r} + \beta u\right|_{r=l} = 0. \tag{8.57}$$

Let us represent the function $u(r, \varphi, t)$ as a product of two functions. The first depends only on r and φ and we denote it as $V(r, \varphi)$, the second depends only on t and is denoted as $T(t)$:

$$u(r, \varphi, t) = V(r, \varphi)T(t). \tag{8.58}$$

Substituting Equation (8.58) in Equation (8.56) and separating variables we obtain

$$\frac{T''(t) + 2\kappa T'(t) + \gamma T(t)}{a^2 T(t)} \equiv \frac{1}{V(r, \varphi)}\left[\frac{\partial^2 V}{\partial r^2} + \frac{1}{r}\frac{\partial V}{\partial r} + \frac{1}{r^2}\frac{\partial^2 V}{\partial \varphi^2}\right] = -\lambda,$$

where λ is a separation constant. Thus, the function $T(t)$ satisfies the ordinary linear homogeneous differential equation of second order

$$T''(t) + 2\kappa T'(t) + (a^2\lambda + \gamma)T(t) = 0, \tag{8.59}$$

and the function $V(r, \varphi)$ satisfies the equation

$$\frac{\partial^2 V}{\partial r^2} + \frac{1}{r}\frac{\partial V}{\partial r} + \frac{1}{r^2}\frac{\partial^2 V}{\partial \varphi^2} + \lambda V = 0, \tag{8.60}$$

with

$$\alpha\frac{\partial V(l, \varphi)}{\partial r} + \beta V(l, \varphi) = 0, \tag{8.61}$$

as a boundary condition. Using physical arguments we also require that the solutions remain finite (everywhere, including point $r = 0$) so that

$$|V(r, \varphi)| < \infty \tag{8.62}$$

and require the solutions to be periodic

$$V(r, \varphi) = V(r, \varphi + 2\pi). \qquad (8.63)$$

For the boundary value problem defined by Equations (8.60) through (8.63) we again may separate the variables, in this case r and φ, using the substitution

$$V(r, \varphi) = R(r)\Phi(\varphi). \qquad (8.64)$$

As shown in Appendix D part 1, for the function $R(r)$ we obtain the Bessel equation, and the function $\Phi(\varphi)$ are sines and cosines. The eigenvalues of the problem are

$$\lambda_{nm} = \left(\frac{\mu_m^{(n)}}{l} \right)^2,$$

and eigenfunctions can be expressed in terms of the Bessel functions

$$V_{nm}^{(1)}(r, \varphi) = J_n \left(\frac{\mu_m^{(n)}}{l} r \right) \cos n\varphi \quad \text{and} \quad V_{nm}^{(2)}(r, \varphi) = J_n \left(\frac{\mu_m^{(n)}}{l} r \right) \sin n\varphi. \qquad (8.65)$$

Different types of boundary conditions lead to different eigenvalues $\mu_m^{(n)}$.

These results completely define λ_{nm}, $V_{nm}^{(1)}$ and $V_{nm}^{(2)}$, the eigenvalues and eigenfunctions of equations for the problem of free oscillations of a circular membrane in the case of homogeneous boundary conditions.

To determine the time evolution of the oscillating membrane we return to Equation (8.59). With $\lambda = \lambda_{nm}$ this equation is

$$T''_{nm}(t) + 2\kappa T'_{nm}(t) + (a^2 \lambda_{nm} + \gamma) T_{nm}(t) = 0. \qquad (8.66)$$

This linear second order equation with constant coefficients has two linearly independent solutions

$$y_{nm}^{(1)}(t) = \begin{cases} e^{-\kappa t} \cos \omega_{nm} t, & \kappa^2 < a^2 \lambda_{nm} + \gamma, \\ e^{-\kappa t} \cosh \omega_{nm} t, & \kappa^2 > a^2 \lambda_{nm} + \gamma, \\ e^{-\kappa t}, & \kappa^2 = a^2 \lambda_{nm} + \gamma, \end{cases}$$

$$y_{nm}^{(2)}(t) = \begin{cases} e^{-\kappa t} \sin \omega_{nm} t, & \kappa^2 < a^2 \lambda_{nm} + \gamma, \\ e^{-\kappa t} \sinh \omega_{nm} t, & \kappa^2 > a^2 \lambda_{nm} + \gamma, \\ t e^{-\kappa t}, & \kappa^2 = a^2 \lambda_{nm} + \gamma, \end{cases} \qquad (8.67)$$

with $\omega_{nm} = \sqrt{|a^2 \lambda_{nm} + \gamma - \kappa^2|}$.

A general solution of Equation (8.66) is a linear combination of these $y_{nm}^{(1)}(t)$ and $y_{nm}^{(2)}(t)$. Collecting the functions $\Phi(\varphi)$, $R(r)$ and $T(t)$ and substituting them into identity (8.58) gives particular solutions to Equation (8.56) in the form of a product of functions satisfying the given boundary conditions:

$$u_{nm}^{(1)}(r, \varphi, t) = T_{nm}^{(1)} V_{nm}^{(1)}(r, \varphi) = \left[a_{nm} y_{nm}^{(1)} + b_{nm} y_{nm}^{(2)} \right] V_{nm}^{(1)}(r, \varphi),$$

$$u_{nm}^{(2)}(r, \varphi, t) = T_{nm}^{(2)} V_{nm}^{(2)}(r, \varphi) = \left[c_{nm} y_{nm}^{(1)} + d_{nm} y_{nm}^{(2)} \right] V_{nm}^{(2)}(r, \varphi).$$

To find solutions to the equation of motion for a membrane satisfying not only the boundary conditions above but also initial conditions let us sum these functions as a series, superimposing all $u_{nm}^{(1)}(r, \varphi, t)$ and $u_{nm}^{(2)}(r, \varphi, t)$:

$$u(r, \varphi, t) = \sum_{n=0}^{\infty} \sum_{m=0}^{\infty} \left\{ \left[a_{nm} y_{nm}^{(1)}(t) + b_{nm} y_{nm}^{(2)}(t) \right] V_{nm}^{(1)}(r, \varphi) \right.$$
$$\left. + \left[c_{nm} y_{nm}^{(1)}(t) + d_{nm} y_{nm}^{(2)}(t) \right] V_{nm}^{(2)}(r, \varphi) \right\}. \quad (8.68)$$

If this series and the series obtained from it by twice differentiating term by term with respect to the variables r, φ and t converges uniformly then its sum will be a solution to Equation (8.56), satisfying boundary condition (8.57).

To satisfy the initial conditions given in Equation (8.55) we require that

$$u|_{t=0} = \phi(r, \varphi) = \sum_{n=0}^{\infty} \sum_{m=0}^{\infty} \left[a_{nm} V_{nm}^{(1)}(r, \varphi) + c_{nm} V_{nm}^{(2)}(r, \varphi) \right] \quad (8.69)$$

and

$$\left. \frac{\partial u}{\partial t} \right|_{t=0} = \psi(r, \varphi) = \sum_{n=0}^{\infty} \sum_{m=0}^{\infty} \left\{ [\omega_{mn} b_{mn} - \kappa a_{mn}] V_{nm}^{(1)}(r, \varphi) \right.$$
$$\left. + [\omega_{mn} d_{mn} - \kappa c_{mn}] V_{nm}^{(2)}(r, \varphi) \right\} \quad (8.70)$$

(like in Section 8.2.1, to treat simultaneously all three cases for κ^2 we replace ω_{nm} by 1 in (8.70) for $\kappa^2 = a^2 \lambda_{nm} + \gamma$).

Multiplying these equations by the area element in polar coordinates, $dS = rdrd\varphi$, integrating and taking into account the orthogonality of the eigenfunctions, we obtain the coefficients a_{nm}, b_{mn}, c_{nm} and d_{nm}:

$$a_{nm} = \frac{1}{||V_{nm}^{(1)}||^2} \int_0^l \int_0^{2\pi} \phi(r, \varphi) V_{nm}^{(1)}(r, \varphi) \, rdrd\varphi,$$

$$b_{nm} = \frac{1}{\omega_{nm}} \left[\frac{1}{||V_{nm}^{(1)}||^2} \int_0^l \int_0^{2\pi} \psi(r, \varphi) V_{nm}^{(1)}(r, \varphi) \, rdrd\varphi + \kappa a_{nm} \right],$$

$$c_{nm} = \frac{1}{||V_{nm}^{(2)}||^2} \int_0^l \int_0^{2\pi} \phi(r, \varphi) V_{nm}^{(2)}(r, \varphi) \, rdrd\varphi,$$

$$d_{nm} = \frac{1}{\omega_{nm}} \left[\frac{1}{||V_{nm}^{(2)}||^2} \int_0^l \int_0^{2\pi} \psi(r, \varphi) V_{nm}^{(2)}(r, \varphi) \, rdrd\varphi + \kappa c_{nm} \right]. \quad (8.71)$$

Equation (8.68) gives the evolution of *free oscillations* of a circular membrane when boundary conditions are homogeneous. It can be considered as the expansion of the (unknown) function $u(r, \varphi, t)$ in a Fourier series using the orthogonal system of functions $V_{nm}(r, \varphi)$. This series converges under sufficiently reasonable assumptions about initial and boundary conditions – it is enough if they are piecewise continuous (see Appendix A).

The particular solutions $u_{nm}(r, \varphi, t) = T_{nm}^{(1)}(t) V_{nm}^{(1)}(r, \varphi) + T_{nm}^{(2)}(t) V_{nm}^{(2)}(r, \varphi)$ are *standing wave* solutions. From this we see that the profile of a standing wave depends on the functions $V_{nm}(r, \varphi)$; the functions $T_{nm}^{(1)}(t)$ and $T_{nm}^{(2)}(t)$ only change the amplitude of the standing wave over time, as was the case for standing waves on a string and the rectangular membrane. Lines on the membrane defined by $V_{nm}(r, \varphi) = 0$ remain at rest for all times and are called *nodal lines* of the standing wave $V_{nm}(r, \varphi)$. Points, where $V_{nm}(r, \varphi)$ reaches a relative maximum or minimum for all times, are called *antinodes* of this standing wave. From the

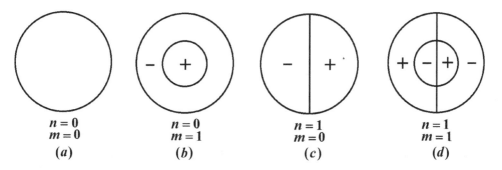

FIGURE 8.9
Drawing of the first few modes of vibrations for the mode $V_{nm}^{(1)}(r, \varphi)$.

above discussion of the Fourier expansion we see that an arbitrary motion of the membrane may be thought of as an infinite sum of these standing waves.

Each mode $u_{nm}(r, \varphi, t)$ possesses a characteristic pattern of nodal lines. The first few of these normal vibration modes for $V_{nm}^{(1)}(r, \varphi) = J_n \left(\frac{\mu_m^{(n)}}{l} r \right) \cos n\varphi$ are sketched in Figure 8.9 with similar pictures for the modes $V_{nm}^{(2)}(r, \varphi)$. In the fundamental mode of vibration corresponding to $\mu_0^{(0)}$, the membrane vibrates as a whole. In the mode corresponding to $\mu_1^{(0)}$ the membrane vibrates in two parts as shown with the part labeled with a plus sign initially above the equilibrium level and the part labeled with a minus sign initially below the equilibrium. The nodal line in this case is a circle which remains at rest as the two sections reverse location. The mode characterized by $\mu_0^{(1)}$ is equal to zero when $\varphi = \pm\pi/2$ and is positive and negative as shown.

8.3.2   Axisymmetric Oscillations of a Membrane

Oscillations of a circular membrane are said to be axisymmetric (or radial) if they do not depend on the polar angle φ (i.e. the deviation of an arbitrary point M from its position of equilibrium at time t depends only on t and the distance between point M and the center of the membrane). Solutions for axisymmetric oscillations will have a simpler form than more general types of oscillations. Physically we see that axisymmetric oscillations will occur when initial displacements and initial velocities do not depend on φ, but rather are functions only of r:

$$u(r, \varphi, t)|_{t=0} = \phi(r), \qquad \frac{\partial u}{\partial t}(r, \varphi, t)\bigg|_{t=0} = \psi(r). \qquad (8.72)$$

In this case all coefficients, a_{nm}, b_{nm}, c_{nm} and d_{nm} with $n \geq 1$ equal zero. We may easily verify this, for example, for a_{nm}

$$a_{nm} = \frac{1}{||V_{nm}^{(1)}||^2} \int_0^l \int_0^{2\pi} \phi(r) J_n \left(\mu_m^{(n)} r/l \right) \cos n\varphi \, r dr d\varphi.$$

Because $\int_0^{2\pi} \cos n\varphi \, d\varphi = 0$ for any integer $n \geq 1$ we have $a_{nm} = 0$.

Similarity $b_{nm} = 0$ for $n \geq 1$, $c_{nm} = 0$, $d_{nm} = 0$ for all n. Thus, *the solution does not contain the functions* $V_{nm}^{(2)}(r, \varphi)$. If $n = 0$ the coefficients a_{0m} and b_{0m} are nonzero and the formulas used to calculate them can be simplified. Putting factors which are independent of φ outside the integral and using $\int_0^{2\pi} d\varphi = 2\pi$, we have

$$a_{0m} = \frac{2\pi}{||V_{0m}^{(1)}||^2} \int_0^l \phi(r) J_0\left(\mu_m^{(0)} r/l\right) r dr. \tag{8.73}$$

Similarly we find

$$b_{0m} = \frac{2\pi}{||V_{0m}^{(1)}||^2} \int_0^l \psi(r) J_0\left(\mu_m^{(0)} r/l\right) r dr. \tag{8.74}$$

Substituting these coefficients into the series in Equation (8.68) we notice that the series reduces from a double series to a single one since all terms in the second sum of this series disappear. Only those terms in the first sum remain for which $n = 0$, making it necessary to sum only on m but not on n. The final result is

$$u(r, \varphi, t) = \sum_{m=0}^{\infty} \left[a_{0m} y_{0m}^{(1)}(t) + b_{0m} y_{0m}^{(2)}(t)\right] J_0\left(\frac{\mu_m^{(0)}}{l} r\right). \tag{8.75}$$

Thus, for axisymmetric oscillations the solution contains only Bessel functions of zero order.

Example 8.5 Find the transverse oscillations of a circular membrane with radius l with a fixed edge. Assume the initial displacement has the form of a paraboloid of rotation, initial velocities are zero and the reaction of the environment is small enough to be neglected.

Solution. We have the following boundary value problem of a circular membrane with fixed edge:

$$\frac{\partial^2 u}{\partial t^2} - a^2 \left[\frac{\partial^2 u}{\partial r^2} + \frac{1}{r}\frac{\partial u}{\partial r}\right] = 0, \quad 0 \leq r < l, \quad 0 \leq \varphi < 2\pi, \quad t > 0,$$

$$u(r, 0) = A\left(1 - \frac{r^2}{l^2}\right), \quad \frac{\partial u}{\partial t}(r, \varphi, 0) = 0, \quad u(l, \varphi, t) = 0.$$

The oscillations of the membrane are axisymmetric since the initial displacement and the initial velocities do not depend on the polar angle φ. Thus, only terms with $n = 0$ are not zero.

Boundary conditions of the problem are of *Dirichlet type*, in which case eigenvalues $\mu_m^{(0)}$ are the solutions of the equation $J_0(\mu) = 0$, and the eigenfunctions are

$$V_{0m}^{(1)}(r, \varphi) = J_0\left(\frac{\mu_m^{(0)}}{l} r\right).$$

The solution $u(r, \varphi, t)$ is given by the series

$$u(r, \varphi, t) = \sum_{m=0}^{\infty} a_{0m} \cos\frac{a\mu_m^{(0)} t}{l} J_0\left(\mu_m^{(0)} r/l\right).$$

The three-dimensional picture shown in Figure 8.10 depicts the two eigenfunctions for the given problem,

$$V_{00}^{(1)}(r, \varphi) = J_0\left(\mu_0^{(0)} r/l\right) \quad \text{and} \quad V_{02}^{(1)}(r, \varphi) = J_0\left(\mu_2^{(0)} r/l\right)$$

(these two eigenfunctions are chosen as examples).

(a)   (b)

FIGURE 8.10
Two eigenfunctions, (a) $V_{01}^{(1)}(r,\varphi)$ and (b) $V_{04}^{(1)}(r,\varphi)$, for the Dirichlet boundary conditions in Example 8.5.

The coefficients a_{0m} are given by Equation (8.73):

$$a_{0m} = \frac{2\pi}{\left\|V_{0m}^{(1)}\right\|^2} \int_0^l A\left(1 - \frac{r^2}{l^2}\right) J_0\left(\mu_m^{(0)}r/l\right) r\,dr.$$

Using the formulas (B.15b and B.14a from Appendix B)

$$\int x^n J_{n-1}(x)dx = x^n J_n(x), \quad J_{n+1}(x) = \frac{2n}{x}J_n(x) - J_{n-1}(x)$$

we calculate integrals (taking into account that $J_0\left(\mu_m^{(0)}\right) = 0$):

$$\int_0^l J_0\left(\frac{\mu_m^{(0)}}{l}\right) r\,dr = \frac{l^2}{\left(\mu_m^{(0)}\right)^2} xJ_1(x)\big|_0^{\mu_m^{(0)}} = \frac{l^2}{\mu_m^{(0)}} J_1\left(\mu_m^{(0)}\right),$$

$$\frac{1}{l^2} \int_0^l r^2 J_0\left(\frac{\mu_m^{(0)}}{l}\right) r\,dr = \frac{1}{l^2}\frac{l^4}{\left(\mu_m^{(0)}\right)^4} \int_0^{\mu_m^{(0)}} x^2 \left[2J_1(x) - xJ_2(x)\right] dx$$

$$= \frac{1}{l^2}\frac{l^4}{\left(\mu_m^{(0)}\right)^4} \left[2\int_0^{\mu_m^{(0)}} x^2 J_1(x)dx - \int_0^{\mu_m^{(0)}} x^3 J_2(x)dx\right] = \left[\frac{l^2}{\mu_m^{(0)}} - \frac{4l^2}{\left(\mu_m^{(0)}\right)^3}\right] J_1\left(\mu_m^{(0)}\right).$$

We may calculate the norms of eigenfunctions taking into account that $J_0'(x) = -J_1(x)$ (see differentiation formulas B.16a). With this, we have (D.17)

$$\left\|V_{0m}^{(1)}\right\|^2 = \pi l^2 \left[J_0'\left(\mu_m^{(0)}\right)\right]^2 = \pi l^2 \left[J_1\left(\mu_m^{(0)}\right)\right]^2.$$

So, the coefficients a_{0m} are given by Equation (8.73):

$$a_{0m} = \frac{8\,A}{\left(\mu_m^{(0)}\right)^3 J_1\left(\mu_m^{(0)}\right)}.$$

Thus,

$$u(r,\varphi,t) = 8A \sum_{m=0}^{\infty} \frac{1}{\left(\mu_m^{(0)}\right)^3 J_1\left(\mu_m^{(0)}\right)} \cos\frac{a\mu_m^{(0)}t}{l} J_0\left(\mu_m^{(0)}r/l\right).$$

Example 8.6 Find the transverse oscillations of a homogeneous circular membrane of radius l with a rigidly fixed edge where the oscillations are initiated by a localized impact, normal to a surface of the membrane. This impact is applied at the point (r_0, φ_0) and supplies an impulse I $(0 < r_0 < l)$ to the membrane. Any initial displacement is absent and the reaction of the environment is negligible.

Solution. The boundary value problem describing the oscillations of the membrane reduces to the solution of the equation

$$\frac{\partial^2 u}{\partial t^2} = a^2 \left[\frac{\partial^2 u}{\partial r^2} + \frac{1}{r}\frac{\partial u}{\partial r} + \frac{1}{r^2}\frac{\partial^2 u}{\partial \varphi^2} \right], \quad 0 \le r < l, \quad 0 \le \varphi < 2\pi, \quad t > 0,$$

under the conditions

$$u(r, \varphi, 0) = 0, \quad \frac{\partial u}{\partial t}(r, \varphi, 0) = \frac{I}{\rho}\delta(r - r_0)\delta(\varphi - \varphi_0), \quad u(l, \varphi, t) = 0.$$

The product $\delta(r - r_0)\delta(\varphi - \varphi_0)$ is a δ-function in two (polar) dimensions.

The boundary condition of the problem is of *Dirichlet type*, so the eigenvalues are given by equation $J_n(\mu) = 0$, the eigenfunctions are given by Equations (8.65), and

$$\left\| V_{nm}^{(1,2)} \right\|^2 = \sigma_n \frac{\pi l^2}{2}\left[J_n'\left(\mu_m^{(n)}\right) \right]^2, \quad \sigma_n = \begin{cases} 2, & \text{if } n = 0, \\ 1, & \text{if } n \ne 0, \end{cases} \text{ (see Appendix B, Equation (B.33)).}$$

The initial displacement of the membrane is zero in which case the solution $u(r, \varphi, t)$ is given by the series

$$u(r, \varphi, t) = \sum_{n=0}^{\infty}\sum_{m=0}^{\infty} [b_{nm}\cos n\varphi + d_{nm}\sin n\varphi] \cdot J_n\left(\mu_m^{(n)}r/l\right)\sin\frac{a\mu_m^{(n)}t}{l}.$$

Next, we calculate the coefficients b_{nm} and d_{nm} in (8.71) to get

$$b_{nm} = \frac{I}{\rho\omega_{nm}\left\| V_{nm}^{(1)} \right\|^2}\int_0^l\int_0^{2\pi}\delta(r - r_0)\delta(\varphi - \varphi_0)\cos n\varphi J_n\left(\mu_m^{(n)}r/l\right)\,r\,dr\,d\varphi$$

$$-\frac{2I\cos n\varphi_0}{\rho\omega_{nm}\sigma_n\pi l^2\left[J_n'\left(\mu_m^{(n)}\right) \right]^2}J_n\left(\mu_m^{(n)}r_0/l\right),$$

and

$$d_{nm} = \frac{I}{\rho\omega_{nm}\left\| V_{nm}^{(2)} \right\|^2}\int_0^l\int_0^{2\pi}\delta(r - r_0)\delta(\varphi - \varphi_0)\sin n\varphi J_n\left(\mu_m^{(n)}r/l\right)\,r\,dr\,d\varphi =$$

$$= \frac{2I\sin n\varphi_0}{\rho\omega_{nm}\sigma_n\pi l^2\left[J_n'\left(\mu_m^{(n)}\right) \right]^2}J_n\left(\mu_m^{(n)}r_0/l\right).$$

Therefore, the evolution of the displacements of points on the membrane is described by series

$$u(r, \varphi, t) = \frac{2I}{\pi\rho l^2}\sum_{n=0}^{\infty}\sum_{m=0}^{\infty}\frac{\cos n(\varphi - \varphi_0)J_n\left(\mu_m^{(n)}r_0/l\right)}{\sigma_n\omega_{nm}\left[J_n'\left(\mu_m^{(n)}\right) \right]^2}J_n\left(\frac{\mu_m^{(n)}}{l}r\right)\sin\frac{a\mu_m^{(n)}t}{l}.$$

Figure 8.11 shows two snapshots of the solution at the times $t = 0.3$ and $t = 4.3$. This solution was obtained for the case $a^2 = 1$, $l = 2$, $r_0 = 1$, $\varphi_0 = \pi$, $I/\rho = 10$.

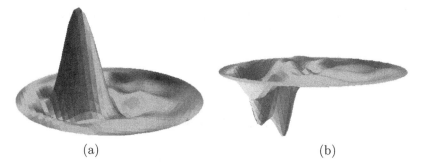

FIGURE 8.11
Graph of the membrane in Example 8.6 at (a) $t = 0.3$, (b) $t = 4.3$.

Example 8.7 The periphery of a flexible circular membrane of radius l is fixed elastically with coefficient h. The initial displacement is zero. Find the transversal vibrations of the membrane if the initial velocities of the membrane are described by the function

$$\psi(r) = A(l - r).$$

Solution. The boundary value problem consists of the solution of the equation

$$\frac{\partial^2 u}{\partial t^2} - a^2 \left[\frac{\partial^2 u}{\partial r^2} + \frac{1}{r} \frac{\partial u}{\partial r} \right] = 0, \quad 0 \leq r < l, \quad 0 \leq \varphi < 2\pi, \quad t > 0,$$

with the conditions

$$u(r, \varphi, 0) = 0, \quad \frac{\partial u}{\partial t}(r, \varphi, 0) = A(l - r), \quad \frac{\partial u}{\partial r} + hu \bigg|_{r=l} = 0.$$

The oscillations of the membrane are axisymmetric since the initial functions do not depend on the polar angle φ; thus only terms with $n = 0$ are not zero.

The boundary condition of the problem is of *mixed type*, in which case eigenvalues $\mu_m^{(0)}$ are given by the roots of the eigenvalue equation $\mu J_0'(\mu) + h J_0(\mu) = 0$. The eigenfunctions are

$$V_{0m}^{(1)}(r, \varphi) = J_0 \left(\mu_m^{(0)} r / l \right), \quad \left\| V_{0m}^{(1)} \right\|^2 = \frac{\pi l^2}{\left(\mu_m^{(n)} \right)^2} \left[\left(\mu_m^{(n)} \right)^2 + l^2 h^2 \right] J_n^2 \left(\mu_m^{(n)} \right)$$

(see Appendix B, Equation (B.35)).

The initial displacement is zero so that the coefficients $a_{0m} = 0$. The coefficients b_{0m} are given by Equation (8.74) which result in

$$b_{0m} = \frac{2\pi}{\|V_{0m}^{(1)}\|^2} \int_0^l A(l - r) J_0 \left(\mu_m^{(0)} r / l \right) r \, dr.$$

The oscillations of the membrane are given by the series in Bessel functions of zero-th order:

$$u(r, \varphi, t) = \sum_{m=0}^{\infty} b_{0m} \sin \frac{a \mu_m^{(0)} t}{l} J_0 \left(\frac{\mu_m^{(0)}}{l} r \right).$$

8.3.3   The Fourier Method for Nonhomogeneous Equations with Homogeneous Boundary Conditions

In this section we are dealing with *nonhomogeneous* Equation (8.53),

$$\frac{\partial^2 u}{\partial t^2} + 2\kappa \frac{\partial u}{\partial t} - a^2 \left(\frac{\partial^2 u}{\partial r^2} + \frac{1}{r} \frac{\partial u}{\partial r} + \frac{1}{r^2} \frac{\partial^2 u}{\partial \varphi^2} \right) + \gamma u = f(r, \varphi, t)$$

satisfying the *homogeneous* boundary condition (8.57)

$$\alpha \frac{\partial u}{\partial r} + \beta \left. u \right|_{r=l} = 0$$

and *nonhomogeneous* initial conditions (8.55)

$$\left. u \right|_{t=0} = \phi(r, \varphi), \quad \left. \frac{\partial u}{\partial t} \right|_{t=0} = \psi(r, \varphi).$$

We begin by searching for a solution in the form of the sum

$$u(r, \varphi, t) = u_1(r, \varphi, t) + u_2(r, \varphi, t), \tag{8.76}$$

where $u_1(r, \varphi, t)$ is the solution to the *homogeneous* equation with *homogeneous* boundary and *nonhomogeneous* initial conditions given by

$$\frac{\partial^2 u_1}{\partial t^2} + 2\kappa \frac{\partial u_1}{\partial t} - a^2 \left(\frac{\partial^2 u_1}{\partial r^2} + \frac{1}{r} \frac{\partial u_1}{\partial r} + \frac{1}{r^2} \frac{\partial^2 u_1}{\partial \varphi^2} \right) + \gamma u_1 = 0, \tag{8.77}$$

$$\alpha \frac{\partial u_1}{\partial r} + \beta \left. u_1 \right|_{r=l} = 0, \tag{8.78}$$

$$\left. u_1 \right|_{t=0} = \phi(r, \varphi), \quad \left. \frac{\partial u_1}{\partial t} \right|_{t=0} = \psi(r, \varphi), \tag{8.79}$$

and $u_2(r, \varphi, t)$ is the solution to the *nonhomogeneous* equation with *zero boundary and initial conditions* given by

$$\frac{\partial^2 u_2}{\partial t^2} + 2\kappa \frac{\partial u_2}{\partial t} - a^2 \left(\frac{\partial^2 u_2}{\partial r^2} + \frac{1}{r} \frac{\partial u_2}{\partial r} + \frac{1}{r^2} \frac{\partial^2 u_2}{\partial \varphi^2} \right) + \gamma u_2 = f(r, \varphi, t), \tag{8.80}$$

$$\alpha \frac{\partial u_2}{\partial r} + \beta u_2 \left.\right|_{r=l} = 0, \tag{8.81}$$

$$\left. u_2 \right|_{t=0} = 0, \quad \left. \frac{\partial u_2}{\partial t} \right|_{t=0} = 0. \tag{8.82}$$

Physically the solution $u_1(r, \varphi, t)$ represents *free oscillations*, i.e. oscillations which occur only due to an initial perturbation. The solution $u_2(r, \varphi, t)$ represents *forced oscillations*, i.e. oscillations which result from the action of external forces when initial perturbations are absent.

Methods for finding the solution $u_1(r, \varphi, t)$ for free oscillations were considered in the previous section; our task here need only be to find the solutions $u_2(r, \varphi, t)$ for forced oscillations. As in the case of free oscillations we search for the solution $u_2(r, \varphi, t)$ in the form of the series

$$u_2(r, \varphi, t) = \sum_{n=0}^{\infty} \sum_{m=0}^{\infty} \left[T_{nm}^{(1)}(t) V_{nm}^{(1)}(r, \varphi) + T_{nm}^{(2)}(t) V_{nm}^{(2)}(r, \varphi) \right], \tag{8.83}$$

where $V_{nm}^{(1)}(r, \varphi)$ and $V_{nm}^{(2)}(r, \varphi)$ are eigenfunctions (8.65) of the corresponding homogeneous boundary value problem, and $T_{nm}^{(1)}(t)$, $T_{nm}^{(2)}(t)$ which we have to find.

Zero boundary conditions given in Equation (8.81) for the function $u_2(r, \varphi, t)$ are satisfied for any choice of $T_{nm}^{(1)}(t)$ and $T_{nm}^{(2)}(t)$ under the restriction of uniform convergence of the series because they are known to be satisfied by the functions $V_{nm}^{(1)}(r, \varphi)$ and $V_{nm}^{(2)}(r, \varphi)$. However, the functions $T_{nm}^{(1)}(t)$ and $T_{nm}^{(2)}(t)$ must also be selected so that the series (8.83) satisfies Equation (8.80) and initial conditions (8.82).

Substituting the series (8.83) into Equation (8.80) we obtain

$$
\sum_{n=0}^{\infty} \sum_{m=0}^{\infty} \left[T_{nm}^{(1)''}(t) + 2\kappa\, T_{nm}^{(1)'}(t) + (a^2\lambda_{nm} + \gamma) T_{nm}^{(1)}(t) \right] V_{nm}^{(1)}(r, \varphi)
$$

$$
+ \sum_{n=0}^{\infty} \sum_{m=0}^{\infty} \left[T_{nm}^{(2)''}(t) + 2\kappa\, T_{nm}^{(2)'}(t) + (a^2\lambda_{nm} + \gamma) T_{nm}^{(2)}(t) \right] V_{nm}^{(2)}(r, \varphi) = f(r, \varphi, t). \quad (8.84)
$$

Next we expand the function $f(r, \varphi, t)$ in a series with functions $V_{nm}^{(1)}(r, \varphi)$ and $V_{nm}^{(2)}(r, \varphi)$ as the basis functions:

$$
f(r, \varphi, t) = \sum_{n=0}^{\infty} \sum_{m=0}^{\infty} \left[f_{nm}^{(1)}(t) V_{nm}^{(1)}(r, \varphi) + f_{nm}^{(2)}(t) V_{nm}^{(2)}(r, \varphi) \right], \quad (8.85)
$$

where

$$
f_{nm}^{(1)}(t) = \frac{1}{||V_{nm}^{(1)}||^2} \int_0^{2\pi} \int_0^l f(r, \varphi, t)\, V_{nm}^{(1)}\, r\, dr\, d\varphi,
$$

$$
f_{nm}^{(2)}(t) = \frac{1}{||V_{nm}^{(2)}||^2} \int_0^{2\pi} \int_0^l f(r, \varphi, t)\, V_{nm}^{(2)}\, r\, dr\, d\varphi. \quad (8.86)
$$

Comparing the series (8.84) and (8.85), we obtain the following equations which will determine the functions $T_{nm}^{(1)}(t)$ and $T_{nm}^{(2)}(t)$:

$$
T_{nm}^{(1)''}(t) + 2\kappa\, T_{nm}^{(1)'}(t) + (a^2\lambda_{nm} + \gamma) T_{nm}^{(1)}(t) = f_{nm}^{(1)}(t),
$$

$$
T_{nm}^{(2)''}(t) + 2\kappa\, T_{nm}^{(2)'}(t) + (a^2\lambda_{nm} + \gamma) T_{nm}^{(2)}(t) = f_{nm}^{(2)}(t). \quad (8.87)
$$

The solution $u_2(r, \varphi, t)$ defined by the series (8.83) satisfies initial conditions (8.82) which imposes on the functions $T_{nm}^{(1)}(t)$ and $T_{nm}^{(2)}(t)$ the conditions

$$
\begin{cases} T_{nm}^{(1)}(0) = 0, \\ T_{nm}^{(1)'}(0) = 0 \end{cases} \quad \text{and} \quad \begin{cases} T_{nm}^{(2)}(0) = 0, \\ T_{nm}^{(2)'}(0) = 0 \end{cases} \quad n, m = 0, 1, 2, \ldots. \quad (8.88)
$$

As for the one-dimensional case, solutions of the Cauchy problems defined in Equations (8.87) and (8.88) for functions $T_{nm}^{(1)}(t)$ and $T_{nm}^{(2)}(t)$ can be written as

$$
T_{nm}^{(1)}(t) = \int_0^t f_{nm}^{(1)}(\tau)\, Y_{nm}(t - \tau)\, d\tau,
$$

$$
T_{nm}^{(2)}(t) = \int_0^t f_{nm}^{(2)}(\tau)\, Y_{nm}(t - \tau)\, d\tau,
$$

where

$$
Y_{nm}(t) = \begin{cases}
\dfrac{1}{\omega_{nm}} e^{-\kappa t} \sin \omega_{nm} t, & \kappa^2 < a^2 \lambda_{nm} + \gamma, \\[2ex]
\dfrac{1}{\omega_{nm}} e^{-\kappa t} \sinh \omega_{nm} t, & \kappa^2 > a^2 \lambda_{nm} + \gamma, \\[2ex]
t e^{-\kappa t}, & \kappa^2 = a^2 \lambda_{nm} + \gamma,
\end{cases}
$$

and $\omega_{nm} = \sqrt{|a^2 \lambda_{nm} + \gamma - \kappa^2|}$. Substituting expressions (8.86) for $f_{nm}^{(1)}(\tau)$ and $f_{nm}^{(2)}(\tau)$, we obtain

$$
T_{nm}^{(1)}(t) = \frac{1}{\left\| V_{nm}^{(1)} \right\|^2} \int_0^t d\tau \int_0^{2\pi} \int_0^l f(r, \varphi, \tau) V_{nm}^{(1)}(r, \varphi) \, Y_{nm}(t - \tau) \, r \, dr \, d\varphi,
$$

$$
T_{nm}^{(2)}(t) = \frac{1}{\left\| V_{nm}^{(2)} \right\|^2} \int_0^t d\tau \int_0^{2\pi} \int_0^l f(r, \varphi, \tau) V_{nm}^{(2)}(r, \varphi) \, Y_{nm}(t - \tau) \, r \, dr \, d\varphi. \tag{8.89}
$$

Substituting these expressions for $T_{nm}^{(1)}(t)$ and $T_{nm}^{(2)}(t)$ in the series (8.83) we obtain solutions to the boundary value problem defined in Equations (8.80) through (8.82), assuming that Equation (8.83) and the series obtained from it by twice differentiating term by term with respect to the variables r, φ and t converge uniformly. Thus, the solution of the problem of forced oscillations with zero boundary conditions is

$$
\begin{aligned}
u(r, \varphi, t) &= u_1(r, \varphi, t) + u_2(r, \varphi, t) \\
&= \sum_{n=0}^{\infty} \sum_{m=0}^{\infty} \Big\{ \Big[T_{nm}^{(1)}(t) + \big(a_{nm} y_{nm}^{(1)}(t) + b_{nm} y_{nm}^{(2)}(t) \big) \Big] \cdot V_{nm}^{(1)}(r, \varphi) \\
&\quad + \Big[T_{nm}^{(2)}(t) + \big(c_{nm} y_{nm}^{(1)}(t) + d_{nm} y_{nm}^{(2)}(t) \big) \Big] \cdot V_{nm}^{(2)}(r, \varphi) \Big\},
\end{aligned} \tag{8.90}
$$

where coefficients a_{nm}, b_{nm}, c_{nm} and d_{nm} were obtained previously in the discussion in Section 8.3.1.

8.3.4 Forced Axisymmetric Oscillations

In the case of axisymmetric oscillations the solution of the nonhomogeneous equation becomes simpler. In this case the initial displacement, initial velocity and function f do not depend on φ and are thus functions of r and t only:

$$
u(r, \varphi, t)|_{t=0} = \phi(r), \quad \frac{\partial u}{\partial t}(r, \varphi, t)\bigg|_{t=0} = \psi(r), \quad \text{and} \quad f(r, \varphi, t) = f(r, t).
$$

For axisymmetric oscillations the solution does not contain the functions $V_{nm}^{(2)}(r, \varphi)$ and the only nonzero coefficients are $f_{0m}^{(1)}$:

$$
f_{0m}^{(1)}(t) = \frac{1}{\left\| V_{0m}^{(1)} \right\|^2} \int_0^l \int_0^{2\pi} f(r, t) J_0 \left(\frac{\mu_m^{(0)}}{l} r \right) r \, dr \, d\varphi = \frac{2\pi}{\left\| V_{0m}^{(1)} \right\|^2} \int_0^l f(r, t) J_0 \left(\frac{\mu_m^{(0)}}{l} r \right) r \, dr.
$$

Substituting these functions into the first of the formulas (8.89) gives $T_{0m}^{(1)}(t)$ and the double series (8.90) reduces to a single series:

$$
u(r, \varphi, t) = \sum_{m=0}^{\infty} \Big\{ T_{0m}^{(1)}(t) + \Big[a_{0m} y_{0m}^{(1)}(t) + b_{0m} y_{0m}^{(2)}(t) \Big] \Big\} \cdot J_0 \left(\frac{\mu_m^{(0)}}{l} r \right).
$$

Example 8.8 Find the transverse oscillations of a homogeneous circular membrane of radius l with a rigidly fixed edge if a pressure proportional to $\cos \omega t$ acts on the membrane. Assume that initial deviations and initial velocities are absent, and that the reaction of the environment is negligibly small.

Solution. The boundary problem modeling the evolution of such oscillations leads to the equation

$$\frac{\partial^2 u}{\partial t^2} - a^2 \left[\frac{\partial^2 u}{\partial r^2} + \frac{1}{r}\frac{\partial u}{\partial r}\right] = A \cos \omega t, \quad A = const.$$

under zero initial and boundary conditions given by

$$u(r,\varphi,0) = 0, \quad \frac{\partial u}{\partial t}(r,\varphi,0) = 0, \quad u(l,\varphi,t) = 0.$$

In this example the oscillations are axisymmetric because the initial conditions are homogeneous and the external pressure is a function of t only. From the previous discussion we have that the eigenvalues $\mu_m^{(0)}$ are roots of equation $J_0(\mu) = 0$, the eigenfunctions are

$$V_{0m}^{(1)}(r,\varphi) = J_0\left(\frac{\mu_m^{(0)}}{l}r\right), \; \left\|V_{0m}^{(1)}\right\|^2 = \pi l^2 \left[J_0'\left(\mu_m^{(0)}\right)\right]^2 = \pi l^2 \left[J_1\left(\mu_m^{(0)}\right)\right]^2$$

(see Appendix B, formula (B.33) and differentiation formulas (B.16)).

Consequently, the solution $u(r,\varphi,t)$ is defined by the series

$$u(r,\varphi,t) = \sum_{m=0}^{\infty} T_{0m}^{(1)}(t) J_0\left(\frac{\mu_m^{(n)}}{l}r\right),$$

where

$$T_{0m}^{(1)}(t) = \frac{1}{\omega_{0m}} \int_0^l f_{0m}^{(1)}(p) \sin \omega_{0m}(t-p)dp, \quad \omega_{0m} = a\frac{\mu_m^{(0)}}{l}.$$

For $f_{0m}^{(1)}(t)$ we have

$$f_{0m}^{(1)}(t) = \frac{A \cos \omega t}{\left\|V_{0m}^{(1)}\right\|^2} 2\pi \int_0^l J_0\left(\frac{\mu_m^{(0)}}{l}\right) r dr.$$

Using the integration formulas (B.15b)

$$\int_0^l J_0\left(\frac{\mu_m^{(0)}}{l}\right) r dr = \frac{l^2}{\left(\mu_m^{(0)}\right)^2} x J_1(x)\big|_0^{\mu_m^{(0)}} = \frac{l^2}{\mu_m^{(0)}} J_1\left(\mu_m^{(0)}\right)$$

we find

$$f_{0m}^{(1)}(t) = \frac{A \cos \omega t}{\left\|V_{0m}^{(1)}\right\|^2} 2\pi \int_0^l J_0\left(\frac{\mu_m^{(0)}}{l}\right) r dr = \frac{2A}{\mu_m^{(0)} J_1\left(\mu_m^{(0)}\right)} \cos \omega t.$$

Using this result we obtain

$$T_{0m}^{(1)}(t) = \frac{2A}{\omega_{0m}\mu_m^{(0)} J_1\left(\mu_m^{(0)}\right)} \int_0^t \cos \omega p \sin \omega_{0m}(t-p)dp$$

$$= -\frac{2A}{\mu_m^{(0)} J_1\left(\mu_m^{(0)}\right)(\omega^2 - \omega_{0m}^2)} [\cos \omega t - \cos \omega_{0m} t].$$

Finally, the deflection of the membrane as a function of time is described by the series

$$u(r, \varphi, t) = 2A \sum_{m=0}^{\infty} \frac{\cos \omega_{0m} t - \cos \omega t}{(\omega^2 - \omega_{0m}^2) \mu_m^{(0)} J_1\left(\mu_m^{(0)}\right)} J_0\left(\frac{\mu_m^{(0)}}{l} r\right).$$

8.3.5 The Fourier Method for Equations with Nonhomogeneous Boundary Conditions

Consider now the general boundary value problem for equations describing the forced oscillations of a circular membrane with nonhomogeneous boundary and initial conditions:

$$\frac{\partial^2 u}{\partial t^2} + 2\kappa \frac{\partial u}{\partial t} - a^2 \left(\frac{\partial^2 u}{\partial r^2} + \frac{1}{r}\frac{\partial u}{\partial r} + \frac{1}{r^2}\frac{\partial^2 u}{\partial \varphi^2}\right) + \gamma u = f(r, \varphi, t),$$

$$\alpha \frac{\partial u}{\partial r} + \beta u \bigg|_{r=l} = g(\varphi, t),$$

$$u|_{t=0} = \phi(r, \varphi), \quad \frac{\partial u}{\partial t}\bigg|_{t=0} = \psi(r, \varphi).$$

We cannot apply the Fourier method directly to this problem because the boundary conditions are nonhomogeneous. First, we should reduce the problem to one with zero boundary conditions.

To do this, we search for the solution of the problem in the form of the sum

$$u(r, \varphi, t) = v(r, \varphi, t) + w(r, \varphi, t),$$

where $v(r, \varphi, t)$ is a new unknown function, and the function $w(r, \varphi, t)$ is chosen so that it satisfies the given nonhomogeneous boundary condition

$$\alpha \frac{\partial w}{\partial r} + \beta w \bigg|_{r=l} = g(\varphi, t)$$

and has the necessary number of continuous derivatives in r, φ and t.

For the function $v(r, \varphi, t)$ we obtain the following boundary value problem:

$$\frac{\partial^2 v}{\partial t^2} + 2\kappa \frac{\partial v}{\partial t} - a^2 \left(\frac{\partial^2 v}{\partial r^2} + \frac{1}{r}\frac{\partial v}{\partial r} + \frac{1}{r^2}\frac{\partial^2 v}{\partial \varphi^2}\right) + \gamma v = \tilde{f}(r, \varphi, t),$$

$$\alpha \frac{\partial v}{\partial r}(l, \varphi, t) + \beta v(l, \varphi, t) = 0,$$

$$v(r, \varphi, t)|_{t=0} = \tilde{\phi}(r, \varphi), \quad \frac{\partial v}{\partial t}(r, \varphi, t)\bigg|_{t=0} = \tilde{\psi}(r, \varphi),$$

where

$$\tilde{f}(r, \varphi, t) = f(r, \varphi, t) - \frac{\partial^2 w}{\partial t^2} - 2\kappa \frac{\partial w}{\partial t} + a^2 \left(\frac{\partial^2 w}{\partial r^2} + \frac{1}{r}\frac{\partial w}{\partial r} + \frac{1}{r^2}\frac{\partial^2 w}{\partial \varphi^2}\right) - \gamma w, \qquad (8.91)$$

$$\tilde{\phi}(r, \varphi) = \phi(r, \varphi) - w(r, \varphi, 0), \qquad (8.92)$$

$$\tilde{\psi}(r, \varphi) = \psi(r, \varphi) - \frac{\partial w}{\partial t}(r, \varphi, 0). \qquad (8.93)$$

The solution to this problem was considered in the previous section.

Reading Exercise: Verify Equations (8.91) through (8.93).

The way to find the auxiliary function was presented in Chapter 7. The results are:

1) For the case of Dirichlet boundary conditions:

a) if $u(l, \varphi, t) = g(\varphi, t)$,

$$w(r, \varphi, t) = \frac{r^2}{l^2} g(\varphi, t) \qquad (8.94)$$

b) if $u(l, \varphi, t) = g(t)$ or $u(l, \varphi, t) = g_0 = \text{const}$

$$w(r, \varphi, t) = g(t) \quad \text{or} \quad w(r, \varphi, t) = g_0. \qquad (8.95)$$

2) For the case of Neumann boundary conditions $\frac{\partial u}{\partial r}(l, \varphi, t) = g(\varphi, t)$

$$w(r, \varphi, t) = \frac{r^2}{2l} \cdot g(\varphi, t) + C, \qquad (8.96)$$

where C is an arbitrary constant.

3) For the case of mixed boundary conditions $\frac{\partial}{\partial r} u(l, \varphi, t) + h u(l, \varphi, t) = g(\varphi, t)$

$$w(r, \varphi, t) = \frac{r^2}{l(2 + hl)} g(\varphi, t). \qquad (8.97)$$

Example 8.9 Find the oscillations of a circular membrane in an environment without resistance and with zero initial conditions where the motion is caused by movement at its edge described by

$$u(l, \varphi, t) = A \sin \omega t, \quad t \geq 0.$$

Solution. The boundary problem modeling the evolution of such oscillations is given by the equation

$$\frac{\partial^2 u}{\partial t^2} - a^2 \left[\frac{\partial^2 u}{\partial r^2} + \frac{1}{r} \frac{\partial u}{\partial r} \right] = 0,$$

$$0 \leq r < l, \quad 0 \leq \varphi < 2\pi, \quad t > 0$$

with zero initial conditions

$$u(r, \varphi, 0) = 0, \quad \frac{\partial u}{\partial t}(r, \varphi, 0) = 0$$

and boundary condition

$$u(l, \varphi, t) = A \sin \omega t.$$

We are searching for the solution in the form of the sum $u(r, \varphi, t) = w(r, \varphi, t) + v(r, \varphi, t)$. Using Equation (8.95), we see that

$$w(r, \varphi, t) = A \sin \omega t.$$

Then we obtain

$$\tilde{f}(r, \varphi, t) = A\omega^2 \sin \omega t,$$

$$\tilde{\phi}(r, \varphi) = 0, \quad \tilde{\psi}(r, \varphi) = -A\omega.$$

The function $\tilde{f}(r, \varphi, t)$ does not depend on the polar angle φ, thus the function $v(r, \varphi, t)$ defines axisymmetric oscillations of the membrane. The solution $v(r, \varphi, t)$ is thus given by the series

$$v(r, \varphi, t) = \sum_{m=0}^{\infty} \left[T_{0m}^{(1)}(t) + b_{0m} \sin \omega_{0m} t \right] \cdot J_0 \left(\frac{\mu_m^{(0)}}{l} r \right),$$

where $\omega_{0m} = a\sqrt{\lambda_{0m}} = a\frac{\mu_m^{(0)}}{l}$, $\mu_m^{(0)}$ are the roots of equation $J_0(\mu) = 0$ and

$$V_{0m}^{(1)}(r, \varphi) = J_0 \left(\frac{\mu_m^{(0)}}{l} r \right), \quad \left\| V_{0m}^{(1)} \right\|^2 = \pi l^2 \left[J_0' \left(\mu_m^{(0)} \right) \right]^2 = \pi l^2 \left[J_1 \left(\mu_m^{(0)} \right) \right]^2.$$

(see Appendix B, formula (B.33) and differentiation formulas (B.16)).

To determine the coefficients b_{0m} we have (see (8.74))

$$b_{0m} = \frac{1}{\omega_{0m} \left\| V_{0m}^{(1)} \right\|^2} \int_0^l \int_0^{2\pi} \left(-A\omega \frac{r^2}{l^2} \right) d\varphi J_0 \left(\frac{\mu_m^{(0)}}{l} r \right) r dr = -\frac{2A\omega \left[\left(\mu_m^{(0)} \right)^2 - 4 \right]}{\omega_{0m} \left(\mu_m^{(0)} \right)^3 J_1 \left(\mu_m^{(0)} \right)}.$$

Next, we determine the function $f_{0m}^{(1)}$ (using the integration formulas (B.15b)):

$$f_{0m}^{(1)}(t) = \frac{1}{\left\| V_{0m}^{(1)} \right\|^2} \int_0^l \int_0^{2\pi} A\omega^2 \sin \omega t J_0 \left(\frac{\mu_m^{(0)}}{l} r \right) r dr d\varphi$$

$$= \frac{A\omega^2 \sin \omega t}{\pi l^2 \left[J_1 \left(\mu_m^{(0)} \right) \right]^2} \cdot 2\pi \cdot \frac{l^2 J_1 \left(\mu_m^{(0)} \right)}{\left(\mu_m^{(0)} \right)^2} = \frac{2A\omega^2 \sin \omega t}{\left(\mu_m^{(0)} \right)^2 J_1 \left(\mu_m^{(0)} \right)}.$$

From this we have

$$T_{0m}^{(1)}(t) = \frac{1}{\omega_{0m}} \int_0^t f_{0m}^{(1)}(\tau) \sin \omega_{0m}(t - \tau) d\tau = \frac{2A\omega^2}{\omega_{0m} \left(\mu_m^{(0)} \right)^2 J_1 \left(\mu_m^{(0)} \right)}$$

$$\times \int_0^t \sin \omega\tau \sin \omega_{0m}(t - \tau) d\tau = \frac{2A\omega^2 \left[\omega \sin \omega_{0m} t - \omega_{0m} \sin \omega t \right]}{\omega_{0m} \left(\mu_m^{(0)} \right)^2 J_1 \left(\mu_m^{(0)} \right) (\omega^2 - \omega_{0m}^2)}.$$

Finally, we may express the solution to the wave equation by the series

$$u(r, \varphi, t) = w(r, \varphi, t) + v(r, \varphi, t)$$

$$= A \sin \omega t + \sum_{m=0}^{\infty} \left[T_{0m}^{(1)}(t) + b_{0m} \sin \omega_{0m} t \right] J_0 \left(\frac{\mu_m^{(0)}}{l} r \right)$$

$$= A \sin \omega t + 2A\omega \sum_{m=0}^{\infty} \frac{\left[\omega_{0m} \sin \omega_{0m} t - \omega \sin \omega t \right]}{\left(\mu_m^{(0)} \right)^2 J_1 \left(\mu_m^{(0)} \right) (\omega^2 - \omega_{0m}^2)} J_0 \left(\frac{\mu_m^{(0)}}{l} r \right).$$

Problems

In problems 1 through 24 we consider transverse oscillations of a rectangular membrane ($0 \leq x \leq l_x$, $0 \leq y \leq l_y$), in problems 25 through 47 a *circular* membrane of radius l. Solve these problems analytically which means the following: formulate the equation and initial and boundary conditions, obtain the eigenvalues and eigenfunctions, write the formulas for coefficients of the series expansion and the expression for the solution of the problem.

The problems discussed refer to membranes, but it should be remembered that many other, similar physical problems are described by the same two-dimensional hyperbolic equation.

You can obtain the pictures of several eigenfunctions and screenshots of the solution and of the auxiliary functions with Maple, Mathematica or software from books [7, 8].

When you choose the parameters of the problem and coefficients of the functions (initial and boundary conditions, and external forces) do not forget that the amplitudes of oscillations should remain small. All the parameters and the variables (time, coordinates, deflection $u(x,t)$), are considered to be dimensionless.

In problems 1 through 5 external forces and resistance of the embedding medium are absent.

Find the free oscillations of the membrane.

1. The membrane is fixed along its edges. The initial conditions are
$$\varphi(x,y) = A \sin \frac{\pi x}{l_x} \sin \frac{\pi y}{l_y}, \quad \psi(x,y) = 0.$$

2. The edge at $x = 0$ of membrane is free and other edges are fixed. The initial conditions are
$$\varphi(x,y) = A \cos \frac{\pi x}{2l_x} \sin \frac{\pi y}{l_y}, \quad \psi(x,y) = 0.$$

3. The edge at $y = 0$ of the membrane is free and the other edges are fixed. The initial conditions are
$$\varphi(x,y) = A \sin \frac{\pi x}{l_x} \cos \frac{\pi y}{2l_y}, \quad \psi(x,y) = 0.$$

4. The edges $x = 0$ and $y = 0$ of membrane are fixed, and edges $x = l_x$ and $y = l_y$ are free. The initial conditions are
$$\varphi(x,y) = 0, \quad \psi(x,y) = Axy \left(l_x - \frac{x}{2} \right) \left(l_y - \frac{y}{2} \right).$$

5. The edges $x = 0$, $x = l_x$ and $y = l_y$ of membrane are fixed, and the edge $y = 0$ is elastically constrained with the coefficient of elasticity $h = 1$. The initial conditions are
$$\varphi(x,y) = Ax \left(1 - \frac{x}{l_x} \right) \sin \frac{\pi y}{l_y}, \quad \psi(x,y) = 0.$$

In problems 6 through 10 external forces are absent and the resistance of the embedding environment is proportional to velocity with proportionality constant κ. Find the transverse oscillations of the membrane.

6. The edges $x = 0$, $y = 0$ and $y = l_y$ of the membrane are fixed and the edge $x = l_x$ is attached elastically with the coefficient of elasticity $h = 1$. The initial conditions are
$$\varphi(x,y) = 0, \quad \psi(x,y) = Axy(l_y - y).$$

7. The edges $x = 0$ and $y = 0$ of the membrane are fixed, the edge $y = l_y$ is free and the edge $x = l_x$ is attached elastically with coefficient of elasticity $h = 1$. The initial conditions are
$$\varphi(x,y) = 0, \quad \psi(x,y) = Ax \left(1 - \frac{x}{l_x} \right) \sin \frac{\pi y}{l_y}.$$

8. The edges $x = l_x$ and $y = l_y$ are fixed, the edge $x = 0$ is free and the edge $y = 0$ is attached elastically with coefficient of elasticity $h = 1$. The initial conditions are

$$\varphi(x,y) = Ay\left(1 - \frac{y}{l_y}\right)\sin\frac{\pi x}{l_x}, \quad \psi(x,y) = 0.$$

9. The edges $y = 0$ and $y = l_y$ are free, the edge $x = 0$ is fixed and the edge $x = l_x$ is attached elastically with coefficient of elasticity $h = 1$. The initial conditions are

$$\varphi(x,y) = Axy^2\left(1 - \frac{x}{l_x}\right), \quad \psi(x,y) = 0.$$

10. The edges $x = 0$ and $y = l_y$ are free, the edge $x = l_x$ is attached elastically with coefficient of elasticity $h = 1$ and the edge $y = 0$ is attached elastically with coefficient of elasticity $h = 1$. The initial conditions are

$$\varphi(x,y) = Ay\left[1 - (x/l_x)^2\right], \quad \psi(x,y) = 0.$$

In problems 11 through 15 at the initial instant of time, $t = 0$, the membrane is set in motion by a blow which applies an impulse, I, at the point (x_0, y_0) $(0 < x_0 < l_x, 0 < y_0 < l_y)$.

For the following cases, find free transverse oscillations of the rectangular membrane assuming the initial displacement is zero, external forcing is absent and the environment causes a resistance proportional to velocity with coefficient $\kappa > 0$.

11. The membrane is fixed along its edges.

12. The edge $x = l_x$ of the membrane is free and the other edges are fixed.

13. The edges $x = 0$ and $y = l_y$ of the membrane are free and the edges $x = l_x$ and $y = 0$ are fixed.

14. The edges $x = l_x$, $y = 0$ and $y = l_y$ of the membrane are fixed and the edge $x = 0$ is attached elastically with coefficient of elasticity $h = 1$.

15. The edges $y = 0$ and $y = l_y$ are free, the edge $x = l_x$ is fixed and the edge $x = 0$ is attached elastically with coefficient of elasticity $h = 1$.

In problems 16 through 20 assume that a force with density $f(x,y,t)$ is acting on the membrane, the initial displacement from equilibrium as well as the initial velocity are zero, and resistance of the embedding medium is absent ($\kappa = 0$).

Solve the equations of motion with the given boundary conditions for each case given below.

16. The membrane is fixed along its boundaries. The external force is

$$f(x,y,t) = Ax\sin\frac{2\pi y}{l_y}(1 - \sin\omega t).$$

17. The edge of the membrane at $x = 0$ is free, and other edges are fixed. The external force is

$$f(x,y,t) = A(l_x - x)\sin\frac{\pi y}{l_y}\sin\omega t.$$

18. The edges $x = l_x$ and $y = l_y$ of the membrane are free and edges $x = 0$ and $y = 0$ are fixed. The external force is

$$f(x,y,t) = Ax\sin\frac{\pi y}{2l_y}\cos\omega t.$$

19. The edges $x = 0$, $y = 0$ and $y = l_y$ of the membrane are fixed and the edge $x = l_x$ is attached elastically with elasticity coefficient $h = 1$. The external force is

$$f(x, y, t) = Axy \cos \frac{\pi y}{2l_y}(1 - \sin \omega t).$$

20. The edges $x = 0$ and $y = l_y$ are attached elastically with elasticity coefficient $h = 1$, the edge $x = l_x$ is fixed, and the edge $y = 0$ is free. The external force is

$$f(x, y, t) = A \sin \frac{\pi x}{l_x} \cos \frac{\pi y}{2l_y} \sin \omega t.$$

In problems 21 through 25 the boundary of the membrane is driven. Assume the initial velocities are zero and the resistance of embedding material is absent ($\kappa = 0$). The initial shape of the membrane is $u(x, y, 0) = \varphi(x, y)$.

Solve the equations of motion with the given boundary conditions for each case given below.

21. The motion of the edge $x = 0$ of the membrane is given by

$$g_1(y, t) = A \sin \frac{\pi y}{l_y} \cos \omega t$$

and the other edges are fixed. Initially the membrane has the shape

$$\varphi(x, y) = A \left(1 - \frac{x}{l_x}\right) \sin \frac{\pi y}{l_y}.$$

22. The motion of the edge $y = l_y$ of the membrane is given by

$$g_4(x, t) = A \sin \frac{\pi x}{l_x} \sin \omega t$$

and the other edges are fixed. Initially the membrane has the shape

$$\varphi(x, y) = Axy \cos \frac{\pi x}{2l_x}.$$

23. Edges $x = 0$ and $x = l_x$ are fixed, the edge $y = 0$ is free and the edge $y = l_y$ moves as
$$g_4(x, t) = A \sin \frac{\pi x}{l_x} \cos \omega t.$$
Initially the membrane has the shape

$$\varphi(x, y) = Ax(l_x - x) \sin \frac{\pi y}{l_y}.$$

24. The edges $x = 0$ and $y = l_y$ are fixed and the edge $y = 0$ is subject to the action of a harmonic force causing displacements

$$g_3(x, t) = A \sin \frac{\pi x}{l_x} \cos \omega t.$$

Initially the membrane has the shape

$$\varphi(x, y) = Ax(y - l_y) \cos \frac{\pi x}{2l_x}.$$

In the following problems consider transverse oscillations of a homogeneous *circular* membrane of radius l, located in the x-y plane.

For all problems the displacement of the membrane, $u(r, \varphi, 0) = \phi(r, \varphi)$, is given at some initial moment of time, $t = 0$, and the membrane is released with initial velocity $u_t(r, \varphi, 0) = \psi(r, \varphi)$.

In problems 25 through 30 external forces and resistance of the environment are absent.

25. The membrane is fixed along its contour. The initial conditions are

$$\phi(r, \varphi) = Ar \left(l^2 - r^2 \right) \sin \varphi, \quad \psi(r, \varphi) = 0.$$

26. The membrane is fixed along its contour. The initial conditions are

$$\phi(r, \varphi) = 0, \quad \psi(r, \varphi) = Ar \left(l^2 - r^2 \right) \cos 4\varphi.$$

27. The edge of the membrane is free. The initial conditions are

$$\phi(r, \varphi) = Ar \left(l - \frac{r}{2} \right) \sin 3\varphi, \quad \psi(r, \varphi) = 0.$$

28. The edge of the membrane is free. The initial conditions are

$$\phi(r, \varphi) = 0, \quad \psi(r, \varphi) = Ar \left(l - \frac{r}{2} \right) \cos 2\varphi.$$

29. The edge of the membrane is fixed elastically with coefficient $h = 1$. The initial conditions are

$$\phi(r, \varphi) = Ar(1 - r) \sin \varphi, \quad \psi(r, \varphi) = 0.$$

30. The edge of the membrane is fixed elastically with coefficient $h = 1$. The initial conditions are

$$\phi(r, \varphi) = 0, \quad \psi(r, \varphi) = Ar(1 - r) \cos \varphi.$$

In problems 31 through 35 external forces are absent but the coefficient of resistance of the environment $\kappa \neq 0$ (resistance is proportional to velocity).

31. The membrane is fixed along its contour. The initial conditions are

$$\phi(r, \varphi) = Ar^3 \left(l^2 - r^2 \right) \sin 3\varphi, \quad \psi(r, \varphi) = 0.$$

32. The membrane is fixed along its contour. The initial conditions are

$$\phi(r, \varphi) = 0, \quad \psi(r, \varphi) = Ar^2 \left(l^2 - r^2 \right) \sin 2\varphi.$$

33. The edge of the membrane is free. The initial conditions are

$$\phi(r, \varphi) = Ar \left(l - \frac{r}{2} \right) \sin 5\varphi, \quad \psi(r, \varphi) = 0.$$

34. The edge of the membrane is free. The initial conditions are

$$\phi(r, \varphi) = 0, \quad \psi(r, \varphi) = Ar \left(l - \frac{r}{2} \right) \cos 4\varphi.$$

35. The edge of the membrane is fixed elastically with coefficient $h = 1$. The initial conditions are

$$\phi(r, \varphi) = Ar(r - 1) \sin 3\varphi, \quad \psi(r, \varphi) = 0.$$

In problems 36 through 38 the membrane is excited at time $t = 0$ by a sharp impact from a hammer, transferring to the membrane an impulse I at a point (r_0, φ_0) where $0 < r_0 < l$ and $0 \le \varphi_0 < 2\pi$.

For the following situations find free transverse oscillations of the membrane assuming the initial displacement is zero, external forcing is absent and the environment causes a resistance proportional to velocity $(\kappa > 0)$.

36. The membrane is fixed along its contour.

37. The edge of the membrane is free.

38. The edge of the membrane is attached elastically with elastic coefficient $h = 1$.

In problems 39 through 43 the initial displacement and initial velocity are zero. The resistance of environment is absent $(\kappa = 0)$. Find the transversal vibrations of the membrane caused by the action of a varying external pressure $f(r, \varphi, t)$ on one side of the membrane surface.

39. The membrane is fixed along its contour. The external force is

$$f(r, \varphi, t) = A \left(\sin \omega t + \cos \omega t \right).$$

40. The membrane is fixed along its contour. The external force is

$$f(r, \varphi, t) = A(l - r) \sin \omega t.$$

41. The edge of the membrane is free. The external force is

$$f(r, \varphi, t) = Ar \cos \omega t.$$

42. The edge of the membrane is fixed elastically with coefficient $h = 1$. The external force is

$$f(r, \varphi, t) = A \left(l^2 - r^2 \right) \sin \omega t.$$

43. The edge of the membrane is fixed elastically with coefficient $h = 1$. The external force is

$$f(r, \varphi, t) = A \left(l - r \right) \cos \omega t.$$

In problems 44 through 47 find the transverse oscillations of a membrane caused by its border being displaced according to the function $g(t)$ (the nonhomogeneous term in the boundary condition). Assume external forces, initial velocities and resistance of the environment are absent $(\kappa = 0)$. The initial displacement is given by $u(r, \varphi, 0) = \phi(r)$.

44. $g(t) = A \sin^2 \omega t, \quad \phi(r) = Br(l - r).$

45. $g(t) = A(1 - \cos \omega t), \quad \phi(r) = B(l^2 - r^2).$

46. $g(t) = A \cos^2 \omega t, \quad \phi(r) = Br\left(\dfrac{r}{2} - l\right).$

47. $g(t) = A(1 - \sin \omega t), \quad \phi(r) = Br(r - l).$

9

Two-Dimensional Parabolic Equations

In this chapter we discuss parabolic equations for a two-dimensional bounded medium. We consider rectangular and circular domains. The presentation is very similar to that for two-dimensional hyperbolic equations in Chapter 8. As before we discuss the heat problem in order to have a specific example at hand but it should be remembered that any other physical problem described by a two-dimensional hyperbolic equation can be solved using the methods discussed below.

9.1 Heat Conduction within a Finite Rectangular Domain

Let a heat-conducting, uniform rectangular plate be placed in the horizontal x-y plane with boundaries given by edges along $x = 0$, $x = l_x$, $y = 0$ and $y = l_y$. The plate is assumed to be thin enough that the temperature is the same at all points with the same x-y coordinates.

Let $u(x, y, t)$ be the temperature of the plate at the point (x, y) at time t. The heat conduction within such a thin uniform rectangular plate is described by the equation

$$\frac{\partial u}{\partial t} = a^2 \left[\frac{\partial^2 u}{\partial x^2} + \frac{\partial^2 u}{\partial y^2} \right] + \xi_x \frac{\partial u}{\partial x} + \xi_y \frac{\partial u}{\partial y} - \gamma u + f(x, y, t),$$
$$0 < x < l_x, \quad 0 < y < l_y, \quad t > 0. \tag{9.1}$$

Here a^2, ξ_x, ξ_y and γ are real constants. In terms of heat exchange, $a^2 = k/c\rho$ is the thermal diffusivity of the material; $\gamma = h/c\rho$ where h is the heat exchange coefficient (for lateral heat exchange with an external medium); $f(x, y, t) = Q(x, y, t)/c\rho$ where Q is the density of the heat source ($Q < 0$ for locations where the heat is absorbed). The terms with coefficients ξ_x and ξ_y describe the heat transfer by the flow with velocity vector $(-\xi_x, -\xi_y)$ due to bulk motion of the surrounding medium. Clearly these coefficients will equal zero for solids but are non-zero for liquids or gases in which bulk movement (advection) of the medium occurs.

The *initial condition* defines the temperature distribution within the plate at time zero:

$$u(x, y, 0) = \varphi(x, y). \tag{9.2}$$

The *boundary conditions* describe the thermal conditions at the boundary at any time t. The boundary conditions can be written in a general form as

$$P_1[u] \equiv \alpha_1 u_x + \beta_1 u|_{x=0} = g_1(y, t), \quad P_2[u] \equiv \alpha_2 u_x + \beta_2 u|_{x=l_x} = g_2(y, t),$$
$$P_3[u] \equiv \alpha_3 u_y + \beta_3 u|_{y=0} = g_3(x, t), \quad P_4[u] \equiv \alpha_4 u_y + \beta_4 u|_{y=l_y} = g_4(x, t), \tag{9.3}$$

where $g_1(y, t), g_2(y, t), g_3(x, t)$ and $g_4(x, t)$ are known functions of time and respective variable and $\alpha_1, \beta_1, \ldots, \alpha_4, \beta_4$ are real constants. As has been discussed previously, physical arguments lead to the sign restrictions $\alpha_1/\beta_1 < 0$ and $\alpha_3/\beta_3 < 0$.

As before, there are three main types of boundary conditions (here and in the following we denote $a = 0$ or l_x and $b = 0$ or $l_y = 6$):

Case I. Boundary condition of the 1st type (*Dirichlet condition*) where we are given the temperature along the y- or x-edge:

$$u(a, y, t) = g(y, t) \quad \text{or} \quad u(x, b, t) = g(x, t).$$

We may also have zero temperature at the edges in which case $g(y, t) \equiv 0$ or $g(x, t) \equiv 0$.

Case II. Boundary condition of the 2nd type (*Neumann condition*) where we are given the heat flow along the y- or x-edge:

$$u_x(a, y, t) = g(y, t) \quad \text{or} \quad u_y(x, b, t) = g(x, t).$$

We may also have a thermally insulated edge in which case $g(y, t) \equiv 0$ or $g(x, t) \equiv 0$.

Case III. Boundary condition of the 3rd type (*mixed condition*) where there is heat exchange with a medium along the y- or x- edge given by

$$u_x(a, y, t) \pm hu(a, y, t) = g(y, t) \quad \text{or} \quad u_y(x, b, t) \pm hu(x, b, t) = g(x, t).$$

We assume that h is a positive constant in which case the positive sign should be chosen in two previous formulas when $a = l_x$, $b = l_y$ and negative when $a = 0$.

The BVP formulated in Equations (9.1)–(9.3) can be reduced to the boundary value problem

$$\frac{\partial v}{\partial t} = a^2 \left[\frac{\partial^2 v}{\partial x^2} + \frac{\partial^2 v}{\partial y^2} \right] - \tilde{\gamma} v + \tilde{f}(x, y, t) \tag{9.4}$$

$$v(x, y, 0) = \tilde{\varphi}(x, y) \tag{9.5}$$

and

$$\alpha_1 v_x + \tilde{\beta}_1 v \Big|_{x=0} = \tilde{g}_1(y, t), \quad \alpha_2 v_x + \tilde{\beta}_2 v \Big|_{x=l_x} = \tilde{g}_2(y, t),$$

$$\alpha_3 v_y + \tilde{\beta}_3 v \Big|_{y=0} = \tilde{g}_3(x, t), \quad \alpha_4 v_y + \tilde{\beta}_4 v \Big|_{y=l_y} = \tilde{g}_4(x, t) \tag{9.6}$$

with the help of the substitution

$$u(x, y, t) = e^{\mu x + \eta y} v(x, y, t), \tag{9.7}$$

where

$$\mu = -\frac{\xi_x}{2a^2} \quad \text{and} \quad \eta = -\frac{\xi_y}{2a^2}.$$

Here

$$\tilde{\gamma} = \gamma + \frac{\xi_x^2 + \xi_y^2}{4a^2},$$

$$\tilde{f}(x, y, t) = e^{-(\mu x + \eta y)} f(x, y, t),$$

$$\tilde{\varphi}(x, y) = e^{-(\mu x + \eta y)} \varphi(x, y),$$

$$\tilde{\beta}_1 = \beta_1 + \mu \alpha_1, \quad \tilde{g}_1(y, t) = e^{-\eta y} g_1(y, t),$$

$$\tilde{\beta}_2 = \beta_2 + \mu \alpha_2, \quad \tilde{g}_2(y, t) = e^{-(\mu l_x + \eta y)} g_2(y, t),$$

$$\tilde{\beta}_3 = \beta_3 + \eta \alpha_3, \quad \tilde{g}_3(x, t) = e^{-\mu x} g_3(x, t),$$

$$\tilde{\beta}_4 = \beta_4 + \eta \alpha_4, \quad \tilde{g}_4(x, t) = e^{-(\mu x + \eta l_y)} g_4(x, t).$$

Reading Exercise: Make the substitution given in Equation (9.7) and verify the results above.

Therefore, below we will only consider Equations (9.1) with $\xi_x = \xi_y = 0$.

First, we should reduce boundary conditions to zero ones. To do that, let us present the solution to the problem defined by Equations (9.1) through (9.3) (with $\xi_x = \xi_y = 0$) as the sum of two functions:

$$u(x, y, t) = v(x, y, t) + w(x, y, t), \tag{9.8}$$

where $v(x, y, t)$ is a new, unknown function and $w(x, y, t)$ is an auxiliary function satisfying boundary conditions (9.3). We shall seek an auxiliary function, $w(x, y, t)$, in the form

$$
\begin{aligned}
w(x, y, t) = &\, g_1(y, t)\overline{X} + g_2(y, t)\overline{\overline{X}} + g_3(x, t)\overline{Y} + g_4(x, t)\overline{\overline{Y}} \\
&+ A(t)\overline{X}\,\overline{Y} + B(t)\overline{X}\,\overline{\overline{Y}} + C(t)\overline{\overline{X}}\,\overline{Y} + D(t)\overline{\overline{X}}\,\overline{Y},
\end{aligned} \tag{9.9}
$$

where $\overline{X}(x)$, $\overline{\overline{X}}(x)$, $\overline{Y}(y)$ and $\overline{\overline{Y}}(y)$ are polynomials of 1st or 2nd order, $A(t)$, etc. are some functions of t. The coefficients of these polynomials will be adjusted in such a way that function $w(x, y, t)$ satisfies boundary conditions given in Equations (9.3). The algorithm to find this function is presented in Appendix C part 2.

Then, the function $v(x, y, t)$ represents heat conduction when heat sources are present within the plate and *boundary conditions are zero*:

$$\frac{\partial v}{\partial t} = a^2 \left[\frac{\partial^2 v}{\partial x^2} + \frac{\partial^2 v}{\partial y^2} \right] - \gamma v + \tilde{f}(x, t), \tag{9.10}$$

$$v(x, y, 0) = \tilde{\varphi}(x, y),$$

$$P_1[v]\big|_{x=0} = 0, \quad P_2[v]\big|_{x=l_x} = 0, \quad P_3[v]\big|_{y=0} = 0, \quad P_4[v]\big|_{y=l_y} = 0,$$

where

$$\tilde{f}(x, y, t) = f(x, y, t) - \frac{\partial w}{\partial t} + a^2 \left(\frac{\partial^2 w}{\partial x^2} + \frac{\partial^2 w}{\partial y^2} \right) - \gamma w,$$

$$\tilde{\varphi}(x, y) = \varphi(x, y) - w(x, y, 0).$$

For this problem the Fourier method can be applied.

To simplify the task even more, we present function $v(x, y, t)$ as the sum of two functions

$$v(x, y, t) = u_1(x, y, t) + u_2(x, y, t), \tag{9.11}$$

where $u_1(x, y, t)$ is the solution of the *homogeneous* equation with the *given initial conditions* and $u_2(x, y, t)$ is the solution of a *nonhomogeneous* equation with *zero initial conditions*. For both functions, u_1 and u_2, *the boundary conditions are zero* (i.e. homogeneous). This step (9.11) does not necessarily have to be done, but it makes the solution more physically transparent.

The solution $u_1(x, y, t)$ represents the case of *free heat exchange*, that is, heat neither generated within nor lost from the plate, but only transferred by conduction. The solution $u_2(x, y, t)$ represents the *non-free heat exchange*, that is, the diffusion of heat due to generation of (or absorption by) internal sources when the initial distribution of temperature is zero. We find these two solutions in the following sub-sections.

9.1.1 The Fourier Method for the Homogeneous Heat Equation (Free Heat Exchange)

Let us first find the solution of the homogeneous equation

$$\frac{\partial u_1}{\partial t} = a^2 \left[\frac{\partial^2 u_1}{\partial x^2} + \frac{\partial^2 u_1}{\partial y^2} \right] - \gamma u_1 \tag{9.12}$$

with the initial condition

$$u_1(x, y, 0) = \varphi(x, y), \tag{9.13}$$

and zero boundary conditions

$$P_1[u_1]|_{x=0} = 0, \quad P_2[u_1]|_{x=l_x} = 0, \quad P_3[u_1]|_{y=0} = 0, \quad P_4[u_1]|_{y=l_y} = 0. \tag{9.14}$$

This describes the case of *free heat exchange* within the plate.

Let us separate time and spatial variables:

$$u_1(x, y, t) = T(t) \cdot V(x, y). \tag{9.15}$$

As a *Reading Exercise* obtain: *a*) the following equation for the function $T(t)$:

$$T'(t) + (a^2\lambda + \gamma)T(t) = 0, \tag{9.16}$$

and *b*) the boundary value problem with homogeneous boundary conditions for the function $V(x, y)$, defined by

$$\frac{\partial^2 V}{\partial x^2} + \frac{\partial^2 V}{\partial y^2} + \lambda V(x, y) = 0, \tag{9.17}$$

$$\alpha_1 V_x(0, y) + \beta_1 V(0, y) = 0, \quad \alpha_2 V_x(l_x, y) + \beta_2 V(l_x, y) = 0,$$
$$\alpha_3 V_y(x, 0) + \beta_3 V(x, 0) = 0, \quad \alpha_4 V_y(0, l_y) + \beta_4 V(0, l_y) = 0, \tag{9.18}$$

where λ is the constant of separation of variables.

Next we can again separate variables:

$$V(x, y) = X(x)Y(y).$$

As a *Reading Exercise*, obtain the following one-dimensional boundary value problems:

$$X''(x) + \lambda_x X(x) = 0,$$
$$\alpha_1 X'(0) + \beta_1 X(0) = 0, \quad \alpha_2 X'(l_x) + \beta_2 X(l_x) = 0, \tag{9.19}$$

and

$$Y''(y) + \lambda_y Y(y) = 0,$$
$$\alpha_3 Y'(0) + \beta_3 Y(0) = 0, \quad \alpha_4 Y'(l_y) + \beta_4 Y(l_y) = 0, \tag{9.20}$$

where the separation of variables constants, λ_x and λ_y, are connected by the relation $\lambda_x + \lambda_y = \lambda$.

If λ_{xn}, λ_{ym}, $X_n(x)$ and $Y_m(y)$ are eigenvalues and eigenfunctions, respectively of the boundary value problems for $X(x)$ and $Y(y)$, then

$$\lambda_{nm} = \lambda_{xn} + \lambda_{ym} \tag{9.21}$$

and

$$V_{nm}(x,y) = X_n(x)Y_m(y) \tag{9.22}$$

are eigenvalues and eigenfunctions of the boundary value problem for $V(x,y)$. The functions $V_{nm}(x,y)$ are orthogonal and their norms are the products

$$\|V_{nm}\|^2 = \|X_n\|^2 \cdot \|Y_m\|^2.$$

Eigenvalues and eigenfunctions of the boundary value problem depend on the types of boundary conditions. Combining different types of boundary conditions one can obtain nine different types of boundary value problems for the solution $X(x)$ and nine different types for the solution $Y(y)$ (see Appendix C part 1). Notice, that Equation (9.17) written in the form

$$\nabla^2 V(x,y) = -\lambda V(x,y),$$

with the Laplace operator

$$\nabla^2 = \frac{\partial^2}{\partial x^2} + \frac{\partial^2}{\partial y^2}$$

allows us to conclude that functions V_{nm} are eigenfunctions, and λ_{nm} are the eigenvalues of this operator for the boundary conditions (9.18).

The eigenvalues can be written as:

$$\lambda_{xn} = \left(\frac{\mu_{xn}}{l_x}\right)^2, \quad \lambda_{ym} = \left(\frac{\mu_{ym}}{l_y}\right)^2, \tag{9.23}$$

where μ_{xn} is the nth root of the equation

$$\tan \mu_x = \frac{(\alpha_1 \beta_2 - \alpha_2 \beta_1) l_x \mu_x}{\mu_x^2 \alpha_1 \alpha_2 + l_x^2 \beta_1 \beta_2}, \tag{9.24}$$

and μ_{ym} is the mth root of the equation

$$\tan \mu_y = \frac{(\alpha_3 \beta_4 - \alpha_4 \beta_3) l_y \mu_y}{\mu_y^2 \alpha_3 \alpha_4 + l_y^2 \beta_3 \beta_4}. \tag{9.25}$$

Similarly, as was obtained in Appendix C part 1 the eigenfunctions are

$$X_n(x) = \frac{1}{\sqrt{\alpha_1^2 \lambda_{xn} + \beta_1^2}} \left[\alpha_1 \sqrt{\lambda_{xn}} \cos \sqrt{\lambda_{xn}} x - \beta_1 \sin \sqrt{\lambda_{xn}} x\right],$$

$$Y_m(y) = \frac{1}{\sqrt{\alpha_3^2 \lambda_{ym} + \beta_3^2}} \left[\alpha_3 \sqrt{\lambda_{ym}} \cos \sqrt{\lambda_{ym}} y - \beta_3 \sin \sqrt{\lambda_{ym}} y\right]. \tag{9.26}$$

Now, having the eigenvalues λ_{nm} and eigenfunctions, $V_{nm}(x,y)$, of the boundary value problem, we may obtain the solution $u_1(x,y,t)$. The solution to equation

$$T'_{nm}(t) + (a^2 \lambda_{nm} + \gamma)T_{nm}(t) = 0 \tag{9.27}$$

is

$$T_{nm}(t) = C_{nm} e^{-(a^2 \lambda_{nm} + \gamma)t}, \tag{9.28}$$

from which we see that the function $u_1(x,y,t)$ can be composed as the infinite series

$$u_1(x,y,t) = \sum_{n=1}^{\infty} \sum_{m=1}^{\infty} T_{nm}(t) V_{nm}(x,y). \tag{9.29}$$

Using the fact that this function must satisfy the initial condition (9.13) and using the orthogonality condition for functions $V_{nm}(x, y)$ we find coefficients C_{nm}:

$$C_{nm} = \frac{1}{\|V_{nm}\|^2} \int_0^{l_x} \int_0^{l_y} \varphi(x, y) V_{nm}(x, y) dx dy. \tag{9.30}$$

It is left to the reader as a *Reading Exercise* to verify Equation (9.30) using the initial conditions (9.13) and the orthogonality condition.

Example 9.1 The initial temperature distribution within a thin uniform rectangular plate $(0 \leq x \leq l_x, 0 \leq y \leq l_y)$ with thermally insulated lateral faces is

$$u(x, y, 0) = Axy(l_x - x)(l_y - y), \quad A = \text{const.}$$

Find the distribution of temperature within the plate at any later time if its boundary is kept a constant zero temperature. Generation (or absorption) of heat by internal sources is absent.

Solution. The problem may be modeled by the solution of the equation

$$\frac{\partial u}{\partial t} = a^2 \left[\frac{\partial^2 u}{\partial x^2} + \frac{\partial^2 u}{\partial y^2} \right], \quad 0 < x < l_x, \ 0 < y < l_y, \ t > 0,$$

under the conditions

$$u(x, y, 0) = \varphi(x, y) = Axy(l_x - x)(l_y - y),$$

$$u(0, y, t) = u(l_x, y, t) = u(x, 0, t) = u(x, l_y, t) = 0.$$

The general solution to this problem can be presented as a sum (9.29)

$$u(x, y, t) = \sum_{n=1}^{\infty} \sum_{m=1}^{\infty} T_{nm} V_{nm}.$$

The boundary conditions of the problem are Dirichlet homogeneous boundary conditions; therefore eigenvalues and eigenfunctions of the problem are

$$\lambda_{nm} = \lambda_{xn} + \lambda_{ym} = \pi^2 \left(\frac{n^2}{l_x^2} + \frac{m^2}{l_y^2} \right), \quad n, m = 1, 2, 3, \dots$$

$$\|V_{nm}\|^2 = \|X_n\|^2 \cdot \|Y_m\|^2 = \frac{l_x l_y}{4}.$$

Applying the Equation (9.30), we obtain

$$C_{nm} = \frac{4}{l_x l_y} \int_0^{l_x} \int_0^{l_y} Axy(l_x - x)(l_y - y) \sin \frac{\pi n x}{l_x} \sin \frac{\pi m y}{l_y} dx dy$$

$$= \begin{cases} \dfrac{64 A l_x^2 l_y^2}{\pi^2 n^2 m^2}, & \text{if } n \text{ and } m \text{ are odd,} \\ 0, & \text{if } n \text{ or } m \text{ are even.} \end{cases}$$

Hence, the distribution of temperatures inside the plate at some instant of time is described by the series

$$u(x, y, t) = \frac{64 A l_x^2 l_y^2}{\pi^6} \sum_{n,m=1}^{\infty} \frac{e^{-\lambda_{nm} a^2 t}}{(2n+1)^2 (2m+1)^2} \sin \frac{(2n+1)\pi x}{l_x} \sin \frac{(2m+1)\pi y}{l_y}.$$

We suggest the reader compare this example and its result with Example 1 from Chapter 8.

9.1.2 The Fourier Method for Nonhomogeneous Heat Equation with Homogeneous Boundary Conditions

As mentioned previously, the solution $u_2(x, y, t)$ represents the *non-free heat exchange* within the plate; that is, the diffusion of heat due to generation (or absorption) of heat by internal sources (or sinks) for the case of an initial distribution of temperatures equal to zero. The solution to the general problem of heat conduction in a plate consists of the sum of the *free heat exchange* solutions, $u_1(x, y, t)$, found in the previous section and the solutions, $u_2(x, y, t)$, which will be discussed in this section.

The function $u_2(x, y, t)$ is a solution of the *nonhomogeneous* equation

$$\frac{\partial u_2}{\partial t} = a^2 \left[\frac{\partial^2 u_2}{\partial x^2} + \frac{\partial^2 u_2}{\partial y^2} \right] - \gamma u_2 + f(x, y, t) \tag{9.31}$$

with *zero initial and boundary conditions*. As above, we can separate time and spatial variables to obtain a general solution in the form

$$u_2(x, y, t) = \sum_{n=1}^{\infty} \sum_{m=1}^{\infty} T_{nm}(t) V_{nm}(x, y), \tag{9.32}$$

where $V_{nm}(x, y)$ are eigenfunctions of the corresponding homogeneous boundary value problem given by Equation (9.29). Here $T_{nm}(t)$ are, as yet, unknown functions of t. Zero boundary conditions for $u_2(x, y, t)$ given by

$$P_1[u_2]|_{x=0} = 0, \quad P_2[u_2]|_{x=l_x} = 0, \quad P_3[u_2]|_{y=0} = 0, \quad P_4[u_2]|_{y=l_y} = 0$$

are valid for any choice of functions $T_{nm}(t)$ (assuming the series converge uniformly) because they are valid for the functions $V_{nm}(x, y)$. We leave it to the reader to obtain these results as a *Reading Exercise*.

We now determine the functions $T_{nm}(t)$ in such a way that the series (9.32) satisfies the nonhomogeneous Equation (9.31) and the homogeneous (zero) initial condition. Substituting the series (9.32) into Equation (9.31) we obtain

$$\sum_{n=1}^{\infty} \sum_{m=1}^{\infty} \left[T'_{nm}(t) + (a^2 \lambda_{nm} + \gamma) T_{nm}(t) \right] V_{nm}(x, y) = f(x, y, t). \tag{9.33}$$

We can expand the function $f(x, y, t)$ in a Fourier series of the functions $V_{nm}(x, y)$ in the rectangular region $[0, l_x; 0, l_y]$ such that

$$f(x, y, t) = \sum_{n=1}^{\infty} \sum_{m=1}^{\infty} f_{nm}(t) V_{nm}(x, y), \tag{9.34}$$

where

$$f_{nm}(t) = \frac{1}{||V_{nm}||^2} \int_0^{l_x} \int_0^{l_y} f(x, y, t) V_{nm}(x, y) dx dy. \tag{9.35}$$

Comparing the two expansions in Equations (9.33) and (9.34) for the same function $f(x, y, t)$ we obtain differential equations for the functions $T_{nm}(t)$:

$$T'_{nm}(t) + (a^2 \lambda_{nm} + \gamma) T_{nm}(t) = f_{nm}(t). \tag{9.36}$$

In order that the solution represented by the series $u_2(x, y, t)$ given in Equation (9.32) satisfies the zero temperature initial condition it is necessary that the functions $T_{nm}(t)$ obey the condition

$$T_{nm}(0) = 0. \tag{9.37}$$

Clearly, the solution of the ordinary differential Equation (9.36) with initial condition (9.37) may be written in the integral form

$$T_{nm}(t) = \int_0^t f_{nm}(\tau) e^{-(a^2 \lambda_{nm} + \gamma)(t-\tau)} d\tau. \tag{9.38}$$

Thus the solution of the nonhomogeneous heat conduction problem for a thin uniform rectangular plate with homogeneous boundary conditions (equal to zero) has the form

$$u(x, y, t) = u_1(x, y, t) + u_2(x, y, t) = \sum_{n=1}^{\infty} \sum_{m=1}^{\infty} \left[T_{nm}(t) + C_{nm} e^{-(a^2 \lambda_{nm} + \gamma)t} \right] V_{nm}(x, y). \tag{9.39}$$

where the functions $T_{nm}(t)$ are defined by Equation (9.38) and the coefficients C_{nm} have been found earlier in Equation (9.30).

Example 9.2 Find the temperature $u(x, y, t)$ of a thin rectangular plate ($0 \leq x \leq l_x$, $0 \leq y \leq l_y$) if its boundary is kept at constant zero temperature, the initial temperature distribution within the plate is zero, and one internal source of heat $Q(t) = A \sin \omega t$ acts at the point (x_0, y_0) in the plate. Assume the plate is thermally insulated over its lateral faces.

Solution. The problem is expressed as

$$\frac{\partial u}{\partial t} = a^2 \left[\frac{\partial^2 u}{\partial x^2} + \frac{\partial^2 u}{\partial y^2} \right] + \frac{A}{c\rho} \sin \omega t \cdot \delta(x - x_0) \delta(y - y_0),$$

under the conditions

$$u(x, y, 0) = 0,$$

$$u(0, y, t) = u(l_x, y, t) = u(x, 0, t) = u(x, l_y, t) = 0.$$

The general solution to this problem can be presented as a sum

$$u(x, y, t) = \sum_{n=1}^{\infty} \sum_{m=1}^{\infty} \left[T_{nm}(t) + C_{nm} e^{-(a^2 \lambda_{nm} + \gamma)t} \right] V_{nm}(x, y).$$

The boundary conditions of the problem are Dirichlet homogeneous boundary conditions, therefore eigenvalues and eigenfunctions of problem are:

$$\lambda_{nm} = \lambda_{xn} + \lambda_{ym} = \pi^2 \left(\frac{n^2}{l_x^2} + \frac{m^2}{l_y^2} \right), \quad n, m = 1, 2, 3, \ldots$$

$$\|V_{nm}\|^2 = \|X_n\|^2 \cdot \|Y_m\|^2 = \frac{l_x l_y}{4}.$$

The initial condition is zero, in which case $C_{nm} = 0$. Applying Equation (9.35), we obtain

$$f_{nm}(t) = \frac{4}{l_x l_y} \int_0^{l_x} \int_0^{l_y} \frac{A}{c\rho} \sin \omega t \cdot \delta(x - x_0) \delta(y - y_0) \sin \frac{\pi n x}{l_x} \sin \frac{\pi m y}{l_y} dx dy$$

$$= \frac{4A}{c\rho l_x l_y} \sin \omega t \sin \frac{\pi n x_0}{l_x} \sin \frac{\pi m y_0}{l_y}.$$

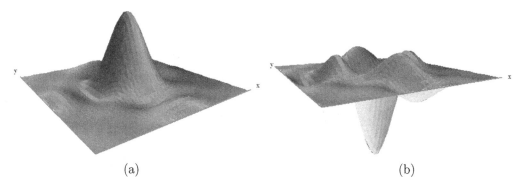

(a)   (b)

FIGURE 9.1
Surface graph of plate temperature at (a) $t = 0.6$, (b) $t = 6.2$ for Example 9.2.

We also have from the above formulas that

$$T_{nm}(t) = \int_0^t f_{nm}(\tau)e^{-\lambda_{nm}a^2(t-\tau)}d\tau = \frac{4A}{c\rho l_x l_y}\sin\frac{\pi n x_0}{l_x}\sin\frac{\pi m y_0}{l_y}$$

$$\times \frac{1}{\omega^2 + (a^2\lambda_{nm})^2}\left[a^2\lambda_{nm}\sin\omega t - \omega\cos\omega t + \omega e^{-\lambda_{nm}a^2 t}\right]$$

so that, finally we obtain

$$u(x,y,t) = \frac{4A}{c\rho l_x l_y}\sum_{n,m=1}^{\infty}\frac{1}{\omega^2 + (a^2\lambda_{nm})^2}\left[a^2\lambda_{nm}\sin\omega t - \omega\cos\omega t + \omega e^{-\lambda_{nm}a^2 t}\right]$$

$$\times \sin\frac{\pi n x_0}{l_x}\sin\frac{\pi m y_0}{l_y}\sin\frac{\pi n x}{l_x}\sin\frac{\pi m y}{l_y}.$$

Figure 9.1 shows snapshots of the solution at times $t = 0.6$ and $t = 6.2$. This solution was obtained for the case $a^2 = 0.25$, $l_x = 4$, $l_y = 6$, $A/c\rho = 120$, $\omega = 5$, $x_0 = 2$ and $y_0 = 3$.

Reading Exercise: Find the periodic in time solution of this problem and compare it with the solution $u(x,y,t)$ obtained above when $t \to \infty$.

Hint: Search the solution in the form $\text{Re}\left[F(x,y)\exp(i\omega t)\right]$, then, for a complex function, $F(x,y)$; you will obtain the Helmholtz equation with zero boundary and initial conditions. A similar problem was discussed for the one-dimensional heat equation.

Example 9.3 A heat-conducting, thin, uniform rectangular plate is thermally insulated over its lateral faces. One side of the plate, at $x = 0$, is thermally insulated and the rest of the boundary is kept at constant zero temperature. The initial temperature distribution within the plate is zero.

Let heat be generated throughout the plate with the intensity of internal sources (per unit mass of the plate) given by

$$Q(x,y,t) = A(l_x - x)\sin\frac{\pi y}{l_y}\sin t.$$

Find the distribution of temperature within the plate when $t > 0$.

Solution. The problem involves finding the solution of the equation

$$\frac{\partial u}{\partial t} = a^2\left[\frac{\partial^2 u}{\partial x^2} + \frac{\partial^2 u}{\partial y^2}\right] + \frac{A}{C\rho}(l_x - x)\sin\frac{\pi y}{l_y}\sin t,$$

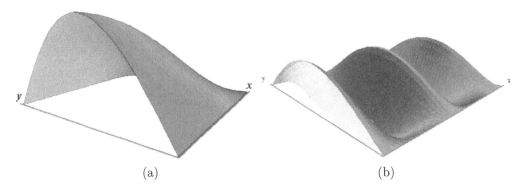

(a)   (b)

FIGURE 9.2
Eigenfunctions (a) $V_{11}(x, y)$ and (b) $V_{51}(x, y)$ for Example 3.

under the conditions

$$u(x, y, 0) = 0,$$

$$\frac{\partial u}{\partial x}(0, y, t) = 0, \quad u(l_x, y, t) = 0, \quad u(x, 0, t) = u(x, l_y, t) = 0.$$

Eigenvalues and eigenfunctions of the problem are given by

$$\lambda_{nm} = \lambda_{xn} + \lambda_{ym} = \pi^2 \left[\frac{(2n - 1)^2}{4l_x^2} + \frac{m^2}{l_y^2} \right], \quad n, m = 1, 2, 3, \ldots,$$

$$V_{nm}(x, y) = X_n(x)Y_m(y) = \cos \frac{(2n - 1)\pi x}{2l_x} \sin \frac{m\pi y}{l_y},$$

$$\|V_{nm}\|^2 = \|X_n\|^2 \cdot \|Y_m\|^2 = \frac{l_x l_y}{4}.$$

The three-dimensional picture shown in Figure 9.2 depicts two eigenfunctions (chosen as examples), $V_{11}(x, y)$ and $V_{51}(x, y)$ for the given problem.

The initial condition is equal to zero, in which case $C_{nm} = 0$. Applying Equation (9.35), we obtain

$$f_{n1}(t) = \frac{4}{l_x l_y} \int_0^{l_x} \int_0^{l_y} f(x, y, t) \cos \frac{(2n - 1)\pi x}{2l_x} \sin \frac{\pi y}{l_y} dx dy = \frac{8Al_x}{\pi^2 (2n - 1)^2} \sin(t),$$

and $f_{nm} = 0$, if $m \neq 1$. Thus we have

$$T_{n1}(t) = \frac{1}{\|v_{n1}\|^2} \int_0^t f_{n1}(\tau) e^{-\lambda_{n1} a^2 (t - \tau)} d\tau$$

$$= \frac{8Al_x}{\pi^2 (2n - 1)^2 \left[1 + (a^2 \lambda_{n1})^2 \right]} \left\{ a^2 \lambda_{n1} \sin(t) - \cos(t) + e^{-\lambda_{n1} a^2 t} \right\},$$

$$T_{nm} = 0, \quad \text{if} \quad m \neq 1.$$

and, finally,

$$u(x, y, t) = \frac{8Al_x}{\pi^2} \sin \frac{\pi y}{l_y} \sum_{n=1}^{\infty} \frac{a^2 \lambda_{n1} \sin t - \cos t + e^{-\lambda_{n1} a^2 t}}{(2n - 1)^2 \left[\omega^2 + (a^2 \lambda_{n1})^2 \right]} \cos \frac{(2n - 1)\pi x}{2l_x}.$$

The BVPs for a rectangular domain in situations with nonhomogeneous boundary conditions are considered in Appendix E part 2.

9.2 Heat Conduction within a Circular Domain

Let a uniform circular plate be placed in the horizontal x-y plane and bounded at the circular periphery by a radius of length l. The plate is assumed to be thin enough so that the temperature is the same at all points with the same x-y coordinates.

In polar coordinates of the Laplace operator is

$$\nabla^2 u = \frac{\partial^2 u}{\partial r^2} + \frac{1}{r}\frac{\partial u}{\partial r} + \frac{1}{r^2}\frac{\partial^2 u}{\partial \varphi^2}$$

and the heat conduction is described by the equation

$$\frac{\partial u}{\partial t} = a^2 \left(\frac{\partial^2 u}{\partial r^2} + \frac{1}{r}\frac{\partial u}{\partial r} + \frac{1}{r^2}\frac{\partial^2 u}{\partial \varphi^2} \right) - \gamma u + f(r, \varphi, t),$$

$$0 \le r < l, \quad 0 \le \varphi < 2\pi, \quad t > 0. \tag{9.40}$$

The *initial condition* defines the temperature distribution within the plate at zero time

$$u(r, \varphi, 0) = \phi(r, \varphi). \tag{9.41}$$

The *boundary condition* describes the thermal condition around the boundary at any time t:

$$P[u]_{r=l} \equiv \alpha \frac{\partial u}{\partial r} + \beta u \Big|_{r=l} = g(\varphi, t). \tag{9.42}$$

It is obvious that function $g(\varphi, t)$ must be a single-valued periodic function in φ of period 2π, that is,

$$g(\varphi + 2\pi, t) = g(\varphi, t).$$

Again we consider three main types of boundary conditions:

i) Boundary condition of the 1st type (*Dirichlet condition*), $u(l, \varphi, t) = g(\varphi, t)$, where the temperature at the boundary is given or is zero in which case $g(\varphi, t) \equiv 0$.

ii) Boundary condition of the 2nd type (*Neumann condition*), $u_r(l, \varphi, t) = g(\varphi, t)$, in which case the heat flow at the boundary is given or the boundary is thermally insulated and $g(\varphi, t) \equiv 0$.

iii) Boundary condition of the 3rd type (*mixed condition*), $u_r(l, \varphi, t) + hu(l, \varphi, t) = g(\varphi, t)$, where the conditions of heat exchange with a medium are specified (here $h = \text{const}$).

In the case of the nonhomogeneous boundary condition we introduce an auxiliary function, $w(r, \varphi, t)$, satisfying the given boundary condition and express the solution to the problem as the sum of two functions:

$$u(r, \varphi, t) = v(r, \varphi, t) + w(r, \varphi, t),$$

where $v(r, \varphi, t)$ is a new, unknown function with zero boundary condition. As in Chapter 8, we will seek function $w(r, \varphi, t)$ in the form

$$w(r, \varphi, t) = (c_0 + c_2 r^2)g(\varphi, t)$$

with constants c_0 and c_2 to be adjusted to satisfy the boundary condition.

As we did a number of times, it helps to present the function $v(r, \varphi, t)$ as the sum

$$v(r, \varphi, t) = u_1(r, \varphi, t) + u_2(r, \varphi, t),$$

where $u_1(r, \varphi, t)$ is the solution of the homogeneous equation (free heat exchange) with the given initial condition and zero boundary condition, and $u_2(r, \varphi, t)$ is the solution of the nonhomogeneous equation (heat exchange involving internal sources) with zero initial and boundary conditions.

9.2.1 The Fourier Method for the Homogeneous Heat Equation

Let us find the solution of the homogeneous equation

$$\frac{\partial u_1}{\partial t} = a^2 \left(\frac{\partial^2 u_1}{\partial r^2} + \frac{1}{r} \frac{\partial u_1}{\partial r} + \frac{1}{r^2} \frac{\partial^2 u_1}{\partial \varphi^2} \right) - \gamma u_1 \tag{9.43}$$

with the initial condition

$$u_1(r, \varphi, 0) = \phi(r, \varphi) \tag{9.44}$$

and zero boundary condition

$$P[u_1]_{r=l} \equiv \alpha \frac{\partial u_1}{\partial r} + \beta u_1 \bigg|_{r=l} = 0. \tag{9.45}$$

Presenting the unknown function as

$$u_1(r, \varphi, t) = T(t)V(r, \varphi), \tag{9.46}$$

substituting this in Equation (9.43) and separating variables we obtain,

$$\frac{T''(t) + \gamma T(t)}{a^2 T(t)} \equiv \frac{1}{V(r, \varphi)} \left[\frac{\partial^2 V}{\partial r^2} + \frac{1}{r} \frac{\partial V}{\partial r} + \frac{1}{r^2} \frac{\partial^2 V}{\partial \varphi^2} \right] = -\lambda,$$

where λ is a separation of variables constant (we know that a choice of minus sign before λ is convenient). Thus, the function $T(t)$ is the solution of the ordinary linear homogeneous differential equation of first order

$$T'(t) + (a^2\lambda + \gamma)T(t) = 0, \tag{9.47}$$

and $V(r, \varphi)$ is the solution to the following boundary value problem:

$$\frac{\partial^2 V}{\partial r^2} + \frac{1}{r} \frac{\partial V}{\partial r} + \frac{1}{r^2} \frac{\partial^2 V}{\partial \varphi^2} + \lambda V = 0, \tag{9.48}$$

$$\alpha \frac{\partial V}{\partial r}(l, \varphi) + \beta V(l, \varphi) = 0. \tag{9.49}$$

Two restrictions on $V(r, \varphi)$ are that it be bounded, $|V(r, \varphi)| < \infty$ and that it be periodic in φ: $V(r, \varphi) = V(r, \varphi + 2\pi)$.

This boundary value problem for the function $V(r, \varphi)$ we have discussed in detail in Appendix D part 1 and in Chapter 7. After separation of variables

$$V(r, \varphi) = R(r)\Phi(\varphi) \tag{9.50}$$

we obtained the Bessel equation for the function $R(r)$ and $\cos n\varphi$ and $\sin n\varphi$ for $\Phi(\varphi)$.

The eigenvalues and eigenfunctions of the boundary value problem for function $R_n(r)$ have the form

$$\lambda_{nm} = \left(\frac{\mu_m^{(n)}}{l}\right)^2, \quad R_{nm}(r) = J_n\left(\frac{\mu_m^{(n)}}{l}r\right), \quad n, m = 0, 1, 2, \ldots, \quad (9.51)$$

where $\mu_m^{(n)}$ is the m-th *positive* root of the equation

$$\alpha\mu J_n'(\mu) + \beta l J_n(\mu) = 0, \quad (9.52)$$

and $J_n(\mu)$ is the Bessel function of the 1st kind.

Collecting the above results we may write the eigenfunctions of the given BVP as

$$V_{nm}^{(1)}(r, \varphi) = J_n\left(\frac{\mu_m^{(n)}}{l}r\right)\cos n\varphi \quad \text{and} \quad V_{nm}^{(2)}(r, \varphi) = J_n\left(\frac{\mu_m^{(n)}}{l}r\right)\sin n\varphi. \quad (9.53)$$

Functions $V_{nm}^{(1,2)}(r, \varphi)$ are eigenfunctions *of the Laplace operator* in polar coordinates in the domain $0 \le r \le l$, $0 \le \varphi < 2\pi$, and λ_{nm} are the corresponding eigenvalues.

With the eigenvalues λ_{nm} we can write the solution of the differential equation

$$T_{nm}'(t) + (a^2\lambda_{nm} + \gamma)T_{nm}(t) = 0 \quad (9.54)$$

as

$$T_{nm}(t) = C_{nm}e^{-(a^2\lambda_{nm}+\gamma)t}. \quad (9.55)$$

From this we see that the general solution for function u_1 is

$$u_1(r, \varphi, t) = \sum_{n=0}^{\infty}\sum_{m=0}^{\infty}\left[a_{nm}V_{nm}^{(1)}(r, \varphi) + b_{nm}V_{nm}^{(2)}(r, \varphi)\right]e^{-(a^2\lambda_{nm}+\gamma)t}. \quad (9.56)$$

The coefficients a_{nm} and b_{nm} are defined using the function which expresses the initial condition and the orthogonality property of functions $V_{nm}^{(1)}(r, \varphi)$ and $V_{nm}^{(2)}(r, \varphi)$:

$$a_{nm} = \frac{1}{\left\|V_{nm}^{(1)}\right\|^2}\int_0^l\int_0^{2\pi}\phi(r, \varphi)V_{nm}^{(1)}(r, \varphi)r\,dr\,d\varphi, \quad (9.57)$$

$$b_{nm} = \frac{1}{\left\|V_{nm}^{(2)}\right\|^2}\int_0^l\int_0^{2\pi}\phi(r, \varphi)V_{nm}^{(2)}(r, \varphi)r\,dr\,d\varphi. \quad (9.58)$$

The norms of eigenfunctions $V_{nm}^{(1)}(r, \varphi)$ and $V_{nm}^{(2)}(r, \varphi)$ can be found in Appendix D part 1.

Example 9.4 The initial temperature distribution within a very long (infinite) cylinder of radius l is

$$\phi(r, \varphi) = u_0\left(1 - \frac{r^2}{l^2}\right), \quad u_0 = const.$$

Find the distribution of temperature within the cylinder if its surface is kept at constant zero temperature. Generation (or absorption) of heat by internal sources is absent.

Solution. The boundary value problem modeling the process of the cooling of an infinite cylinder is

$$\frac{\partial u}{\partial t} = a^2 \left[\frac{\partial^2 u}{\partial r^2} + \frac{1}{r} \frac{\partial u}{\partial r} \right],$$

$$u(r, \varphi, 0) = u_0 \left(1 - \frac{r^2}{l^2} \right), \quad u(l, \varphi, t) = 0.$$

The initial temperature does not depend on the polar angle φ; thus, only terms with $n = 0$ are not zero and the solution includes only functions $V_{0m}^{(1)}(r, \varphi)$ given by the first of Equations (9.53). The solution $u(r, \varphi, t)$ is therefore given by the series

$$u(r, \varphi, t) = \sum_{m=0}^{\infty} a_{0m} e^{-a^2 \lambda_{0m} t} J_0 \left(\frac{\mu_m^{(0)}}{l} r \right).$$

The boundary condition of the problem is of *Dirichlet type*, so eigenvalues $\mu_m^{(n)}$ are given by the roots of equation

$$J_n(x) = 0,$$

where $x = \mu r / l$. As it was mentioned in Appendix D part 1 for simplicity, such equations are often written just as $J_n(\mu) = 0$.

The coefficients a_{0m} are given by Equation (9.57):

$$a_{0m} = \frac{2\pi}{||V_{0m}^{(1)}||^2} \int_0^l u_0 \left(1 - \frac{r^2}{l^2} \right) J_0 \left(\mu_m^{(0)} r / l \right) r dr.$$

Thus, the distribution of temperature within the cylinder is given by the series in Bessel functions of zero-th order:

$$u(r, \varphi, t) = 8u_0 \sum_{m=0}^{\infty} \frac{e^{-a^2 \lambda_{0m} t}}{\left(\mu_m^{(0)} \right)^3 J_1 \left(\mu_m^{(0)} \right)} J_0 \left(\frac{\mu_m^{(0)}}{l} r \right).$$

We see that, due to the exponential nature of the coefficients the final temperature of the cylinder after a long time will be zero. This is due to dissipation of energy to the surrounding space and could have been anticipated from the physical configuration of the problem.

Example 9.5 The initial temperature distribution within a very long (infinite) cylinder of radius l is

$$u(r, \varphi, 0) = u_0 = \text{const.}$$

Find the distribution of temperature within the cylinder if it is subjected to convective heat transfer according to Newton's law at its surface and the temperature of the medium is zero.

Solution. The boundary value problem modeling the process of the cooling of an infinite cylinder is

$$\frac{\partial u}{\partial t} = a^2 \left[\frac{\partial^2 u}{\partial r^2} + \frac{1}{r} \frac{\partial u}{\partial r} \right],$$

$$u(r, \varphi, 0) = u_0,$$

$$\frac{\partial u}{\partial r}(l, \varphi, t) + hu(l, \varphi, t) = 0.$$

The boundary condition of the problem is of *mixed type* so eigenvalues of problem $\lambda_{nm} = (\mu_m^{(n)}/l)^2$ are the roots of the equation

$$\mu J_n'(\mu) + hl J_n(\mu) = 0, \text{ and}$$

the eigenfunctions are given by (9.53).

The coefficients a_{nm} and b_{nm} are given by Equations (9.57) and (9.58):

$$a_{nm} = \frac{u_0}{||V_{nm}^{(1)}||^2} \int_0^{2\pi} \cos n\varphi d\varphi \int_0^l r J_n\left(\frac{\mu_m^{(n)}}{l}r\right) dr \ ,$$

$$b_{nm} = \frac{u_0}{||V_{nm}^{(2)}||^2} \int_0^{2\pi} \sin n\varphi d\varphi \int_0^l r J_n\left(\frac{\mu_m^{(n)}}{l}r\right) dr.$$

The initial temperature does not depend on the polar angle φ; thus, only terms with $n = 0$ are not zero. Obviously $b_{nm} = 0$ for all n. Let us calculate coefficient a_{0m}. First, taking into account the relation $\mu J_0'(\mu) + hl J_0(\mu) = 0$, we may write

$$\left\|V_{0m}^{(1)}\right\|^2 = \pi \left[l^2 h^2 + \left(\mu_m^{(0)}\right)^2\right] \frac{l^2}{\left(\mu_m^{(0)}\right)^2} J_0^2\left(\mu_m^{(0)}\right).$$

From this we have

$$a_{0m} = \frac{2u_0 \mu_m^{(0)}}{\left[\left(\mu_m^{(0)}\right)^2 + h^2 l^2\right] J_0^2\left(\mu_m^{(0)}\right)} J_1\left(\mu_m^{(0)}\right).$$

Using the above relations we may write the distribution of temperature within the cylinder as the series in Bessel functions of zero order:

$$u(r,\varphi,t) = 2u_0 \sum_{m=0}^{\infty} \frac{\mu_m^{(0)} J_1\left(\mu_m^{(0)}\right) e^{-a^2 \lambda_{0m} t}}{\left[\left(\mu_m^{(0)}\right)^2 + h^2 l^2\right] J_0^2\left(\mu_m^{(0)}\right)} J_0\left(\frac{\mu_m^{(0)}}{l}r\right).$$

As in the previous example, dissipation of energy to the environment brings the final temperature of the plate to zero after long time periods.

9.2.2 The Fourier Method for the Nonhomogeneous Heat Equation

Function $u_2(r,\varphi,t)$ represents the *non-free heat exchange* within the plate; that is, the diffusion of heat due to generation (or absorption) of heat by internal sources when the initial distribution of temperature is zero. The function $u_2(r,\varphi,t)$ is the solution of the *nonhomogeneous* equation

$$\frac{\partial u_2}{\partial t} = a^2 \left(\frac{\partial^2 u_2}{\partial r^2} + \frac{1}{r}\frac{\partial u_2}{\partial r} + \frac{1}{r^2}\frac{\partial^2 u_2}{\partial \varphi^2}\right) - \gamma u_2 + f(r,\varphi,t) \tag{9.59}$$

with *initial and boundary conditions equal to zero*.

After the separation of variables the general solution to this equation clearly is

$$u_2(r,\varphi,t) = \sum_{n=0}^{\infty}\sum_{m=0}^{\infty} \left[T_{nm}^{(1)}(t)V_{nm}^{(1)}(r,\varphi) + T_{nm}^{(2)}(t)V_{nm}^{(2)}(r,\varphi)\right], \tag{9.60}$$

where $V_{nm}^{(1)}(r, \varphi)$ and $V_{nm}^{(2)}(r, \varphi)$ are eigenfunctions of the corresponding homogeneous boundary value problem and $T_{nm}^{(1)}(t)$ and $T_{nm}^{(2)}(t)$ are unknown functions of t.

The zero boundary condition for $u_2(r, \varphi, t)$

$$P[u_2]_{r=l} \equiv \alpha \frac{\partial u_2}{\partial r} + \beta u_2 \bigg|_{r=l} = 0$$

is valid for any choice of functions $T_{nm}^{(1)}(t)$ and $T_{nm}^{(2)}(t)$ because it is valid for the functions $V_{nm}^{(1)}(r, \varphi)$ and $V_{nm}^{(2)}(r, \varphi)$.

Substitution of the series (9.60) into Equation (9.59) gives

$$\sum_{n=0}^{\infty} \sum_{m=0}^{\infty} \left[\frac{dT_{nm}^{(1)}}{dt} + (a^2 \lambda_{nm} + \gamma) T_{nm}^{(1)}(t) \right] V_{nm}^{(1)}(r, \varphi)$$

$$+ \sum_{n=0}^{\infty} \sum_{m=0}^{\infty} \left[\frac{dT_{nm}^{(2)}}{dt} + (a^2 \lambda_{nm} + \gamma) T_{nm}^{(2)}(t) \right] V_{nm}^{(2)}(r, \varphi) = f(r, \varphi, t). \quad (9.61)$$

Next expand function $f(r, \varphi, t)$ within a circle of radius l as

$$f(r, \varphi, t) = \sum_{n=0}^{\infty} \sum_{m=0}^{\infty} \left[f_{nm}^{(1)}(t) V_{nm}^{(1)}(r, \varphi) + f_{nm}^{(2)}(t) V_{nm}^{(2)}(r, \varphi) \right], \quad (9.62)$$

where, using the orthogonality of this set of functions, the coefficients are

$$f_{nm}^{(1)}(t) = \frac{1}{\left\| V_{nm}^{(1)} \right\|^2} \int_0^l \int_0^{2\pi} f(r, \varphi, t) V_{nm}^{(1)}(r, \varphi) r \, dr \, d\varphi, \quad (9.63)$$

$$f_{nm}^{(2)}(t) = \frac{1}{\left\| V_{nm}^{(1)} \right\|^2} \int_0^l \int_0^{2\pi} f(r, \varphi, t) V_{nm}^{(2)}(r, \varphi) r \, dr \, d\varphi. \quad (9.64)$$

Comparing the expansions (9.61) and (9.62), we obtain differential equations for determining the functions $T_{nm}^{(1)}(t)$ and $T_{nm}^{(2)}(t)$:

$$\frac{dT_{nm}^{(1)}}{dt} + (a^2 \lambda_{nm} + \gamma) T_{nm}^{(1)}(t) = f_{nm}^{(1)}(t),$$

$$\frac{dT_{nm}^{(2)}}{dt} + (a^2 \lambda_{nm} + \gamma) T_{nm}^{(2)}(t) = f_{nm}^{(2)}(t). \quad (9.65)$$

In addition, these functions are necessarily subject to the initial conditions:

$$T_{nm}^{(1)}(0) = 0 \quad \text{and} \quad T_{nm}^{(2)}(0) = 0. \quad (9.66)$$

The solutions of the differential Equations (9.65) with initial conditions (9.66) can be presented in the form of integral relations

$$T_{nm}^{(1)}(t) = \int_0^t f_{nm}^{(1)}(\tau) e^{-(a^2 \lambda_{nm} + \gamma)(t - \tau)} d\tau, \quad (9.67)$$

$$T_{nm}^{(2)}(t) = \int_0^t f_{nm}^{(2)}(\tau) e^{-(a^2 \lambda_{nm} + \gamma)(t - \tau)} d\tau. \quad (9.68)$$

Thus, we have that the solution of the nonhomogeneous equation with zero initial and boundary conditions can be written as

$$u(x, y, t) = u_1(x, y, t) + u_2(x, y, t)$$

$$= \sum_{n=0}^{\infty} \sum_{m=0}^{\infty} \left\{ \left[T_{nm}^{(1)}(t) + a_{nm} e^{-(a^2 \lambda_{nm} + \gamma)t} \right] V_{nm}^{(1)}(x, y) \right.$$

$$\left. + \left[T_{nm}^{(2)}(t) + b_{nm} e^{-(a^2 \lambda_{nm} + \gamma)t} \right] V_{nm}^{(2)}(x, y) \right\} \tag{9.69}$$

with coefficients a_{nm} and b_{nm} given by formulas (9.57), (9.58).

Example 9.6 Find the temperature within a thin circular plate of radius l if its boundary is kept at constant zero temperature, the initial temperature distribution within the plate is zero, and one internal source of heat $Q(t) = A \sin \omega t$ acts at the point (r_0, φ_0) of the plate where $0 \le r_0 < l$, $0 \le \varphi_0 < 2\pi$. The plate is thermally insulated over its lateral faces.

Solution. The problem is expressed as

$$\frac{\partial u}{\partial t} = a^2 \left[\frac{\partial^2 u}{\partial r^2} + \frac{1}{r} \frac{\partial u}{\partial r} \right] + \frac{A}{c\rho} \delta(r - r_0) \delta(\varphi - \varphi_0) \sin \omega t$$

under the conditions

$$u(r, \varphi, 0) = 0, \quad u(l, \varphi, t) = 0.$$

The boundary condition of the problem is of *Dirichlet type*, so eigenvalues are given by the equation $J_n(\mu) = 0$, the eigenfunctions are given by formulas (9.53), the eigenfunctions squared norms can be calculated using Equations (D.17) and (D.18) from Appendix D part 1:

$$\left\| V_{nm}^{(1)} \right\|^2 = \left\| V_{nm}^{(2)} \right\|^2 = \sigma_n \pi \frac{l^2}{2} \left[J_n' \left(\mu_m^{(n)} \right) \right]^2, \quad \sigma_n = \begin{cases} 2, & \text{if } n = 0, \\ 1, & \text{if } n > 0. \end{cases}$$

The initial temperature of the plate is zero, so we have $a_{nm} = 0$ and $b_{nm} = 0$ for all n, m.

Let us next find $f_{nm}^{(1)}(t)$ and $f_{nm}^{(2)}(t)$:

$$f_{nm}^{(1)}(t) = \frac{A}{c\rho} \frac{2}{\sigma_n \pi l^2 \left[J_n' \left(\mu_m^{(n)} \right) \right]^2} \sin \omega t \cos n\varphi_0 J_n \left(\frac{\mu_m^{(n)}}{l} r_0 \right),$$

$$f_{nm}^{(2)}(t) = \frac{A}{c\rho} \frac{2}{\sigma_n \pi l^2 \left[J_n' \left(\mu_m^{(n)} \right) \right]^2} \sin \omega t \sin n\varphi_0 J_n \left(\frac{\mu_m^{(n)}}{l} r_0 \right).$$

From this we have

$$T_{nm}^{(1)}(t) = \frac{A}{c\rho} \frac{2}{\sigma_n \pi l^2 \left[J_n' \left(\mu_m^{(n)} \right) \right]^2} \cos n\varphi_0 J_n \left(\frac{\mu_m^{(n)}}{l} r_0 \right) \cdot I(t),$$

$$T_{nm}^{(2)}(t) = \frac{A}{c\rho} \frac{2}{\sigma_n \pi l^2 \left[J_n' \left(\mu_m^{(n)} \right) \right]^2} \sin n\varphi_0 J_n \left(\frac{\mu_m^{(n)}}{l} r_0 \right) \cdot I(t),$$

where we have introduced

$$I(t) = \frac{1}{\left[\omega^2 + (a^2 \lambda_{nm})^2 \right]} \left\{ a^2 \lambda_{nm} \sin \omega t - \omega \cos \omega t + \omega e^{-\lambda_{nm} a^2 t} \right\}.$$

FIGURE 9.3
Temperature of plate at (a) $t = 0.5$ and (b) $t = 1$ for Example 9.6.

Therefore, the evolution of temperature within the plate is described by the series

$$u(r, \varphi, t) = \sum_{n=0}^{\infty} \sum_{m=0}^{\infty} \left[T_{nm}^{(1)} \cos(n\varphi) + T_{nm}^{(2)} \sin(n\varphi) \right] J_n \left(\frac{\mu_m^{(n)}}{l} r \right)$$

$$= \frac{2A}{c\rho\pi l^2} \sum_{n=0}^{\infty} \sum_{m=0}^{\infty} \frac{I(t)}{\sigma_n \left[J_n' \left(\mu_m^{(n)} \right) \right]^2} J_n \left(\frac{\mu_m^{(n)}}{l} r_0 \right) J_n \left(\frac{\mu_m^{(n)}}{l} r \right) \cos n(\varphi - \varphi_0).$$

Figure 9.3 shows two snapshots of the solution at the times $t = 0.5$ and $t = 1$. This solution was obtained for the case when $a^2 = 0.25$, $l = 2$, $r_0 = 1$, $\varphi_0 = 1$, $A/c\rho = 100$ and $\omega = 5$.

Reading Exercise: Discuss the above result for the equilibrium (final) state when the source of heat is placed in the center of the plate at $r_0 = 0$. Show that in this case the problem can be reduced to a one-dimensional ordinary differential equation.

Example 9.7 A heat-conducting, thin, uniform, circular plate of radius l is thermally insulated over its lateral faces. The boundary of the plate is kept at constant zero temperature, and the initial temperature distribution within the plate is zero. Let heat be generated throughout the plate with the intensity of internal sources (per unit mass of the membrane) given by

$$Q(t) = A \cos \omega t.$$

Find the distribution of temperature within the plate when $t > 0$.

Solution. The problem may be expressed by the equation

$$\frac{\partial u}{\partial t} = a^2 \left[\frac{\partial^2 u}{\partial r^2} + \frac{1}{r} \frac{\partial u}{\partial r} \right] + A \cos \omega t,$$

under the conditions

$$u(r, \varphi, 0) = 0, \quad u(l, \varphi, t) = 0.$$

The boundary condition of the problem is of *Dirichlet type*, so eigenvalues $\mu_m^{(n)}$ are given by equation $J_n(\mu) = 0$. The initial temperature of the plate is zero and the intensity of internal sources depends only on time t, so the solution $u(r, t)$ includes only functions $V_{0m}^{(1)} = J_0 \left(\mu_m^{(0)} r/l \right)$ and is given by the series

$$u(r, t) = \sum_{m=0}^{\infty} T_{0m}^{(1)}(t) J_0 \left(\frac{\mu_m^{(n)}}{l} r \right),$$

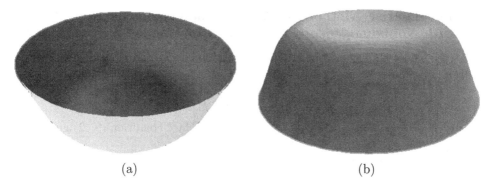

(a)   (b)

FIGURE 9.4
Temperature of plate at (a) $t = 1$ and (b) $t = 3$ for Example 9.7.

where

$$T_{0m}^{(1)}(t) = \int_0^t f_{0m}^{(1)}(\tau)e^{-(a^2\lambda_{0m}+\gamma)(t-\tau)}d\tau.$$

Taking into account that (see Appendix B)

$$\left\|V_{0m}^{(1)}\right\|^2 = \pi l^2 \left[J_0'\left(\mu_m^{(0)}\right)\right]^2 = \pi l^2 \left[J_1\left(\mu_m^{(0)}\right)\right]^2,$$

$$\int_0^l J_0\left(\frac{\mu_m^{(0)}}{l}r\right)r\,dr = \frac{l^2}{\left[\mu_m^{(0)}\right]^2}\int_0^{\mu_m^{(0)}} xJ_0(x)dx = \frac{l^2}{\mu_m^{(0)}}J_1\left(\mu_m^{(0)}\right),$$

we find $f_{0m}^{(1)}(t)$ in the form

$$f_{0m}^{(1)}(t) = \frac{A\cos\omega t}{\left\|V_{0m}^{(1)}\right\|^2}2\pi\int_0^l J_0\left(\frac{\mu_m^{(0)}}{l}r\right)r\,dr = \frac{2A}{\mu_m^{(0)}J_1\left(\mu_m^{(0)}\right)}\cos\omega t.$$

Then

$$T_{0m}^{(1)}(t) = \frac{2A}{\mu_m^{(0)}J_1\left(\mu_m^{(0)}\right)}\int_0^t \cos\omega\tau \cdot e^{-a^2\lambda_{0m}(t-\tau)}d\tau$$

$$= \frac{2A}{\mu_m^{(0)}J_1\left(\mu_m^{(0)}\right)\left[\omega^2 + (a^2\lambda_{0m})^2\right]}\left[a^2\lambda_{0m}\cos\omega t + \omega\sin\omega t - a^2\lambda_{0m}e^{-a^2\lambda_{0m}t}\right].$$

Therefore, the evolution of temperature within the plate is described by series

$$u(r,t) = \sum_{m=0}^{\infty} T_{0m}^{(1)}(t)J_0\left(\frac{\mu_m^{(n)}}{l}r\right).$$

Figure 9.4 shows two snapshots of the solution at the times $t = 1$ and $t = 3$. This solution was obtained for the case when $a^2 = 0.25$, $l = 2$, $A = 100$ and $\omega = 5$.

Reading Exercise: Discuss the above result for the equilibrium (final) state. Show that in this case the problem can be reduced to a one-dimensional ordinary differential equation.

9.2.3　The Fourier Method for the Nonhomogeneous Heat Equation with Nonhomogeneous Boundary Conditions

Finally, we consider the general boundary problem for heat conduction, Equation (9.40), given by

$$\frac{\partial u}{\partial t} = a^2 \left(\frac{\partial^2 u}{\partial r^2} + \frac{1}{r} \frac{\partial u}{\partial r} + \frac{1}{r^2} \frac{\partial^2 u}{\partial \varphi^2} \right) - \gamma u + f(r, \varphi, t)$$

with nonhomogeneous initial condition (9.41) and boundary condition (9.42) given by

$$u(r, \varphi, 0) = \phi(r, \varphi),$$

$$P[u]_{r=l} \equiv \alpha \frac{\partial u}{\partial r} + \beta u \bigg|_{r=l} = g(\varphi, t).$$

To reduce the nonhomogeneous boundary condition to homogeneous, we introduce an auxiliary function $w(r, \varphi, t)$ which satisfies the given nonhomogeneous boundary condition. As always, we will search for the solution of the problem as

$$u(r, \varphi, t) = v(r, \varphi, t) + w(r, \varphi, t),$$

where $v(r, \varphi, t)$ is a function satisfying the homogeneous boundary condition, and will seek the auxiliary function $w(r, \varphi, t)$ in the form

$$w(r, \varphi, t) = \left(c_0 + c_1 r + c_2 r^2 \right) g(\varphi, t).$$

The constants will be adjusted to satisfy the boundary condition. Because

$$\frac{1}{r} \frac{\partial w}{\partial r}(r, \varphi, t) = \left(\frac{c_1}{r} + 2c_2 \right) g(\varphi, t)$$

and $r = 0$ is a regular point, the coefficient $c_1 \equiv 0$ and the auxiliary function reduce to

$$w(r, \varphi, t) = \left(c_0 + c_2 r^2 \right) g(\varphi, t),$$

where c_0 and c_2 are real constants.

Example 9.8 The initial temperature distribution within a very long (infinite) cylinder of radius l is

$$u(r, \varphi, 0) = u_0 = \text{const.}$$

Find the distribution of temperature within the cylinder if a constant heat flow,

$$\frac{\partial u}{\partial r}(l, \varphi, t) = Q = \frac{q}{\kappa},$$

is supplied to the surface of the cylinder from the outside starting at time $t = 0$. Generation (or absorption) of heat by internal sources is absent.

Solution. This is the BVP

$$\frac{\partial u}{\partial t} = a^2 \left[\frac{\partial^2 u}{\partial r^2} + \frac{1}{r} \frac{\partial u}{\partial r} \right],$$

$$u(r, \varphi, 0) = u_0, \quad \frac{\partial u}{\partial r}(l, \varphi, t) = Q.$$

An auxiliary function satisfying the given boundary condition is

$$w(r, \varphi, t) = \frac{Q}{2l} r^2.$$

The solution to the problem should be sought in the form

$$u(r, \varphi, t) = w(r, \varphi, t) + v(r, \varphi, t),$$

where the function $v(r, \varphi, t)$ is the solution to the boundary value problem with homogeneous boundary conditions and

$$\tilde{f}(r, \varphi, t) = \frac{2a^2}{l}Q, \quad \tilde{\phi}(r, \varphi) = u_0 - \frac{Q}{2l}r^2.$$

These functions (defined in Section 9.1) do not depend on the polar angle φ, which is why the solution for function $v(r, t)$ contains only the Bessel function of zero-th order:

$$v(r, t) = \sum_{m=0}^{\infty} \left[T_{0m}^{(1)}(t) + a_{0m}e^{-a^2\lambda_{0m}t} \right] J_0 \left(\frac{\mu_m^{(0)}}{l}r \right).$$

The boundary condition of the problem is of *Neumann type*, so eigenvalues $\mu_m^{(n)}$ are given by the equation

$$J_n'(\mu) = 0.$$

The eigenfunctions and their norms are (D.16), (D.17) (Appendix D part 1)

$$V_{0m}^{(1)} = J_0 \left(\frac{\mu_m^{(0)}}{l}r \right), \quad \left\| V_{0m}^{(1)} \right\|^2 = \pi l^2 J_0^2 \left(\mu_m^{(0)} \right), \quad \left\| V_{00}^{(1)} \right\|^2 = \pi l^2,$$

in which case we have

$$a_{0m} = \frac{1}{\left\| V_{0m}^{(1)} \right\|^2} \int_0^{2\pi} \int_0^l \left(u_0 - \frac{Q}{2l}r^2 \right) J_0 \left(\frac{\mu_m^{(0)}}{l}r \right) r dr d\varphi = -\frac{2Ql}{\left(\mu_m^{(0)} \right)^2 J_0 \left(\mu_m^{(0)} \right)},$$

$$a_{00} = \frac{1}{\left\| V_{00}^{(1)} \right\|^2} \int_0^{2\pi} \int_0^l \left(u_0 - \frac{Q}{2l}r^2 \right) r dr d\varphi = u_0 - \frac{Ql}{4}.$$

Using the Equation (9.63) we obtain

$$f_{0m}^{(1)}(t) = \frac{2a^2Q}{l \left\| V_{0m}^{(1)} \right\|^2} \int_0^{2\pi} d\varphi \int_0^l J_0 \left(\frac{\mu_m^{(0)}}{l}r \right) r dr = \frac{4a^2Q}{l\mu_m^{(0)} J_0^2 \left(\mu_m^{(0)} \right)} J_1 \left(\mu_m^{(0)} \right) = 0,$$

$$f_{00}^{(1)}(t) = \frac{2a^2Q}{l \left\| V_{00}^{(1)} \right\|^2} \int_0^{2\pi} d\varphi \int_0^l r dr = \frac{2a^2Q}{l},$$

and with Equations (9.67), (9.68) we have

$$T_{0m}^{(1)}(t) = \int_0^t f_{0m}^{(1)}(\tau)e^{-a^2\lambda_{0m}(t-\tau)}d\tau = 0, \quad T_{00}^{(1)}(t) = \int_0^t f_{00}^{(1)}(\tau)d\tau = \frac{2a^2Q}{l}t.$$

Hence, the distribution of temperature within the cylinder at some instant of time is described by the series (9.69):

$$u(r, t) = \frac{Q}{2l}r^2 + u_0 - \frac{Ql}{4} + \frac{2a^2Q}{l}t - 2Ql \sum_{m=1}^{\infty} \frac{e^{-a^2\lambda_{0m}t}}{(\mu_m^{(0)})^2 J_0(\mu_m^{(0)})} J_0 \left(\frac{\mu_m^{(0)}}{l}r \right).$$

Reading Exercise: Discuss the role and the origin of each term in this solution.

9.3 Heat Conduction in an Infinite Medium

Consider the homogeneous heat-conduction equation in three dimensions

$$\frac{\partial u}{\partial t} = a^2 \nabla^2 u. \tag{9.70}$$

For a finite medium the separation of variables procedure, $u(\boldsymbol{r}, t) = X(\boldsymbol{r})T(t)$, gives the discrete spectrum of eigenvalues λ_n due to boundary conditions. For infinite medium problems, when there is no boundary condition, let us express the function $u(\boldsymbol{r}, t)$ as a Fourier integral with respect to the coordinates:

$$u(\boldsymbol{r}, t) = \frac{1}{(2\pi)^3} \int u_{\boldsymbol{k}}(t) e^{i\boldsymbol{k}\boldsymbol{r}} d^3\boldsymbol{k}, \qquad d^3k = dk_x dk_y dk_z, \tag{9.71}$$

where the Fourier coefficients are

$$u_{\boldsymbol{k}}(t) = \int u(\boldsymbol{r}, t) e^{-i\boldsymbol{k}\boldsymbol{r}} d\boldsymbol{r}, \qquad d\boldsymbol{r} = dx dy dz. \tag{9.72}$$

Substituting expression (9.71) into Equation (9.70), we obtain

$$\frac{1}{(2\pi)^3} \int \left(\frac{du_{\boldsymbol{k}}}{dt} + k^2 a^2 u_{\boldsymbol{k}} \right) e^{i\boldsymbol{k}\boldsymbol{r}} d^3k = 0.$$

Therefore, for each Fourier component, $u_{\boldsymbol{k}}(t)$, Equation (9.70) gives

$$\frac{\partial u_{\boldsymbol{k}}}{\partial t} + k^2 a^2 u_{\boldsymbol{k}} = 0,$$

from which we obtain

$$u_{\boldsymbol{k}}(t) = u_{0\mathrm{k}} e^{-k^2 a^2 t}. \tag{9.73}$$

It is clear that the coefficients $u_{0\boldsymbol{k}}$ are determined by the initial temperature distribution (initial condition) given by

$$u(\boldsymbol{x}, 0) = u_0(\boldsymbol{x}) \equiv \varphi(\boldsymbol{r}). \tag{9.74}$$

From Equations (9.73) and (9.74) we have

$$u_{0\boldsymbol{k}}(t) = \int \varphi(\boldsymbol{r}') e^{-i\boldsymbol{k}\boldsymbol{r}'} d\boldsymbol{r}'. \tag{9.75}$$

Thus, the temperature distribution as function of coordinates and time is

$$u(\boldsymbol{r}, t) = \frac{1}{(2\pi)^3} \int \varphi(\boldsymbol{r}') e^{-\boldsymbol{k}^2 a^2 t} e^{i\boldsymbol{k}(\boldsymbol{r}-\boldsymbol{r}')} d\boldsymbol{r}' d^3k. \tag{9.76}$$

The integral over $d^3\boldsymbol{k}$ is the product of three integrals; each of them is

$$\int_{-\infty}^{\infty} e^{-\alpha k_i^2} \cos \beta k_i dk_i = \left(\frac{\pi}{\alpha} \right)^{1/2} e^{-\beta^2/4\alpha}, \quad i = x, y, z \tag{9.77}$$

(the integrals with sines in place of cosines are zero since the sine function is odd). Finally, we obtain the following formula which gives the complete solution of the problem since it determines the temperature within the medium for any moment in time if the initial temperature distribution is known:

$$u(\boldsymbol{r}, t) = \frac{1}{8(\pi a^2 t)^{3/2}} \int \varphi(\boldsymbol{r}') \exp\left[-\frac{(\boldsymbol{r} - \boldsymbol{r}')^2}{4a^2 t}\right] d\boldsymbol{r}'. \tag{9.78}$$

If the initial temperature distribution is a function of only one coordinate, x (for example the case of a thin infinite bar), then performing an integration over y and z in Equation (9.78), yields

$$u(x, t) = \frac{1}{2\sqrt{\pi a^2 t}} \int_{-\infty}^{\infty} \varphi(x') \exp\left[-\frac{(x - x')^2}{4a^2 t}\right] dx'. \tag{9.79}$$

Reading Exercise. Check the step between Equations (9.78) and (9.79) and show that the result corresponds to the one-dimensional case discussed in Chapter 6.

Let us consider a useful illustration of the result in Equation (9.78) – the case when the temperature at $t = 0$ is zero everywhere except the origin of the coordinate system where it is infinite. We assume also that the total amount of energy (or heat) is proportional to $\int \varphi(\boldsymbol{r}) d\boldsymbol{r}$. For a point source at the origin we may write

$$\varphi(\boldsymbol{r}) = A\delta(\boldsymbol{r}), \quad A = \text{const},$$

where $\delta(\boldsymbol{r}) = \delta(x)\delta(y)\delta(z)$ is a three-dimensional delta function. From Equation (9.78) we immediately obtain

$$u(\boldsymbol{r}, t) = A\frac{1}{8(\pi a^2 t)^{3/2}} \exp\left[-\frac{r^2}{4a^2 t}\right]. \tag{9.80}$$

At the origin the temperature decreases as $t^{-3/2}$ and there is a corresponding temperature rise in the surrounding space. The size of the space where the temperature substantially differs from zero is determined by the exponential in Equation (9.80). That is, it is given by $l \approx a\sqrt{t}$ and l increases as the square root of the time. If at $t = 0$ the heat was concentrated in the plane at $x = 0$, or more generally if it does not depend on the x and y axis, then

$$u(x, t) = A\frac{1}{2\sqrt{\pi a^2 t}} \exp\left[-\frac{x^2}{4a^2 t}\right] \tag{9.81}$$

which follows from Equation (9.79).

The solution in Equation (9.80) with the delta-function as the initial condition is Green's function $G(\boldsymbol{r} - \boldsymbol{r}', t)$ for homogeneous heat conduction. In terms of Green's function, the solution in Equation (9.78) can be written as

$$u(\boldsymbol{r}, t) = \int \varphi(\boldsymbol{r}') G(\boldsymbol{r} - \boldsymbol{r}', t) d\boldsymbol{r}',$$

where

$$G(\boldsymbol{r} - \boldsymbol{r}', t) = \frac{1}{8(\pi a^2 t)^{3/2}} \exp\left[-\frac{(\boldsymbol{r} - \boldsymbol{r}')^2}{4a^2 t}\right] \tag{9.82}$$

We refer the reader to Chapter 6 where the properties of Green's functions were described; as was mentioned there, they also hold for the three-dimensional case. Using the three-dimensional Green's function it should be clear how to generalize the material presented in this section to, for example, the nonhomogeneous heat conduction equation given by

$$\frac{\partial u}{\partial t} = a^2 \nabla^2 u + f(\vec{r}, t).$$

9.4   Heat Conduction in a Semi-Infinite Medium

We discuss several cases.

Case I. Let us consider a three-dimensional medium located at $x > 0$ and begin with the case where constant temperature is maintained on the boundary plane located at $x = 0$. We will take this constant temperature as zero, i.e. the boundary and initial conditions are

$$u(x, y, z, t = 0) = \varphi(x, y, z), \quad u|_{x=0} = 0. \tag{9.83}$$

To apply the methods previously developed for an infinite medium we first imagine that the medium is extended to the left from $x = 0$ and an initial temperature distribution is defined for $x < 0$ by the same function, φ, taken with a minus sign. Thus the initial distribution for infinite space is an odd function of x:

$$\varphi(-x, y, z) = -\varphi(x, y, z). \tag{9.84}$$

From Equation (9.83) it follows that $\varphi(0, y, z) = 0$. From symmetry it is clear that this boundary condition will be valid for $t > 0$.

Now we can solve the Equation (9.70) for an infinite medium with an initial temperature distribution that satisfies Equation (9.84). This solution is given by the general Equation (9.78), in which we divide the range of integration over x into two parts, from $-\infty$ to 0 and from 0 to $+\infty$. Using Equation (9.84) we have

$$u(\boldsymbol{r}, t) = \frac{1}{8(\pi a^2 t)^{3/2}} \int_{-\infty}^{\infty} \int_{-\infty}^{\infty} \int_{0}^{\infty} \varphi(\boldsymbol{r}') \exp\left[-\frac{(y - y')^2 + (z - z')^2}{4a^2 t}\right]$$

$$\times \left\{ \exp\left[-\frac{(x - x')^2}{4a^2 t}\right] - \exp\left[-\frac{(x + x')^2}{4a^2 t}\right] \right\} dx' dy' dz'. \tag{9.85}$$

If the initial temperature distribution is a function of only x, Equation (9.85) gives

$$u(x, t) = \frac{1}{2\sqrt{\pi a^2 t}} \int_{0}^{\infty} \varphi(x') \left\{ \exp\left[-\frac{(x - x')^2}{4a^2 t}\right] - \exp\left[-\frac{(x + x')^2}{4a^2 t}\right] \right\} dx'. \tag{9.86}$$

Example 9.9 The temperature is maintained equal to zero on the (boundary) plane at $x = 0$. The initial temperature is constant everywhere for $x > 0$, i.e. $\varphi(x) = u_0$.

Solution. Performing the substitutions $\xi = \frac{x' \mp x}{2a\sqrt{t}}$ in the two integrals in Equation (9.86), with a minus sign in the first and a plus sign in the second, we obtain

$$u(x, t) = \frac{u_0}{2} \left[\mathrm{erf}\left(\frac{x}{2a\sqrt{t}}\right) - \mathrm{erf}\left(-\frac{x}{2a\sqrt{t}}\right) \right],$$

where $\mathrm{erf}(x) = \frac{2}{\sqrt{\pi}} \int_{0}^{x} e^{-\xi^2} d\xi$ is the *error function* ($\mathrm{erf}(\infty) = 1$). Since $\mathrm{erf}(-x) = -\mathrm{erf}(x)$, we obtain

$$u(x, t) = u_0 \, \mathrm{erf}\left(\frac{x}{2a\sqrt{t}}\right). \tag{9.87}$$

This result could be written immediately since it follows from the properties of the error function. It is easy to check that differentiating Equation (9.87) twice with respect to x and once with respect to t we obtain the equation $u_t = a^2 u_{xx}$, thus Equation (9.87) satisfies both the initial and boundary conditions.

Reading Exercise. The reader is encouraged to check the preceding statement.

Thus we see that the function $u(x,t) = A \ \mathrm{erf}(\frac{x}{2a\sqrt{t}})$, where A is an arbitrary constant, is a solution of equation $u_t = a^2 u_{xx}$.

Reading Exercises.

   a) Generalize the previous result to the three-dimensional case.

   b) Prove that if a semi-infinite bar is initially at zero temperature and the end at $x = 0$ is kept at temperature u_0, the temperature at time t is

$$u(x,t) = u_0 \left[1 - \mathrm{erf}\left(\frac{x}{2a\sqrt{t}}\right)\right].$$

According to Equation (9.87) the temperature propagates into space at a rate proportional to $\sqrt{t}$. The result (9.87) depends on the single dimensionless parameter $x/2a\sqrt{t}$.

Case II. Let us consider the case of a thermally insulated boundary plane at $x = 0$. That is, a boundary with no heat flux through it. The boundary and initial conditions are

$$u(x,y,z,t=0) = \varphi(x), \qquad \left.\frac{\partial u}{\partial x}\right|_{x=0} = 0. \qquad (9.88)$$

As in the previous example, imagine the medium to extend on both sides of the plane at $x = 0$, but in this case extend the initial temperature distribution, $\varphi(x,y,z)$, as an even function of x:

$$\varphi(-x,y,z) = \varphi(x,y,z), \qquad (9.89)$$

for which

$$\frac{\partial \varphi}{\partial x}(x,y,z) = -\frac{\partial \varphi}{\partial x}(-x,y,z) \quad \text{and} \quad \frac{\partial \varphi}{\partial x}(0,y,z) = 0 \quad \text{for} \quad x = 0.$$

From symmetry it is clear that this condition will be satisfied for all $t > 0$. Repeating the calculations above but using Equation (9.89) instead of Equation (9.84), we obtain the general solution which differs from Equations (9.85) and (9.86) by replacing the subtraction of two terms by the summation of two terms:

$$u(\mathbf{r},t) = \frac{1}{8(\pi a^2 t)^{3/2}} \int_{-\infty}^{\infty} \int_{-\infty}^{\infty} \int_{0}^{\infty} \varphi(\mathbf{r}') \exp\left[-\frac{(y-y')^2 + (z-z')^2}{4a^2 t}\right]$$

$$\times \left\{\exp\left[-\frac{(x-x')^2}{4a^2 t}\right] + \exp\left[-\frac{(x+x')^2}{4a^2 t}\right]\right\} dx' dy' dz', \qquad (9.90)$$

$$u(x,t) = \frac{1}{2\sqrt{\pi a^2 t}} \int_{0}^{\infty} \varphi(x') \left\{\exp\left[-\frac{(x-x')^2}{4a^2 t}\right] + \exp\left[-\frac{(x+x')^2}{4a^2 t}\right]\right\} dx', \qquad (9.91)$$

Case III. Assume that a heat flux enters the medium through its boundary plane at $x = 0$ for a medium located at $x > 0$; i.e. the boundary condition is

$$-\kappa \left.\frac{\partial u}{\partial x}\right|_{x=0} = q(t), \qquad (9.92)$$

with $q(t)$ as a given flux function. Because q does not depend on coordinates y and z the problem reduces to the one-dimensional case. The initial condition is

$$u(x, y, z, t = 0) = 0. \tag{9.93}$$

To begin we first solve an auxiliary problem with $q(t) = \delta(t)$. This problem is equivalent to the problem that led to Equation (9.81), i.e. to the problem of heat propagation in an infinite medium from a point source which produces a given amount of heat. Indeed Equation (9.92) means that a unit of energy enters through each unit area of the plane at $x = 0$ at the instant $t = 0$. In this problem where the initial condition is $u = \frac{2}{\rho c}\delta(x)$ for $t = 0$, an amount of heat $\int \rho c u dx = 2$ is concentrated in the same area at time $t = 0$. From symmetry we may argue that half of this energy flows in the $x > 0$ direction, the other half in the $x < 0$ direction. Since the solutions of both problems are identical, from Equation (9.81) we obtain

$$u(x, t) = \frac{1}{\kappa}\frac{a}{\sqrt{\pi t}}\exp\left[-\frac{x^2}{4a^2 t}\right]. \tag{9.94}$$

The heat conduction equation is linear so that for arbitrary $q(t)$ instead of $\delta(t)$ the general solution of Equation (9.70) with the conditions in Equations (9.92) and (9.93) is

$$u(x, t) = \frac{1}{\kappa}\int_{-\infty}^{t}\frac{a}{\sqrt{\pi(t - t')}}q(t')\exp\left[-\frac{x^2}{4a^2(t - t')}\right]dt'. \tag{9.95}$$

Reading Exercise. Check in detail the derivation of Equations (9.94) and (9.95).

In particular the temperature on the plane at $x = 0$ varies according to

$$u(0, t) = \frac{1}{\kappa}\int_{-\infty}^{t}\frac{a}{\sqrt{\pi(t - t')}}q(t')dt'. \tag{9.96}$$

Using Equation (9.95) we can solve the problem in which the temperature on the plane $x = 0$ is the given function of time,

$$u|_{x=0} = g(t), \tag{9.97}$$

and the initial temperature is constant (which can be taken as zero):

$$u(-x, y, z, t = 0) = 0. \tag{9.98}$$

Notice that if $u(x, t)$ satisfies Equation (9.70) then so does its derivative, $\frac{\partial u}{\partial x}$. Differentiating Equation (9.95) with respect to x, we obtain

$$-\kappa\frac{\partial u}{\partial x} = \int_{-\infty}^{t}\frac{xq(t')}{2a\sqrt{\pi(t - t')^3}}\exp\left[-\frac{x^2}{4a^2(t - t')}\right]dt'.$$

According to Equation (9.92) $q(t)$ has the same value at $x = 0$. Writing $u(x, t)$ instead of $-\kappa\frac{\partial u}{\partial x}$ and using $g(t)$ instead of $q(t)$, we obtain the solution of the problem as

$$u(x, t) = \frac{x}{2a\sqrt{\pi}}\int_{-\infty}^{t}\frac{g(t')}{\sqrt{(t - t')^3}}\exp\left[-\frac{x^2}{4a^2(t - t')}\right]dt'. \tag{9.99}$$

Reading Exercise. A radioactive gas is diffusing into the atmosphere from contaminated soil (the boundary of which we can locate at $x = 0$).

a) Provide the arguments to show that the density of the gas in the air, $\rho(x,t)$, is described by the boundary value problem

$$\frac{\partial \rho}{\partial t} = D\frac{\partial^2 \rho}{\partial x^2} - \lambda\rho, \quad \frac{\partial \rho}{\partial x}\bigg|_{x=0} = -\kappa, \quad \rho(x,0) = 0, \ (D, \lambda, \quad \kappa = \text{const}).$$

b) Show that the solution to this problem is

$$\rho(x,t) = \kappa\sqrt{\frac{D}{\pi}} \int_0^t \frac{1}{\sqrt{t'}} \exp\left[-\lambda t' - \frac{x^2}{4Dt'}\right] dt'.$$

Case IV. Let us consider an important particular case when the temperature varies periodically in time on the boundary plane at $x = 0$:

$$u(x = 0, y, z, t) = u_0 \cos \omega t. \tag{9.100}$$

This problem is equivalent to the classical problem (G. Stokes) about waves within an incompressible fluid generated by an infinite, rigid, flat surface harmonically oscillating in its own plane $(y\text{-}z)$. If we investigate the process at a time which is sufficiently long from the initial moment, the influence of the initial condition is practically negligible. Thus this is a problem without an initial condition and we seek a stationary solution. Formally we can choose the zero initial condition

$$u(x, y, z, t = 0) = \varphi(x, y, z) = 0.$$

Assume the fluid surface is at $x > 0$ and the plane oscillates along the y-axis, i.e. velocities in the fluid have only a y-component. The fluid velocity satisfies the Navier–Stokes equation which, for this geometry, reduces to a one-dimensional heat conduction equation

$$u_t = a^2 u_{xx},$$

where $u(x,t)$ is the y-component of fluid velocity and $a^2 = \nu$ is the dynamic viscosity of the fluid.

It is convenient to write the boundary condition as the real part of the complex expression $u = \text{Re}\left\{u_0 e^{-i\omega t}\right\}$. In the following while performing intermediate (linear) operations, we omit the symbol Re and take the real part of the final result. Thus we write the boundary condition as

$$u(0, t) = u_0 e^{-i\omega t}. \tag{9.101}$$

It is natural to seek a solution periodic in x and t given by

$$u(x, t) = u_0 e^{i(kx - \omega t)}, \tag{9.102}$$

which satisfies condition (9.101). Substitution of Equation (9.102) into the equation $u_t = a^2 u_{xx}$ gives

$$i\omega = a^2 k^2, \quad k = (1+i)/\delta, \quad \delta = \sqrt{2a^2/\omega}. \tag{9.103}$$

Thus

$$u(x, t) = u_0 e^{-x/\delta} \exp\left[i(x/\delta - \omega t)\right]. \tag{9.104}$$

(the choice of the sign before the root $\sqrt{i} = +(1+i)/2$ in the last of Equations (9.104) is determined by the physical requirement that the velocity should be bounded as x increases).

From this discussion we see that transverse waves can exist in fluids with non-zero viscosity where the velocity of the fluid is perpendicular to the wave propagation direction.

The oscillations damp quickly as the distance from the surface increases. The constant δ is called the penetration depth. At a distance δ from the surface the wave's amplitude decreases e times, in other words it decreases $e^{2\pi} \approx 540$ times during one wavelength. The penetration depth, δ, decreases with increasing frequency and increases with increasing viscosity. In a more general case, when a plane wave is moving according to some function $u(x = 0, t) = u_0(t)$ instead of the simple harmonic motion of Equation (9.101), the solution is given by the formula (9.99).

Similarly the temperature propagation inside a body when the temperature changes periodically according to Equation (9.100) on the boundary at $x = 0$, is described by the same Equation (9.104). Periodically changing surface temperature propagates from the boundary into a body in the form of a temperature wave with the amplitude decreasing exponentially with the depth (Fourier's first law).

An analogous phenomenon exists when alternating current flows in a metal conductor. Alternating current does not flow through a conductor with a uniform cross-sectional profile but concentrates close to the conductor surface (the so-called skin-effect). Inside a conductor the displacement current is insignificant in comparison to conduction current and the charge density is zero in which case the equations for the electrical and magnetic fields inside a homogeneous conductor become

$$\frac{\partial \vec{E}}{\partial t} = a^2 \nabla^2 \vec{E}, \quad \frac{\partial \vec{H}}{\partial t} = a^2 \nabla^2 \vec{H}, \tag{9.105}$$

where $a^2 = c^2/(4\pi\mu\lambda)$. Here c is the speed of light, μ is the magnetic susceptibility, λ is the electric conductivity and $\rho = 1/\lambda$ is the resistivity of the medium.

Let us consider the same geometry as in the previous problem, i.e. a conductor is placed at $x > 0$ where the x-axis is directed inside the conductor and an external electric field is directed along the y-axis, which is parallel to the conductor's surface. It is clear from symmetry that for a big surface (formally an infinite plane) the field depends on x (and on time), but does not depend on y and z. If the electric field changes along the plane according to Equation (9.100) or (9.101) then inside the conductor it is given by expression (9.104). Taking its real part we obtain the equation for the electric field inside the conductor as

$$E(x, t) = E_0 e^{-x/\delta} \cos\left[i(x/\delta - \omega t)\right], \tag{9.106}$$

where E_0 is the amplitude of the electric field on the surface of the conductor. The field (and current density $j = \lambda E$) are concentrated close to the surface in a layer of thickness δ. For example a cooper conductor ($\mu \approx 1$) with an applied field, E with a wavelength of 3000 m (radio frequency) has a penetration depth of $\delta \approx 0.2$mm. In the case of a direct current, $\omega = 0$ and thus $\delta \to \infty$; i.e. a direct current is evenly distributed across the cross-section of a conductor. A magnetic field is described by an equation identical to the one for the electric field. From these arguments we see that high-frequency electromagnetic fields do not penetrate deeply into a conductor but concentrate near its surface.

Problems

In problems 1 through 19 we consider a heat conduction within a *rectangular* plate ($0 \leq x \leq l_x$, $0 \leq y \leq l_y$), in problems 20 through 40 within a *circular* plate of radius l. Solve these problems analytically which means the following: formulate the equation and initial and boundary conditions, obtain the eigenvalues and eigenfunctions, write the formulas for coefficients of the series expansion and the expression for the solution of the problem.

You can obtain the pictures of several eigenfunctions and screenshots of the solution and of the auxiliary functions with Maple, Mathematica or software from books [7,8].

In problems 1 through 5 we consider rectangular plates which are thermally insulated over their lateral surfaces. In the initial time, $t = 0$, the temperature distribution is given by $u(x, y, 0) = \varphi(x, y)$. There are no heat sources or absorbers inside the membrane. Find the distribution of temperature within the membrane at any later time.

1. The boundary of the plate is kept at constant zero temperature. The initial temperature of the plate membrane is given as

$$\varphi(x, y) = A \sin \frac{\pi x}{l_x} \sin \frac{\pi y}{l_y}.$$

2. The edge at $x = 0$ of plate is thermally insulated and other edges are kept at zero temperature. The initial temperature of the plate is given as

$$\varphi(x, y) = A \cos \frac{\pi x}{2l_x} \sin \frac{\pi y}{l_y}.$$

3. The edge at $y = 0$ of the plate is thermally insulated and other edges are kept at zero temperature. The initial temperature of the plate is given as

$$\varphi(x, y) = A \sin \frac{\pi x}{l_x} \cos \frac{\pi y}{2l_y}.$$

4. The edges $x = 0$ and $y = 0$ of the plate are kept at zero temperature, and the edges $x = l_x$ and $y = l_y$ are thermally insulated. The initial temperature of the plate is given as

$$\varphi(x, y) = A x y \left(l_x - x\right) \left(l_y - y\right).$$

5. The edges $x = 0$, $x = l_x$ and $y = l_y$ of plate are kept at zero temperature, and the edge $y = 0$ is subjected to convective heat transfer with the environment which has a temperature of zero. The initial temperature of the plate is given as

$$\varphi(x, y) = A x y \left(l_x - x\right) \left(l_y - y\right).$$

In problems 6 through 10 we consider a rectangular plate which is thermally insulated over its lateral surfaces. The initial temperature distribution within the plate is zero, and one internal source of heat acts at the point (x_0, y_0) of the plate. The value of this source is $Q(t)$. Find the temperature within the plate.

6. The edges $x = 0$, $y = 0$ and $y = l_y$ of the plate are kept at zero temperature and the edge $x = l_x$ is subjected to convective heat transfer with the environment which has a temperature of zero. The value of the internal source is

$$Q(t) = A \cos \omega t.$$

7. The edges $x = 0$ and $y = 0$ of the plate are kept at zero temperature, the edge $y = l_y$ is thermally insulated and the edge $x = l_x$ is subjected to convective heat transfer with the environment which has a temperature of zero. The value of the internal source is

$$Q(t) = A \sin \omega t.$$

8. The edges $y = 0$ and $y = l_y$ are thermally insulated, the edge $x = 0$ is kept at zero temperature and the edge $x = l_x$ is subjected to convective heat transfer with the environment which has a temperature of zero. The value of the internal source is

$$Q(t) = Ae^{-t}\sin\omega t.$$

9. The edges $x = 0$ and $y = l_y$ are thermally insulated, the edge $x = l_x$ is kept at zero temperature and the edge $y = 0$ is subjected to convective heat transfer with the environment which has a temperature of zero. The value of the internal source is

$$Q(t) = Ae^{-t}\cos\omega t.$$

10. Find the heat distribution in a thin rectangular plate if it is subjected to heat transfer according to Newton's law at its edges. The temperature of the medium is $u_{md} = \text{const}$, the initial temperature of the plate is zero, and there is a constant source of heat, Q, uniformly distributed over the plate.

Hint: The problem is formulated as follows:

$$\frac{\partial u}{\partial t} = a^2\left[\frac{\partial^2 u}{\partial x^2} + \frac{\partial^2 u}{\partial y^2}\right] + Q, \quad 0 < x < l_x, \quad 0 < y < l_y, \quad t > 0,$$

$$u(x, y, 0) = 0,$$

$$\left.\frac{\partial u}{\partial x} - h(u - u_{md})\right|_{x=0} = 0, \quad \left.\frac{\partial u}{\partial x} + h(u - u_{md})\right|_{x=l_x} = 0,$$

$$\left.\frac{\partial u}{\partial y} - h(u - u_{md})\right|_{y=0} = 0, \quad \left.\frac{\partial u}{\partial y} + h(u - u_{md})\right|_{y=l_y} = 0.$$

For problems 11 through 13 consider a rectangular plate which is thermally insulated over its lateral surfaces. The edges of the plate are kept at the temperatures described by the function of $u(x, y, t)|_\Gamma$ given below. The initial temperature distribution within the plate is $u(x, y, 0) = u_0 = \text{const}$. Find heat distribution in the plate if there are no heat sources or absorbers inside the plate.

11. $u|_{x=0} = u|_{y=0} = u_1, \quad u|_{x=l_x} = u|_{y=l_y} = u_2,$

12. $u|_{x=0} = u|_{x=l_x} = u_1, \quad u|_{y=0} = u|_{y=l_y} = u_2.$

13. $u|_{x=0} = u|_{x=l_x} = u|_{y=0} = u_1, \quad u|_{y=l_y} = u_2.$

For problems 14 through 16 a thin homogeneous plate with sides of length π lies in the x-y plane. The edges of the plate are kept at the temperatures described by the function of $u(x, y, t)|_\Gamma$, given below. Find the temperature in the plate if initially the temperature has a constant value A and there are no heat sources or absorbers inside the plate.

14. $u|_{x=0} = u|_{x=\pi} = 0, \quad u|_{y=0} = u|_{y=\pi} = x^2 - x.$

15. $u|_{x=0} = u|_{x=\pi} = 0, \quad u|_{y=0} = x^2, \quad u|_{y=\pi} = 0.5x.$

16. $u|_{x=0} = 0, \quad u|_{x=\pi} = \cos y, \quad u|_{y=0} = u|_{y=\pi} = 0.$

In problems 17 through 19 an infinitely long rectangular cylinder has its central axis along the z-axis and its cross-section is a rectangle with sides of length π. The sides of the cylinder are kept at the temperature described by functions

$u(x, y, t)|_\Gamma$, given below. Find the temperature within the cylinder if initially the temperature is $u(x, y, 0) = Axy$ and there are no heat sources or absorbers inside the cylinder.

17. $u|_{x=0} = 3y^2$, $u|_{x=\pi} = 0$, $u|_{y=0} = u|_{y=\pi} = 0$.

18. $u|_{x=0} = u|_{x=\pi} = y^2$, $u|_{y=\pi} = 0$, $u|_{y=0} = x$.

19. $u|_{x=0} = u|_{x=\pi} = \cos 2y$, $u|_{y=0} = u|_{y=\pi} = 0$.

In problems 20 through 22 we consider circular plates of radius l which are thermally insulated over their lateral surfaces. In the initial time, $t = 0$, the temperature distribution is given by $u(r, \varphi, 0) = \phi(r, \varphi)$. There are no heat sources or absorbers inside the membrane. Find the distribution of temperature within the plate at any later time.

20. The boundary of the plate is kept at constant zero temperature. The initial temperature of the plate is given as

$$\phi(r, \varphi) = Ar \left(l^2 - r^2 \right) \sin \varphi.$$

21. The boundary of the plate is thermally insulated. The initial temperature of the plate is given as

$$\phi(r, \varphi) = u_0 r \cos 2\varphi.$$

22. The boundary of the membrane is subjected to convective heat transfer with the environment which has a temperature of zero. The initial temperature of the plate is given as

$$\phi(r, \varphi) = u_0 \left(1 - r^2/l^2 \right).$$

In problems 23 through 25 we consider a very long (infinite) cylinder of radius l. The initial temperature distribution within the cylinder is given by $u(r, \varphi, 0) = \phi(r, \varphi)$. There are no heat sources or absorbers inside the cylinder. Find the distribution of temperature within the cylinder at any later time.

23. The surface of the cylinder is kept at constant temperature $u = u_0$. The initial temperature distribution within the cylinder is given by

$$\phi(r, \varphi) = Ar \left(l^2 - r^2 \right) \sin \varphi.$$

24. The constant heat flow $\partial u/\partial r(l, \varphi, t) = Q$ is supplied to the surface of the cylinder from outside. The initial temperature distribution within the cylinder is given by

$$\phi(r, \varphi) = Ar \left(l - \frac{r}{2} \right) \sin 3\varphi.$$

25. The surface of the cylinder is subjected to convective heat transfer with the environment which has a temperature $u = u_{md}$. The initial temperature distribution within the cylinder is given by

$$\phi(r, \varphi) = u_0 \sin 4\varphi.$$

In problems 26 through 28 we consider a circular plate of radius l which is thermally insulated over its lateral surfaces. The initial temperature distribution within the plate is zero, and one internal source of heat acts at the point (r_0, φ_0) of the plate. The value of this source is $Q(t)$. Find the temperature within the plate.

26. The edge of the plate is kept at zero temperature. The value of the internal source is
$$Q(t) = A \cos \omega t.$$

27. The edge of the plate is thermally insulated. The value of the internal source is
$$Q(t) = A \sin \omega t.$$

28. The edge of the plate is subjected to convective heat transfer with the environment which has a temperature of zero. The value of the internal source is
$$Q(t) = A(\sin \omega t + \cos \omega t).$$

In problems 29 through 32 we consider a circular plate of radius l which is thermally insulated over its lateral surfaces. The initial temperature distribution within the plate is zero. Heat is generated uniformly throughout the plate; the intensity of internal sources (per unit area of the plate) is $Q(t)$. Find the temperature distribution within the plate.

29. The edge of the plate is kept at zero temperature. The intensity of the internal sources is
$$Q(t) = A \cos \omega t.$$

30. The edge of the plate is kept at zero temperature. The intensity of the internal sources is
$$Q(t) = A(l - r) \sin \omega t.$$

31. The edge of the membrane is thermally insulated. The intensity of the internal sources is
$$Q(t) = A \sin \omega t.$$

32. The edge of the plate is thermally insulated. The intensity of the internal sources is
$$Q(t) = A \left(l^2 - r^2\right) \sin \omega t.$$

In problems 33 through 35 we consider a very long (infinite) cylinder of radius l. The initial temperature of the cylinder is $u_0 = $ const. Find the temperature distribution within the cylinder.

33. The surface of the cylinder is kept at the temperature described by the function $u(l, \varphi, t) = A \sin \omega t.$

34. The heat flow at the surface is governed by $\frac{\partial}{\partial r} u(l, \varphi, t) = A \cos \omega t.$

35. The temperature exchange with the environment with zero temperature is governed according to Newton's law $\frac{\partial u}{\partial r} + u\big|_{r=l} = A(1 - \cos \omega t).$

Problems 36 through 39 are related to Section 9.4.

36. The temperature distribution in the earth takes place with a phase displacement. The time Δt between the occurrence of the temperature maximum (minimum) at depth x is described by formula $\left(\sqrt{1/2a^2\omega}\right) x$ (Fourier's second law). Derive this formula.

37. For two temperature distributions with periods T_1 and T_2, the corresponding depths x_1 and x_2 in which the relative temperature changes coincide are connected by the equation $x_2 = (\sqrt{T_2/T_1})x_1$ (Fourier's third law). Obtain this formula. Apply it for the daily and yearly variations to compare the depths of penetration.

38. The estimation for thermal diffusivity of the Earth is $a^2 \approx 0.4 \cdot 10^{-6} m^2/s$. How much time does it take for (the maximum) temperature to reach a 4 m depth?

39. An infinitely long rectangular cylinder, $0 \le x \le l_x$, $0 \le y \le l_y$, with the central axis along the z-axis is placed in a coil. At $t = 0$ a current in the coil turns on and the coil starts to generate an oscillation magnetic field outside the cylinder directed along the z-axis:

$$u(x, y, t) = H_0 \sin \omega t, \quad H_0 = \text{const}, \quad 0 < t < \infty.$$

Find the magnetic field inside the cylinder.

Hint: The problem is formulated as follows:

$$\frac{\partial u}{\partial t} = a^2 \left[\frac{\partial^2 u}{\partial x^2} + \frac{\partial^2 u}{\partial y^2} \right], \quad 0 < x < l_x, \quad 0 < y < l_y, \quad t > 0,$$

$$u(x, y, 0) = 0, \quad u|_{x=0} = u|_{x=l_x} = u|_{y=0} = u|_{y=l_y} = H_0 \sin \omega t.$$

Answer:

$$u(x, y, t) = H_0 \sin(\omega t) + \sum_{n=1}^{\infty} \sum_{m=1}^{\infty} T_{nm}(t) \sin \frac{n \pi x}{l_x} \sin \frac{m \pi y}{l_y},$$

where

$$T_{nm}(t) = -\frac{4 H_0 \omega \left[1 - (-1)^n \right] \left[1 - (-1)^m \right]}{n m \pi^2 \left[\omega^2 + (a^2 \lambda_{nm})^2 \right]}$$

$$\times \left\{ a^2 \lambda_{nm} \cos(\omega t) + \omega \sin(\omega t) - a^2 \lambda_{nm} e^{-\lambda_{nm} a^2 t} \right\}.$$

$$\lambda_{nm} = \lambda_{xn} + \lambda_{ym} = \pi^2 \left[\frac{n^2}{l_x^2} + \frac{m^2}{l_y^2} \right], \quad n, m = 1, 2, 3, \dots$$

10

Nonlinear Equations

In previous chapters, except Chapter 2, we dealt with linear PDEs. The present section is devoted to some famous nonlinear PDEs that have numerous applications in physics. All these equations are *integrable* in the sense that finding their solution can be reduced to solving some *linear* problems.

10.1 Burgers Equation

In Chapter 2 we considered the *non-viscous Burgers equation*,

$$u_t + uu_x = 0 \tag{10.1}$$

(another notation of a variable was used), which describes the propagation of a nonlinear sound wave in the non-viscous gas. We have seen that the nonlinearity can lead to an unbounded growth of the wave steepness $|u_x|$ during a finite time interval. In Chapter 6 we considered the heat equation,

$$u_t = \mu u_{xx} \tag{10.2}$$

(another notation for the coefficient was used), which describes different dissipative processes (heat transfer, diffusion, viscosity etc.). The linear term with the second derivative tends to diminish any spatial inhomogeneities.

When a nonlinear sound wave propagates in a viscous and heat-conducting gas, both terms are present in the governing equation,

$$u_t + uu_x = \mu u_{xx}. \tag{10.3}$$

The exact expression for the positive coefficient μ, which is determined by the gas shear and volume viscosities and heat conductivity, can be found in books on fluid dynamics (see, e.g., [4]). Equation (10.3) is called *Burgers equation*.

10.1.1 Kink Solution

Let us consider Equation (10.3) in an infinite region, $-\infty < x < \infty$, and assume that

$$u(\infty, t) = u_1, \quad u(-\infty, t) = u_2, \quad u_2 > u_1. \tag{10.4}$$

One can assume that under the action of both factors, the nonlinearity which makes the wave steeper and the linear term that makes it flatter, the wave will tend to a certain balanced stationary shape. Let us find the particular solution in the form of *traveling wave*

$$u(x, t) = U(X), \quad X = x - ct. \tag{10.5}$$

Substituting (10.5) into (10.3), we obtain the ODE

$$\mu U'' + cU' - UU' = 0, \quad -\infty < X < \infty, \tag{10.6}$$

which has to be solved with boundary conditions

$$U(\infty) = u_1, \quad U(-\infty) = u_2. \tag{10.7}$$

Integrating (10.6), we find:

$$\mu U' + cU - \frac{U^2}{2} = C, \tag{10.8}$$

where C is a constant which is obtained from the boundary conditions,

$$C = cu_1 - \frac{u_1^2}{2} = cu_2 - \frac{u_2^2}{2}. \tag{10.9}$$

Relations (10.9) give

$$c = \frac{u_1 + u_2}{2}, \quad C = \frac{u_1 u_2}{2}. \tag{10.10}$$

Substituting (10.10) into (10.8), we find that

$$\frac{dU}{dX} = \frac{1}{2\mu}(U - u_1)(U - u_2). \tag{10.11}$$

Solving (10.11), (10.7), we obtain

$$U = \frac{u_1 + u_2 \exp[-(u_2 - u_1)(X - x_0)/2\mu]}{1 + \exp[-(u_2 - u_1)(X - x_0)/2\mu]},$$

where x_0 is an arbitrary constant, hence

$$u(x,t) = \frac{u_1 + u_2 \exp[-(u_2 - u_1)(X - x_0)/2\mu + (u_2^2 - u_1^2)t/4\mu]}{1 + \exp[-(u_2 - u_1)(X - x_0)/2\mu + (u_2^2 - u_1^2)t/4\mu]}. \tag{10.12}$$

In the literature, one calls this solution "shock", "step" or "kink".

Solution (10.12) can be also written as

$$u(x,t) = \frac{u_1 + u_2}{2} - \frac{u_2 - u_1}{2} \tanh\left[\frac{u_2 - u_1}{4\mu}\left(x - x_0 - \frac{u_1 + u_2}{2}t\right)\right]. \tag{10.13}$$

Note that the bigger the mean value $(u_1 + u_2)/2$, the faster the wave; the bigger the amplitude $(u_2 - u_1)/2$, the steeper the wave.

10.1.2 Symmetries of the Burger's Equation

The features of the family of solutions obtained above can be easily explained using the symmetry properties of Equation (10.3).

Galilean Symmetry

Let us perform the following change of variables:

$$v = u - C, \quad X = x - Ct, \quad T = t,$$

where C is a real number ("*Galilean transformation*"). We find:

$$u = v + C, \quad u_t = v_T - Cv_X, \quad u_x = v_X, \quad u_{XX} = v_{XX}. \qquad (10.14)$$

Substituting (10.14) into (10.3), we obtain

$$v_T + vv_X = \mu v_{XX}, \qquad (10.15)$$

which coincides with (10.3) up to the renaming of variables. Thus, if Equation (10.13) has a certain solution $u(x,t) = f(x,t)$, then (10.15) has a solution $v(X,T) = f(X,T)$. That means that

$$u(x,t) = f(x - Ct, t) + C \qquad (10.16)$$

is a solution of (10.3) for any C. Thus, any solution $u(x,t) = f(x,t)$ is a member of a set of solutions (10.16); addition of a constant to the solution leads to a corresponding change of the velocity. Obviously, family (10.13) possesses that property.

Scaling Symmetry

Let us perform now a *scaling transformation*:

$$X = \alpha x, \quad T = \beta t, \quad u(x,t) = \gamma v(X,T), \qquad (10.17)$$

hence

$$\frac{\partial}{\partial x} = \alpha \frac{\partial}{\partial X}, \quad \frac{\partial}{\partial t} = \beta \frac{\partial}{\partial T}. \qquad (10.18)$$

Substituting (10.17) into (10.3), we find:

$$\gamma\beta v_T + \gamma^2 \alpha v v_X = \gamma\alpha^2 \mu v_{XX}.$$

Hence, we obtain an equation equivalent to (10.3), if we take

$$\beta = \alpha^2, \quad \gamma = \alpha.$$

With that choice of coefficients, if there exists a solution $u(x,t) = f(x,t)$, then there exists also solution $v(X,T) = f(X,T)$, i.e.,

$$u(x,t) = \alpha f(\alpha X, \alpha^2 T). \qquad (10.19)$$

Family (10.12) has that property: if u_1 and u_2 are taken α times larger, the wave becomes α times steeper, and the temporal rate becomes α^2 times larger.

Translational Symmetries

Also, the Burgers equation is invariant to the transformation of coordinates

$$X = x + C_1, \quad T = t + C_2,$$

where C_1 and C_2 are arbitrary real numbers, hence solution $u(x,t) = f(x,t)$ generates a set of solutions $u(x,t) = f(x + C_1, t + C_2)$.

Reflection Symmetry

Equation (10.3) is symmetric to the simultaneous transformation $X = -x$, $v = -u$. Thus, if the sign of u is changed, the wave moves in the opposite direction.

10.2   General Solution of the Cauchy Problem

Let us consider now the general initial value problem,

$$u_t + u u_x = \mu u_{xx}, \quad -\infty < x < \infty, \quad t > 0; \tag{10.20}$$

$$u(x,0) = u_0(x), \quad -\infty < x < \infty. \tag{10.21}$$

Below we will transform this problem to a linear one and find its solution explicitly.

First, let us introduce function ψ such that $u = \psi_x$. That means that

$$\psi = \int_{x_0}^{x} u(x_1) dx_1 + f(t),$$

where x_0 is an arbitrary number and $f(t)$ is an arbitrary function. The equation for ψ,

$$\psi_{xt} + \psi_x \psi_{xx} - \mu \psi_{xxx} = 0,$$

can be integrated:

$$\psi_t + \frac{1}{2} \psi_x^2 - \mu \psi_{xx} = C(t). \tag{10.22}$$

Because ψ is determined up to an arbitrary function of t, we can choose $C(t) = 0$.

Let us apply now the transformation

$$\psi = -2\mu \ln \varphi, \quad \varphi > 0, \tag{10.23}$$

called the *Hopf-Cole transformation*. Substituting (10.23) into (10.22) (with $C(t) = 0$) and taking into account that $\varphi \neq 0$, $\mu \neq 0$, we obtain *the heat equation*

$$\varphi_t = \mu \varphi_{xx}, \quad -\infty < x < \infty, \quad t > 0. \tag{10.24}$$

Equation (10.24) has to be solved with the initial condition

$$\varphi(x,0) = \varphi_0(x), \quad -\infty < x < \infty, \tag{10.25}$$

where

$$\varphi_0(x) = \exp\left[-\frac{1}{2\mu} \int_{x_0}^{x} u_0(x_1) dx_1\right]. \tag{10.26}$$

As we have seen in Section 6.8, the solution of the initial value problem (10.24), (10.25) is

$$\varphi(x,t) = \frac{1}{\sqrt{4\pi\mu t}} \int_{-\infty}^{\infty} dx_1 \varphi_0(x_1) \exp\left[-\frac{(x-x_1)^2}{4\mu t}\right]. \tag{10.27}$$

Substituting expression (10.26) into (10.27), we find that

$$\varphi(x,t) = \frac{1}{\sqrt{4\pi\mu t}} \int_{-\infty}^{\infty} dx_1 \exp\left[-\frac{1}{2\mu} G(x, x_1, t)\right], \tag{10.28}$$

where

$$G(x, x_1, t) = \int_{x_0}^{x_1} u_0(x_2) dx_2 + \frac{(x-x_1)^2}{2t}. \tag{10.29}$$

The original function is

$$u(x,t) = \psi_x = -2\mu \frac{\varphi_x(x,t)}{\varphi(x,t)}. \tag{10.30}$$

Taking into account that

$$\frac{\partial G}{\partial x} = \frac{x - x_1}{t},$$

we obtain the following exact solution of the initial value problem (10.20), (10.21):

$$u(x,t) = \frac{\int_{-\infty}^{\infty} dx_1 \frac{x - x_1}{t} \exp\left[-\frac{G(x, x_1, t)}{2\mu}\right]}{\int_{-\infty}^{\infty} dx_1 \exp\left[-\frac{G(x, x_1, t)}{2\mu}\right]}, \tag{10.31}$$

10.2.1 Interaction of Kinks

Let us return to the "kink solution" (10.12) and find what solution of the linear equation (10.24) corresponds to it. Multiplying the numerator and the denominator of (10.12) by

$$\exp\left[-\frac{u_1(x - x_0)}{2\mu} + \frac{u_1^2 t}{4\mu}\right],$$

we find that

$$u(x,t) = \frac{u_1 \exp[-u_1(x - x_0)/2\mu + u_1^2 t/4\mu] + u_2 \exp[-u_2(x - x_0)/2\mu + u_2^2 t/4\mu]}{\exp[-u_1(x - x_0)/2\mu + u_1^2 t/4\mu] + \exp[-u_2(x - x_0)/2\mu + u_2^2 t/4\mu]}.$$

This expression can be obtained according to Equation (10.30) from the following expression for φ,

$$\varphi = C\left\{\exp\left[-\frac{u_1(x - x_0)}{2\mu} + \frac{u_1^2 t}{4\mu}\right] + \exp\left[-\frac{u_2(x - x_0)}{2\mu} + \frac{u_2^2 t}{4\mu}\right]\right\},$$

where C is an arbitrary constant. We can see that

$$\varphi = \varphi_1 + \varphi_2,$$

where

$$\varphi_j = \exp\left(-\frac{u_j x}{2\mu} + \frac{u_1^2 t}{4\mu} - b_j\right), \quad b_j = -\ln C - u_j x_0, \quad j = 1, 2. \tag{10.32}$$

Note that each φ_j is indeed a solution of the heat equation, (10.24), but expressions (10.32), which are unbounded at infinity, are quite different from those that we obtained in the context of the heat transfer in Chapter 6.

Because Equation (10.24) is linear, any superposition

$$\varphi = \sum_{j=1}^{N} \exp\left(-\frac{u_j x}{2\mu} + \frac{u_j^2 t}{4\mu} - b_j\right), \tag{10.33}$$

where $u_i \neq u_j$ if $i \neq j$, generates an exact solution (10.30) of the Burgers equation. One can show that solution (10.33) describes the interaction of $N - 1$ kinks.

Let us consider the case $n = 3$ in more detail. Assume $u_1 < u_2 < u_3$. By means of transformations $\varphi \to C\varphi$, $x \to x + a$, $t \to t + b$, we can assign arbitrary values to b_j, $j = 1, 2, 3$. Let us take $b_1 = b_2 = 0$, $b_3 = (u_3 - u_2)/2\mu$. Then

$$\varphi(x, 0) = \varphi_1(x, 0) + \varphi_2(x, 0) + \varphi_3(x, 0), \tag{10.34}$$

where

$$\varphi_1(x, 0) = \exp\left(-\frac{u_1 x}{2\mu}\right), \quad \varphi_2(x, 0) = \exp\left(-\frac{u_2 x}{2\mu}\right), \quad \varphi_3(x, 0) = \exp\left(-\frac{u_3(x + 1) - u_2}{2\mu}\right).$$

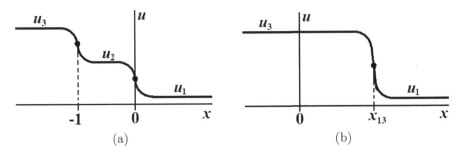

FIGURE 10.1
The shape of $u(x,t)$: (a) at $t = 0$; (b) at $t > t_*$.

For the sake of simplicity, assume that $\mu \ll 1$. In that case, for any two real numbers r_1 and r_2, if $r_1 < r_2$ and $r_2 - r_1 = O(1)$, then $\exp(r_1/\mu) \ll \exp(r_2/\mu)$. Then, as the rule, only one of the terms in the right-hand side of (10.34) is significant, while two other terms can be neglected, except the vicinity of the point $x = -1$, where $\varphi_2 = \varphi_3$, and the vicinity of the point $x = 0$, where $\varphi_1 = \varphi_2$. Specifically, for $x < -1$, $\varphi \approx \varphi_3$, hence $u \approx u_3$; for $-1 < x < 0$, $\varphi \approx \varphi_2$, hence $u \approx u_2$; for $x > 0$, $\varphi \approx \varphi_1$. Near the point $x = 1$, where $\varphi_2 = \varphi_3 \gg \varphi_1$, $\varphi \approx \varphi_2 + \varphi_3$, which corresponds to a kink between values u_3 and u_2; near the point $x = 0$, where $\varphi_1 = \varphi_2 \gg \varphi_3$, $\varphi \approx \varphi_1 + \varphi_2$, which corresponds to a kink between values u_2 and u_1 (see Figure 10.1(a)).

For $t > 0$, we have

$$\varphi_1 = \exp\left(-\frac{u_1 x}{2\mu} + \frac{u_1^2 t}{4\mu}\right), \quad \varphi_2 = \exp\left(-\frac{u_2 x}{2\mu} + \frac{u_2^2 t}{4\mu}\right), \quad \varphi_3 = \exp\left(-\frac{u_3(x+1) - u_2}{2\mu} + \frac{u_3^2 t}{4\mu}\right).$$

The coordinate of the right kink center $x_{12}(t)$ is determined by the equation $\varphi_1(x_{12}) = \varphi_2(x_{12})$, which gives

$$x_{12} = \frac{u_1 + u_2}{2} t.$$

The coordinate of the left kink center $x_{23}(t)$ is determined by the equation $\varphi_2(x_{23}) = \varphi_3(x_{23})$, which gives

$$x_{23} = -1 + \frac{u_2 + u_3}{2} t.$$

Because $u_3 > u_1$, the left kink moves faster than the right one, and both kinks collide at $t = t_*$,

$$t_* = \frac{2}{u_3 - u_1}.$$

For $t > t_*$, φ_2 is smaller than φ_1 and φ_2 everywhere, therefore, $\varphi \approx \varphi_1 + \varphi_3$, which corresponds to a kink between u_3 and u_1 (see Figure 10.1(b)). Thus, the collision of two kinks leads to their merging and formation of a kink with the center coordinate $x_{13}(t)$ determined by the equation $\varphi_1(x_{13}) = \varphi_3(x_{13})$; one finds that

$$x_{13} = \frac{u_1 + u_3}{2} t - \frac{u_3 - u_2}{u_3 - u_1}$$

(see Figure 10.2). Note that the kink velocities satisfy inequalities

$$\frac{dx_{12}}{dt} < \frac{dx_{13}}{dt} < \frac{dx_{23}}{dt}.$$

Similar phenomena of kink merging take place for any N, i.e., for any initial number of kinks.

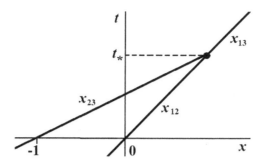

FIGURE 10.2
Trajectories of kink centers.

10.3   Korteweg-de Vries Equation

The non-viscous Burgers equation (10.1) describes long non-dispersive waves in different physical contexts. Besides the sound waves, it governs the propagation of nonlinear gravity waves in "shallow water", i.e., in the case where the wavelength is large with respect to the liquid layer depth. For gravity waves, the main physical factor that prevents the development of a singularity is *dispersion* rather than dissipation (see Section 5.12). For long waves, the dispersion is described by a term with the third spatial derivative; thus the wave propagation is governed by the *Korteweg-de Vries (KdV) equation*,

$$u_t + uu_x + \kappa u_{xxx} = 0.$$

We can change the coefficient κ in an arbitrary way using an appropriate scaling transformation of the kind $u \to Cu$, $x \to ax$, $t \to bt$. Later on, we use the standard choice, $\kappa = 6$:

$$u_t + u_{xxx} + 6uu_x = 0, \quad -\infty < x < \infty, \quad -\infty < t < \infty. \tag{10.35}$$

Physically meaningful solutions should be bounded.

10.3.1   Symmetry Properties of the KdV Equation

First, let us discuss the symmetry properties of the KdV equation which are rather similar to those of the Burgers equation (see Subsection 10.1.2). The KdV equation is invariant to coordinate translation. Due to the Galilean symmetry, any solution $u(x,t) = f(x,t)$ generates the family of solutions,

$$u(x,t;C) = f(x - 6Ct, t) + C, \tag{10.36}$$

where C is an arbitrary number.

   The major difference from the Burgers equation is in the scaling property. The scaling transformation

$$X = \alpha x, \quad T = \beta t, \quad u(x,t) = \gamma v(X,T)$$

gives

$$\gamma \beta v_T + \gamma \alpha^3 v_{XXX} + \gamma^2 \alpha v v_X = 0;$$

therefore, the original KdV equation is reproduced if

$$\beta = \alpha^3, \quad \gamma = \alpha^2.$$

The family of solutions generated by the scaling transformation of the solution $u(x,t) = f(x,t)$ is

$$u(x,t;\alpha) = \alpha^2 f(\alpha x,\ \alpha^3 t).\tag{10.37}$$

10.3.2 Cnoidal Waves

Let us consider now traveling wave solutions

$$u(x,t) = U(X),\quad X = x - ct,\tag{10.38}$$

that satisfy the ODE

$$U''' - cU' + 6UU' = 0,\quad -\infty < X < \infty.\tag{10.39}$$

Obviously, any constant function is a solution of Equation (10.39). We are interested in finding non-constant solutions.

Integrating (10.39), we obtain

$$U'' - cU + 3U^2 = C,\quad -\infty < X < \infty,\tag{10.40}$$

where C is an arbitrary constant. Let us multiply both sides of that equation by U':

$$U'U'' - cUU' + 3U^2U' - CU' = 0.\tag{10.41}$$

The left-hand side of Equation (10.41) is a full derivative,

$$\frac{d}{dX}\left[\frac{(U')^2}{2} + U^3 - \frac{c}{2}U^2 - CU\right] = 0,$$

hence

$$\frac{(U')^2}{2} + U^3 - \frac{c}{2}U^2 - CU = D,\tag{10.42}$$

where D is an arbitrary constant. Equation (10.42) can be written as

$$\frac{(U')^2}{2} + V(U) = 0,\tag{10.43}$$

where

$$V(U) = (U - U_1)(U - U_2)(U - U_3)$$

is a cubic polynomial. The roots U_1, U_2 and U_3 of the polynomial satisfy the relation

$$U_1 + U_2 + U_3 = \frac{c}{2}.\tag{10.44}$$

Note that a cubic polynomial with real coefficients can have either three real roots or only one real root and two more complex-conjugate roots.

To understand the properties of the solutions of Equation (10.45), it is convenient to use a "mechanical interpretation" of that equation. It is formally equivalent to the energy conservation law for a fictitious particle with the mass equal to 1 (so that its kinetic energy is $(U')^2/2$) moving in a potential $V(U)$. The energy of the particle is equal to 0. We are interested in particle trajectories that do not tend to infinity. One can see that if the polynomial $V(U)$ has three real roots, $U_1 \geq U_2 \geq U_3$, the particle can oscillate between points U_1 and U_2 (see Figure 10.3(a)). That corresponds to a spatially periodic solution $U(X)$ with

$$\max_X U(X) = U_1,\quad \min_X U(X) = U_2.$$

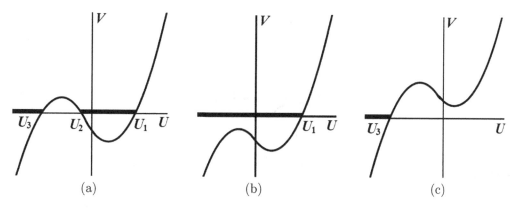

FIGURE 10.3
Plots of the potential $V(U)$: (a) bounded oscillations between U_1 and U_2; (b) no bounded solutions; (c) no bounded solutions.

If there is only one real root (see Figure 10.3(b), 10.3(c)), the particle with energy 0 always escapes to infinity, which is not acceptable.

Below we assume that the polynomial $V(U)$ has three different real roots, $U_1 > U_2 > U_3$. If $U_1 = U_2$, there are no bounded solutions except the constant ones. The case $U_2 = U_3$ is considered in the next subsection.

Equation (10.43) can be integrated using elliptic functions. Taking into account that $U_2 \leq U(X) \leq U_1$, we define a new variable $\psi(X)$ by the relation

$$U(X) = U_2 + (U_1 - U_2)\cos^2 \psi(X). \tag{10.45}$$

Substituting (10.45) into (10.43), we find

$$2(\psi')^2 = (U_1 - U_3) - (U_1 - U_2)\sin^2 \psi(X),$$

hence

$$\frac{dX}{d\psi} = \sqrt{\frac{2}{U_1 - U_3}} \frac{1}{\sqrt{1 - m\sin^2 \psi}},$$

where parameter

$$m = \frac{U_1 - U_2}{U_1 - U_3}, \quad 0 < m < 1. \tag{10.46}$$

We find that

$$X = \sqrt{\frac{2}{U_1 - U_3}} F(\psi|m) + X_0,$$

where

$$F(\psi|m) = \int_0^\psi \frac{d\theta}{\sqrt{1 - m\sin^2 \theta}}$$

is the *incomplete elliptic integral of the first kind*. The function inverse to the incomplete elliptic integral,

$$\psi = \text{am}(\xi|m), \quad \xi = \sqrt{\frac{U_1 - U_3}{2}}(X - X_0), \tag{10.47}$$

is called *Jacobi amplitude*. Substituting (10.47) into (10.45), we find

$$U(X) = U_2 + (U_1 - U_2)\,\text{cn}^2(\xi|m),$$

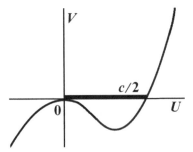

FIGURE 10.4
Plot of the potential in the case of solution.

where $\mathrm{cn}(\xi|m) \equiv \cos \mathrm{am}(\xi|m)$ is the *Jacobi elliptic cosine*. Because

$$c = 2\left(U_1 + U_2 + U_3\right),$$

we obtain the following expression for a nonlinear spatially periodic (*cnoidal*) wave:

$$u(x,t) = U_2 + (U_1 - U_2)$$

$$\times \mathrm{cn}^2\left[\sqrt{\frac{U_1 - U_3}{2}}\left[x - X_0 - 2\left(U_1 + U_2 + U_3\right)t\right]\left|\frac{U_1 - U_2}{U_1 - U_3}\right.\right]. \tag{10.48}$$

While linear periodic waves have a sinusoidal shape, cnoidal waves (10.48) have sharper crests and flatter troughs.

10.3.3   Solitons

Let us consider now the special case $U_3 = U_2 \equiv U_\infty$ which corresponds to a *solitary wave* (*soliton*) with

$$\lim_{X \to \pm\infty} U(X) = U_\infty.$$

Because of the Galilean symmetry (10.36), it is sufficient to find a solution with $U_\infty - 0$; all other solutions of this kind can be obtained by transformation (10.36).
Because of relation (10.44), $U_1 = c/2$ (see Figure 10.4). Then (10.43) gives:

$$\frac{(U')^2}{2} = U^2\left(\frac{c}{2} - U\right),$$

hence

$$\frac{dX}{du} = \pm\frac{1}{U\sqrt{c - 2U}}.$$

Denoting

$$v = \pm\sqrt{c/2 - U},$$

we find

$$dX/dv = \frac{\sqrt{2}}{v^2 - c/2} = \frac{1}{\sqrt{c}}\left(\frac{1}{v - \sqrt{c/2}} - \frac{1}{v + \sqrt{c/2}}\right),$$

hence

$$X - X_0 = \frac{1}{\sqrt{c}}\ln\frac{\sqrt{c/2} - v}{\sqrt{c/2} + v},$$

and
$$v = \sqrt{c/2} \tanh\left[\frac{\sqrt{c}}{2}\left(X - X_0\right)\right].$$

Thus,
$$U = \frac{c}{2} - v^2 = \frac{c}{2}\cosh^{-2}\left[\frac{\sqrt{c}}{2}\left(X - X_0\right)\right].$$

Finally, we obtain the following family of *solitary waves (solitons)*:
$$u(x, t; c) = \frac{c}{2}\cosh^{-2}\left[\frac{\sqrt{c}}{2}\left(x - ct - X_0\right)\right].$$

Denote $c = 4\kappa^2$, then
$$u(x, t; \kappa) = 2\kappa^2\cosh^{-2}\left[\kappa\left(x - 4\kappa^2 t - X_0\right)\right] \tag{10.49}$$

In accordance with (10.37), a higher soliton is narrower, and it moves with a higher velocity.

10.3.4 Bilinear Formulation of the KdV Equation

Below we describe an approach that will allow us to construct *multisoliton solutions*.

Let us start with the transformation somewhat similar to the Hopf-Cole transformation described in Section 10.1. Define $u = \psi_x$, then
$$\psi_{xt} + \psi_{xxxx} + 6\psi_x\psi_{xx} = 0$$

and
$$\psi_t + \psi_{xxx} + 3\psi_x^2 = f(t). \tag{10.50}$$

Because ψ is defined up to an arbitrary function of t, we can choose $f(t) = 0$. Now define
$$\psi = 2(\ln\phi)_x, \quad \varphi > 0. \tag{10.51}$$

Substituting (10.51) into (10.50) (with $f(t) = 0$), we obtain a *bilinear form* (i.e., all the terms are quadratic in φ and its derivatives):
$$\varphi\left(\varphi_t + \varphi_{xxx}\right)_x - \varphi_x\left(\varphi_t + \varphi_{xxx}\right) + 3\left(\varphi_{xx}^2 - \varphi_x\varphi_{xxx}\right) = 0. \tag{10.52}$$

The bilinear KdV equation can be written in a more compact form, if we introduce the following bilinear operator acting on a *pair of functions*:
$$D_x^m D_t^n\left[f(x, t), g(x, t)\right] \equiv \left[\left(\frac{\partial}{\partial x} - \frac{\partial}{\partial x'}\right)^m\left(\frac{\partial}{\partial t} - \frac{\partial}{\partial t'}\right)^n f(x, t)g(x', t')\right]_{x'=x, t'=t}$$

(the derivatives are calculated for x, t, x' and t' as independent variables, then x and t replace x' and t', correspondingly). With that notation,
$$D_x D_t(\varphi, \varphi) = \left[\left(\frac{\partial^2}{\partial x\partial t} - \frac{\partial^2}{\partial x\partial t'} - \frac{\partial^2}{\partial x'\partial t} + \frac{\partial^2}{\partial x'\partial t'}\right)\varphi(x, t)\varphi(x', t')\right]_{x'=x, t'=t}$$
$$= 2\left(\varphi\varphi_{xt} - \varphi_x\varphi_t\right),$$
$$D_x^4(\varphi, \varphi) = \left(\frac{\partial}{\partial x} - \frac{\partial}{\partial x'}\right)^4 \varphi(x, t)\varphi(x', t')\bigg|_{x'=x, t'=t} = 2\varphi\varphi_{xxxx} - 8\varphi_{xxx}\varphi_x + 6\varphi_{xx}^2,$$

hence Equation (10.52) can be written as
$$\left(D_x D_t + D_x^4\right)(\varphi, \varphi) = 0. \tag{10.53}$$

Let us list some important properties of operator $D_x^m D_t^n$.

1. $D_x^m D_t^n (f,1) = \left(\dfrac{\partial}{\partial x} - \dfrac{\partial}{\partial x'} \right)^m \left(\dfrac{\partial}{\partial t} - \dfrac{\partial}{\partial t'} \right)^n f(x,t) = \dfrac{\partial^m}{\partial x^m} \dfrac{\partial^n}{\partial t^n} f(x,t).$

2. A change of notations cannot change the expression, hence

$$\left(\frac{\partial}{\partial x} - \frac{\partial}{\partial x'} \right)^m \left(\frac{\partial}{\partial t} - \frac{\partial}{\partial t'} \right)^n f(x,t) g(x',t') \bigg|_{x'=x, t'=t}$$

$$= \left(\frac{\partial}{\partial x'} - \frac{\partial}{\partial x} \right)^m \left(\frac{\partial}{\partial t'} - \frac{\partial}{\partial t} \right)^n f(x',t') g(x,t) \bigg|_{x'=x, t'=t}$$

That relation gives

$$D_x^m D_t^n (f,g) = (-1)^{m+n} D_x^m D_t^n (g,f).$$

3. Therefore,
$$D_x^m D_t^n (f,f) = (-1)^{m+n} D_x^m D_t^n (f,f);$$

that means that
$$D_x^m D_t^n (f,f) = 0$$

if $m+n$ is odd.

4. Applying the operator $D_x^m D_t^n$ to exponential functions, we find:

$$D_x^m D_t^n \left[\exp(k_1 x - \omega_1 t), \exp(k_2 x - \omega_2 t) \right]$$
$$= (k_1 - k_2)^m (\omega_2 - \omega_1)^n \exp\left[(k_1 + k_2) x - (\omega_1 + \omega_2) t \right].$$

If $k_1 = k_2$ or $\omega_1 = \omega_2$, this expression is equal to zero.

10.3.5 Hirota's Method

Our goal is to find solutions of Equation (10.53) in a systematic way. To get a hint about how to proceed with finding solutions, let us find φ corresponding to soliton solutions (10.49). We can present (10.49) in the form

$$u = 2\kappa \frac{\partial}{\partial x} \left[\tanh \kappa \left(x - 4\kappa^2 t - X_0 \right) \right]$$

and choose

$$\psi = 2\kappa \left[\tanh \kappa \left(x - 4\kappa^2 t - X_0 \right) - 1 \right] = 2 \frac{\partial}{\partial x} \left[\ln \cosh \kappa \left(x - 4\kappa^2 t - X_0 \right) - \kappa x \right].$$

That gives

$$\varphi = C(t) \cosh \left[\kappa \left(x - 4\kappa^2 t - X_0 \right) \right] e^{-\kappa x}$$
$$= \frac{1}{2} C(t) \left[\exp\left(-4\kappa^3 t - \kappa X_0 \right) + \exp\left(-2\kappa x + 4\kappa^3 t + \kappa X_0 \right) \right],$$

where $C(t)$ is an arbitrary function of t. Choosing

$$C(t) = 2 \exp\left(4\kappa^3 t + \kappa X_0 \right),$$

we find

$$\varphi = 1 + \exp\eta, \quad \eta = -2\kappa \left(x - X_0 \right) + 8\kappa^3 t. \tag{10.54}$$

Let us check directly that function (10.54) satisfies Equation (10.53):

$$\begin{aligned}
\left(D_x D_t + D_x^4\right)(1 + \exp\eta, 1 + \exp\eta) &= \left(D_x D_t + D_x^4\right)(1,1) \\
&+ \left(D_x D_t + D_x^4\right)(1, \exp\eta) + \left(D_x D_t + D_x^4\right)(\exp\eta, 1) \\
&+ \left(D_x D_t + D_x^4\right)(\exp\eta, \exp\eta).
\end{aligned} \tag{10.55}$$

The first term and the fourth term in (10.55) vanish due to property 4 ($k_1 = k_2 = 0$ and $k_1 = k_2 = -2\kappa$, correspondingly). The second term and the third term, which are equal because of property 2, can be calculated according to property 1 as

$$\frac{\partial}{\partial x}\left(\frac{\partial}{\partial t} + \frac{\partial^3}{\partial t^3}\right)\exp\eta = \frac{\partial}{\partial x}\left[8\kappa^3 - (2\kappa)^3\right]\exp\eta = 0.$$

Note that

$$\varphi = 1 + \varepsilon \exp\eta = 1 + \exp\left[-2\kappa(x - X_0) + 8\kappa^3 t + \ln\varepsilon\right]$$

is also a solution of (10.53) for any ε.

Following Hirota, we search solutions of Equation (10.53) in the form

$$\varphi = \sum_{n=1}^{M} \varepsilon^n \varphi_n, \tag{10.56}$$

where ε is a *formal* expansion parameter. Let us substitute expansion (10.56) into (10.53),

$$\left(D_t D_x + D_x^4\right)\left(\varphi_0 + \varepsilon\varphi_1 + \varepsilon^2\varphi_2 + \ldots + \varepsilon^m + \ldots, \varphi_0 + \varepsilon\varphi_1 + \varepsilon^2\varphi_2 + \ldots + \varepsilon^m + \ldots\right) = 0,$$

and collect the terms with the same power of ε.

At the zeroth order in ε, we obtain equation

$$\left(D_t D_x + D_x^4\right)(\varphi_0, \varphi_0) = 0$$

and choose solution $\varphi_0 = 1$.

At the first order in ε, we obtain *a linear problem* for φ_1,

$$\left(D_t D_x + D_x^4\right)(\varphi_0, \varphi_1) + \left(D_t D_x + D_x^4\right)(\varphi_1, \varphi_0) = 0, \tag{10.57}$$

and find its solution. At the second order in ε, we obtain and solve the linear problem for φ_2,

$$\left(D_t D_x + D_x^4\right)(\varphi_0, \varphi_2) + \left(D_t D_x + D_x^4\right)(\varphi_2, \varphi_0) = -\left(D_t D_x + D_x^4\right)(\varphi_1, \varphi_1). \tag{10.58}$$

Generally, at the mth order in ε, we obtain the linear problem for φ_m,

$$\begin{aligned}
&\left(D_t D_x + D_x^4\right)(\varphi_0, \varphi_m) + \left(D_t D_x + D_x^4\right)(\varphi_m, \varphi_0) \\
&= -\left(D_t D_x + D_x^4\right)(\varphi_1, \varphi_{m-1}) - \ldots - \left(D_t D_x + D_x^4\right)(\varphi_{m-1}, \varphi_1).
\end{aligned}$$

It is important that the sum (10.56) has to contain only a finite number M of nonzero terms; in that case the described algorithm gives an exact solution of Equation (10.53).

The crucial point is the choice of the solution of the homogeneous linear Equation (10.57). If we choose it as *one* exponential function,

$$\varphi_1 = \exp\eta, \quad \eta = -2\kappa(x - x_0) + 8\kappa^3 t,$$

then Equation (10.58) becomes

$$2 \left(\frac{\partial^2}{\partial x \partial t} + \frac{\partial^4}{\partial x^4} \right) \varphi_2(x,t) = -\left(D_t D_x + D_x^4 \right) (\exp \eta, \exp \eta).$$

The right-hand side of that equation is equal to zero due to property 4; thus we can choose $\varphi_2 = 0$. At the third order, the right-hand side is

$$-2 \left(D_x D_t + D_x^4 \right) (\varphi_1, \varphi_2) = 0$$

because $\varphi_2 = 0$, hence $\varphi_3 = 0$. Step by step, we find that $\varphi_m = 0$ for all higher values of m. Taking $\varepsilon = 1$, we obtain the *one-soliton* solution (10.54).

10.3.6 Multisoliton Solutions

Because Equation (10.57) is linear, we can take its solution as an arbitrary sum of exponential functions. Let us choose a sum of *two* exponential functions,

$$\varphi_1 = \exp \eta_1 + \exp \eta_2,$$

where

$$\eta_n = -2\kappa_n \left(x - x_n^0 \right) + 8\kappa_n^3 t, \quad n = 1, 2; \quad \kappa_1 \neq \kappa_2.$$

It is convenient to define

$$x_n(t) = x_n^0 + 4\kappa_n^2 t, \tag{10.59}$$

then

$$\eta_n = -2\kappa_n \left[x - x_n(t) \right]. \tag{10.60}$$

The right-hand side of Equation (10.58) is non-zero:

$$- \left(D_t D_x + D_x^4 \right) (\exp \eta_1 + \exp \eta_2, \exp \eta_1 + \exp \eta_2) = -2 \left(D_t D_x + D_x^4 \right) (\exp \eta_1, \exp \eta_2).$$

Using property 4, we obtain:

$$2 \left(\frac{\partial^2}{\partial x \partial t} + \frac{\partial^4}{\partial x^4} \right) \varphi_2(x,t) = 96 \kappa_1 \kappa_2 \left(\kappa_1 - \kappa_2 \right)^2 \exp \left(\eta_1 + \eta_2 \right).$$

Substituting the ansatz

$$\varphi_2 = K \exp \left(\eta_1 + \eta_2 \right),$$

we find that

$$K = \left(\frac{\kappa_1 - \kappa_2}{\kappa_1 + \kappa_2} \right)^2,$$

hence

$$\varphi_2 = \exp(\eta_1 + \eta_2 + A_{12}),$$

$$A_{12} = \ln \left[\left(\frac{\kappa_1 - \kappa_2}{\kappa_1 + \kappa_2} \right)^2 \right] < 0. \tag{10.61}$$

At the third order, a direct calculation of the right-hand side

$$-2 \left(D_t D_x + D_x^4 \right) \left[\exp \eta_1 + \exp \eta_2, \exp \left(\eta_1 + \eta_2 + A_{12} \right) \right]$$

shows that it is equal to zero. Indeed, according to property 4,

$$\left(D_t D_x + D_x^4 \right) \left[\exp \eta_1, \exp \left(\eta_1 + \eta_2 \right) \right] = \exp \left(2\eta_1 + \eta_2 \right) \left[2\kappa_2 \left(-8\kappa_2^3 \right) + 16\kappa_4 \right] = 0.$$

Similarly,

$$\left(D_t D_x + D_x^4\right) [\exp \eta_2, \exp (\eta_1 + \eta_2)] = 0.$$

Thus, we can choose $\varphi_3 = 0$.

At the fourth order, the right-hand side of the equation

$$-\left(D_t D_x + D_x^4\right) (\varphi_2, \varphi_2) = 0$$

due to property 4, hence we can choose $\phi_4 = 0$.

One can see that if $\varphi_3 = \varphi_4 = 0$, the right-hand side of the equation for φ_5 is equal to zero, hence $\varphi_5 = 0$ etc. Finally, only φ_1 and φ_2 are different from zero. Taking $\varepsilon = 1$, we obtain the exact solution,

$$\varphi(x, t) = 1 + \exp \eta_1 + \exp \eta_2 + \exp (\eta_1 + \eta_2 + A_{12}), \qquad (10.62)$$

where η_1, η_2 and A_{12} are determined by Equations (10.61)-(10.63).

Formula (10.62) determines the solution of the original KdV equation (10.35),

$$u(x, t) = 2 \frac{\partial^2}{\partial x^2} (\ln \varphi(x, t)). \qquad (10.63)$$

To understand the physical meaning of solution (10.63), let us consider it in two limits, for $t < 0$, $|t| \gg 1$ ("far ago") and $t > 0$, $t \gg 1$ ("far future"). Below we assume that $\kappa_1 > \kappa_2 > 0$.

For $t < 0$ and sufficiently large $|t|$, $x_1(t) < x_2(t)$, $x_2(t) - x_1(t) \gg 1$. Let us consider the behavior of the solution in different regions of x. First, consider x around $x_1(t)$. In that region $\exp \eta_1 = O(1)$, $\exp \eta_2 \gg 1$, therefore

$$\varphi \approx \exp \eta_2 [1 + \exp (\eta_1 + A_{12})].$$

Calculating u according to formula (10.63), we obtain

$$u(x, t) \approx 2\kappa_1^2 \cosh^{-2} [\kappa_1 (x - x_1(t) + \Delta_1)],$$

where

$$\Delta_1 = -\frac{A_{12}}{2\kappa_1} > 0.$$

Thus, for $|x - x_1(t)| = O(1)$ the solution is close to a soliton solution with the center coordinate $x = x_1 - \Delta_1$.

If x is near $x_2(t)$, then $\exp \eta_1 \ll 1$, $\exp \eta_2 = O(1)$, therefore

$$\varphi \approx 1 + \exp \eta_2,$$

which leads to

$$u(x, t) \approx 2\kappa_2^2 \cosh^{-2} [\kappa_2 (x - x_2(t))],$$

i.e., we observe another soliton with the center coordinate $x_2(t)$. In all other regions of x, $u(x, t)$ is small.

Thus, at large negative t the solution describes two solitons on the large distance of each other. The faster soliton moving with velocity $4\kappa_1^2$ is located near the point $x = x_1(t) - \Delta_1$, and the slower soliton moving with velocity $4\kappa_2^2$ is located near the point $x = x_2(t)$.

With the growth of t, the faster soliton collides with the slower soliton. The collision is described by the full expressions (10.62) and (10.63).

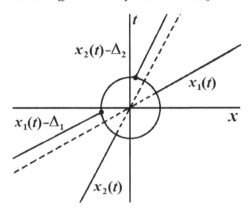

FIGURE 10.5
Interaction of KdV solutions.

Let us consider now the limit $t > 0$, $t \gg 1$. In that case, $x_1(t) \gg x_2(t)$. For x near $x_2(t)$, $\exp \eta_1 \gg 1$, $\exp \eta_2 = O(1)$, thus

$$\varphi \approx \exp \eta_1 \left[1 + \exp \left(\eta_2 + A_{12} \right) \right]$$

and

$$u(x, t) \approx 2\kappa_2^2 \cosh^{-2} \left[\kappa_2 \left(x - x_2(t) + \Delta_2 \right) \right], \quad \Delta_2 = -\frac{A_{12}}{2\kappa_2} > 0,$$

i.e. the trajectory of the slower soliton center is shifted backward to $x = x_2(t) - \Delta_2$. For x near $x_1(t)$, $\exp \eta_1 = O(1)$, $\exp \eta_2 \ll 1$, thus

$$\varphi \approx 1 + \exp \eta_1,$$

which leads to

$$u(x, t) \approx 2\kappa_1^2 \cosh^{-2} \left[\kappa_1 \left(x - x_1(t) \right) \right],$$

thus, the trajectory of the center coordinate is $x = x_1(t)$.

We see that after collision, both solitons are intact, and they keep their shape, amplitude and velocity. The only change is the shift of the center coordinate: the faster soliton is shifted forward by Δ_1, and the slower soliton is shifted backward by Δ_2 (see Figure 10.5).

Solution (10.62), (10.63) is called a *two-soliton solution*.

If we take the solution of (10.57) with three exponential functions, the Hirota's expansion allows us to obtain the *three-soliton solution*:

$$\varphi = 1 + \exp \eta_1 + \exp \eta_2 + \exp \eta_3 + \exp \left(\eta_1 + \eta_2 + A_{12} \right) + \exp \left(\eta_1 + \eta_3 + A_{13} \right)$$
$$+ \exp \left(\eta_2 + \eta_3 + A_{23} \right) + \exp \left(\eta_1 + \eta_2 + \eta_3 + A_{12} + A_{13} + A_{23} \right),$$

where

$$A_{ij} = \ln \left(\frac{\kappa_i - \kappa_j}{\kappa_i + \kappa_j} \right)^2.$$

Generally, the Hirota's expansion allows us to obtain the *multisoliton solutions* with arbitrary numbers of solitons.

The multisoliton solutions are, of course, only particular solutions of the KdV equation. However, their role is crucial in the KdV dynamics. The general Cauchy problem with initial condition $u(x, 0) = u_0(x)$ can be solved by means of the method of *inverse scattering transform* that is beyond the subject of the present book [5]. One can show that in the case where $u_0(x)$ tends to zero sufficiently fast as $x \to \pm\infty$, the solution tends either to 0 or to one of the soliton solutions described above. The number of solitons is determined by the initial conditions.

10.4 Nonlinear Schrödinger Equation

In Subsection 7.16.1, we have seen that the evolution of the *envelope function*, which describes the spatio-temporal modulation of a nearly unidirectional monochromatic linear wave, is governed by the linear Schrödinger equation. When the wave propagates in a *nonlinear medium*, e.g., when the refraction index depends on the light intensity, the equation for the envelope function becomes nonlinear as well. By means of asymptotic methods, one can derive a universal equation for the envelope function,

$$i\Psi_t + \beta\nabla^2\Psi + \gamma|\Psi|^2\Psi = 0, \qquad (10.64)$$

which is called *the nonlinear Schrödinger equation (NSE)* [5]. Besides the envelope function for waves of different physical nature, including fiber optics and water waves, the NSE provides an approximate description of the dynamics of the macroscopic wave functions in the superfluidity theory, where it is called *the Gross-Pitaevskii equation*. The variable t can be a temporal or a spatial variable, depending on the physical context. Below we consider only the case where $\Psi = \Psi(x,t)$ and $\nabla^2 = d^2/dx^2$.

The simplest class of solutions has the form

$$\Psi = R\exp\left(iKx - i\Omega t\right),$$

with

$$\Omega = \beta K^2 - \gamma R^2.$$

Thus, the frequency of the nonlinear wave depends not only on the wavenumber K (dispersion) but also on its amplitude R (nonlinear shift of frequency).

10.4.1 Symmetry Properties of NSE

First, let us consider the symmetry properties of NSE. Assume that Equation (10.64) has a solution $\Psi(x,t) = f(x,t)$. Then the following functions will also be solutions of that equation (below C is an arbitrary constant):

1. $\Psi_1(x,t) = f(x+C,t)$ (translation in space).
2. $\Psi_2(x,t) = f(x,t+C)$ (translation in time).
3. $\Psi_3(x,t) = f(x,t)\exp\left(iC\right)$ (phase invariance). The phase change for the envelope function corresponds to the translation of the carrying wave. Note that the change of the sign of the solution is a particular case of the phase transformation with $C = \pi$.
4. $\Psi_4(x,t) = f(-x,t)$ (reflection). In a contradistinction to the Burgers equation and the KdV equation, propagation of a wave is possible in both directions. Also, standing oscillations of the kind $\Psi = R\exp(-i\Omega t)$ with constant R are possible.
5. $\Psi_5(x,t) = f^*(x,-t)$ (time reversal). Note that the properties listed above are inherited from the linear Schrödinger equation.

6. $\Psi_6(x,t) = af(ax, a^2 t)$, where a is an arbitrary real number. That property can be obtained in the same way as that has been done for the Burgers equation and the KdV equation.

7. The Galilean invariance for the NSE has a more complex form than for the Burgers equation and the KdV equation:

$$\Psi_7(x,t) = f(x - vt, t) \exp\left[i(Kx - \Omega t)\right]. \tag{10.65}$$

Substituting (10.65) into (10.64), one can find that

$$K = \frac{v}{2\beta}, \quad \Omega = \frac{v^2}{4\beta},$$

i.e.

$$\Psi_7(x,t) = f(x - vt, t) \exp\left[i\left(\frac{v}{2\beta}x - \frac{v^2}{4\beta}t\right)\right]. \tag{10.66}$$

Note that in the case of the quantum-mechanical linear Schrödinger equation, this transformation (with $\beta = \hbar/2m$) corresponds to the transformation of the momentum and energy of a particle by the change of the reference frame, i.e., by the Galilean transformation in its original meaning.

10.4.2   Solitary Waves

The dynamics depend qualitatively on the sign of γ/β, which cannot be changed by a scaling transformation. If $\gamma/\beta > 0$, this is the *focusing NSE*, if $\gamma/\beta < 0$, this is the *defocusing NSE*. As an example, let us consider a solution in the form of solitary waves. Because of the Galilean invariance, it is sufficient to calculate the solution in the form of a standing wave,

$$\Psi(x,t) = R(x)\exp(-i\Omega t). \tag{10.67}$$

If that solution is found, the waves moving with a definite velocity can be found by means of the Galilean transformation (10.66). Substituting (10.67) into (10.64), we obtain the ODE,

$$\beta R'' + \Omega R + \gamma R^3 = 0. \tag{10.68}$$

Acting like in Subsection 10.2.2, we multiply Equation (10.68) by R' and obtain

$$\frac{d}{dx}\left(\frac{1}{2}\beta(R')^2 + \frac{1}{2}\Omega R^2 + \frac{1}{4}\gamma R^4\right) = 0,$$

hence

$$\frac{1}{2}\left(R'\right)^2 + V(R) = E, \quad V(R) = \frac{\Omega}{2\beta}R^2 + \frac{\gamma}{4\beta}R^4. \tag{10.69}$$

The constant E plays the role of the energy of a fictitious particle with the mass equal to 1 moving in the potential $V(R)$.

A solitary wave solution tending to a constant value at infinity is possible if the function $V(R)$ is non-monotonic, i.e., if $\Omega/2\beta$ and $\gamma/4\beta$ have different signs. Let us consider the cases of focusing and defocusing NSE separately.

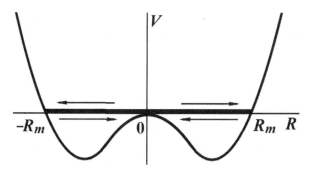

FIGURE 10.6
Potential in the case of focusing NSE.

Bright solitons

In the case of the focusing NSE ($\gamma/4\beta > 0$, $\Omega/2\beta < 0$), the corresponding potential $V(R)$ is shown in Figure 10.6.
The solitary wave with

$$\lim_{x \to \pm\infty} R(x) = 0$$

corresponds to $E = 0$; for any other admissible values of E, $R(x)$ is a periodic function. For $E = 0$, Equation (10.71) can be written as

$$\left(R'\right)^2 = \frac{\gamma}{2\beta} R^2 \left(R_m^2 - R^2\right), \tag{10.70}$$

where

$$R_m^2 = -\frac{-2\Omega}{\gamma}. \tag{10.71}$$

Let us introduce the new variable $y = \sqrt{R_m^2 - R^2}$, then

$$\frac{dy}{dx} = \pm\sqrt{\frac{\gamma}{2\beta}} \left(R_m^2 - y^2\right). \tag{10.72}$$

Integrating Equation (10.72), we find

$$x - x_0 = \pm\frac{1}{2R_m} \ln\left[\frac{R_m + y}{R_m - y}\right],$$

where x_0 is an arbitrary number, thus

$$y = \pm R_m \tanh\left[\sqrt{\frac{\gamma}{2\beta}} R_m \left(x - x_0\right)\right]$$

and

$$R = \pm R_m \cosh^{-1}\left[\sqrt{\frac{\gamma}{2\beta}} R_m \left(x - x_0\right)\right].$$

According to (10.71), $\Omega = -\gamma R_m^2/2$. Thus we obtain the following family of *bright solitons*:

$$\Psi(x,t) = \pm R_m \cosh^{-1}\left[\sqrt{\frac{\gamma}{2\beta}} R_m \left(x - x_0\right)\right] \exp\left[i\frac{\gamma}{2} R_m^2 t\right],$$

which is invariant with respect to the scaling transformation discussed above. This family can be further extended by means of the phase transformation and the Galilean transformation. Finally, we find:

$$\Psi(x,t) = R_m \cosh^{-1}\left[\sqrt{\frac{\gamma}{2\beta}}R_m(x - vt - x_0)\right] \exp\left[i\frac{\gamma}{2}R_m^2(t - t_0)\right]$$

$$\times \exp\left[i\left(\frac{v}{2\beta}x - \frac{v^2}{4\beta}t\right)\right]. \tag{10.73}$$

The obtained family of solutions is a member of a wider class of exact solutions of the focusing NSE equation, which includes multisoliton solutions and so called breathers. Like in the case of the KdV equation discussed in Section 10.2, the initial value problem for the one-dimensional NSE equation is solvable by means of the inverse scattering transform method (see [6]).

Dark Solitons

Let us consider now the defocusing NSE ($\gamma/4\beta < 0$, $\Omega/2\beta > 0$). The shape of the potential $V(R)$ (see (10.74)) is shown in Figure 10.7. The potential has maxima in the points $R = \pm R_m$,

$$R_m^2 = -\frac{\Omega}{\gamma}. \tag{10.74}$$

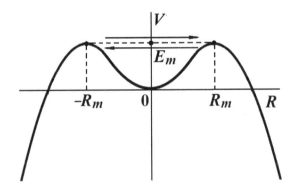

FIGURE 10.7
Potential in the case of defocusing NSE.

Using the mechanical interpretation of the problem, we come to the conclusion that the equation has unbounded solutions and bounded periodic solutions for $E < E_m = V(R_m)$ and only unbounded solutions for $E > E_m$.

For $E = E_m$, the equation has two bounded non-periodic solutions corresponding to the particle motion between to maxima, rightward, with

$$\lim_{x \to \pm\infty} R(x) = \pm R_m,$$

or leftward, with

$$\lim_{x \to \pm\infty} R(x) = \mp R_m.$$

Let us find these solutions.

Taking into account that

$$\Omega = -\gamma R_m^2 \tag{10.75}$$

(see (10.74)), we find that

$$E_m - V(R_m) = \frac{1}{4} \left| \frac{\gamma}{\beta} \right| R_m^4. \tag{10.76}$$

Substituting (10.75) and (10.76) into (10.69), we obtain equation

$$\frac{1}{2}(R')^2 = \frac{1}{4} \left| \frac{\gamma}{\beta} \right| \left(R_m^2 - R^2 \right)^2,$$

hence

$$R' = \pm \sqrt{\frac{1}{2} \left| \frac{\gamma}{\beta} \right|} \left(R_m^2 - R^2 \right). \tag{10.77}$$

Integrating (10.77), we obtain

$$\sqrt{\frac{1}{2} \left| \frac{\gamma}{\beta} \right|} (x - x_0) = \frac{1}{2R_m} \ln \left(\frac{R_m + R}{R_m - R} \right),$$

therefore,

$$R = \pm R_m \tanh \left[\sqrt{\frac{1}{2} \left| \frac{\gamma}{\beta} \right|} (x - x_0) \right],$$

where x_0 is an arbitrary number.

Finally, we obtain

$$\Psi(x,t) = \pm R_m \tanh \left[\sqrt{\frac{1}{2} \left| \frac{\gamma}{\beta} \right|} (x - x_0) \right] \exp(i\gamma R_m^2 t). \tag{10.78}$$

Solution (10.78) is called *a dark soliton*, because in the case of a light wave, the intensity of light $I(x,t) = |\Psi(x,t)|^2$ is equal to zero in the center of the solitary wavy, at $x = x_0$. Family (10.78) can be extended by means of the phase transformation and the Galilean transformation.

Problems

1. Find the solution of the Burger's equation

$$u_t + u u_x = u_{xx}, \quad -\infty < x < \infty, \quad t > 0$$

with initial condition

$$u(x,0) = u_0(x), \quad -\infty < x < \infty,$$

where

$$u_0(x) = u_-, \quad x < 0; \quad u_0(x) = u_+, \quad x \geq 0; \quad u_- > u_+.$$

2. Find one-soliton and two-soliton solutions by means of Hirota's method for the Kadomtsev-Petviashvili equation

$$(u_t + u_{xxx} + 6u u_x)x + \alpha u_{yy} = 0, \quad -\infty < x < \infty, \quad -\infty < t < \infty.$$

3. The sine-Gordon equation

$$u_{tt} - u_{xx} + \sin u, \quad -\infty < x < \infty, \quad -\infty < t < \infty$$

is given.

A. Find the one-soliton solutions in the form $u = u(x - vt)$, $|v| < 1$, that satisfies boundary conditions $u(\infty, t) = u(-\infty, t) + 2\pi$ (kink) or $u(\infty, t) = u(-\infty, t) - 2\pi$ (antikink).

B. Find the one-soliton and two-soliton solutions by means of Hirota's method.

   Hint: using the change of the variable $u = 2i \ln(f^*/f)$, where f is a complex function and f^* is its complex-conjugate, transform the sine-Gordon equation to the form

   $$\left(D_x^2 - D_t^2\right)(f, f) + \frac{1}{2}\left(f^*\right)^2 = C(x, t);$$

   choose C in such a way that $f = 1$ is a solution; apply Hirota's expansion $f = 1 + \varepsilon f_1 + \varepsilon^2 f_2 + \ldots$; take $f_1 = i \exp \eta$, $\eta = kx - \omega t + c$ for one-soliton solutions and $f_1 = i \exp \eta_1 + i \exp \eta_2$, $\eta_i = k_i x - \omega_i t + c_i$, $i = 1, 2$ for two-soliton solutions.

4. Find the one-soliton solutions of the focusing NSE

   $$iu_t + u_{xx} + |u|^2 u = 0, \quad -\infty < x < \infty, \quad -\infty < t < \infty; \quad |u(-\infty, t)| = |u(\infty, t)| = 0$$

   using Hirota's method.

   Hint: substitute $u = g/f$, where f is a real function, and transform the equation to the system

   $$\left(iD_t + D_x^2\right)(g, f) = 0, \quad D_x^2(f, f) = |g|^2.$$

5. Using Hirota's method, find the solution of the following boundary-value problem:

   $$u_{xt} = \sin u + 2 \sin \frac{u}{2}, \quad -\infty < x < \infty, \quad -\infty < t < \infty; \quad u(-\infty, 0) = 0, \quad u(\infty, 0) = 4\pi.$$

   Hint: substitute $u = 4 \arctan(f/g)$; assume $D_x D_t(f, g) = \mu f g$; use Hirota's expansion in the form: $g = 1 + \varepsilon^2 g_2 + \varepsilon^4 g_4 + \ldots$, $f = \varepsilon f_1 + \varepsilon^3 f_3 + \ldots$.

6. Find the family of *grey soliton* solutions of the defocusing NSE in the form

   $$\Psi(x, t) = [R(x - vt) + iQ] \exp(-i\Omega t),$$

   where v, Q, $R(\infty)$ and $R(-\infty)$ are some constants.

A

Fourier Series, Fourier and Laplace Transforms

A.1 Periodic Processes and Periodic Functions

In the sciences and in technology very often we encounter periodic phenomena. It means that some processes repeat after some time interval T, called the period. Alternating electric currents, an object in circular motion and wave phenomena are examples of physical processes which are periodic. Such processes can be associated with mathematical functions periodic in time, t, which have the property

$$\varphi(t + T) = \varphi(t).$$

The simplest periodic function is the sine (or cosine) function, $A\sin(\omega t + \alpha)$ (or $A\cos(\omega t + \alpha)$), where ω is the *angular frequency* related to the period by the relationship $\omega = 2\pi/T$(quantity $f = 1/T$ is called frequency, constant α is called phase).

With these simple periodic functions more complex periodic functions can be constructed as was noted by the French mathematician Joseph Fourier. For example if we add the functions

$$y_0 = A_0, \quad y_1 = A_1\sin(\omega t + \alpha_1), \quad y_2 = A_2\sin(2\omega t + \alpha_2),$$
$$y_3 = A_3\sin(3\omega t + \alpha_3), \quad \ldots \tag{A.1}$$

with multiple frequencies $\omega, 2\omega, 3\omega, \ldots$, i.e. with the periods $T, T/2, T/3, \ldots$ we obtain a periodic function (with period T), which, when graphed, has an appearance very distinct from the graphs of any of the functions in Equation (A.1).

It is natural to also investigate the reverse problem. Is it possible to resolve a given arbitrary periodic function, $\varphi(t)$, with period T, into a sum of simple functions such as those in Equation (A.1)? As we shall see, for a very wide class of functions the answer to this question is positive, but to do so may require an infinite sequence of the functions in Equation (A.1). In that case the periodic function $\varphi(t)$ can be resolved into the infinite *trigonometric series*

$$\varphi(t) = A_0 + A_1\sin(\omega t + \alpha_1) + A_2\sin(2\omega t + \alpha_2) + \ldots$$
$$= A_0 + \sum_{n=1}^{\infty} A_n\sin(n\omega t + \alpha_n), \tag{A.2}$$

where A_n and α_n are constants, and $\omega = 2\pi/T$. Each term in Equation (A.2) is called a *harmonic* and the decomposition of periodic functions into harmonics is called *harmonic analysis*.

In many cases it is useful to introduce the dimensionless variable

$$x = \omega t = \frac{2\pi t}{T}$$

and to work with the functions

$$f(x) = \varphi\left(\frac{x}{\omega}\right)$$

which are also periodic but with the *standard period* 2π: $f(x + 2\pi) = f(x)$. Using this shorthand, Equation (A.2) becomes

$$f(x) = A_0 + A_1 \sin(x + \alpha_1) + A_2 \sin(2x + \alpha_2) + \ldots$$

$$= A_0 + \sum_{n=1}^{\infty} A_n \sin(nx + \alpha_n). \tag{A.3}$$

With the trigonometric identity $\sin(\alpha + \beta) = \sin\alpha\cos\beta + \cos\alpha\sin\beta$ and the notation

$$A_0 = 2a_0, \quad A_n \sin\alpha_n = a_n, \quad A_n \cos\alpha_n = b_n, \quad n = 1, 2, 3, \ldots,$$

we obtain a standardized form for the harmonic analysis of a periodic function $f(x)$ as

$$f(x) = \frac{a_0}{2} + (a_1\cos x + b_1\sin x) + (a_2\cos 2x + b_2\sin 2x) + \ldots$$

$$= \frac{a_0}{2} + \sum_{n=1}^{\infty}(a_n\cos nx + b_n\sin nx), \tag{A.4}$$

which is referred to as the *trigonometric Fourier expansion*.

A.2 Fourier Formulas

To determine the limits of validity for the representation in Equation (A.4) of a given function $f(x)$ with period 2π and to find the coefficients a_n and b_n we follow the approach that was originally elaborated by Fourier. We first assume that the function $f(x)$ can be integrated over the interval $[-\pi, \pi]$. If $f(x)$ is discontinuous at any point we assume that the integral of $f(x)$ converges and in this case we also assume that the integral of the absolute value of the function, $|f(x)|$, converges. A function with these properties is said to be *absolutely integrable*. Integrating the expression (A.4) term by term we obtain

$$\int_{-\pi}^{\pi} f(x)dx = \pi a_0 + \sum_{n=1}^{\infty}\left[a_n \int_{-\pi}^{\pi}\cos nxdx + b_n \int_{-\pi}^{\pi}\sin nxdx\right].$$

Since $\int_{-\pi}^{\pi}\cos nxdx = \int_{-\pi}^{\pi}\sin nxdx = 0$, all the terms in the sum are zero and we obtain

$$a_0 = \frac{1}{\pi}\int_{-\pi}^{\pi} f(x)dx. \tag{A.5}$$

To find coefficients a_n we multiply Equation (A.4) by $\cos mx$ and then integrate term by term over the interval $[-\pi, \pi]$:

$$\int_{-\pi}^{\pi} f(x)\cos mxdx = a_0 \int_{-\pi}^{\pi}\cos mxdx + \sum_{n=1}^{\infty}\left[a_n \int_{-\pi}^{\pi}\cos nx\cos mxdx + b_n \int_{-\pi}^{\pi}\sin nx\cos mxdx\right].$$

For any n and m we also have

$$\int_{-\pi}^{\pi}\sin nx\cos mxdx = \frac{1}{2}\int_{-\pi}^{\pi}[\sin(n+m)x + \sin(n-m)x]\,dx = 0 \tag{A.6}$$

and if $n \neq m$ we obtain

$$\int_{-\pi}^{\pi} \cos nx \cos mx dx = \frac{1}{2} \int_{-\pi}^{\pi} [\cos(n+m)x + \cos(n-m)x] \, dx = 0. \qquad (A.7)$$

Using these formulas along with the identity $\int_{-\pi}^{\pi} \cos^2 mx dx = \pi$, we see that all the integrals in the sum are zero except the one with the coefficient a_m. We thus have

$$a_m = \frac{1}{\pi} \int_{-\pi}^{\pi} f(x) \cos mx dx, \quad m = 1, 2, 3, \ldots. \qquad (A.8)$$

The usefulness of introducing the factor $1/2$ in the first term in Equation (A.4) is now apparent since it allows the same formulas to be used for all a_n, including $n = 0$.

Similarly, multiplying Equation (A.4) by $\sin mx$ and using, along with Equation (A.6), two other simple integrals $\int_{-\pi}^{\pi} \sin nx \sin mx dx = 0$ if $n \neq m$, and $\int_{-\pi}^{\pi} \sin^2 mx dx = \pi$, we obtain the second coefficient

$$b_m = \frac{1}{\pi} \int_{-\pi}^{\pi} f(x) \sin mx dx, \quad m = 1, 2, 3, \ldots. \qquad (A.9)$$

Reading Exercise: Obtain the same result as in Equations (A.6) and (A.7) using Euler's formula

$$e^{imx} = \cos mx + i \sin mx.$$

Equations (A.6) and (A.7) also indicate that the system of functions

$$1, \ \cos x, \ \sin x, \ \cos 2x, \ \sin 2x, \ \ldots, \ \cos nx, \ \sin nx, \ldots \qquad (A.10)$$

is *orthogonal* on $[-\pi, \pi]$.

It is important to notice that the above system is not orthogonal on the reduced interval $[0, \pi]$ because for n and m with different parity (one odd and the other even) we have

$$\int_{0}^{\pi} \sin nx \cos mx dx \neq 0.$$

However, the system consisting of cosine functions only

$$1, \ \cos x, \ \cos 2x, \ \ldots, \ \cos nx, \ \ldots \qquad (A.11)$$

is orthogonal on $[0, \pi]$ and the same is true for

$$\sin x, \ \sin 2x, \ \ldots, \ \sin nx, \ \ldots \qquad (A.12)$$

A second observation, which we will need later, is that on an interval $[0, l]$ of arbitrary length l, both systems of functions

$$1, \ \cos \frac{\pi x}{l}, \ \cos \frac{2\pi x}{l}, \ \ldots, \ \cos \frac{n\pi x}{l}, \ \ldots \qquad (A.13)$$

and

$$\sin \frac{\pi x}{l}, \ \sin \frac{2\pi x}{l}, \ \ldots, \ \sin \frac{n\pi x}{l}, \ \ldots \qquad (A.14)$$

are orthogonal.

Reading Exercise: Prove the above three statements.

Equations (A.5), (A.8), and (A.9) are known as the *Fourier coefficients* and the series (A.4) with these definitions is called the *Fourier series*. Equation (A.4) is also referred to as the *Fourier expansion* of the function $f(x)$.

Notice that for the function $f(x)$ having period 2π, the integral

$$\int_{\alpha}^{\alpha+2\pi} f(x)dx$$

does not depend on the value of α. As a result we may also use the following expressions for the Fourier coefficients:

$$a_m = \frac{1}{\pi} \int_0^{2\pi} f(x)\cos mx\, dx \quad \text{and} \quad b_m = \frac{1}{\pi} \int_0^{2\pi} f(x)\sin mx\, dx. \tag{A.15}$$

It is important to realize that to obtain the results above we used a term by term integration of the series which is justified only if the series converges uniformly. Until we know for sure that the series converges we can only say that the series (A.4) *corresponds* to the function $f(x)$ which usually is denoted as

$$f(x) \sim \frac{a_0}{2} + \sum_{n=1}^{\infty} (a_n \cos nx + b_n \sin nx).$$

At this point we should remind the reader what is meant by *uniform convergence*. The series $\sum_{n=1}^{\infty} f_n(x)$ converges to the sum $S(x)$ uniformly on the interval $[a, b]$ if, for any arbitrarily small $\varepsilon > 0$ we can find a number N such that for all $n \geq N$ the remainder of the series $|\sum_{n=N}^{\infty} f_n(x)| \leq \varepsilon$ for all $x \in [a, b]$. This indicates that the series approaches its sum uniformly with respect to x.

The most important features of a uniformly converging series are:

i) If $f_n(x)$ for any n is a continuous function, then $S(x)$ is also a continuous function;

ii) The equality $\sum_{n=1}^{\infty} f_n(x) = S(x)$ can be integrated term by term along any interval within the interval $[a, b]$;

iii) If the series $\sum_{n=1}^{\infty} f_n'(x)$ converges uniformly then its sum is equal to $S'(x)$; i.e. the formula $\sum_{n=1}^{\infty} f_n(x) = S(x)$ can be differentiated term by term.

There is a simple and very practical criterion for convergence established by Karl Weierstrass that says that if $|f_n(x)| < c_n$ for each term $f_n(x)$ in the series defined on the interval $x \in [a, b]$ (i.e. $f_n(x)$ is limited by c_n), where $\sum_{n=1}^{\infty} c_n$ is a converging numeric series, then the series

$$\sum_{n=1}^{\infty} f_n(x)$$

converges uniformly on $[a, b]$. For example, the numeric series $\sum_{n=1}^{\infty} \frac{1}{n^2}$ is known to converge, so any trigonometric series with terms such as $\sin nx/n^2$ or similar will converge uniformly for all x because $|\sin nx/n^2| \leq 1/n^2$.

A.3   Convergence of Fourier Series

In this section we study the range of validity of Equation (A.4) with Fourier coefficients given by Equations (A.5), (A.8), and (A.9). To start, it is clear that if the function $f(x)$ is finite on $[-\pi, \pi]$ then the Fourier coefficients are bounded. This is easily verified, for instance for a_n since

$$|a_n| = \frac{1}{\pi} \left| \int_{-\pi}^{\pi} f(x) \cos nx dx \right| \leq \frac{1}{\pi} \int_{-\pi}^{\pi} |f(x)| \cdot |\cos nx| \, dx \leq \frac{1}{\pi} \int_{-\pi}^{\pi} |f(x)| \, dx \qquad \text{(A.16)}$$

The same result is valid in cases where $f(x)$ is not finite but is absolutely integrable, i.e. the integral of its absolute value converges:

$$\int_{-\pi}^{\pi} |f(x)| \, dx < \infty. \qquad \text{(A.17)}$$

The *necessary condition* that any series converges is that its terms tend to zero as $n \to \infty$. Because the absolute values of sine and cosine functions are bounded, the necessary condition that the trigonometric series in Equation (A.4) converges is that coefficients of expansion a_n and b_n tend to zero as $n \to \infty$. This condition is valid for functions that are integrable (or absolutely integrable in the case of functions which are not finite) which is clear from the following lemma.

Riemann's lemma

If the function $f(t)$ is absolutely integrable on $[a, b]$, then

$$\lim_{\alpha \to \infty} \int_a^b f(t) \sin \alpha t dt = 0 \quad \text{and} \quad \lim_{\alpha \to \infty} \int_a^b f(t) \cos \alpha t dt = 0. \qquad \text{(A.18)}$$

We will not prove this rigorously but its sense should be obvious. In the case of very fast oscillations the sine and cosine functions change their sign very quickly as $\alpha \to \infty$. Thus these integrals vanish for "reasonable" (i.e. absolutely integrable) functions $f(t)$ because they do not change substantially as the sine (and cosine) alternate with opposite signs in their semi-periods.

Thus, for absolutely integrable functions the necessary condition of convergence of Fourier series is satisfied. Before we discuss the problem of convergence of Fourier series in more detail, let us notice that practically any interesting function for applications can be expanded in a converging Fourier series.

It is important to know how quickly the terms in (A.4) decrease as $n \to \infty$. If they decrease rapidly, the series converges rapidly. In this case, using very few terms we have a good trigonometric approximation for $f(x)$ and the partial sum of the series, $S_n(x)$, is a good approximation to the sum $S(x) = f(x)$. If the series converges more slowly, a larger number of terms is needed to have a sufficiently accurate approximation.

Assuming that the series (A.4) converges, the speed of its convergence to $f(x)$ depends on the behavior of $f(x)$ over its period, or, in the case of non-periodic functions, on the way it is extended from the interval $[a, b]$ to the entire axis x, as we will discuss below. Convergence is most rapid for very smooth functions (functions which have continuous derivatives of higher order). Discontinuities in the derivative of the function, $f'(x)$, substantially reduce the rate of convergence whereas discontinuities in $f(x)$ reduce the convergence rate even more with the result that many terms in the Fourier series must be used to approximate the function $f(x)$ with the necessary precision. This should be fairly obvious since the "smoothness" of $f(x)$ determines the rate of decreasing of the coefficients a_n and b_n.

It can be shown [7, 8], that the coefficients decrease

a) faster than $1/n^2$ (for example $1/n^3$) when $f(x)$ and $f'(x)$ are continuous but $f''(x)$ has a discontinuity;

b) at about the same rate as $1/n^2$ when $f(x)$ is continuous but $f'(x)$ has discontinuities; and

c) at about the same rate as $1/n$ if $f(x)$ is not continuous.

It is important to note that in the first two cases the series converges uniformly which follows from the Weierstrass criterion, because each term of Equation (A.4) is bounded by the corresponding term in the converging numeric series $\sum_{n=1}^{\infty} \frac{1}{n^2} < \infty$.

The following very important theorem describes the convergence of the Fourier series given in Equation (A.4) for a function $f(x)$ at a point x_0 where $f(x)$ is continuous or where it may have a discontinuity (the proof can be found in books [7, 8]).

The Dirichlet theorem

If the function $f(x)$ with period 2π is piecewise continuous in $[-\pi, \pi]$ and has a finite number of points of discontinuity in this interval, then its Fourier series converges to $f(x_0)$ when x_0 is a continuity point, and to

$$S(x_0) = \frac{f(x_0 + 0) + f(x_0 - 0)}{2}$$

if x_0 is a point of discontinuity.

At the ends of the interval $[-\pi, \pi]$ the Fourier series converges to

$$\frac{f(-\pi + 0) + f(\pi - 0)}{2}.$$

A function $f(x)$ defined on $[a, b]$ is called *piecewise continuous* if:

i) It is continuous on $[a, b]$ except perhaps at a finite number of points;

ii) If x_0 is one such point then the left and right limits of $f(x)$ at x_0 exist and are finite;

iii) Both the limit from the right of $f(x)$ at a and the limit from the left at b exist and are finite.

Stated more briefly, for the Fourier series of a function $f(x)$ to converge, this function should be piecewise continuous with a finite number of discontinuities.

A.4 Fourier Series for Non-periodic Functions

We assumed above that the function $f(x)$ is defined on the entire x-axis and has period 2π. But very often we need to deal with non-periodic functions defined only on the interval $[-\pi, \pi]$. The theory discussed above can still be used if we extend $f(x)$ periodically from $(-\pi, \pi)$ to all x. In other words, we assign the same values of $f(x)$ to all the intervals $(\pi, 3\pi)$, $(3\pi, 5\pi)$, ..., $(-3\pi, \pi)$, $(-5\pi, -3\pi)$, ... and then use Equations (A.8) and (A.9) for the Fourier coefficients of this new function which is periodic. If $f(-\pi) = f(\pi)$ we can include the end points, $x = \pm\pi$ and the Fourier series converges to $f(x)$ everywhere on $[-\pi, \pi]$. Over the entire axis the expansion gives a periodic extension of the function $f(x)$

given originally on $[-\pi, \pi]$. In many cases $f(-\pi) \neq f(\pi)$ and the Fourier series at the ends of the interval $[-\pi, \pi]$ converges to

$$\frac{f(-\pi) + f(\pi)}{2}$$

which differs from both $f(-\pi)$ and $f(\pi)$.

The rate of convergence of the Fourier series depends on the discontinuities of the function and derivatives of the function after its extension to the entire axis. Some extensions do not increase the number of discontinuities of the original function whereas others do increase this number. In the latter case the rate of convergence is reduced. In Example A.2 below the function is extended to the entire axis as an even function and remains continuous so that the coefficients of the Fourier series decrease as $1/n^2$. In Example A.3 the function is extended as an odd function and has discontinuities at $x = k\pi$ (integer k) in which case the coefficients decrease slower, as $1/n$.

A.5   Fourier Expansions on Intervals of Arbitrary Length

Suppose that a function $f(x)$ is defined on some interval $[-l, l]$ of arbitrary length $2l$ (where $l > 0$). Using the substitution

$$x = \frac{ly}{\pi}, \quad -\pi \le y \le \pi,$$

we obtain the function $f\left(\frac{yl}{\pi}\right)$ of the variable y on the interval $[-\pi, \pi]$ which can be expanded using the standard Equations (A.4), (A.8) and (A.9) as

$$f\left(\frac{yl}{\pi}\right) = \frac{a_0}{2} + \sum_{n=1}^{\infty} (a_n \cos ny + b_n \sin ny),$$

with

$$a_n = \frac{1}{\pi} \int_{-\pi}^{\pi} f\left(\frac{yl}{\pi}\right) \cos ny\, dy \quad \text{and} \quad b_n = \frac{1}{\pi} \int_{-\pi}^{\pi} f\left(\frac{yl}{\pi}\right) \sin ny\, dy.$$

Returning to the variable x we obtain

$$f(x) = \frac{a_0}{2} + \sum_{n=1}^{\infty} \left(a_n \cos \frac{n\pi x}{l} + b_n \sin \frac{n\pi x}{l}\right) \tag{A.19}$$

with

$$a_n = \frac{1}{l} \int_{-l}^{l} f(x) \cos \frac{n\pi x}{l} dx, \quad n = 0, 1, 2, \ldots,$$

$$b_n = \frac{1}{l} \int_{-l}^{l} f(x) \sin \frac{n\pi x}{l} dx, \quad n = 1, 2, \ldots. \tag{A.20}$$

If the function is given, but not on the interval $[-l, l]$, and instead on an arbitrary interval of length $2l$, for instance $[0, 2l]$, the formulas for the coefficients of the Fourier series (A.19) become

$$a_n = \frac{1}{l} \int_{0}^{2l} f(x) \cos \frac{n\pi x}{l} dx \quad \text{and} \quad b_n = \frac{1}{l} \int_{0}^{2l} f(x) \sin \frac{n\pi x}{l} dx. \tag{A.21}$$

In both cases, the series in Equation (A.19) gives a periodic function with the period $T = 2l$.

If the function $f(x)$ is given on an interval $[a, b]$ (where a and b may have the same or opposite sign, that is, the interval $[a, b]$ can include or exclude the point $x = 0$), different periodic continuations onto the entire x-axis may be made (see Figure A.1). As an example, consider the periodic continuation $F(x)$ of the function $f(x)$, defined by the condition

$$F(x + n(b - a)) = f(x), \quad n = 0, \pm 1, \pm 2, \ldots \quad \text{for all} \quad x.$$

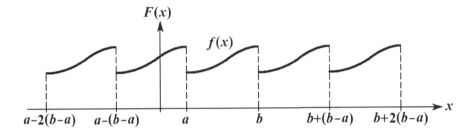

FIGURE A.1
Arbitrary function $f(x)$ defined on the interval $[a, b]$ extended to the x-axis as the function $F(x)$.

In this case the Fourier series is given by Equation (A.19) where $2l = b - a$. Clearly, instead of Equations (A.20) the following formulas for the Fourier coefficients should be used:

$$a_n = \frac{2}{b - a} \int_a^b f(x) \cos \frac{2n\pi x}{b - a} dx, \quad b_n = \frac{2}{b - a} \int_a^b f(x) \sin \frac{2n\pi x}{b - a} dx. \tag{A.22}$$

The series in Equation (A.19) gives a periodic function with the period $T = 2l = b - a$; however the original function was defined only on the interval $[a, b]$ and is not periodic in general.

A.6 Fourier Series in Cosine or in Sine Functions

Suppose that $f(x)$ is an *even* function on $[-\pi, \pi]$ so that $f(x) \sin nx$ is odd. For this case

$$b_n = \frac{1}{\pi} \int_{-\pi}^{\pi} f(x) \sin nx dx = 0$$

since the integral of an odd function over a symmetric interval equals zero. Coefficients a_n can be written as

$$a_n = \frac{1}{\pi} \int_{-\pi}^{\pi} f(x) \cos nx dx = \frac{2}{\pi} \int_0^{\pi} f(x) \cos nx dx, \tag{A.23}$$

since the integrand is even. Thus, for even functions, $f(x)$ we may write

$$f(x) = \frac{a_0}{2} + \sum_{n=1}^{\infty} a_n \cos nx. \tag{A.24}$$

Similarly, if $f(x)$ is an *odd* function we have

$$a_n = \frac{1}{\pi} \int_{-\pi}^{\pi} f(x) \cos nx\, dx = 0 \quad \text{and} \quad b_n = \frac{2}{\pi} \int_0^{\pi} f(x) \sin nx\, dx, \qquad \text{(A.25)}$$

in which case we have

$$f(x) = \sum_{n=1}^{\infty} b_n \sin nx. \qquad \text{(A.26)}$$

Thus, an even on $[-\pi, \pi]$ function is expanded in the set (A.11)

$$1, \ \cos x, \ \cos 2x, \ \ldots, \ \cos nx, \ \ldots$$

The odd on $[-\pi, \pi]$ function is expanded in the set (A.12)

$$\sin x, \ \sin 2x, \ \ldots, \ \sin nx, \ \ldots$$

Any function can be presented as a sum of even and odd functions with set (A.10),

$$f(x) = f_1(x) + f_2(x),$$

where

$$f_1(x) = \frac{f(x) + f(-x)}{2} \quad \text{and} \quad f_2(x) = \frac{f(x) - f(-x)}{2},$$

in which case $f_1(x)$ can be expanded into a cosine Fourier series and $f_2(x)$ into a sine series.

If the function $f(x)$ is defined only on the interval $[0, \pi]$ we can extend it to the interval $[-\pi, 0)$. This extension may be made in different ways corresponding to different Fourier series. In particular, such an extension can make $f(x)$ even or odd on $[-\pi, \pi]$ which leads to cosine or sine series with period 2π. In the first case on the interval $[-\pi, 0)$ we have

$$f(-x) = f(x) \qquad \text{(A.27)}$$

and in the second case

$$f(-x) = -f(x). \qquad \text{(A.28)}$$

The points $x = 0$ and $x = \pi$ need special consideration because the sine and cosine series behave differently at these points. If $f(x)$ is continuous at these points, because of Equations (A.24) and (A.27) the cosine series converges to $f(0)$ at $x = 0$ and to $f(\pi)$ at $x = \pi$. The situation is different for the sine series, however. At $x = 0$ and $x = \pi$ the sum of the sine series in Equation (A.26) is zero thus the series is equal to the functions $f(0)$ and $f(\pi)$, respectively, only when these values are zero.

If $f(x)$ is given on the interval $[0, l]$ (where $l > 0$), the cosine is

$$\frac{a_0}{2} + \sum_{n=1}^{\infty} a_n \cos \frac{n\pi x}{l} \quad \text{with} \quad a_n = \frac{2}{l} \int_0^l f(x) \cos \frac{n\pi x}{l} dx, \qquad \text{(A.29)}$$

and sine series is

$$\sum_{n=1}^{\infty} b_n \sin \frac{n\pi x}{l} \quad \text{with} \quad b_n = \frac{2}{l} \int_0^l f(x) \sin \frac{n\pi x}{l} dx. \qquad \text{(A.30)}$$

To summarize the above discussion, we see that the Fourier series provides a way to obtain an *analytic formula* for functions defined by different formulas on different intervals by combining these intervals into a larger one. Such analytic formulas replace a discontinuous function by a continuous Fourier series expansion which is often more convenient in a given application. As we have seen above there are often many different choices of how to extend the original function, defined initially on an interval, to the entire axis. The specific choice of extension depends on the application to which the expansion is to be used. Many examples and problems demonstrating these points will be presented in the examples at the end of each of the following sections and the problems at the end of this chapter.

A.7   Examples

All the functions given below are differentiable or piecewise differentiable and can be represented by the Fourier series.

Example A.1 Find the cosine series for $f(x) = x^2$ on the interval $[-\pi, \pi]$.

Solution. The coefficients are

$$\frac{1}{2}a_0 = \frac{1}{\pi}\int_0^\pi x^2 dx = \frac{\pi^2}{3},$$

$$a_n = \frac{2}{\pi}\int_0^\pi x^2 \cos nx\, dx = \frac{2}{\pi}x^2 \frac{\sin nx}{n}\Big|_0^\pi - \frac{4}{n\pi}\int_0^\pi x \sin nx\, dx = (-1)^n \frac{4}{n^2}.$$

Thus

$$x^2 = \frac{\pi^2}{3} + 4\sum_{n=1}^\infty (-1)^n \frac{\cos nx}{n^2}, \quad -\pi \le x \le \pi. \tag{A.31}$$

In the case where $x = \pi$ we obtain a famous expansion,

$$\frac{\pi^2}{6} = \sum_{n=1}^\infty \frac{1}{n^2}. \tag{A.32}$$

Example A.2 Let the function $f(x) = x$ on the interval $[0, \pi]$. Find the cosine series.

Solution. Figure A.2 gives an even periodic continuation of $f(x) = x$ from $[0, \pi]$ onto the entire axis. For coefficients we have

$$\frac{1}{2}a_0 = \frac{1}{\pi}\int_0^\pi x\, dx = \frac{\pi}{2}, \quad a_n = \frac{2}{\pi}\int_0^\pi x \cos nx\, dx = 2\frac{\cos n\pi - 1}{n^2\pi} = 2\frac{(-1)^n - 1}{n^2\pi}, \quad n > 0,$$

that is,

$$a_{2k} = 0, \quad a_{2k-1} = -\frac{4}{(2k-1)^2\pi}, \quad k = 1, 2, 3, \ldots,$$

and thus,

$$x = \frac{\pi}{2} - \frac{4}{\pi}\sum_{k=1}^\infty \frac{\cos(2k-1)x}{(2k-1)^2}, \quad 0 \le x \le \pi. \tag{A.33}$$

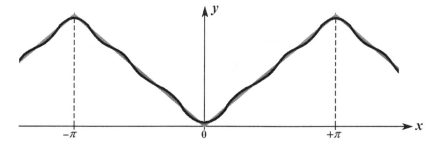

FIGURE A.2
Original function extended from $[0, \pi]$ to the x-axis as even function, plotted together with the partial sum of the first five terms.

Figure A.2 shows the graph of the partial sum

$$y = S_5(x) = \frac{\pi}{2} - \frac{4}{\pi}\left(\cos x + \frac{1}{3^2}\cos 3x + \frac{1}{5^2}\cos 5x\right)$$

together with the graph of the extended function.

Example A.3 Find the Fourier series for $f(x) = \frac{\pi - x}{2}$ on the interval $(0, 2\pi)$.

Solution. The coefficients are

$$a_0 = \frac{1}{\pi}\int_0^{2\pi}\frac{\pi - x}{2}dx = 0, \quad a_n = \frac{1}{\pi}\int_0^{2\pi}\frac{\pi - x}{2}\cos nxdx = 0, \quad b_n = \frac{1}{\pi}\int_0^{2\pi}\frac{\pi - x}{2}\sin nxdx = \frac{1}{n}.$$

This contains the interesting result

$$\frac{\pi - x}{2} = \sum_{n=1}^{\infty}\frac{\sin nx}{n}, \quad 0 < x < 2\pi. \tag{A.34}$$

This equation is not valid at $x = 0$ and $x = 2\pi$ because the sum of the series equals zero. The equality is also violated beyond $(0, 2\pi)$.

Notice that this series converges more slowly than Example A.2, thus we need more terms to obtain the same deviation from the original function. Also, this series does not converge uniformly (to understand why, attempt to differentiate it term by term and note what happens).

For $x = \frac{\pi}{2}$ we have another interesting result that was obtained by Leibnitz by other means:

$$\frac{\pi}{4} = 1 - \frac{1}{3} + \frac{1}{5} - \frac{1}{7} + \dots \tag{A.35}$$

And for $x = \frac{\pi}{6}$ we obtain another representation of π:

$$\frac{\pi}{4} = 1 + \frac{1}{5} - \frac{1}{7} - \frac{1}{11} + \frac{1}{13} + \frac{1}{17} - \dots \tag{A.36}$$

A.8   The Complex Form of the Trigonometric Series

For a real function $f(x)$ with period 2π the Fourier expansion (A.4) can be rewritten in complex form. From Euler's formula

$$e^{iax} = \cos ax + i \sin ax \tag{A.37}$$

we have

$$\cos nx = \frac{1}{2}\left(e^{inx} + e^{-inx}\right), \quad \sin nx = \frac{i}{2}\left(e^{-inx} - e^{inx}\right),$$

from which we obtain

$$f(x) = \frac{a_0}{2} + \sum_{n=1}^{\infty}\left[\frac{1}{2}\left(a_n - b_n i\right)e^{inx} + \frac{1}{2}\left(a_n + b_n i\right)e^{-inx}\right].$$

Using the notations

$$c_0 = \frac{1}{2}a_0, \quad c_n = \frac{1}{2}\left(a_n - b_n i\right), \quad c_{-n} = \frac{1}{2}\left(a_n + b_n i\right),$$

we have

$$f(x) = \sum_{n=-\infty}^{\infty} c_n e^{inx}. \tag{A.38}$$

With the Fourier coefficients a_n and b_n (A.5), (A.8), and (A.9) it is easy to see that the coefficients c_n can be written as

$$c_n = \frac{1}{2\pi}\int_{-\pi}^{\pi} f(x)e^{-inx}dx, \quad n = 0, \pm 1, \pm 2, \ldots . \tag{A.39}$$

It is clear that for functions with period $2l$, the Equations (A.38) and (A.39) have the form

$$f(x) = \sum_{n=-\infty}^{\infty} c_n e^{\frac{in\pi x}{l}} \quad \text{with} \quad c_n = \frac{1}{2l}\int_{-l}^{l} f(x)e^{-\frac{in\pi x}{l}}dx, \quad n = 0, \pm 1, \pm 2, \ldots . \tag{A.40}$$

For periodic functions in time t with a period T, the same formulas can be written as

$$f(t) = \sum_{n=-\infty}^{\infty} c_n e^{\frac{2in\pi t}{T}} \tag{A.41}$$

and

$$c_n = \frac{1}{T}\int_{-T/2}^{T/2} f(t)e^{-\frac{2in\pi x}{T}}dt, \quad n = 0, \pm 1, \pm 2, \ldots . \tag{A.42}$$

Several useful properties of these results can be easily verified:

i)   Because $f(x)$ is real, c_n and c_{-n} are complex conjugate and we have $c_{-n} = c_n^*$;

ii)   If $f(x)$ is even, all c_n are real;

iii)   If $f(x)$ is odd, $c_0 = 0$ and all c_n are pure imaginary.

Example A.4 Represent the function $f(x) = \begin{cases} 0, & -\pi < x \le 0, \\ 1, & 0 < x \le \pi, \end{cases}$ by a complex Fourier series.

Solution. The coefficients are

$$c_0 = \frac{1}{2\pi} \int_0^\pi dx = \frac{1}{2}, \quad c_n = \frac{1}{2\pi} \int_0^\pi e^{-inx} dx = \frac{1 - e^{-in\pi}}{2\pi ni} = \begin{cases} 0, & n = \text{even}, \\ \dfrac{1}{\pi ni}, & n = \text{odd}. \end{cases}$$

Thus

$$f(x) = \frac{1}{2} + \frac{1}{\pi i} \sum_{\substack{n = -\infty \\ n = \text{odd}}}^{+\infty} \frac{1}{n} e^{inx}.$$

Reading Exercise: Using Euler's formula check that from this expression follows

$$Im f(x) = 0 \quad (\text{as should be}) \quad \text{and} \quad Re f(x) = \frac{1}{2} + \frac{2}{\pi} \sum_{n=1,3,\dots}^\infty \frac{\sin nx}{n}.$$

The same result can be obtained if we apply the real form of the Fourier series from the beginning.

Example A.5 Find the Fourier series of the function $f(x) = e^{-x}$ on the interval $(-\pi, \pi)$.

Solution. First use the complex Fourier series with coefficients

$$c_n = \frac{1}{2\pi} \int_{-\pi}^\pi e^{-x} e^{-inx} dx = \frac{1}{2\pi} \int_{-\pi}^\pi e^{-(1+in)x} dx = \frac{e^\pi e^{in\pi} - e^{-\pi} e^{-in\pi}}{2\pi(1 + in)}.$$

Then with $e^{\pm in\pi} = \cos n\pi \pm i \sin n\pi = (-1)^n$ we have $c_n = \frac{(-1)^n (e^\pi - e^{-\pi})}{2\pi(1+in)}$, thus

$$e^{-x} = \sum_{n=-\infty}^\infty c_n e^{\frac{in\pi x}{l}} = \frac{e^\pi - e^{-\pi}}{2\pi} \sum_{n=-\infty}^\infty \frac{(-1)^n e^{inx}}{1 + in}.$$

In the interval $(-\pi, \pi)$ this series converges to e^{-x} and at points $x = \pm\pi$ its sum is $(e^\pi + e^{-\pi})/2$.

Reading Exercise: Apply Euler's formula and check that this series in real form becomes

$$e^{-x} = \frac{e^\pi - e^{-\pi}}{\pi} \left[\frac{1}{2} + \sum_{n=1}^\infty \frac{(-1)^n}{1 + n^2} (\cos nx + n \sin nx) \right].$$

The same result is obtained if we apply the real form of the Fourier series from the beginning.

A.9   Fourier Series for Functions of Several Variables

In this section we extend the previous ideas to generate the Fourier series for functions of two variables, $f(x, y)$, which have period 2π in both the variables x and y. Analogous to the development of Equation (A.38) we write a double Fourier series for the function $f(x, y)$ as

$$f(x, y) = \sum_{n,m=-\infty}^{+\infty} \alpha_{nm} e^{i(nx+my)} \tag{A.43}$$

in the domain $(D) = (-\pi \leq x \leq \pi, -\pi \leq y \leq \pi)$.

The coefficients α_{nm} can be obtained by multiplying Equation (A.43) by $e^{-i(nx+my)}$ and integrating over the domain (D), performing this integration for the series term by term. Because the e^{inx} form a complete set of orthogonal functions on $[-\pi, \pi]$ (and the same for e^{imy}), we obtain

$$\alpha_{nm} = \frac{1}{4\pi^2} \iint_{(D)} f(x, y) e^{-i(nx+my)} dxdy, \quad n, m = 0, \pm 1, \pm 2, \ldots. \tag{A.44}$$

The previous two formulas give the Fourier series for $f(x, y)$ in complex form. For the real Fourier series instead of Equation (A.43) we have

$$f(x, y) = \sum_{n,m=0}^{+\infty} [a_{nm} \cos nx \cos my + b_{nm} \cos nx \sin my$$
$$+ c_{nm} \sin nx \cos my + d_{nm} \sin nx \sin my], \tag{A.45}$$

where

$$a_{00} = \frac{1}{4\pi^2} \iint_{(D)} f(x, y) dxdy, \qquad a_{n0} = \frac{1}{2\pi^2} \iint_{(D)} f(x, y) \cos nx dxdy,$$

$$a_{0m} = \frac{1}{2\pi^2} \iint_{(D)} f(x, y) \cos my dxdy, \qquad a_{nm} = \frac{1}{\pi^2} \iint_{(D)} f(x, y) \cos nx \cos my dxdy,$$

$$b_{0m} = \frac{1}{2\pi^2} \iint_{(D)} f(x, y) \sin my dxdy, \qquad b_{nm} = \frac{1}{\pi^2} \iint_{(D)} f(x, y) \cos nx \sin my dxdy,$$

$$c_{n0} = \frac{1}{2\pi^2} \iint_{(D)} f(x, y) \sin nx dxdy, \qquad c_{nm} = \frac{1}{\pi^2} \iint_{(D)} f(x, y) \sin nx \cos my dxdy,$$

$$d_{nm} = \frac{1}{\pi^2} \iint_{(D)} f(x, y) \sin nx \sin my dxdy \qquad \text{for} \quad n, m = 1, 2, 3, \ldots$$

$$\tag{A.46}$$

A.10   Generalized Fourier Series

Consider expansions similar to trigonometric Fourier series using a set of orthogonal functions as a basis for the expansion. Recall that two complex functions, $\varphi(x)$ and $\psi(x)$, of a real variable x are said to be orthogonal on the interval $[a, b]$ (which can be an infinite interval) if

$$\int_a^b \varphi(x)\psi^*(x)dx = 0, \tag{A.47}$$

where $\psi^*(x)$ is the complex conjugate of $\psi(x)$ (when $\psi(x)$ is real $\psi^* = \psi$).

Let us expand some function $f(x)$ into a set of orthogonal functions $\{\varphi_n(x)\}$:

$$f(x) = c_1\varphi_1(x) + c_2\varphi_2(x) + \ldots + c_n\varphi_n(x) + \ldots = \sum_{n=1}^{\infty} c_n\varphi_n(x). \tag{A.48}$$

Multiplying by $\varphi_n(x)$, integrating and using the orthogonality condition, we obtain the coefficients

$$c_n = \frac{\int_a^b f(x)\varphi_n^*(x)dx}{\int_a^b \varphi_n(x)\varphi_n^*(x)dx} = \frac{1}{\lambda_n} \int_a^b f(x)\varphi_n^*(x)dx, \tag{A.49}$$

where

$$\lambda_n = \int_a^b |\varphi_n(x)|^2 (x)dx$$

are real numbers – squared *norms* of functions $\varphi_n(x)$.

Series (A.48) with coefficients (A.49) is called *generalized Fourier series*.

If the set $\{\varphi_n(x)\}$ is normalized, $\lambda_n = 1$, and the previous formula becomes

$$c_n = \int_a^b f(x)\varphi_n^*(x)dx. \tag{A.50}$$

In the case of trigonometric Fourier series (A.4), the orthogonal functions $\varphi_n(x)$ are

$$1, \; \cos x, \; \sin x, \; \cos 2x, \; \sin 2x, \; \ldots, \; \cos nx, \; \sin nx, \ldots \tag{A.51}$$

This set is *complete* on the interval $[-\pi, \pi]$. Rather than the standard interval $[-\pi, \pi]$, we may also wish to consider any interval of length 2π, or the interval $[-l, l]$ where instead of the argument x in Equation (A.51) we have $\frac{n\pi x}{l}$, etc.

Other sets of orthogonal functions are a system of sines (A.12) or a system of cosines (A.11) on the interval $[0, \pi]$.

With the set of exponential functions

$$\ldots, \; e^{-i2x}, \; e^{-ix}, \; 1, \; e^{ix}, \; e^{i2x}, \ldots \tag{A.52}$$

orthogonal on $[-\pi, \pi]$, we have the expansion (A.38). Another complex set of functions which we use for expansion (A.41)

$$\ldots, \; e^{-\frac{i2\pi x}{l}}, \; e^{-\frac{i\pi x}{l}}, \; 1, \; e^{-\frac{i\pi x}{l}}, \; e^{\frac{i2\pi x}{l}}, \ldots \tag{A.53}$$

is complete and orthogonal on $[-l, l]$. Notice that for this set $\lambda_n = b - a$ and the functions $\{\varphi_n(x)/\lambda_n\}$ are normalized, i.e. have norms equal to one.

Let us multiply Equation (A.48) by its complex conjugated, $f^*(x) = \sum_{n=1}^{\infty} c_n^* \varphi_n^*(x)$, and integrate over the interval $[a, b]$ (or the entire axis). This gives, due to the orthogonality of the functions $\{\varphi_n(x)\}$,

$$\int_a^b |f|^2 (x)dx = \sum_{n=1}^{\infty} |c_n|^2 \int_a^b |\varphi_n(x)|^2 \, dx = \sum_{n=1}^{\infty} |c_n|^2 \lambda_n. \tag{A.54}$$

Equation (A.54) is known as the *completeness equation* or *Parsevale's equality*. If this equation is satisfied the set of functions $\{\varphi_n(x)\}$ is *complete*. Equation (A.54) is an extension of the Pythagorean theorem to a space with an infinite number of dimensions; the square of the diagonal of an (infinite dimensional) parallelepiped is equal to the sum of the squares of all its sides.

The completeness of set $\{\varphi_n(x)\}$ means that any function $f(x)$ (for which $\int_a^b |f(x)|^2 \, dx < \infty$) can be expanded in this set (formula(A.48)) and no other functions except $\{\varphi_n(x)\}$ need to be included.

When the norms of the functions equal unity, i.e. $\lambda_n = 1$, Equation (A.54) has its simplest form:

$$\int_a^b |f(x)|^2 \, dx = \sum_{n=1}^{\infty} |c_n|^2 . \tag{A.55}$$

Note that from formula (A.55) it follows that $c_n \to 0$ as $n \to \infty$.

For the trigonometric Fourier series (A.4) on $[-\pi, \pi]$, Equation (A.55) becomes

$$\int_{-\pi}^{\pi} f^2(x)dx = \frac{1}{2}\pi a_0^2 + \pi \sum_{n=1}^{\infty} \left(a_n^2 + b_n^2\right), \tag{A.56}$$

and for a series on the interval $[-l, l]$ we have

$$\int_{-l}^{l} f^2(x)dx = \frac{1}{2}la_0^2 + l \sum_{n=1}^{\infty} \left(a_n^2 + b_n^2\right). \tag{A.57}$$

A.11   The Gibbs Phenomenon

In this section we take a closer look at the behavior of the Fourier series of a function $f(x)$ near a point of discontinuity (a finite jump) of the function. At these points the series cannot converge uniformly and, in addition, partial sums exhibit specific defects.

Let us begin with an example. The Fourier series for the function

$$f(x) = \begin{cases} -\pi/2, & \text{if} \quad -\pi < x < 0, \\ 0, & \text{if} \quad x = 0, \pm\pi, \\ \pi/2, & \text{if} \quad 0 < x < \pi, \end{cases}$$

is

$$2 \sum_{n=1}^{\infty} \frac{\sin(2n-1)x}{2n-1} = 2\left[\sin x + \frac{\sin 3x}{3} + \frac{\sin 5x}{5} + \ldots\right]. \tag{A.58}$$

This expansion gives (an odd) continuation of the function $f(x)$ from the interval $(-\pi/2, \pi/2)$ to the entire x-axis. Because of the periodicity we can restrict the analysis to the interval $(0, \pi/2)$. The partial sums, shown in Figure A.3, like the original function $f(x)$, have jumps at points $x = 0$ and $x = \pi$.

We may isolate these discontinuity points within infinitely small regions, $[0, \varepsilon)$ and $(\pi - \varepsilon, \pi]$, so that on the rest of the interval, $[\varepsilon, \pi - \varepsilon]$, this series converges uniformly. In the figure this corresponds to the fact that the graphs of the partial sums, for large enough n, are very close to the line $y = \pi/2$ along the interval $[\varepsilon, \pi - \varepsilon]$. Close to the points $x = 0$ and $x = \pi$ it is clear that the uniformity of the approximation of $f(x)$ with partial sums is violated because of the jump discontinuity in $f(x)$.

Next we point out another phenomenon that can be observed near the points $x = 0$ and $x = \pi$. Near $x = 0$, approaching the origin from the right, the graphs of the partial sums (shown in Figure A.3) oscillate about the line $y = \pi/2$. The significant thing to note is that the amplitudes of these oscillations do not diminish to zero as $n \to \infty$. On the contrary, the height of the first bump (closest to $x = 0$) approaches the value of $\delta = 0.281$ above the $y = \pi/2$ line. This corresponds to an additional $\delta : (\pi/2) = 18\%$ of the height of the partial

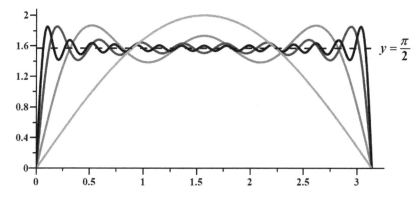

FIGURE A.3
The first partial sums of the expansion of (A.58) demonstrating the Gibb's phenomena –
S_1, S_3, S_{15}, S_{30}.

sum above the "expected" value. The situation is similar when x approaches the value π from the left. Such a defect of the convergence was first found by Josian Gibbs and is known as the *Gibbs phenomenon*. In general if the function $f(x)$ has a finite jump $|D|$ at some point x, the maximum elevation of the partial sum value near x when $n \to \infty$ is bigger than $|D|$ by

$$2|D|\delta/\pi$$

i.e. by about 11.46%.

A.12 Fourier Transforms

A Fourier series is a representation of a function which uses a *discrete* system of orthogonal functions. This idea may be expanded to a *continuous* set of orthogonal functions. The corresponding expansion in this case is referred to as a *Fourier transform*.

Lets start with the complex form of the Fourier series for function $f(x)$ on the interval $[-l, l]$

$$f(x) = \sum_{n=-\infty}^{\infty} c_n e^{\frac{in\pi x}{l}} \tag{A.59}$$

with coefficients

$$c_n = \frac{1}{2l} \int_{-l}^{l} f(x) e^{-\frac{in\pi x}{l}} dx, \quad n = 0, \pm 1, \pm 2, \ldots. \tag{A.60}$$

In physics terminology, Equation (A.60) gives a discrete spectrum of function $f(x)$ with *wave numbers* $k_n = \frac{n\pi}{l}$. Here $c_n e^{ik_n x}$ is a harmonic with complex amplitude c_n defined by Equation (A.60) or

$$c_n = \frac{1}{2l} \int_{-l}^{l} f(x) e^{-ik_n x} dx. \tag{A.61}$$

Suppose now that l is very large, thus the distance between two neighboring wave numbers, $\Delta k = \frac{\pi}{l}$, is very small. Using the notation

$$\hat{f}(k) = \int_{-\infty}^{\infty} f(x)e^{-ikx}dx \tag{A.62}$$

we may write Equation (A.61) in the form

$$c_n = \frac{1}{2\pi}\int_{-\infty}^{\infty} f(x)e^{-ik_n x}dx \cdot \frac{\pi}{l} = \frac{1}{2\pi}\hat{f}(k_n)\Delta k. \tag{A.63}$$

Using this definition Equation (A.59) can be written as

$$f(x) = \sum_n c_n e^{ik_n x} = \frac{1}{2\pi}\sum_n \hat{f}(k_n)e^{ik_n x}\Delta k, \quad -l < x < l. \tag{A.64}$$

In the limit $l \to \infty$ this becomes the integral

$$f(x) = \frac{1}{2\pi}\int_{-\infty}^{\infty} \hat{f}(k)e^{ikx}dk, \quad -\infty < x < \infty. \tag{A.65}$$

In this limit the wave number takes all the values from $-\infty$ to ∞, i.e. when $l \to \infty$ the spectrum is continuous. The amplitudes are distributed continuously and for each infinitesimal interval from k to $k + dk$ there is an infinitesimal amplitude

$$dc = \frac{1}{2\pi}\hat{f}(k)dk. \tag{A.66}$$

With this equation as a definition, $\hat{f}(k)$ is called the *spectral density* of $f(x)$.

Equations (A.62) and (A.65) define the *Fourier transform*. Equation (A.62) is called the direct Fourier transform and Equation (A.65) is referred to as the inverse Fourier transform. These formulas are valid if the function $f(x)$ is absolutely integrable on $(-\infty, \infty)$:

$$\int_{-\infty}^{\infty} |f(x)|dx < \infty. \tag{A.67}$$

It should be noted that there are different ways to deal with the factor $1/2\pi$ in the formulas for direct and inverse transforms. Often, this factor is placed in the direct transform formula while other authors split this factor into two identical factors, $1/\sqrt{2\pi}$, one in each equation. Using the definition given by Equations (A.62) and (A.65) has the advantage that the Fourier transform of the Dirac *delta function*

$$\delta(x) = \frac{1}{2\pi}\int_{-\infty}^{\infty} e^{ikx}dk \tag{A.68}$$

equals one, as can be seen by comparing Equations (A.65) and (A.68). Here we remind the reader that the most useful property of the delta function is

$$\int_{-\infty}^{\infty} f(x')\delta(x - x')dx' = f(x). \tag{A.69}$$

The delta function defined with the coefficient in Equation (A.68) obeys the normalization condition

$$\int_{-\infty}^{\infty} \delta(x - x')dx' = 1. \tag{A.70}$$

The step or Heaviside function is defined as

$$H(x) = \begin{cases} 1, & x \geq 0, \\ 0, & x < 0, \end{cases}$$

and is related to the delta function by the relation

$$\frac{d}{dx}H(x) = \delta(x). \tag{A.71}$$

Reading Exercise: Prove the following two properties of the delta function:

$$\delta(-x) = \delta(x) \quad \text{and} \quad \delta(ax) = \frac{1}{|a|}\delta(x). \tag{A.72}$$

For many practical applications it is useful to present Fourier transform formulas for another pair of physical variables, time and frequency. Using Equations (A.62) and (A.65) we may write the direct and inverse transforms as

$$\hat{f}(\omega) = \int_{-\infty}^{\infty} f(t)e^{-i\omega t}dt \quad \text{and} \quad f(t) = \frac{1}{2\pi}\int_{-\infty}^{\infty} \hat{f}(\omega)e^{i\omega t}d\omega. \tag{A.73}$$

Fourier transform equations are easy to generalize to cases of higher dimensions. For instance, for an application with spatial variables represented as vectors, Equations (A.62) and (A.65) become

$$\hat{f}(\vec{k}) = \int_{-\infty}^{\infty} f(\vec{x})e^{-i\vec{k}\vec{x}}d\vec{x} \quad \text{and} \quad f(\vec{x}) = \frac{1}{2\pi}\int_{-\infty}^{\infty} \hat{f}(\vec{k})e^{i\vec{k}\vec{x}}d\vec{k}. \tag{A.74}$$

Next we briefly discuss *Fourier transforms of even or odd functions*. If the function $f(x)$ is even we have

$$\hat{f}(k) = \int_{-\infty}^{\infty} f(x)\cos kx\,dx - i\int_{-\infty}^{\infty} f(x)\sin kx\,dx = 2\int_{0}^{\infty} f(x)\cos kx\,dx.$$

From here we see that $\hat{f}(k)$ is also even and with Equation (A.65) we obtain

$$f(x) = \frac{1}{\pi}\int_{0}^{\infty} \hat{f}(k)\cos kx\,dk. \tag{A.75}$$

These formulas give what is known as the *Fourier cosine transform*. Similarly if $f(x)$ is odd we obtain the *Fourier sine transform*

$$i\hat{f}(k) = 2\int_{0}^{\infty} f(x)\sin kx\,dx, \quad f(x) = \frac{1}{\pi}\int_{0}^{\infty} i\hat{f}(k)\sin kx\,dk. \tag{A.76}$$

In this case usually $i\hat{f}(k)$ (rather than $\hat{f}(k)$) is called the Fourier transform. We leave it to the reader to obtain Equations (A.76) as a *Reading Exercise*.

If the function $f(x)$ is given on the interval $0 < x < \infty$ it can be extended to $-\infty < x < 0$ in either an even or odd way and we may use either sine or cosine transforms.

Example A.6 Let $f(x) = 1$ on $-1 < x < 1$ and zero outside this interval. This is an even function thus with the cosine Fourier transform we have

$$\hat{f}(k) = 2\left(\int_{0}^{1} 1 \cdot \cos kx\,dx + \int_{1}^{\infty} 0 \cdot \cos kx\,dx\right) = \frac{2\sin k}{k}.$$

The inverse transform gives

$$f(x) = 2 \int_0^\infty \frac{\sin k}{\pi k} \cos kx\, dk. \qquad (A.77)$$

As in the case for the regular Fourier series, if we substitute some value, x_0, into the formula for the inverse transform we obtain the value $f(x_0)$ at the point where this function is continuous. The equation

$$[f(x_0 + 0) + f(x_0 - 0)]/2$$

gives the value at a point where it has a finite discontinuity. For instance, substituting $x = 0$ in (A.77) gives

$$1 = 2 \int_0^\infty \frac{\sin k}{\pi k}\, dk,$$

from which we obtain the interesting result

$$\int_0^\infty \frac{\sin k}{k}\, dk = \frac{\pi}{2}. \qquad (A.78)$$

Example A.7 Investigate the connection between a Gaussian function and its Fourier transform.

Solution. The Fourier transform of a Gaussian, $f(x) = e^{-ax^2}$, $a > 0$ is given by

$$\hat{f}(k) = \int_{-\infty}^\infty e^{-ax^2} e^{-ikx}\, dx.$$

Since

$$\hat{f}'(k) = \int_{-\infty}^\infty (-ix)e^{-ax^2} e^{-ikx}\, dx = \frac{i}{2a} \int_{-\infty}^\infty \frac{d}{dx}(e^{-ax^2}) e^{-ikx}\, dx$$

$$= -\frac{k}{2a} \int_{-\infty}^\infty e^{-ax^2} e^{-ikx}\, dx = -\frac{k}{2a}\hat{f}(k)$$

$\hat{f}(k)$ can be obtained as a solution of a simple differential equation (by separation of variables):

$$\hat{f}(k) = \hat{f}(0)e^{-k^2/4a}.$$

Here $\hat{f}(0) = \int_{-\infty}^\infty e^{-ax^2}\, dx$. With the substitution $z = x\sqrt{a}$ we obtain $\hat{f}(0) = \frac{1}{\sqrt{a}}\int_{-\infty}^\infty e^{-z^2}\, dz$. Because it is well known that $\int_{-\infty}^\infty e^{-z^2}\, dz = \sqrt{\pi}$ we have that $\hat{f}(0) = \sqrt{\frac{\pi}{a}}$ and $\hat{f}(k) = \sqrt{\frac{\pi}{a}}e^{-k^2/4a}$. Thus we have obtained a remarkable result: the Fourier transform of a Gaussian is also a Gaussian. Both functions are bell-shaped and their widths are determined by the value of a. If a is small, then $f(x)$ is a broadly spread Gaussian and its Fourier transform is sharply peaked near $k = 0$. On the other hand, if $f(x)$ is a narrowly peaked Gaussian function corresponding to a being large, its Fourier transform is broadly spread. An application of this result is the uncertainty principle in quantum mechanics where momentum and position probabilities may be represented as Fourier transforms of each other with the result that a narrow uncertainty in the probable location of an object results in a broad uncertainty in the momentum of the object and vice versa.

Reading Exercises:
    Here are several important properties of Fourier transform proofs which we leave to the reader.

i) Prove that the Fourier transform of $f(-x)$ is equal to $\hat{f}(-k)$.

ii) Prove that the Fourier transform of $f'(x)$ is equal to $ik\hat{f}(k)$. *Hint:* The Fourier transform of $f'(x)$ is $\int_{-\infty}^{\infty} f'(x)e^{-ikx}dx$; differentiate it by parts and take into account that $\int_{-\infty}^{\infty} |f(x)|dx < \infty$.

iii) Prove that the Fourier transform of $f(x - x_0)$ (a shift of origin) is equal to $e^{-kx_0}\hat{f}(k)$.

iv) Prove that the Fourier transform of $f(\alpha x)$ (where α is constant) is equal to $\frac{1}{\alpha}\hat{f}\left(\frac{k}{\alpha}\right)$. This property shows that if we stretch the size of an "object" along the x-axis, then the size of the Fourier "image" compresses by the same factor. This means that it is not possible to localize a function in both "x-and k-spaces" which is a mathematical expression representing the *uncertainty principle* in quantum mechanics.

v) Prove Parsevale's equality for the Fourier transform

$$\int_{-\infty}^{\infty} |f(x)|^2 dx = 2\pi \int_{-\infty}^{\infty} |\hat{f}(k)|^2 dk.$$

A.13 Laplace Transforms

A Laplace transform $L[f(x)]$ of a real function of a real variable $f(x)$ is defined as follows:

$$\hat{f}(p) = L[f(x)] = \int_0^{\infty} e^{-px} f(x)dx, \qquad (A.79)$$

where p is, generally, a complex parameter. Function $\hat{f}(p)$ is often called the *image* of the *original* function $f(x)$.

The right side of Equation (A.79) is called the *Laplace integral*. For its convergence in the case where p is real, it is necessary that $p > 0$, when p is complex; then for convergence it is necessary that $\operatorname{Re} p > 0$. Also for convergence a growth of function $f(x)$ should be restricted. We will consider only functions $f(x)$ that are increasing slower than some exponential function as $x \to \infty$. This means that for any x there exist positive constants M and a, such that

$$|f(x)| \leq Me^{ax}.$$

Clearly, for the convergence of the integral it should be $\operatorname{Re} p > a$.
And obviously, function $f(x)$ should be regular on $(0, \infty)$, for instance, for $f(x) = 1/x^{\beta}$, $\beta > 0$, the LT does not exist.

To determine the original function from the image, $\hat{f}(p)$, one has to perform the *inverse Laplace transform*, which is denoted as

$$f(x) = L^{-1}\left[\hat{f}(p)\right]. \qquad (A.80)$$

As an example, find Laplace transforms of two functions:

Let $f(x) = 1$, then $L[f(x)] = \int_0^{\infty} e^{-px}dx = -\frac{1}{p}e^{-px}\,|_0^{\infty} = \frac{1}{p}$;
Let $f(x) = e^{ax}$, then

$$L[f(x)] = \int_0^{\infty} e^{-px}e^{at}dx = \int_0^{\infty} e^{(a-p)x}dx = \frac{1}{a-p}e^{(a-p)x}\,|_0^{\infty} = \frac{1}{p-a}.$$

TABLE A.1

Laplace transforms of some functions.

$f(x)$	$L[f(x)]$	Convergence condition
1	$\dfrac{1}{p}$	$\text{Re } p > 0$
x^n	$\dfrac{n!}{p^{n+1}}$	$n \geq 0$ is integer, $\text{Re } p > 0$
x^a	$\dfrac{\Gamma(a+1)}{p^{a+1}}$	$a > -1$, $\text{Re } p > 0$
e^{ax}	$\dfrac{1}{p-a}$	$\text{Re } p > \text{Re } a$
$\sin cx$	$\dfrac{c}{p^2 + c^2}$	$\text{Re } p > |\text{Im } c|$
$\cos cx$	$\dfrac{p}{p^2 + c^2}$	$\text{Re } p > |\text{Im } c|$
$\sinh bx$	$\dfrac{b}{p^2 - b^2}$	$\text{Re } p > |\text{Re } b|$
$\cosh bx$	$\dfrac{p}{p^2 - b^2}$	$\text{Re } p > |\text{Re } b|$
$x^n e^{ax}$	$\dfrac{n!}{(p-a)^{n+1}}$	$\text{Re } p > \text{Re } a$
$x \sin cx$	$\dfrac{2pc}{(p^2 + c^2)^2}$	$\text{Re } p > |\text{Im } c|$
$x \cos cx$	$\dfrac{p^2 - c^2}{(p^2 + c^2)^2}$	$\text{Re } p > |\text{Im } c|$
$e^{ax} \sin cx$	$\dfrac{c}{(p-a)^2 + c^2}$	$\text{Re } p > (\text{Re } a + |\text{Im } c|)$
$e^{ax} \cos cx$	$\dfrac{p-a}{(p-a)^2 + c^2}$	$\text{Re } p > (\text{Re } a + |\text{Im } c|)$
$\text{erf}(x)$	$\dfrac{e^{p^2/4} [1 - erf(p/2)]}{p}$	$\text{Re } p > 0$

It is seen that for the convergence of the integral, $\text{Re } p$ should be larger than a. Laplace transforms of some functions can be found in Table A.1.

Properties of the Laplace Transform

Property 1. A Laplace transform is linear:

$$L[C_1 f_1(x) + C_2 f_2(x)] = C_1 L[f_1(x)] + C_2 L[f_2(x)]. \tag{A.81}$$

This follows from the linearity of the integral (A.76).

An inverse transform is also linear:

$$L^{-1}\left[C_1 \hat{f}_1(p) + c_2 \hat{f}_2(p)\right] = C_1 L^{-1}\left[\hat{f}_1(p)\right] + C_2 L^{-1}\left[\hat{f}_2(p)\right]. \tag{A.82}$$

Property 2. Let $L[f(x)] = \hat{f}(p)$. Then

$$L[f(kx)] = \frac{1}{k}\hat{f}\left(\frac{p}{k}\right). \tag{A.83}$$

This can be proven using the change of a variable, $x = t/k$:

$$\int_0^\infty e^{-px} f(kx)dx = \frac{1}{k}\int_0^\infty e^{-\frac{p}{k}t}f(t)dt = \frac{1}{k}\hat{f}\left(\frac{p}{k}\right).$$

Property 3. Let $L[f(x)] = \hat{f}(p)$, and

$$f_a(x) = \begin{cases} 0, & x < a, \\ f(x-a), & x \geq a, \end{cases}$$

where $a > 0$. Then

$$\hat{f}_a(p) = e^{-pa}\hat{f}(p). \tag{A.84}$$

This property is also known as the *Delay theorem*.

Property 4. Let $L[f(x)] = \hat{f}(p)$. Then for any complex constant c:

$$L[e^{cx}f(x)] = \int_0^\infty e^{-px}e^{cx}f(x)dx = \hat{f}(p-c). \tag{A.85}$$

This property is also known as the *Shift theorem*.

Property 5. First define *convolution* of functions $f_1(x)$ and $f_2(x)$ as function $f(x)$ given by

$$f(x) = \int_0^x f_1(x-t)f_2(t)dt. \tag{A.86}$$

Convolution is symbolically denoted as:

$$f(x) = f_1(x) * f_2(x).$$

The property states that if $f(x)$ is a convolution of functions $f_1(x)$ and $f_2(x)$, then:

$$\hat{f}(p) = \hat{f}_1(p)\hat{f}_2(p), \tag{A.87}$$

in other words

$$L[f(x)] = L\left[\int_0^x f_1(x-t)f_2(t)dt\right] = \hat{f}_1(p)\hat{f}_2(p) \tag{A.88}$$

(the integral should converge absolutely), where

$$\hat{f}_1(p) = \int_0^\infty f_1(x)e^{-px}dx, \quad \hat{f}_2(p) = \int_0^\infty f_2(x)e^{-px}dx, \hat{f}(p) = \int_0^\infty f(x)e^{-px}dx.$$

Property 6. The Laplace transform of a derivative is:

$$L[f'(x)] = p\hat{f}(p) - f(0). \tag{A.89}$$

Indeed, integration by parts gives:

$$\int_0^\infty e^{-px} f'(x)dx = f(x)e^{-px} \mid_0^\infty + p\int_0^\infty e^{-px} f(x)dx = -f(0) + p\hat{f}(p).$$

Analogously, for the second derivative we obtain:

$$L\left[f''(x)\right] = p^2 \hat{f}(p) - pf(0) - f'(0). \tag{A.90}$$

And for the nth derivative:

$$L\left[f^{(n)}(x)\right] = p^n \hat{f}(p) - p^{n-1}f(0) - p^{n-2}f'(0) - \ldots - f^{(n-1)}(0). \tag{A.91}$$

Property 7. If $L\left[f(x)\right] = \hat{f}(p)$, then the Laplace transform of the integral can be represented as

$$L\left[\int_0^t f(t)dt\right] = \frac{1}{p}\hat{f}(p). \tag{A.92}$$

This property can be proven by writing the Laplace transform of the integral as a double integral and then interchanging the order of the integrations.

Property 8. The Laplace transform of the delta function, $\delta(x)$, is:

$$L\left[\delta(x)\right] = \int_{-0}^{+0} e^{-px}\delta(x)dx + \int_{+0}^{\infty} e^{-px}\delta(x)dx = \int_{-0}^{+0} \delta(x)dx + 0 = 1. \tag{A.93}$$

A.14   Applications of Laplace Transform for ODE

One of the applications of the Laplace transform is the solution of initial value problems (IVPs) for differential equations. After the transform has been found, it must be *inverted*, that is, the original function (the solution of a differential equation) must be obtained. Often this step needs a calculation of integrals of a complex variable, but in many situations, such as, for example, the case of linear equations with constant coefficients, the inverted transforms just can be found in the LT Table. Some partial differential equations also can be solved using the method of Laplace transform.

Consider the IVP for the 2nd-order linear equation with constant coefficients:

$$ax'' + bx' + cx = f(t), \tag{A.94}$$

$$x(0) = \beta, \quad x'(0) = \gamma. \tag{A.95}$$

Let us apply the Laplace transform to the both sides of Equation (A.94). Using linearity of the transform and Property 6 gives the *algebraic* equation for the transform function $\hat{x}(p)$:

$$a(p^2\hat{x} - \beta p - \gamma) + b(p\hat{x} - \beta) + c\hat{x} = \hat{f}(p). \tag{A.96}$$

We have used the initial conditions (A.95). Solving this algebraic equation, one finds:

$$\hat{x}(p) = \frac{\hat{f}(p) + a\beta p + a\gamma + b\beta}{ap^2 + bp + c}. \tag{A.97}$$

Note that when the initial conditions are zero, the transform takes a simple form:

$$\hat{x}(p) = \frac{\hat{f}(p)}{ap^2 + bp + c}. \tag{A.98}$$

Next, inverting the transform, gives the function $x(t)$ – the solution of the IVP.

Example A.8 Solve differential equation

$$x'' + 9x = 6\cos 3t$$

with zero initial conditions.

Solution. Applying the LT to both sides of the equation and taking into account that $x(0) = 0$, $x'(0) = 0$, gives:

$$p^2 \hat{x}(p) + 9\hat{x}(p) = \frac{6p}{p^2 + 9}.$$

Then,

$$\hat{x}(p) = \frac{6p}{(p^2 + 9)^2} = \frac{2 \cdot 3p}{(p^2 + 3^2)^2}.$$

The original function $x(t)$ is read out directly from the Laplace transform table:

$$x(t) = t\sin 3t.$$

This function is the solution of the IVP.

Example A.9 Solve the IVP

$$x'' + 4x = e^t, \quad x(0) = 4, \quad x'(0) = -3.$$

Solution.

Applying the LT to both sides of the equation and taking into the account the initial conditions, we obtain:

$$p^2 \hat{x}(p) - 4p + 3 + 4\hat{x}(p) = \frac{1}{p - 1}.$$

Solving for $\hat{x}(p)$ gives

$$\hat{x}(p) = \frac{4p^2 - 7p + 4}{(p^2 + 4)(p - 1)}.$$

Next, using the partial fractions we can write

$$\frac{4p^2 - 7p + 4}{(p^2 + 4)(p - 1)} = \frac{A}{p - 1} + \frac{Bp + C}{p^2 + 4}.$$

From there

$$4p^2 - 7p + 4 = Ap^2 + 4A + Bp^2 + Cp - Bp - C,$$

and equating the coefficients of the second, first, and zeroth degrees of p, we have

$$4 = A + B, \quad -7 = C - B, \quad 4 = 4A - C.$$

The solution to this system of equations is

$$A = 1/5, \quad B = 19/5, \quad C = -16/5.$$

Then

$$x(t) = \frac{1}{5} L^{-1} \left[\frac{1}{p - 1} \right] + \frac{19}{5} L^{-1} \left[\frac{p}{p^2 + 4} \right] - \frac{16}{5} L^{-1} \left[\frac{1}{p^2 + 4} \right].$$

Using the inverse transform from the Table gives *the solution of the IVP*:

$$x(t) = (e^t + 19\cos 2t - 16\sin 2t)/5.$$

It is easy to check that this solution satisfies the equation and the initial conditions.

Problems

Expand the function into a trigonometric Fourier series using

    a) the *general* method of expansion (cosine & sine series)

    b) the *even terms only* method of expansion (cosine series)

    c) the *odd terms only* method of expansion (sine series):

1. $f(x) = x^2$ on the interval $[0, \pi]$.
2. $f(x) = e^{ax}$ on the interval $[0, \pi]$.
3. $f(x) = \cos ax$ on the interval $[-\pi, \pi]$.
4. $f(x) = \sin ax$ on the interval $[-\pi, \pi]$.
5. $f(x) = \pi - 2x$ on the interval $[0, \pi]$.
6. Prove the Fourier transform of $f(x) = e^{-a|x|}$ equals $\frac{2a}{a^2+k^2}$ where a is a positive constant.
7. Show that cosine and sine transforms of the function $f(x) = 1/\sqrt{x}$ are equal to the original function $1/\sqrt{x}$.

Solve the IVPs for ordinary differential equations using the Laplace transform.

1. $x'' + x = 4\sin t$, $x(0) = 0$, $x'(0) = 0$.
2. $x'' + 16x = 3\cos 2t$, $x(0) = 0$, $x'(0) = 0$.
3. $x'' + 4x = e^t$, $x(0) = 0$, $x'(0) = 0$.
4. $x'' + 25x = e^{-2t}$, $x(0) = 1$, $x'(0) = 0$.
5. $x'' + x' = te^t$, $x(0) = 0$, $x'(0) = 0$.
6. $x'' + 2x' + 2x = te^{-t}$, $x(0) = 0$, $x'(0) = 0$.
7. $x'' - 9x = e^{3t}\cos t$, $x(0) = 0$, $x'(0) = 0$.
8. $x'' - x = 2e^t - t^2$ $x(0) = 0$, $x'(0) = 0$.

B

Bessel and Legendre Functions

In this chapter we discuss the Bessel and Legendre equations and their solutions: Bessel and Legendre functions. These famous functions are widely used in the sciences, in particular Bessel functions serve as a set of basis functions for BVP with circular or cylindrical symmetry, the Legendre functions serve as a set of basis functions for BVP in spherical coordinates.

B.1 Bessel Equation

In many problems one often encounters a differential equation

$$r^2 y''(r) + r y'(r) + (\lambda r^2 - p^2) y(r) = 0, \tag{B.1}$$

where p is given fixed value and λ is an additional parameter. Equation (B.1) is called the *Bessel equation of order* p. In application of the Bessel equation we will, in addition to Equation (B.1), encounter boundary conditions. For example the function $y(r)$, defined on a closed interval $[0,\ l]$, could be restricted to a specified behavior at the points $r = 0$ and $r = l$. For instance at $r = l$ the value of the solution, $y(l)$, could be prescribed, or its derivative $y'(l)$, or their linear combination $\alpha y'(l) + \beta y(l)$. Generally, Equation (B.1) has nontrivial solutions which correspond to a given set of boundary conditions only for certain values of the parameter λ, which are called *eigenvalues*. The goal then becomes to find eigenvalues for these boundary conditions and the corresponding solutions, $y(r)$, which are called *eigenfunctions*.

The Bessel equation (B.1) is often presented in the form which we use in Chapters 7-9:

$$y'' + \frac{y'}{r} + \left(\lambda - \frac{p^2}{r^2} \right) y = 0, \tag{B.2}$$

where we meet the Bessel functions as the eigenfunctions of the Laplace operator in polar coordinates.

To start solving Equation (B.1) make the change of variables $x = \sqrt{\lambda} r$ to yield

$$x^2 y''(x) + x y'(x) + (x^2 - p^2) y(x) = 0, \tag{B.3}$$

or

$$y'' + \frac{y'}{x} + \left(1 - \frac{p^2}{x^2} \right) y = 0. \tag{B.4}$$

Let us first consider integer values of the parameter p. Try to find the solution in the form of power series in x and let a_0 be the first non-zero coefficient of the series, then

$$y = a_0 x^m + a_1 x^{m+1} + a_2 x^{m+2} + \ldots + a_n x^{m+n} + \ldots \tag{B.5}$$

where $m \geq 0$, $a_0 \neq 0$ and, by construction, m is an integer.

Substituting the series (B.5) into Equation (B.4) and equating coefficients of powers of x to zero we obtain (the details can be found in books [7, 8]):

$$y = a_0 2^p p! \left[\frac{\left(\frac{x}{2}\right)^p}{p!} - \frac{\left(\frac{x}{2}\right)^{p+2}}{1!(p+1)!} + \frac{\left(\frac{x}{2}\right)^{p+4}}{2!(p+2)!} - \ldots + (-1)^k \frac{\left(\frac{x}{2}\right)^{p+2k}}{k!(p+k)!} + \ldots \right].$$

The series in the square brackets is absolutely convergent for all values of x which is easy to confirm using the D'Alembert criterion: $\lim\limits_{k \to \infty} |a_{k+1}/a_k| = 0$. Due to the presence of factorials in the denominator this series converges very fast; its sum is called *Bessel function* of order p and it is denoted as $J_p(x)$:

$$J_p(x) = \frac{\left(\frac{x}{2}\right)^p}{p!} - \frac{\left(\frac{x}{2}\right)^{p+2}}{1!(p+1)!} + \frac{\left(\frac{x}{2}\right)^{p+4}}{2!(p+2)!} - \ldots + (-1)^k \frac{\left(\frac{x}{2}\right)^{p+2k}}{k!(p+k)!} + \ldots \qquad (B.6)$$

that is

$$J_p(x) = \sum_{k=0}^{\infty} \frac{(-1)^k}{k!\,(p+k)!} \left(\frac{x}{2}\right)^{p+2k}. \qquad (B.7)$$

Thus we have obtained the solution of Equation (B.4) in the case that the function $y(x)$ is finite at $x = 0$. (The constant coefficient $a_0 2^p p!$ in the series for $y(x)$ can be dropped because the Bessel equation is homogeneous.) Note that this solution was obtained for *integer, nonnegative values of p only.*

Now let us generalize the above case for *arbitrary real p.* To do this it is necessary to replace the integer-valued function $p!$ by the Gamma function $\Gamma(p+1)$, which is defined for arbitrary real values of p. The definition and main properties of the Gamma function are listed in Appendix B (section B.5). Using the Gamma function, the Bessel function of order p, where p is real, can be defined by a series which is built analogously to the series in Equation (B.6):

$$a_2 = -\frac{a_0}{2(2+2p)} = -\frac{a_0}{2^2(1+p)} = -\frac{a_0 \Gamma(p+1)}{2^2 \Gamma(p+2)},$$

$$a_4 = -\frac{a_2}{2^3(p+2)} = \frac{a_0}{2!\,2^4(p+1)(p+2)} = \frac{a_0 \Gamma(p+1)}{2!\,2^4 \Gamma(p+3)},$$

$$\cdots\cdots\cdots\cdots\cdots\cdots\cdots$$

$$a_{2k} = \frac{(-1)^k a_0 \Gamma(p+1)}{2^{2k} k! \Gamma(p+k+1)!},$$

$$\cdots\cdots\cdots\cdots\cdots\cdots\cdots$$

$$J_p(x) = \sum_{k=0}^{\infty} \frac{(-1)^k}{\Gamma(k+1)\Gamma(k+p+1)} \left(\frac{x}{2}\right)^{p+2k}$$

$$= \left(\frac{x}{2}\right)^p \sum_{k=0}^{\infty} \frac{(-1)^k}{\Gamma(k+1)\Gamma(k+p+1)} \left(\frac{x}{2}\right)^{2k}, \qquad (B.8)$$

which *converges for any p.* In particular, *replacing p by $-p$* we obtain

$$J_{-p}(x) = \left(\frac{x}{2}\right)^{-p} \sum_{k=0}^{\infty} \frac{(-1)^k}{\Gamma(k+1)\Gamma(k-p+1)} \left(\frac{x}{2}\right)^{2k}. \qquad (B.9)$$

Since Bessel equation (B.4) contains p^2, functions $J_p(x)$ and $J_{-p}(x)$ are solutions of the equation for the same p. If p *is non-integer*, these solutions are linearly independent, since

TABLE B.1

Positive roots of $J_0(x)$, $J_1(x)$, $J_2(x)$.

Roots / Function	μ_1	μ_2	μ_3	μ_4	μ_5
$J_0(x)$	2.4048	5.5201	8.6537	11.7915	14.9309
$J_1(x)$	3.8317	7.0156	10.1735	13.3237	16.4706
$J_2(x)$	5.136	8.417	11.620	14.796	17.960

the first terms in Equations (B.8) and (B.9) contain different powers of x; x^p and x^{-p}, respectively, then the general solution of Equation (B.4) can be written in the form

$$y = C_1 J_p(x) + C_2 J_{-p}(x). \tag{B.10}$$

Point $x = 0$ must be excluded from the domain of definition of the function (B.9), since x^{-p} for $p > 0$ diverges at this point.

The functions $J_p(x)$ are bounded as $x \to 0$. In fact the functions $J_p(x)$ are continuous for all x since they are the sum of a converging power series. For non-integer values of p this follows from the properties of the Gamma function and the series (B.8).

Reading Exercise: 1) Show that $J_{-n}(x) = (-1)^n J_n(x)$ for $n = 1, 2, 3 \ldots$;

2) Show that $\lim\limits_{x \to 0} \dfrac{J_1(x)}{x} = \dfrac{1}{2}$.

For integer values of p $J_p(x)$ and $J_{-p}(x)$ that are not linearly independent, the general solution can be formed from the linear combination of the functions $J_p(x)$ and $N_p(x)$ as

$$y = C_1 J_p(x) + C_2 N_p(x). \tag{B.11}$$

The functions $N_p(x)$ are singular at $x = 0$, thus if the physical formulation of the problem requires regularity of the solution at zero, the coefficient C_2 in the solution in Equation (B.10) must be zero.

We conclude this section presenting several first Bessel functions $J_n(x)$ of integer order. The first few terms of the expansion in Equation (B.6) near zero for the first three functions are:

$$J_0(x) = 1 - \frac{x^2}{2^2} + \frac{x^4}{2^4 \cdot 2! \cdot 2!} - \ldots, \quad J_1(x) = \frac{x}{2} - \frac{x^3}{2^3 \cdot 2!} + \frac{x^5}{2^5 \cdot 2! \cdot 3!} - \ldots$$

$$J_2(x) = \frac{x^2}{2^2 \cdot 2!} - \frac{x^4}{2^4 \cdot 3!} + \frac{x^6}{2^6 \cdot 2! \cdot 4!} - \ldots, \quad J_3(x) = \ldots \ldots \tag{B.12}$$

Note that the $J_n(x)$ are even if n is integer and odd if n is non-integer functions of x (although in general, problems with physical boundary values have $x \geq 0$, and we do not need to consider the behavior of the function for $x < 0$). For future reference we present the useful fact that at $x = 0$ we have $J_0(0) = 1$ and $J_n(0) = 0$ for $n \geq 1$. Figure B.1 shows graphs of functions $J_0(x)$, $J_1(x)$ and $J_2(x)$. Table B.1 lists a few first roots of Bessel functions of orders 0, 1 and 2.

A convenient notation for roots of equation $J_n(x) = 0$ is $\mu_k^{(n)}$, where n stands for the order of the Bessel function and k stands for the root number. Often we will write equation $J_n(x) = 0$ in the form $J_n(\mu) = 0$. In cases when the value of n is clear we will omit the upper index in μ_k (like in Table B.1).

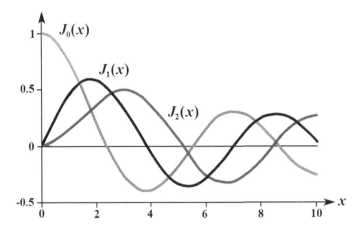

FIGURE B.1
Graphs of functions $J_0(x)$, $J_1(x)$ and $J_2(x)$.

B.2   Properties of Bessel Functions

The properties of functions $J_p(x)$ listed below follow from the expansion in Equation (B.6).

1. All Bessel functions are defined and continuous on the real axis and have derivatives of all orders. This is because any Bessel function can be expanded in a power series which converges for all x, and the sum of the power series is a continuous function which has derivatives of all orders.

2. For integer $p = n$ Bessel functions of even orders are even functions (since their expansion contains only even powers of the argument). Bessel functions of odd orders are odd functions.

3. Each Bessel function has an infinite number of real roots. Roots located on the positive semi-axis can be marked by integer numbers in the increasing order. Zeros of $J_p(x) = 0$ fall between the zeros of $J_{p+1}(x) = 0$.

4. Behavior of Bessel functions in the vicinity of zero is given by the first terms of the series in Equation (B.6); for large x the asymptotic formula may be used

$$J_p(x) \approx \sqrt{\frac{2}{\pi x}} \cos\left(x - \frac{p\pi}{2} - \frac{\pi}{4}\right). \tag{B.13}$$

With increasing x the accuracy of this formula quickly increases. When $J_p(x)$ is replaced by the right hand side of Equation (B.13), the error is very small for large x and has the same order as $x^{-3/2}$.

From Equation (B.1) it follows, in particular, that the function $J_p(x)$ has roots that are close (for large x) to the roots of the equation

$$\cos\left(x - \frac{p\pi}{2} - \frac{\pi}{4}\right) = 0;$$

thus the difference between two adjacent roots of the function $J_p(x)$ tends to π when roots tend to infinity. A graph of $J_p(x)$ has the shape of a curve which depicts decaying oscillation; the "wavelength" is almost constant (close to π), and the amplitude decays inversely proportional to the square root of x. In fact we have $\lim\limits_{x\to\infty} J_p(x) = 0$.

5. Recurrence formulas

$$J_{p+1}(x) = \frac{2p}{x}J_p(x) - J_{p-1}(x), \tag{B.14a}$$

$$J_{p+1}(x) = \frac{p}{x}J_p(x) - J_p'(x), \tag{B.14b}$$

$$J_p'(x) = -\frac{p}{x}J_p(x) + J_{p-1}(x). \tag{B.14c}$$

6. Integration formulas

$$\int x^{-p}J_{p+1}(x)dx = -x^{-p}J_p(x), \tag{B.15a}$$

$$\int x^p J_{p-1}(x)dx = x^p J_p(x), \tag{B.15b}$$

$$\int_0^x zJ_0(z)dz = xJ_1(x), \tag{B.15c}$$

$$\int_0^x z^3 J_0(z)dz = 2x^2 J_0(x) + x(x^2 - 4)J_1(x). \tag{B.15d}$$

7. Differentiation formulas

$$\frac{d}{dx}[x^{-p}J_p(x)] \equiv -x^{-p}J_{p+1}(x), \quad p = 0, 1, 2, \ldots \tag{B.16a}$$

$$\frac{d}{dx}[x^p J_p(x)] \equiv x^p J_{p-1}(x), \quad p = 1, 2, 3, \ldots \tag{B.16b}$$

In particular, $J_0'(x) = -J_1(x), \quad J_1'(x) = J_0(x) - \dfrac{J_1(x)}{x}$.

These identities are easily established by operating on the series which defines the function.

In many physical problems with spherical symmetry one encounters Bessel functions of half-integer orders where $p = (2n+1)/2$ for $n = 0, 1, 2, \ldots$ For instance, solving Equation (B.4) with $p = 1/2$ and $p = -1/2$ by using the series expansion

$$y(x) = x^{1/2}\sum_{k=0}^{\infty} a_k x^k \quad (a_0 \neq 0), \tag{B.17}$$

we obtain

$$J_{1/2}(x) = \left(\frac{2}{\pi x}\right)^{1/2}\sum_{k=0}^{\infty}(-1)^k \frac{x^{2k+1}}{(2k+1)!} \tag{B.18}$$

and

$$J_{-1/2}(x) = \left(\frac{2}{\pi x}\right)^{1/2}\sum_{k=0}^{\infty}(-1)^k \frac{x^{2k}}{(2k)!}. \tag{B.19}$$

By comparing expansions in Equations (B.18) and (B.19) to the McLaren series expansions in $\sin x$ and $\cos x$, we obtain

$$J_{1/2} = \left(\frac{2}{\pi x}\right)^{1/2}\sin x. \tag{B.20}$$

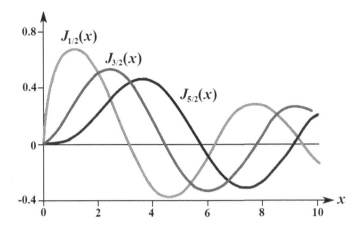

FIGURE B.2
Graphs of functions $J_{1/2}(x)$, $J_{3/2}(x)$ and $J_{5/2}(x)$.

$$J_{-1/2} = \left(\frac{2}{\pi x}\right)^{1/2} \cos x. \tag{B.21}$$

Note that $J_{1/2}(x)$ is bounded for all x and function $J_{-1/2}(x)$ diverges at $x = 0$. Recall that Equation (B.8) gives an expansion of $J_p(x)$ which is valid for any value of p. Figure B.2 shows graphs of functions $J_{1/2}(x)$, $J_{3/2}(x)$ and $J_{5/2}(x)$; Figure B.3 shows graphs for $J_{-1/2}(x)$, $J_{-3/2}(x)$ and $J_{-5/2}(x)$.

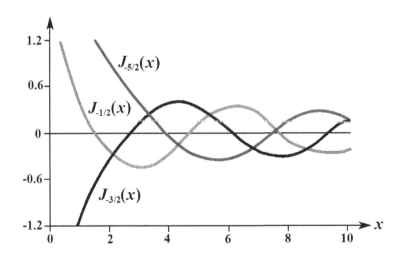

FIGURE B.3
Graphs of $J_{-1/2}(x)$, $J_{-3/2}(x)$ and $J_{-5/2}(x)$.

Reading Exercise:
Using the recurrence Equations (B.14) and the expression for $J_{1/2}(x)$, obtain the functions $J_{3/2}(x)$, $J_{-3/2}(x)$, $J_{5/2}(x)$ and $J_{-5/2}(x)$. For instance, the answer for $J_{3/2}(x)$ is

$$J_{3/2} = \left(\frac{2}{\pi x}\right)^{1/2} \left(\frac{\sin x}{x} - \cos x\right). \tag{B.22}$$

TABLE B.2

Positive roots of the functions $N_0(x)$, $N_1(x)$ and $N_2(x)$.

Roots Function	μ_1	μ_2	μ_3	μ_4	μ_5
$N_0(x)$	0.8936	3.9577	7.0861	10.2223	13.3611
$N_1(x)$	2.1971	5.4297	8.5960	11.7492	14.8974
$N_2(x)$	3.3842	6.7938	10.0235	13.2199	16.3789

Let us now briefly consider Bessel functions of the second kind, or Neumann functions $N_p(x)$. Table B.2 lists a few roots of Neumann functions with $p = 1, 2, 3$. Neumann functions for non-integer p can be obtained as

$$N_p(x) = \frac{J_p(x) \cos p\pi - J_{-p}(x)}{\sin p\pi}. \tag{B.23}$$

It is easy to check, that $N_p(x)$ satisfies the Bessel equation (B.4). Figure B.4 shows graphs of Neumann functions with $p = 1, 2, 3$. In case of integer order n, this function is defined as $N_n(x) = \lim_{p \to n} N_p(x)$.

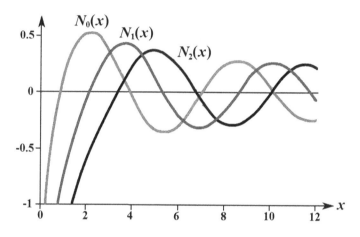

FIGURE B.4

Graphs of $N_0(x)$, $N_1(x)$ and $N_2(x)$.

For integer n

$$N_{-n}(x) = (-1)^n N_n(x). \tag{B.24}$$

Other useful properties are that as $x \to 0$ the functions $N_n(x)$ diverge logarithmically; as $x \to \infty$, $N_n(x) \to 0$, oscillating with decaying amplitude. At large x we have the asymptotic form

$$N_n(x) \approx \sqrt{\frac{2}{\pi x}} \sin\left(x - \frac{n\pi}{2} - \frac{\pi}{4}\right). \tag{B.25}$$

B.3 Boundary Value Problems and Fourier-Bessel Series

In applications it is often necessary to solve Bessel equations (B.1) accompanied by boundary condition(s). For instance function $y(x)$ defined on the interval $[0, l]$ is finite at $x = 0$ and at $x = l$ obeys a homogeneous boundary condition $\alpha y'(l) + \beta y(l) = 0$ (notice, that starting from here we will use letter x instead of r considering equation (B.1)). When $\alpha = 0$ this mixed boundary condition becomes the Dirichlet type, when $\beta = 0$ it is the Neumann type. The Bessel equation and boundary condition(s) form the Sturm-Liouville problem, thus it has the nontrivial solutions only for nonnegative discrete eigenvalues λ_k:

$$0 \leq \lambda_1 < \lambda_2 < \ldots < \lambda_k < \ldots. \tag{B.26}$$

The corresponding eigenfunctions are:

$$J_p\left(\sqrt{\lambda_1}x\right), J_p\left(\sqrt{\lambda_2}x\right), \ldots, J_p\left(\sqrt{\lambda_k}x\right), \ldots \tag{B.27}$$

(if function $y(x)$ has not to be finite at $x = 0$, we can also consider the set of eigenfunctions $N_p\left(\sqrt{\lambda_k}x\right)$). The boundary condition at $x = l$ gives

$$\alpha\sqrt{\lambda}\, J_p'\left(\sqrt{\lambda}l\right) + \beta J_p\left(\sqrt{\lambda}l\right) = 0.$$

Setting $\sqrt{\lambda}l \equiv \mu$ we obtain a transcendental equation for μ:

$$\alpha\mu J_p'(\mu) + \beta l J_p(\mu) = 0, \tag{B.28}$$

which has an infinite number of roots which we label as $\mu_k^{(p)}$. From there the eigenvalues are $\lambda_k = (\mu_k/l)^2$; we see that we need only *positive roots*, $\mu_m^{(n)}$, because negative roots do not give new values of λ_k.

For example, for the Dirichlet problem $\mu_k^{(p)}$ are the positive roots of the Bessel function $J_p(x)$.

Thus, for fixed p we have the set of eigenfunctions (index p in $\mu_k^{(p)}$ is dropped)

$$J_p\left(\frac{\mu_1}{l}x\right), \quad J_p\left(\frac{\mu_2}{l}x\right), \quad \ldots, \quad J_p\left(\frac{\mu_k}{l}x\right), \quad \ldots. \tag{B.29}$$

As follows from the Sturm-Liouville theory, these functions form a *complete set* and are pair-wise orthogonal (with weight x) on the interval $[0, l]$:

$$\int_0^l J_p\left(\frac{\mu_k}{l}x\right) J_p\left(\frac{\mu_j}{l}x\right) x\,dx = 0, \quad i \neq j. \tag{B.30}$$

A Fourier series expansion (or generalized Fourier series) of an arbitrary function $f(x)$ using the set of functions (B.29) is called *Fourier-Bessel series* and is given by the expression

$$f(x) = \sum_{k=0}^{\infty} c_k J_p\left(\frac{\mu_k}{l}x\right). \tag{B.31}$$

The orthogonality property allows us to find the coefficients of this series. We multiply Equation (B.31) by $J_p\left(\frac{\mu_k}{l}x\right)$ and integrate term by term with weight x. This gives an expression for the coefficient as

$$c_k = \frac{\int_0^l f(x) J_p\left(\frac{\mu_k}{l}x\right) x\,dx}{\int_0^l x\left[J_p\left(\frac{\mu_k}{l}x\right)\right]^2 dx}. \tag{B.32}$$

The *squared norm* $\|R_{pk}\|^2 = \int_0^l x J_p^2\left(\frac{\mu_k^{(p)}}{l}x\right) dx$ is:

1) For the *Dirichlet* boundary condition $\alpha = 0$ and $\beta = 1$, in which case eigenvalues are obtained from the equation

$$J_p(\mu) = 0$$

and we have

$$\|R_{pk}\|^2 = \frac{l^2}{2}\left[J_p'\left(\mu_k^{(p)}\right)\right]^2. \tag{B.33}$$

2) For the *Neumann* boundary condition $\alpha = 1$ and $\beta = 0$, in which case eigenvalues are obtained from the equation

$$J_p'(\mu) = 0$$

and we have

$$\|R_{pk}\|^2 = \frac{l^2}{2\left(\mu_k^{(p)}\right)^2}\left[\left(\mu_k^{(p)}\right)^2 - p^2\right] J_p^2\left(\mu_k^{(p)}\right). \tag{B.34}$$

3) For the *mixed* boundary condition $\alpha = 1$ and $\beta = h$, in which case eigenvalues are obtained from the equation

$$\mu J_p'(\mu) + hl J_p(\mu) = 0$$

and we have

$$\|R_{pk}\|^2 = \frac{l^2}{2}\left[1 + \frac{l^2 h^2 - p^2}{\left(\mu_k^{(p)}\right)^2}\right] J_p^2\left(\mu_k^{(p)}\right). \tag{B.35}$$

In Chapters 7-9 we use solutions of different BVP in the form

$$V_{nm}^{(1)} = J_n\left(\frac{\mu_m^{(n)}}{l}r\right)\cos n\varphi, \quad V_{nm}^{(2)} = J_n\left(\frac{\mu_m^{(n)}}{l}r\right)\sin n\varphi.$$

Clearly, $\|V_{nm}^{(1,2)}\|^2 = \sigma_n \pi \|R_{nm}\|^2$, where $\sigma_n = 2$ for $n = 0$ and $\sigma_n = 1$ for $n \neq 0$.

The completeness of the set of functions $J_p\left(\frac{\mu_k}{l}x\right)$ on the interval $(0,l)$ means that for any square integrable on $[0,\ l]$ function $f(x)$ the following is true:

$$\int_0^l xf^2(x)dx = \sum_k \left\|J_p\left(\frac{\mu_k}{l}x\right)\right\|^2 c_k^2. \tag{B.36}$$

This is Parseval's equality for the Fourier-Bessel series. It has the same significance of completeness as in the case of the trigonometric Fourier series with sines and cosines as the basis functions where the weight function equals one instead of x as in the Bessel series.

Regarding the convergence of the series (B.31), we note that the sequence of the partial sums of the series, $S_n(x)$, converges on the interval $(0,l)$ on average (i.e. in the mean) to $f(x)$ (with weight x), which may be written as

$$\int_0^l [f(x) - S_n(x)]^2 x dx \to 0, \quad \text{if} \quad n \to \infty.$$

This property is true for any function $f(x)$ from the class of piecewise-continuous functions because the orthogonal set of functions (B.29) is complete on the interval $[0,\ l]$. For such functions, $f(x)$, the series (B.31) converges absolutely and uniformly. We present the following theorem without the proof which states a somewhat stronger result about the convergence of the series in Equation (B.31) than convergence in the mean:

Theorem

If the function $f(x)$ is piecewise-continuous on the interval $(0, l)$, then the Fourier-Bessel series converges to $f(x)$ at the points where the function $f(x)$ is continuous, and to

$$\frac{1}{2}\left[f(x_0 + 0) + f(x_0 - 0)\right],$$

if x_0 is a point of finite discontinuity of the function $f(x)$.

Below we consider several examples of the expansion of functions into the Fourier-Bessel series using the functions $J_p(x)$.

Example B.1 Let us expand the function $f(x) = A$, $A = \text{const}$, in a series using the Bessel functions $X_k(x) = J_0\left(\mu_k^{(0)} x/l\right)$ on the interval $[0, l]$, where $\mu_k^{(0)}$ are the positive roots of the equation $J_0(\mu) = 0$.

Solution. First, we calculate the norm, $\|X_k\|^2 = \left\|J_0\left(\mu_k^{(0)} x/l\right)\right\|^2$ using the relation $J_0'(x) = -J_1(x)$ to obtain

$$\left\|J_0\left(\mu_k^{(0)} x/l\right)\right\|^2 = \frac{l^2}{2}\left[J_0'\left(\mu_k^{(0)}\right)\right]^2 = \frac{l^2}{2} J_1^2\left(\mu_k^{(0)}\right).$$

Using the substitution $z = \mu_k^{(0)} x/l$ and second of relations (B.14b) we may calculate the integral

$$\int_0^l J_0\left(\frac{\mu_k^{(0)}}{l} x\right) x\, dx = \frac{l^2}{\left(\mu_k^{(0)}\right)^2} \int_0^{\mu_k^{(0)}} J_0(z)\, z\, dz = \frac{l^2}{\left(\mu_k^{(0)}\right)^2}\left[z J_1(z)\right]_0^{\mu_k^{(0)}} = \frac{l^2}{\mu_k^{(0)}} J_1\left(\mu_k^{(0)}\right).$$

For the coefficients c_k of the expansion (B.31) we have

$$c_k = \frac{\int_0^l A\, J_0\left(\mu_k^{(0)} x/l\right)\, x\, dx}{\left\|J_0\left(\mu_k^{(0)} x/l\right)\right\|^2} = \frac{2A}{l^2\left[J_1\left(\mu_k^{(0)}\right)\right]^2} \frac{l^2}{\mu_k^{(0)}} J_1\left(\mu_k^{(0)}\right) = \frac{2A}{\mu_k^{(0)} J_1\left(\mu_k^{(0)}\right)}.$$

Thus, the expansion is

$$f(x) = 2A \sum_{k=0}^{\infty} \frac{1}{\mu_k^{(0)} J_1\left(\mu_k^{(0)}\right)} J_0\left(\frac{\mu_k^{(0)}}{l} x\right).$$

Figure B.5 shows the function $f(x) = 1$ and the partial sum of its Fourier-Bessel series when $l = 1$. From this figure it is seen that the series converges very slowly (see Figure B.5d) and even when 50 terms are kept in the expansion (Figure B.5c) the difference from $f(x) = 1$ can easily be seen. This is because at the endpoints of the interval the value of the function $f(x) = 1$ and the functions $J_0\left(\mu_k^{(0)} x\right)$ are different. The obtained expansion does not converge well near the endpoints.

Example B.2 Let us modify the boundary condition in the previous problem. We expand the function $f(x) = A$, $A = \text{const}$, given on the interval $[0, l]$, in a Fourier series in Bessel functions $X_k(x) = J_0\left(\mu_k^{(0)} x/l\right)$, where $\mu_k^{(0)}$ are now the positive roots of the equation $J_0'(\mu) = 0$.

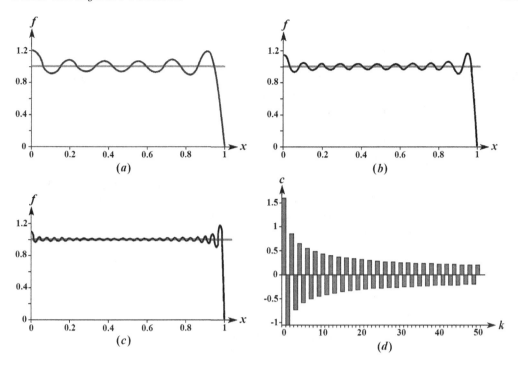

FIGURE B.5

The function $f(x) = 1$ and the partial sum of its Fourier-Bessel series. The graph of $f(x)$ is shown in grey, the graph of the series by the black line. a) 11 terms are kept in the series ($N = 10$); b) $N = 20$; c) $N = 50$; d) values of the coefficients c_k of the series.

Solution. For *Neumann boundary condition*, $J_0'(\mu) = 0$, for $k = 0$ we have

$$\mu_0^{(0)} = 0, \quad X_0(x) = J_0(0) = 1, \quad \|X_0\|^2 = \|J_0(0)\|^2 = \frac{l^2}{2},$$

The first coefficient of the expansion (B.31) is

$$c_0 = \frac{A}{\|J_0(0)\|^2} \int_0^l J_0(0)x\,dx = \frac{2A}{l^2} \int_0^l x\,dx = A.$$

The next coefficients c_k can be evaluated by using the substitution $z = \mu_k^{(0)} x/l$ and using the integration formula

$$\int_0^l J_0\left(\frac{\mu_k^{(0)}}{l}x\right)x\,dx = \frac{l^2}{\left(\mu_k^{(0)}\right)^2}\,[zJ_1(z)]_0^{\mu_k^{(0)}} = \frac{l^2}{\mu_k^{(0)}}J_1\left(\mu_k^{(0)}\right).$$

Applying the relation $J_0'(x) = -J_1(x)$ and then recalling that $J_0'\left(\mu_k^{(0)}\right) = 0$ we find

$$c_k = \frac{2A}{\mu_k^{(0)}}J_1\left(\mu_k^{(0)}\right) = 0, \quad \text{when} \quad k > 0.$$

Thus, we obtain simple expansion, $f(x) = c_0 J_0\left(\mu_0^{(0)} x/l\right) = A$. In fact, this means that the given function is actually one of the functions from the set of eigenfunctions used for eigenfunction expansion.

B.4   Spherical Bessel Functions

In this section we briefly consider *spherical Bessel functions* which are related to the solutions of certain boundary value problems in spherical coordinates.

Consider the following equation:

$$\frac{d^2 R(x)}{dx^2} + \frac{2}{x}\frac{dR(x)}{dx} + \left[1 - \frac{l(l+1)}{x^2}\right]R(x) = 0. \tag{B.37}$$

Parameter l takes discrete non-negative integer values: $l = 0, 1, 2, \ldots$ Equation (B.37) is called the *spherical Bessel equation*. It differs from the cylindrical Bessel equation, Equation (B.4) by the coefficient 2 in the second term. Equation (B.37) can be transformed to a Bessel cylindrical equation by the substitution $R(x) = y(x)/\sqrt{x}$.

Reading Exercise: Check that equation for $y(x)$ is

$$\frac{d^2 y(x)}{dx^2} + \frac{1}{x}\frac{dy(x)}{dx} + \left[1 - \frac{(l+1/2)^2}{x^2}\right]y(x) = 0. \tag{B.38}$$

If we introduce $s = l + 1/2$ in Equation (B.38) we recognize this equation as the Bessel equation which has the general solution

$$y(x) = C_1 J_s(x) + C_2 N_s(x), \tag{B.39}$$

where $J_s(x)$ and $N_s(x)$ are (cylindrical) Bessel and Neumann functions. Because $s = l + 1/2$ these functions are of half-integer order. Inverting the transformation we have that the solution, $R(x)$, to Equation (B.37) is

$$R(x) = C_1 \frac{J_{l+1/2}(x)}{\sqrt{x}} + C_2 \frac{N_{l+1/2}(x)}{\sqrt{x}}. \tag{B.40}$$

If we consider a regular at $x = 0$ solution, the coefficient $C_2 \equiv 0$.

The *spherical Bessel function* $j_l(x)$ is defined to be a solution finite at $r = 0$ thus it is a multiple of $J_{l+1/2}(x)/\sqrt{x}$. The coefficient of proportionality is usually chosen to be $\sqrt{\pi/2}$ so that

$$j_l(x) = \sqrt{\frac{\pi}{2x}} J_{l+1/2}(x). \tag{B.41}$$

For $l = 0$, $J_{1/2}(x) = \sqrt{\frac{2}{\pi x}} \sin x$, thus

$$j_0(x) = \frac{\sin x}{x}. \tag{B.42}$$

Analogously we may define the *spherical Neumann functions* as

$$n_l(x) = \sqrt{\frac{\pi}{2x}} N_{l+1/2}(x), \tag{B.43}$$

from where (using Equation (B.22))

$$n_0(x) = \sqrt{\frac{\pi}{2x}} J_{-1/2}(x) = -\frac{\cos x}{x}. \tag{B.44}$$

Expressions for the first few terms of the functions $j_l(x)$ and $n_l(x)$ are

$$j_1(x) = \frac{\sin x}{x^2} - \frac{\cos x}{x} + \ldots, \quad j_2(x) = \left(\frac{3}{x^3} - \frac{1}{x}\right)\sin x - \frac{3}{x^2}\cos x + \ldots, \tag{B.45}$$

$$n_1(x) = -\frac{\cos x}{x^2} - \frac{\sin x}{x} + \ldots, \quad n_2(x) = -\left(\frac{3}{x^3} - \frac{1}{x}\right)\cos x - \frac{3}{x^2}\sin x + \ldots \tag{B.46}$$

The spherical Bessel functions with $l = 0, 1, 2$ are sketched in Figures B.6 and B.7.

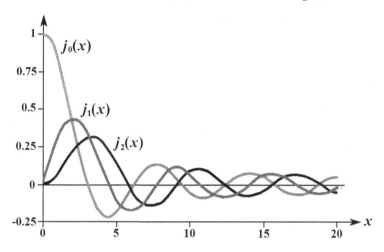

FIGURE B.6
Graphs of functions $j_0(x)$, $j_1(x)$ and $j_2(x)$.

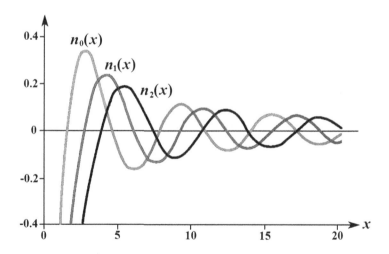

FIGURE B.7
Graphs of functions $n_0(x)$, $n_1(x)$ and $n_2(x)$.

The following *recurrence relations* are valid (here the symbol f is written to stand for j or n):

$$f_{l-1}(x) + f_{l+1}(x) = (2l + 1)x^{-1}f_l(x), \tag{B.47}$$

$$lf_{l-1}(x) - (l + 1)f_{l+1}(x) = (2l + 1)\frac{d}{dx}f_l(x). \tag{B.48}$$

Differentiation formulas:

$$\frac{d}{dx}\left[x^{l+1}j_l(x)\right] = x^{l+1}j_{l-1}(x), \quad \frac{d}{dx}\left[x^{-l}j_l(x)\right] = -x^{-1}j_{l+1}(x). \tag{B.49}$$

Asymptotic values:

$$j_l(x) \sim \frac{1}{x}\cos\left[x - \frac{\pi}{2}(l+1)\right], \quad n_l(x) \sim \frac{1}{x}\sin\left[x - \frac{\pi}{2}(l+1)\right] \quad \text{as} \quad x \to \infty \tag{B.50}$$

(the last expression has good precision for $x \gg l(l+1)$).

B.5   The Gamma Function

In this section we develop the essential properties of the Gamma function. One of the most important applications of the Gamma function is that it allows us to find factorials of positive numbers which are not integers. The gamma function is defined by the integral

$$\Gamma(x) = \int_0^\infty t^{x-1}e^{-t}dt, \quad x > 0. \tag{B.51}$$

Here x is an arbitrary, real, non-negative number. In the case that x is an integer and $x \geq 2$, the integral in Equation (B.51) can be evaluated by parts and we have, after the substitution $t^{x-1} = u$, $e^{-t}dt = dv$:

$$\Gamma(x) = t^{x-1}e^{-t}\Big|_0^\infty + (x-1)\int_0^\infty t^{x-2}e^{-t}dt = (x-1)\int_0^\infty t^{x-2}e^{-t}dt. \tag{B.52}$$

The obtained integral is equal to $\Gamma(x-1)$, i.e. we may write

$$\Gamma(x) = (x-1)\Gamma(x-1). \tag{B.53}$$

Substituting into Equation (B.53) we have

$$\Gamma(x-1) = (x-2)\Gamma(x-2),$$

$$\Gamma(x-2) = (x-3)\Gamma(x-3),$$

etc., and thus we obtain the general expression

$$\Gamma(x) = (x-1)(x-2)\ldots\Gamma(1), \tag{B.54}$$

where

$$\Gamma(1) = \int_0^\infty e^{-t}dt = 1. \tag{B.55}$$

Substituting Equation (B.55) into Equation (B.54) we have

$$\Gamma(x) = (x-1)! \quad \text{for} \quad x = 2, 3, \ldots. \tag{B.56}$$

We derived Equation (B.56) for integer values $x \geq 2$, but it is possible to generalize it to define *factorials of any numbers*. First, we verify that Equation (B.56) is valid for $x = 1$.

Let $x = 1$ in Equation (B.56) in which case we have $\Gamma(1) = 0! = 1$ which does agree with Equation (B.55). Thus, for integer values of the argument, $n = 1, 2, 3, \ldots$

$$\Gamma(1) = 1, \quad \Gamma(2) = 1, \ldots, \Gamma(n) = (n-1)! \tag{B.57}$$

Now consider non-integer values of x. For $x = 1/2$, taking into account definition (B.51) and with the substitution $t = z^2$, we obtain

$$\Gamma(1/2) = \int_0^\infty t^{-1/2} e^{-t} dt = 2 \int_0^\infty e^{-z^2} dz = \sqrt{\pi}. \tag{B.58}$$

Now using Equation (B.53), we find

$$\Gamma(3/2) = (1/2)\Gamma(1/2) = \sqrt{\pi}/2. \tag{B.59}$$

Reading Exercise: Show that for any integer $n \geq 1$,

$$\Gamma\left(n + \frac{1}{2}\right) = \frac{1 \cdot 3 \cdot 5 \cdot \ldots (2n-1)}{2^n} \Gamma\left(\frac{1}{2}\right). \tag{B.60}$$

We can also generalize definition (B.51) to negative values of x using Equation (B.52). First replace x by $x + 1$ which gives

$$\Gamma(x) = \frac{\Gamma(x+1)}{x}. \tag{B.61}$$

We may use this equation to find, for example, $\Gamma(-1/2)$ in the following way:

$$\Gamma\left(-\frac{1}{2}\right) = \frac{\Gamma(1/2)}{-1/2} = -2\sqrt{\pi}. \tag{B.62}$$

It is clear from this that using Equations (B.51) and (B.62) we can find a value of $\Gamma(x)$ for all values of x except 0 and negative integers.

The function $\Gamma(x)$ diverges at $x = 0$ as is seen from Equation (B.51). Then, from Equation (B.61) we see that $\Gamma(-1)$ is not defined because it involves $\Gamma(0)$. Thus, $\Gamma(x)$ does not exist for negative integer values of x. From Equation (B.61) it is obvious (taking into account that $\Gamma(1) = 1$) that at all these values of x the function $\Gamma(x)$ has simple poles. A graph of $\Gamma(x)$ is plotted in Figure B.8. Table B.3 contains the values of Gamma function for interval $[1, 2]$.

Equation (B.61) allows us to find the value of $\Gamma(x)$ for any non-negative integer x using the value of $\Gamma(x)$ on the interval $1 \leq x \leq 2$. For example, $\Gamma(3.4) = 3.4 \cdot \Gamma(2.4) = 3.4 \cdot 2.4 \cdot \Gamma(1.4)$. Based on this fact, a table of values for $\Gamma(x)$ only need include the interval $[1, 2]$ as values of x. The minimum value of $\Gamma(x)$ is reached at $x = 1.46116321\ldots$

Equation (B.57), which is now valid for all non-negative integer values of x, can be written as

$$\Gamma(x+1) = x! \tag{B.63}$$

On the other hand, from Equation (B.51) we have

$$\Gamma(x+1) = \int_0^\infty t^x e^{-t} dt. \tag{B.64}$$

The integrand $t^x e^{-t}$ has a sharp peak at $t = x$ that allows us to obtain a famous approximation formula, known as Stirling's approximation, for $x!$ which works very well for large x:

$$x! \sim (2\pi x)^{1/2} x^x e^{-x}. \tag{B.65}$$

This formula agrees very well with the precise value of $x!$ even for values of x which are not very large. For instance, for $x = 10$ the relative error of Equation (B.65) is less than 0.8%. Most of the applications of Equation (B.65) belong to statistical physics where it is often necessary to evaluate factorials of very large numbers.

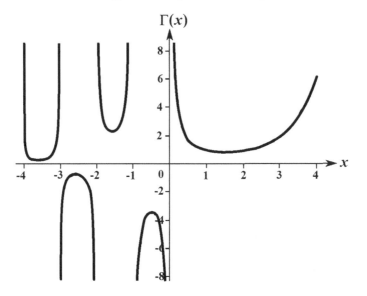

FIGURE B.8
Graph of the Gamma function, $\Gamma(x)$.

TABLE B.3
Values of $\Gamma(x)$ for $x \in [1,\, 2]$.

x	$\Gamma(x)$	x	$\Gamma(x)$	x	$\Gamma(x)$
1	1	1.35	0.8911514420	1.7	0.9086387329
1.05	0.9735042656	1.4	0.8872638175	1.75	0.9190625268
1.1	0.9513507699	1.45	0.8856613803	1.8	0.9313837710
1.15	0.9330409311	1.5	0.8862269255	1.85	0.9456111764
1.2	0.9181687424	1.55	0.8888683478	1.9	0.9617658319
1.25	0.9064024771	1.6	0.8935153493	1.95	0.9798806513
1.3	0.8974706963	1.65	0.9001168163	2	1

B.6 Legendre Equation and Legendre Polynomials

In applications one often encounters an eigenvalue problem containing a second order linear homogeneous differential equation

$$\left(1 - x^2\right) y'' - 2xy' + \lambda y = 0, \quad -1 \le x \le 1, \tag{B.66}$$

where λ is a real parameter. Equation (B.66) can be rewritten in *Sturm-Liouville form* given by

$$\frac{d}{dx}\left[\left(1 - x^2\right)\frac{dy}{dx}\right] + \lambda y = 0, \quad -1 \le x \le 1 \tag{B.67}$$

and is called *the Legendre equation*. Such an equation frequently arises after a separation of variables procedure in spherical coordinates (in those problems the variable x is $x = \cos\theta$, where θ is a meridian angle) in many problems of mathematical physics. Prominent examples include heat conduction in a spherical domain, vibrations of spherical solids and shells as well as boundary value problems for the electric potential in spherical coordinates (see [7, 8]).

Because this problem is a particular example of a Sturm-Liouville problem we can expect that the eigenvalues are *nonnegative real discrete* λ_n, and the eigenfunctions corresponding to different eigenvalues are *orthogonal* on the interval $[-1, 1]$ with the weight $r(x) = 1$:

$$\int_{-1}^{1} y_n(x)y_m(x)dx = 0, \quad n \neq m. \tag{B.68}$$

Let us solve Equation (B.67) on the interval $x \in [-1, 1]$ assuming that the function $y(x)$ is finite at the points $x = -1$ and $x = 1$. Let us search for a solution in the form of a power series in x:

$$y(x) = \sum_{n=0}^{\infty} a_n x^n. \tag{B.69}$$

Substitution of Equation (B.69) into Equation (B.67) results in the following equality:

$$\sum_{n=2}^{\infty} n(n-1)a_n x^{n-2} - \sum_{n=2}^{\infty} n(n-1)a_n x^n - \sum_{n=2}^{\infty} 2na_n x^n + \lambda \sum_{n=2}^{\infty} a_n x^n = 0.$$

Changing the index of summations in the first term from n to $n + 2$ to yield

$$\sum_{n=0}^{\infty} (n+2)(n+1)a_{n+2} x^n$$

allows us to group all the terms with $n \geq 2$ leaving the terms with $n = 0$ and $n = 1$ which we write separately and we have

$$(6a_3 - 2a_1 + \lambda a_1) x + 2a_2 + \lambda a_0$$

$$+ \sum_{n=2}^{\infty} \left[(n+2)(n+1) a_{n+2} - \left(n^2 + n - \lambda \right) a_n \right] x^n = 0.$$

By setting coefficients of each power of x to zero we obtain an infinite system of equations for the coefficients a_n:

$$n = 0 \quad 2a_2 + \lambda a_0 = 0, \tag{B.70}$$

$$n = 1 \quad 6a_3 - 2a_1 + \lambda a_1 = 0, \tag{B.71}$$

$$n \geq 2 \quad (n+2)(n+1) a_{n+2} - \left(n^2 + n - \lambda \right) a_n = 0. \tag{B.72}$$

Equation (B.72) is *the recurrence formula* for coefficients.
From Equation (B.70) we have

$$a_2 = -\frac{\lambda}{2}a_0. \tag{B.73}$$

Using Equations (B.73) and (B.72) we obtain

$$a_4 = \frac{6 - \lambda}{3 \cdot 4} a_2 = \frac{-\lambda(6 - \lambda)}{4!} a_0,$$

$$a_6 = \frac{20 - \lambda}{5 \cdot 6} a_4 = \frac{-\lambda(6 - \lambda)(20 - \lambda)}{6!} a_0,$$

etc. Each coefficient a_{2n} with even index is multiplied by a_0 and depends on n and the parameter λ.

Similarly let us proceed with the odd terms, a_{2k+1}. From Equation (B.71) we have

$$a_3 = \frac{2 - \lambda}{6} a_1 = \frac{2 - \lambda}{3!} a_1.$$

Then from the recurrence formula (B.72) we obtain

$$a_5 = \frac{12 - \lambda}{4 \cdot 5} a_3 = \frac{(2 - \lambda)(12 - \lambda)}{5!} a_1,$$

$$a_7 = \frac{30 - \lambda}{6 \cdot 7} a_5 = \frac{(2 - \lambda)(12 - \lambda)(30 - \lambda)}{7!} a_1,$$

etc. Each coefficient, a_{2k+1}, with odd index is multiplied by a_1 and depends on n and λ. Substituting the obtained coefficients in Equation (B.69) we have

$$y(x) = \sum_{n=0}^{\infty} a_n x^n = a_0 \left[1 - \frac{\lambda}{2} x^2 - \frac{\lambda(6 - \lambda)}{4!} x^4 - \dots \right]$$

$$+ a_1 \left[x + \frac{2 - \lambda}{3!} x^3 + \frac{(2 - \lambda)(12 - \lambda)}{5!} x^5 + \dots \right]. \tag{B.74}$$

In Equation (B.74) there are two sums, one of which contains coefficients with even indexes and another with odd indexes. As a result, we obtain two *linearly independent* solutions of Equation (B.67); one contains even powers of x, the other odd:

$$y^{(1)}(x) = \sum_{n=0}^{\infty} a_{2n} x^{2n} = a_0 \left[1 - \frac{\lambda}{2} x^2 - \frac{\lambda(6 - \lambda)}{4!} x^4 - \dots \right], \tag{B.75}$$

$$y^{(2)}(x) = \sum_{n=0}^{\infty} a_{2n+1} x^{2n+1} = a_1 \left[x + \frac{2 - \lambda}{3!} x^3 + \frac{(2 - \lambda)(12 - \lambda)}{5!} x^5 + \dots \right]. \tag{B.76}$$

Now from the recurrence relation (B.72) for the coefficients,

$$a_{n+2} = \frac{(n^2 + n - \lambda)}{(n + 2)(n + 1)} a_n, \tag{B.77}$$

we can state an important fact. The series in Equation (B.69) converges on an open interval $-1 < x < 1$, as can be seen from a ratio test, $\lim_{n \to \infty} \left| \frac{a_{n+2} x^{n+2}}{a_n x^n} \right| = x^2$, but diverges at the points $x = \pm 1$. Therefore, this series cannot be used as an acceptable solution of the differential equation on the entire interval $-1 \leq x \leq 1$ unless it *terminates as a polynomial* with a finite number of terms. This can occur if the numerator in Equation (B.77) is zero for some index value, $n_{\max}$, such that

$$\lambda = n_{\max}(n_{\max} + 1).$$

This gives $a_{n_{\max}+2} = 0$ and consequently $a_{n_{\max}+4} = 0$, $a_{n_{\max}+6} = 0$, ... thus $y(x)$ will contain a finite number of terms and thus turn out to be a polynomial of degree $n_{\max}$. In order not to overcomplicate the notation, from here on we will denote $n_{\max}$ as n. We may conclude from this discussion that λ can take only *nonnegative integer* values:

$$\lambda = n(n + 1). \tag{B.78}$$

Let us consider several particular cases. If $n = 0$ (which means that the highest degree of the polynomial is 0), $a_0 \neq 0$ and $a_2 = 0$, $a_4 = 0$, etc. The value of λ for this case is

$\lambda = 0$, and we have $y^{(1)}(x) = a_0$. If $n = 1$, $a_1 \neq 0$ and $a_3 = a_5 = \ldots = 0$ then $\lambda = 2$ and $y^{(2)}(x) = a_1 x$. If $n = 2$, the highest degree of the polynomial is 2; we have $a_2 \neq 0$, $a_4 = a_6 = \ldots = 0$, $\lambda = 6$ and from the recurrence relation we obtain $a_2 = -3a_0$. This results in $y^{(1)}(x) = a_0 \left(1 - 3x^2\right)$. If $n = 3$, $a_3 \neq 0$ and $a_5 = a_7 = \ldots = 0$, $\lambda = 12$ and from the recurrence relation we obtain $a_3 = -5/3a_1$ and as the result, $y^{(2)}(x) = a_1 \left(1 - 5x^3/3\right)$. Constants a_0 and a_1 remain arbitrary unless we impose some additional requirement. A convenient requirement is that the solutions (the polynomials) obtained in this way should have the value 1 when $x = 1$.

The polynomials obtained above are denoted as $P_n(x)$ and called *the Legendre polynomials*. The first few, which we derived above, are

$$P_0(x) = 1, \quad P_1(x) = x, \quad P_2(x) = \frac{1}{2}\left(3x^2 - 1\right), \quad P_3(x) = \frac{1}{2}\left(5x^3 - 3x\right). \quad \text{(B.79)}$$

Let us list two more (which the reader may derive as a *Reading Exercise* using the above relationships):

$$P_4(x) = \frac{1}{8}\left(35x^4 - 30x^2 + 3\right), \quad P_5(x) = \frac{1}{8}\left(63x^5 - 70x^3 + 15x\right). \quad \text{(B.80)}$$

Reading Exercise: Obtain $P_n(x)$ for $n = 6, 7$.
Reading Exercise: Show by direct substitution that $P_2(x)$ and $P_3(x)$ satisfy Equation (B.67).
As we see, $P_n(x)$ are even functions for even values of n, and odd functions for odd n. The functions $y^{(1)}(x)$ and $y^{(2)}(x)$, bounded on the closed interval $-1 \leq x \leq 1$, are

$$y^{(1)}(x) = P_{2n}(x), \quad y^{(2)}(x) = P_{2n+1}(x). \quad \text{(B.81)}$$

Rodrigues' formula allows us to calculate the Legendre polynomials:

$$P_n(x) = \frac{1}{2^n n!} \frac{d^n}{dx^n} \left(x^2 - 1\right)^n \quad \text{(B.82)}$$

(here a zero-order derivative means the function itself).
Reading Exercise: Using Rodrigues' formula show that

$$P_n(-x) = (-1)^n P_n(x), \quad \text{(B.83)}$$

$$P_n(-1) = (-1)^n. \quad \text{(B.84)}$$

Let us state few more useful properties of Legendre polynomials:

$$P_n(1) = 1, \quad P_{2n+1}(0) = 0, \quad P_{2n}(0) = (-1)^n \frac{1 \cdot 3 \cdot \ldots \cdot (2n-1)}{2 \cdot 4 \cdot \ldots \cdot 2n}. \quad \text{(B.85)}$$

The Legendre polynomial $P_n(x)$ has n real and simple (i.e. not repeated) roots, all lying in the interval $-1 < x < 1$. Zeroes of the polynomials $P_n(x)$ and $P_{n+1}(x)$ alternate as x increases. In Figure B.9 the first four polynomials, $P_n(x)$, are shown and their properties, as listed in Equations (B.82) through (B.84), are reflected in these graphs. The following *recurrence formula* relates three polynomials

$$(n+1)P_{n+1}(x) - (2n+1)xP_n(x) + nP_{n-1}(x) = 0. \quad \text{(B.86)}$$

This formula gives a simple (and the most practical) way to obtain the Legendre polynomials of any order, one by one, starting with $P_0(x) = 1$ and $P_1(x) = x$.

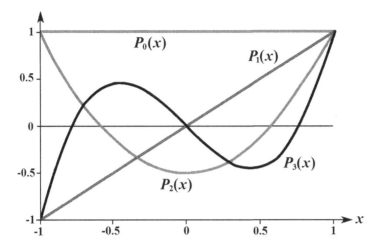

FIGURE B.9
First four polynomials, $P_n(x)$.

The following recurrence relations for Legendre polynomials are often useful:

$$P'_{n-1}(x) - xP'_n(x) + nP_n(x) = 0, \tag{B.87}$$

$$P'_n(x) - xP'_{n-1}(x) - nP_{n-1}(x) = 0. \tag{B.88}$$

To summarize, the solution of the Sturm-Liouville problem for Equation (B.67) with boundary conditions stating that the solution is bounded on the closed interval $-1 \leq x \leq 1$ is a set of Legendre polynomials, $P_n(x)$, which are the eigenfunctions of the Sturm-Liouville operator. The eigenvalues are $\lambda = n(n+1)$, $n = 0, 1, 2, \ldots$ As a solution of the Sturm-Liouville problem, the Legendre polynomials, $P_n(x)$, form a *complete orthogonal* set of functions on the closed interval $[-1, 1]$, a property we will find very useful in the applications considered below.

It the points $x = \pm 1$ are excluded from a domain the solution in the form of an infinite series is also acceptable. In this case logarithmically diverges at $x = \pm 1$ functions $Q_n(x)$ are also the solutions of Legendre equation (for details see books [7, 8]).

B.7   Fourier-Legendre Series in Legendre Polynomials

The Legendre polynomials are orthogonal on the interval $[-1, 1]$:

$$\int_{-1}^1 P_n(x)P_m(x)dx = 0, \quad m \neq n. \tag{B.89}$$

The norm squared of Legendre polynomials are (see [7, 8])

$$\|P_n\|^2 = \int_{-1}^1 P_n^2(x)dx = \frac{2}{2n+1}. \tag{B.90}$$

Equations (B.89) and (B.92) can be combined and written as

$$\int_{-1}^{1} P_n(x)P_m(x)dx = \begin{cases} 0, & m \neq n, \\ \dfrac{2}{2n+1}, & m = n. \end{cases} \quad \text{(B.91)}$$

The Legendre polynomials form a complete set of functions on the interval $[-1, 1]$, thus $\{P_n(x)\}$, $n = 0, 1, 2, \ldots$ provide a basis for an eigenfunction expansion for functions $f(x)$, bounded on the interval $[-1, 1]$:

$$f(x) = \sum_{n=0}^{\infty} c_n P_n(x). \quad \text{(B.92)}$$

Due to the orthogonality of the functions $P_n(x)$ with different indexes, the coefficients c_n are

$$c_n = \frac{1}{\|P_n\|^2} \int_{-1}^{1} f(x)P_n(x)dx = \frac{2n+1}{2} \int_{-1}^{1} f(x)P_n(x)dx. \quad \text{(B.93)}$$

As we know from general theory discussed previously in Chapter 4, the sequence of the partial sums of this series, $S_N(x)$, converges on the interval $(-1, 1)$ on average (i.e. in the mean) to $f(x)$, which may be written as

$$\int_{-1}^{1} [f(x) - S_N(x)]^2 \, dx \to 0 \quad \text{as} \quad N \to \infty. \quad \text{(B.94)}$$

The function $f(x)$ should be square integrable, that is we require that the integral $\int_{-1}^{1} f^2(x)dx$ exists.

For an important class of piecewise-continuous functions the series in Equation (B.92) converges absolutely and uniformly. The following theorem states a stronger result about the convergence of the series (B.92) than convergence in the mean.

Theorem

If the function $f(x)$ is piecewise-continuous on the interval $(-1, 1)$, then the Fourier-Legendre series converges to $f(x)$ at the points where the function $f(x)$ is continuous and

$$\frac{1}{2} [f(x_0 + 0) + f(x_0 - 0)] \quad \text{(B.95)}$$

if x_0 is a point of finite discontinuity of the function $f(x)$.

Because the Legendre polynomials form a complete set, for any square integrable function, $f(x)$, we have

$$\int_{-1}^{1} f^2(x)dx = \sum_{n=0}^{\infty} \|P_n\|^2 c_n^2 = \sum_{n=0}^{\infty} \frac{2}{2n+1} c_n^2. \quad \text{(B.96)}$$

This is Parserval's equality (the completeness equation) for the Fourier-Legendre series. Clearly, for a partial sum on the right we have Bessel's inequality

$$\int_{-1}^{1} f^2(x)dx \geq \sum_{n=0}^{N} \frac{2}{2n+1} c_n^2. \quad \text{(B.97)}$$

Below we consider several examples of the expansion of functions into the Fourier-Legendre series. In some cases the coefficients can be found analytically; otherwise we may calculate them numerically using Maple, Mathematica, or software from [7, 8].

Example B.3 Expand function $f(x) = A$, $A = \text{const}$, in a Fourier-Legendre series in $P_n(x)$ on the interval $-1 \le x \le 1$.

Solution. The series is

$$A = c_0 P_0(x) + c_1 P_1(x) + \cdots,$$

where coefficients c_n are

$$c_n = \frac{1}{\|P_n^2(x)\|} \int_{-1}^{1} A P_n(x) dx = \frac{(2n+1)A}{2} \int_{-1}^{1} P_n(x) dx.$$

From this formula it is clear that the only nonzero coefficient is $c_0 = A$.

Example B.4 Expand the function $f(x) = x$ in a Fourier-Legendre series in $P_n(x)$ on the interval $-1 \le x \le 1$.

Solution. The series is

$$x = c_0 P_0(x) + c_1 P_1(x) + \cdots,$$

where c_n are

$$c_n = \frac{1}{\|P_n^2(x)\|} \int_{-1}^{1} x P_n(x) dx = \frac{2n+1}{2} \int_{-1}^{1} x P_n(x) dx.$$

Clearly the only nonzero coefficient is

$$c_1 = \frac{3}{2} \int_{-1}^{1} x P_1(x) dx = \frac{3}{2} \int_{-1}^{1} x^2 dx = 1.$$

As in the previous example this result is apparent because one of the polynomials, $P_1(x)$ in this example, coincides with the given function, $f(x) = x$.

Example B.5 Expand the function $f(x)$ given by

$$f(x) = \begin{cases} 0, & -1 < x < 0, \\ 1, & 0 < x < 1 \end{cases}$$

in a Fourier-Legendre series.

Solution. The expansion $f(x) = \sum_{n=0}^{\infty} c_n P_n(x)$ has coefficients

$$c_n = \frac{1}{\|P_n\|^2} \int_{-1}^{1} f(x) P_n(x) dx = \frac{2n+1}{2} \int_{0}^{1} P_n(x) dx.$$

The first several are

$$c_0 = \frac{1}{2} \int_{0}^{1} dx = \frac{1}{2}, \quad c_1 = \frac{3}{2} \int_{0}^{1} x dx = \frac{3}{4}, \quad c_2 = \frac{5}{2} \int_{0}^{1} \frac{1}{2} \left(3x^2 - 1 \right) dx = 0.$$

Continuing, we find for the given function $f(x)$,

$$f(x) = \frac{1}{2} P_0(x) + \frac{3}{4} P_1(x) - \frac{7}{16} P_3(x) + \frac{11}{32} P_5(x) + \cdots.$$

The series converges slowly because of discontinuity of a given function $f(x)$ at point $x = 0$ (see Figure B.10).

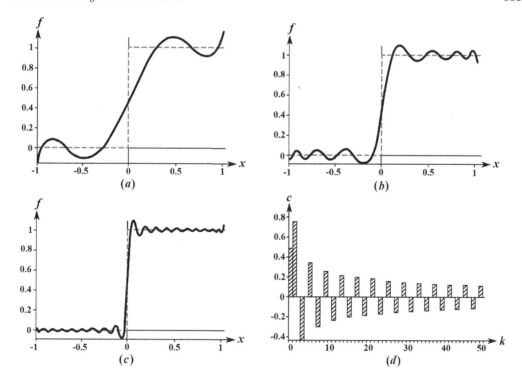

FIGURE B.10
The function $f(x)$ and the partial sum of its Fourier-Legendre series. The graph of $f(x)$ is shown by the dashed line, and the graph of the series is shown by the solid line. $(N+1)$ terms are kept in the series. (a) $(N = 5)$. (b) $N = 15$. (c) $N = 50$. (d) Values of the coefficients c_n of the series.

B.8 Associated Legendre Functions

In this section we consider a generalization of Equation (B.66):

$$\left(1 - x^2\right) y'' - 2xy' + \left(\lambda - \frac{m^2}{1 - x^2}\right) y = 0, \quad -1 \le x \le 1, \tag{B.98}$$

where m is a specified number. Like Equation (B.67), Equation (B.98) has nontrivial solutions bounded at $x = \pm 1$ only for the values of $\lambda = n(n + 1)$. In mathematical physics problems the values of m are *integer*, also the values of m and n are related by inequality $|m| \le n$. Equation (B.98) is called the *associated Legendre equation of order* m. In Sturm-Liouville form this equation can be written as

$$\frac{d}{dx}\left[\left(1 - x^2\right)\frac{dy}{dx}\right] + \left(\lambda - \frac{m^2}{1 - x^2}\right) y = 0, \quad -1 \le x \le 1. \tag{B.99}$$

To solve Equation (B.99) we can use a solution of Equation (B.67). First, we will discuss positive values of m. Let us introduce a new function, $z(x)$, to replace $y(x)$ in Equation (B.99) using the definition

$$y(x) = \left(1 - x^2\right)^{\frac{m}{2}} z(x). \tag{B.100}$$

Substituting Equation (B.100) into Equation (B.99) we obtain

$$\left(1 - x^2\right) z'' - 2(m+1)xz' + [\lambda - m(m+1)] z = 0. \tag{B.101}$$

If $m = 0$, Equation (B.101) reduces to Equation (7.1), thus its solutions are Legendre polynomials $P_n(x)$.

Let us solve Equation (B.101) expanding $z(x)$ in a power series:

$$z = \sum_{k=0}^{\infty} a_k x^k. \tag{B.102}$$

With this we have

$$z' = \sum_{k=1}^{\infty} k a_k x^{k-1} = \sum_{k=0}^{\infty} k a_k x^{k-1},$$

$$z'' = \sum_{k=2}^{\infty} k(k-1) a_k x^{k-2} = \sum_{k=0}^{\infty} (k+2)(k+1) a_{k+2} x^k,$$

$$x^2 z'' = \sum_{k=2}^{\infty} k(k-1) a_k x^k = \sum_{k=0}^{\infty} k(k-1) a_k x^k.$$

Substituting these into Equation (B.101) we obtain

$$\sum_{k=0}^{\infty} \left\{ (k+2)(k+1) a_{k+2} + [\lambda - (k+m)(k+m+1)] a_k \right\} x^k = 0.$$

Functions x^k are linearly independent, thus the coefficients of each power of x^k must be zero which leads to a recurrence relation for coefficients a_k:

$$a_{k+2} = -\frac{\lambda - (k+m)(k+m+1)}{(k+2)(k+1)} a_k. \tag{B.103}$$

Reading Exercise: Using this recurrence relation check that the series in Equation (B.102) converges for $-1 < x < 1$ and diverges at the ends of the intervals, $x = \pm 1$.

Below we will discuss only solutions which are regular on the closed interval, $-1 \leq x \leq 1$. This means that the series (B.102) should *terminate as a polynomial* of some maximum degree. Denoting this degree as q we obtain $a_q \neq 0$ and $a_{q+2} = 0$ so that if $\lambda = (q+m)(q+m+1)$, $q = 0, 1, \ldots$, then $a_{q+2} = a_{q+4} = \ldots = 0$. Introducing $n = q + m$, because q and m are nonnegative integers, we have $n = 0, 1, 2, \ldots$ and $n \geq m$. Thus, we see that $\lambda = n(n+1)$ as in the case of Legendre polynomials. Clearly if $n = 0$ the value of $m = 0$, thus $\lambda = 0$ and the function $z(x) = a_0$ and $y(x) = P_0(x)$.

From the above discussion we obtain that $z(x)$ is an even or odd polynomial of degree $(n - m)$:

$$z(x) = a_{n-m} x^{n-m} + a_{n-m-2} x^{n-m-2} + \ldots + \begin{cases} a_0, \\ a_1 x. \end{cases} \tag{B.104}$$

Let us present several examples for $m = 1$. If $n = 1$, then $q = 0$, thus $z(x) = a_0$. If $n = 2$ then $q = 1$, thus $z(x) = a_1 x$. If $n = 3$, $q = 2$ and $z(x) = a_0 + a_2 x^2$, and from the recurrence formula we have $a_2 = -5a_0$.

Reading Exercises:

1. Find $z(x)$ for $m = 1$ and $n = 4$.
2. Find $z(x)$ for $m = 2$ and $n = 4$.
3. For all above examples check that (keeping the lowest coefficients arbitrary) $z(x) = \frac{d^m}{dx^m} P_n(x)$.

Given that $\lambda = n(n+1)$ we can obtain a solution of Equation (B.101) using the solution of the Legendre equation, Equation (B.67). Let us differentiate Equation (B.101) with respect to the variable x:

$$\left(1 - x^2\right) \left(z'\right)'' - 2\left[(m+1) + 1\right] x \left(z'\right)' + \left[n(n+1) - (m+1)(m+2)\right] z' = 0. \qquad (B.105)$$

It is seen that if in this equation we replace z' by z and $(m+1)$ by m, the obtained equation becomes Equation (B.101). In other words, if $P_n(x)$ is a solution of Equation (B.101) for $m = 0$, then $P_n'(x)$ is a solution of Equation (B.105) for $m = 1$. Repeating this we obtain that $P_n''(x)$ is a solution for $m = 2$, $P_n'''(x)$ is a solution for $m = 3$, etc. For arbitrary integer m, where $0 \le m \le n$, a solution of Equation (B.101) is the function $\frac{d^m}{dx^m} P_n(x)$, thus

$$z(x) = \frac{d^m}{dx^m} P_n(x), \quad 0 \le m \le n. \qquad (B.106)$$

With Equations (B.105) and (B.100) we have a solution of Equation (B.99) given by

$$y(x) = \left(1 - x^2\right)^{\frac{m}{2}} \frac{d^m}{dx^m} P_n(x), \quad 0 \le m \le n. \qquad (B.107)$$

The functions defined in Equation (B.106) are called the *associated Legendre functions* and denoted as $P_n^m(x)$:

$$P_n^m(x) = \left(1 - x^2\right)^{\frac{m}{2}} \frac{d^m}{dx^m} P_n(x). \qquad (B.108)$$

Notice that $\frac{d^m}{dx^m} P_n(x)$ is a polynomial of degree $n - m$, thus

$$P_n^m(-x) = (-1)^{n-m} P_n^m(x), \qquad (B.109)$$

which is referred to as the parity property. From Equation (B.107) it is directly seen that $P_n^m(x) = 0$ for $|m| > n$ because in this case m-th order derivatives of polynomial $P_n(x)$ of degree n are equal to zero. The graphs of several $P_n^m(x)$ are plotted in Figure B.11. Thus, from the above discussion, we see that Equation (B.98) has eigenvalues λ

$$m(m + 1), \quad (m + 1)(m + 2), \quad (m + 2)(m + 3), \qquad (B.110)$$

with corresponding eigenfunctions, bounded on $[-1, 1]$, are

$$P_m^m(x), \quad P_{m+1}^m(x), \quad P_{m+2}^m(x), \quad \dots \qquad (B.111)$$

Equation (B.98) (or (B.99)) does not change when the sign of m changes. Therefore, a solution of Equation (B.51) for positive m is also a solution for negative values $-|m|$. Thus, we can define $P_n^m(x)$ as equal to $P_n^{|m|}(x)$ *for* $-n \le m \le n$.

$$P_n^{-|m|}(x) = P_n^{|m|}(x). \qquad (B.112)$$

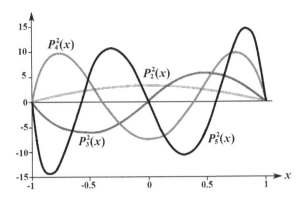

FIGURE B.11
Graphs of $P_2^2(x)$, $P_3^2(x)$, $P_4^2(x)$ and $P_5^2(x)$.

The first several associated Legendre functions $P_n^1(x)$ for $m = 1$ are

$$P_1^1(x) = \sqrt{1 - x^2} \cdot [P_1(x)]' = \sqrt{1 - x^2},$$

$$P_2^1(x) = \sqrt{1 - x^2} \cdot [P_2(x)]' = \sqrt{1 - x^2} \cdot 3x \,,$$

$$P_3^1(x) = \sqrt{1 - x^2} \cdot [P_3(x)]' = \sqrt{1 - x^2} \cdot \frac{3}{2}\left(5x^2 - 1\right)$$

and the first several associated Legendre functions, $P_n^2(x)$, for $m = 2$ are

$$P_2^2(x) = \left(1 - x^2\right) \cdot [P_2(x)]'' = \left(1 - x^2\right) \cdot 3,$$

$$P_3^2(x) = \left(1 - x^2\right) \cdot [P_3(x)]'' = \left(1 - x^2\right) \cdot 15x,$$

$$P_4^2(x) = \left(1 - x^2\right) \cdot [P_4(x)]'' = \left(1 - x^2\right) \cdot \frac{15}{2}\left(7x^2 - 1\right).$$

Associated Legendre function $P_n^m(x)$ has $(n - m)$ simple (not repeating) real roots on the interval $1 < x < 1$.

The following *recurrence formula* is often useful:

$$(2n + 1)xP_n^m(x) - (n - m + 1)P_{n+1}^m(x) - (n + m)P_{n-1}^m(x) = 0. \qquad \text{(B.113)}$$

B.9   Fourier-Legendre Series in Associated Legendre Functions

Functions $P_n^m(x)$ *for fixed value of* $|m|$ (the upper index of the associated Legendre functions) and all possible values of the lower index,

$$P_m^m(x), \quad P_{m+1}^m(x), \quad P_{m+2}^m(x), \quad \dots \qquad \text{(B.114)}$$

form an *orthogonal* (with respect to the weight function $r(x) = 1$) and *complete* set of functions on the interval $[-1, 1]$. In other words, for each value of m there is an orthogonal and complete set of Equations (B.112). This follows from the fact that these functions are also the solutions of a Sturm-Liouville problem. Thus the set of Equations (B.112) for any given m is a basis for an eigenfunction expansion for functions bounded on $[-1, 1]$ and we may write

$$f(x) = \sum_{k=0}^{\infty} c_k P_{m+k}^m(x). \tag{B.115}$$

The formula for the coefficients c_k $(k = 0, 1, 2, \ldots)$ follows from the orthogonality of the functions in Equation (B.112):

$$c_k = \frac{1}{\left\| P_{m+k}^m \right\|^2} \int_{-1}^1 f(x) P_{m+k}^m(x) dx = \frac{[2(m+k) + 1]\, k!}{2\,(2m+k)!} \int_{-1}^1 f(x) P_{m+k}^m(x) dx. \tag{B.116}$$

As previously, the sequence of the partial sums $S_N(x)$ of series (B.113) converges on the interval $(-1, 1)$ in the mean to the square integrable function $f(x)$, i.e.

$$\int_{-1}^1 [f(x) - S_N(x)]^2\, dx \to 0 \quad \text{as} \quad N \to \infty.$$

For piecewise-continuous functions the same theorem as in the previous section is valid.

Example B.6 Expand the function

$$f(x) = \begin{cases} 1 + x, & -1 \le x < 0, \\ 1 - x, & 0 \le x \le 1 \end{cases}$$

in terms of associated Legendre functions $P_n^m(x)$ of order $m = 2$.

Solution. The series is

$$f(x) = c_0 P_2^2(x) + c_1 P_3^2(x) + c_2 P_4^2(x) + c_3 P_5^2(x) \ldots,$$

where coefficients c_k are

$$c_k = \frac{2k+5}{2} \frac{k!}{(k+4)!} \left[\int_{-1}^0 (1+x) P_{k+2}^2(x) dx + \int_0^1 (1-x) P_{k+2}^2(x) dx \right].$$

Because $f(x)$ is an even function of x, $c_1 = c_3 = c_5 = \ldots = 0$.
The first two coefficients with even index are

$$c_0 = \frac{5}{48} \left[\int_{-1}^0 (1+x)3\left(1 - x^2\right) dx + \int_0^1 (1-x)3\left(1 - x^2\right) dx \right] = \frac{25}{96},$$

$$c_2 = \frac{1}{80} \left[\int_{-1}^0 (1+x)\frac{15}{2}\left(1 - x^2\right)\left(7x^2 - 1\right) dx + \int_0^1 (1-x)\frac{15}{2}\left(1 - x^2\right)\left(7x^2 - 1\right) dx \right] = -\frac{1}{80}.$$

The graph of $f(x)$ and the partial sum of respective Fourier-Legendre series is shown on Figure B.12.

B.10   Airy Functions

The Airy functions are the particular solution of the equation (Airy equation):

$$y'' - xy = 0.$$

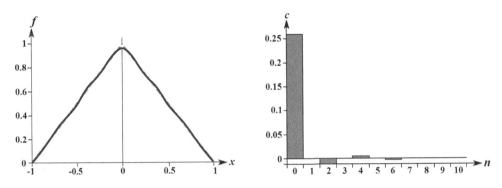

FIGURE B.12
The function $f(x)$ and the partial sum of its Fourier-Legendre series in terms of associated Legendre functions $P_n^2(x)$. a) The graph of $f(x)$ is shown by the dashed line, the graph of the partial sum with $N = 10$ terms of series by the solid line; b) values of the coefficients c_k of the series.

This simple equation has a turning point where the character of the solution *changes from oscillatory to exponential*.

There are Airy functions of the first kind, $Ai(x)$, and the second kind, $Bi(x)$; when $x \to -\infty$ both oscillate (with equal decaying amplitude, the phase shift between them is $\pi/2$, see Figure B.13). For $x \to +\infty$, function $Ai(x)$ exponentially decays and function $Bi(x)$ exponentially increases.

For real x functions $Ai(x)$ and $Bi(x)$ can be presented by the integrals

$$Ai(x) = \frac{1}{\pi} \int_0^\infty \cos\left(\frac{t^3}{3} + xt\right) dt,$$

$$Bi(x) = \frac{1}{\pi} \int_0^\infty \left[\exp\left(-\frac{t^3}{3} + xt\right) + \sin\left(\frac{t^3}{3} + xt\right)\right] dt.$$

This can be easily proven - differentiating these integrals we obtain the Airy equation in both cases.

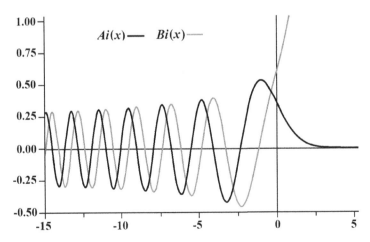

FIGURE B.13
Functions $Ai(x)$ (black) and $Bi(x)$ (gray).

Relation to Bessel functions of different kinds

For $x > 0$ the Airy functions are related to modified Bessel functions:

$$Ai(x) = \frac{1}{\pi}\sqrt{\frac{1}{3}x}K_{1/3}\left(\frac{2}{3}x^{3/2}\right),$$

$$Bi(x) = \sqrt{\frac{1}{3}x}\left[I_{1/3}\left(\frac{2}{3}x^{3/2}\right) + I_{-1/3}\left(\frac{2}{3}x^{3/2}\right)\right],$$

where $I_{\pm 1/3}$ and $K_{1/3}$ are solutions of the equation $x^2 y'' + xy' - (x^2 + 1/9)y = 0$.
For $x < 0$ the Airy functions are related to Bessel functions:

$$Ai(-x) = \frac{1}{3}\sqrt{x}\left[J_{1/3}\left(\frac{2}{3}x^{3/2}\right) + J_{-1/3}\left(\frac{2}{3}x^{3/2}\right)\right],$$

$$Bi(-x) = \sqrt{\frac{1}{3}x}\left[J_{-1/3}\left(\frac{2}{3}x^{3/2}\right) - J_{1/3}\left(\frac{2}{3}x^{3/2}\right)\right],$$

where $J_{\pm 1/3}$ are solutions of the equation $x^2 y'' + xy' - (x^2 + 1/9)y = 0$.

Problems

Bessel Equation

1. Find the eigenfunctions of the Sturm-Liouville problems for the Bessel equation on $[0, l]$ assuming that function $y(x)$ is finite at $x = 0$ – this a boundary condition at $x = 0$. Equations below are the same as Equation (B.1) where x is used for r.

   (a) $x^2 y'' + xy' + (\lambda x^2 - 1)y = 0$, $y(1) = 0$;
   (b) $x^2 y'' + xy' + (\lambda x^2 - 4)y = 0$, $y(1) = 0$;
   (c) $x^2 y'' + xy' + \lambda x^2 y = 0$, $y'(2) = 0$;
   (d) $x^2 y'' + xy' + (\lambda x^2 - 9)y = 0$, $y(3) + 2y'(3) = 0$.

2. Expand the function $f(x)$, given on the interval $[0, 1]$, in a Fourier series in Bessel functions of the first kind, $X_k(x) = J_0(\mu_k x)$, where $\mu_k^{(0)}$ are positive roots of the equation $J_0(\mu) = 0$ (in Problems 2 through 8 coefficients of expansion you can find with Maple, Mathematica, or with the software from books [7, 8]), if:

   (a) $f(x) = \sin \pi x$;
   (b) $f(x) = x^2$;
   (c) $f(x) = \sin^2 \pi x$;
   (d) $f(x) = 1 - x^2$;
   (e) $f(x) = \cos \dfrac{\pi x}{2}$.

3. Expand the function $f(x)$, given on the interval $[0, 1]$, in a Fourier series in Bessel functions $X_k(x) = J_1(\mu_k x)$, where $\mu_k^{(1)}$ are positive roots of the equation $J_1(\mu) = 0$, if:

   (a) $f(x) = x$;
   (b) $f(x) = \sin \pi x$;
   (c) $f(x) = \sin^2 \pi x$;
   (d) $f(x) = x(1 - x)$;

(e) $f(x) = x(1 - x^2)$.

4. Expand the function $f(x)$, given on the interval $[0, 1]$, in a Fourier series in Bessel functions $X_k(x) = J_0(\mu_k x)$, where $\mu_k^{(0)}$ are positive roots of the equation $J_0'(\mu) = 0$, if:

   (a) $f(x) = x(1 - x)$;
   (b) $f(x) = x(1 - x^3)$;
   (c) $f(x) = x(1 - x^2)$;
   (d) $f(x) = x^3$.

5. Expand the function

$$f(x) = A\left(1 - \frac{x^2}{l^2}\right), \quad A = \text{const},$$

   given on the interval $[0, l]$, in Fourier series in Bessel functions $X_k(x) = J_0(\mu_k^{(0)} x/l)$, where $\mu_k^{(0)}$ are positive roots of the equation $J_0(\mu) = 0$.

6. Expand the function

$$f(x) = Ax, \quad A = \text{const},$$

   given on the interval $[0, l]$, in a Fourier series in Bessel functions $X_k(x) = J_1(\mu_k^{(1)} x/l)$, where $\mu_k^{(1)}$ are positive roots of the equation $J_1'(\mu) = 0$.

7. Expand the function

$$f(x) = Ax^2, \quad A = \text{const},$$

   given on the interval $[0, l]$, in Fourier series in Bessel functions of the first kind $X_k(x) = J_0(\mu_k^{(0)} x/l)$, where $\mu_k^{(0)}$ are positive roots of the equation $J_0'(\mu) = 0$.

8. Expand the function

$$f(x) = \begin{cases} x^2, & 0 \le x < 1, \\ x, & 1 \le x < 2, \end{cases}$$

   given on the interval $[0, 2]$, in Fourier series in Bessel functions of the first kind $X_k(x) = J_2\left(\mu_k^{(2)} x/l\right)$ $(l = 2)$, where $\mu_k^{(2)}$ are positive roots of the equation $\mu J_2'(\mu) + h l J_2(\mu) = 0$.

Fourier-Legendre functions

For each of Problems 1 through 8 expand the function $f(x)$ in the Fourier-Legendre functions $P_n^m(x)$ on $[-1, 1]$. Do expansion for a) $m = 0$ – in this case the functions $P_n^m(x)$ are the Legendre polynomials $P_n(x)$; b) for $m = 1$; c) for $m = 2$.

Write the formulas for coefficients of the series expansion and the expression for the Fourier-Legendre series. If the integrals in the coefficients are not easy to evaluate, try to evaluate them numerically.

Using Maple or Mathematica, or software from books [7, 8], obtain the pictures of several orthonomal functions $P_{m+k}^m(x)$, plot the graphs of the given function $f(x)$ and of the partial sums $S_N(x)$ and build the histograms of coefficients c_k of the series.

1. $f(x) = 2x - 1$.
2. $f(x) = 1 - x^2$.

3. $f(x) = \begin{cases} -1, & -1 < x < 0, \\ 1, & 0 < x < 1. \end{cases}$

4. $f(x) = \begin{cases} 0, & -1 < x < 0, \\ x, & 0 < x < 1. \end{cases}$

5. $f(x) = \cos \dfrac{\pi x}{2}$ for $-1 \le x \le 1$.

6. $f(x) = \begin{cases} 0, & -1 < x < 0, \\ \sqrt{1-x}, & 0 < x < 1. \end{cases}$

7. $f(x) = \sin \pi x$.

8. $f(x) = e^x$.

C

Sturm-Liouville Problem and Auxiliary Functions for One and Two Dimensions

C.1 Eigenvalues and Eigenfunctions of 1D Sturm-Liouville Problem for Different Types of Boundary Conditions

The one-dimensional Sturm-Liouville boundary value problem for eigenvalues and eigenfunctions is formulated as:

Find values of parameter λ for which there exist nontrivial (not identically equal to zero) solutions of the boundary value problem:

$$X'' + \lambda X = 0, 0 < x < l,$$

$$P_1[X] \equiv \alpha_1 X' + \beta_1 X|_{x=0} = 0, \quad |\alpha_1| + |\beta_1| \neq 0,$$

$$P_2[X] \equiv \alpha_2 X' + \beta_2 X|_{x=l} = 0, \quad |\alpha_2| + |\beta_2| \neq 0.$$

The eigenfunctions of this Sturm-Liouville problem are

$$X_n(x) = \frac{1}{\sqrt{\alpha_1^2 \lambda_n + \beta_1^2}} \left[\alpha_1 \sqrt{\lambda_n} \cos \sqrt{\lambda_n} x - \beta_1 \sin \sqrt{\lambda_n} x \right].$$

These eigenfunctions are orthogonal. Their square norms are

$$\|X_n\|^2 = \int_0^l X_n^2(x) dx = \frac{1}{2} \left[l + \frac{(\beta_2 \alpha_1 - \beta_1 \alpha_2)(\lambda_n \alpha_1 \alpha_2 - \beta_1 \beta_2)}{(\lambda_n \alpha_1^2 + \beta_1^2)(\lambda_n \alpha_2^2 + \beta_2^2)} \right].$$

The eigenvalues are

$$\lambda_n = \left(\frac{\mu_n}{l} \right)^2,$$

where μ_n is the nth root of the equation

$$\tan \mu = \frac{(\alpha_1 \beta_2 - \alpha_2 \beta_1) l \mu}{\mu^2 \alpha_1 \alpha_2 + l^2 \beta_1 \beta_2}.$$

Below we consider all possible cases (combinations) of boundary conditions.

1. **Boundary conditions** ($\alpha_1 = \alpha_2 = 0$, $\beta_1 = \beta_2 = 1$):

$$\begin{cases} X(0) = 0 & - \text{ Dirichlet condition,} \\ X(l) = 0 & - \text{ Dirichlet condition.} \end{cases}$$

Eigenvalues: $\lambda_n = \left[\frac{\pi n}{l} \right]^2, \quad n = 1, 2, \ldots, \infty.$

Eigenfunctions: $X_n(x) = \sin \frac{\pi n}{l} x, \quad \|X_n\|^2 = \frac{l}{2}.$

2. **Boundary conditions** ($\alpha_1 = 0$, $\beta_1 = 1$, $\alpha_2 = 0$, $\beta_2 = 1$):

$$\begin{cases} X(0) = 0 & \text{– Dirichlet condition,} \\ X'(l) = 0 & \text{– Neumann condition.} \end{cases}$$

Eigenvalues: $\lambda_n = \left[\dfrac{\pi(2n+1)}{2l}\right]^2$, $\quad n = 0, 1, 2, \ldots, \infty$.

Eigenfunctions: $X_n(x) = \sin\dfrac{\pi(2n+1)}{2l}x$, $\quad ||X_n||^2 = \dfrac{l}{2}$.

3. **Boundary conditions** ($\alpha_1 = 0$, $\beta_1 = 1$, $\alpha_2 = 1$, $\beta_2 = h_2$):

$$\begin{cases} X(0) = 0 & \text{– Dirichlet condition,} \\ X'(l) + h_2 X(l) = 0 & \text{– mixed condition.} \end{cases}$$

Eigenvalues: $\lambda_n = \left[\dfrac{\mu_n}{l}\right]^2$, where μ_n is n-th root of the equation $\tan\mu = -\dfrac{\mu}{h_2 l}$, $n = 0, 1, 2, \ldots, \infty$.

Eigenfunctions: $X_n(x) = \sin\sqrt{\lambda_n}x$, $\quad ||X_n||^2 = \dfrac{1}{2}\left(l + \dfrac{h_2}{\lambda_n + h_2^2}\right)$.

4. **Boundary conditions** ($\alpha_1 = 1$, $\beta_1 = 0$, $\alpha_2 = 0$, $\beta_2 = 1$):

$$\begin{cases} X'(0) = 0 & \text{– Neumann condition,} \\ X(l) = 0 & \text{– Dirichlet condition.} \end{cases}$$

Eigenvalues: $\lambda_n = \left[\dfrac{\pi(2n+1)}{2l}\right]^2$, $\quad n = 0, 1, 2, \ldots, \infty$.

Eigenfunctions: $X_n(x) = \cos\dfrac{\pi(2n+1)}{2l}x$, $\quad ||X_n||^2 = \dfrac{l}{2}$.

5. **Boundary conditions** ($\alpha_1 = \alpha_2 = 1$, $\beta_1 = \beta_2 = 0$):

$$\begin{cases} X'(0) = 0 & \text{– Neumann condition,} \\ X'(l) = 0 & \text{– Neumann condition.} \end{cases}$$

Eigenvalues. $\lambda_n = \left[\dfrac{\pi n}{l}\right]^2$, $\quad n = 0, 1, 2, \ldots, \infty$.

Eigenfunctions: $X_n(x) = \cos\dfrac{\pi n}{l}x$, $\quad ||X_n||^2 = \begin{cases} l, & n = 0 \\ l/2, & n > 0 \end{cases}$.

6. **Boundary conditions** ($\alpha_1 = 1$, $\beta_1 = 0$, $\alpha_2 = 1$, $\beta_2 = h_2$):

$$\begin{cases} X'(0) = 0 & \text{– Neumann condition,} \\ X'(l) + h_2 X(l) = 0 & \text{– mixed condition.} \end{cases}$$

Eigenvalues: $\lambda_n = \left[\dfrac{\mu_n}{l}\right]^2$, where μ_n is n-th root of the equation $\tan\mu = \dfrac{h_2 l}{\mu}$, $n = 0, 1, 2, \ldots, \infty$.

Eigenfunctions: $X_n(x) = \cos\sqrt{\lambda_n}x$, $\quad ||X_n||^2 = \dfrac{1}{2}\left(l + \dfrac{h_2}{\lambda_n + h_2^2}\right)$.

7. **Boundary conditions** ($\alpha_1 = 1$, $\beta_1 = h_1$, $\alpha_2 = 1$, $\beta_2 = 0$):

$$\begin{cases} X'(0) - h_1 X(0) = 0 & \text{– mixed condition,} \\ X'(l) = 0 & \text{– Neumann condition.} \end{cases}$$

Eigenvalues: $\lambda_n = \left[\dfrac{\mu_n}{l}\right]^2$, where μ_n is n-th root of the equation $\tan\mu = \dfrac{h_1 l}{\mu}$,
$n = 0, 1, 2, \ldots, \infty$.

Eigenfunctions: $X_n(x) = \dfrac{1}{\sqrt{\lambda_n + h_1^2}}\left[\sqrt{\lambda_n}\cos\sqrt{\lambda_n}x + h_1\sin\sqrt{\lambda_n}x\right]$,

$$\|X_n\|^2 = \frac{1}{2}\left(l + \frac{h_1}{\lambda_n + h_1^2}\right).$$

8. **Boundary conditions** ($\alpha_1 = 1$, $\beta_1 = h_1$, $\alpha_2 = 1$, $\beta_2 = h_2$):

$$\begin{cases} X'(0) - h_1 X(0) = 0 & \text{– mixed condition,} \\ X'(l) + h_2 X(l) = 0 & \text{– mixed condition.} \end{cases}$$

Eigenvalues: $\lambda_n = \left[\dfrac{\mu_n}{l}\right]^2$, where μ_n is n-th root of the equation

$$\tan\mu = \frac{(h_1 + h_2)l\mu}{\mu^2 - h_1 h_2 l^2}, \quad n = 0, 1, 2, \ldots, \infty.$$

Eigenfunctions: $X_n(x) = \dfrac{1}{\sqrt{\lambda_n + h_1^2}}\left[\sqrt{\lambda_n}\cos\sqrt{\lambda_n}x + h_1\sin\sqrt{\lambda_n}x\right]$,

$$\|X_n\|^2 = \frac{1}{2}\left(l + \frac{(h_1 + h_2)(\lambda_n + h_1 h_2)}{(\lambda_n + h_1^2)(\lambda_n + h_2^2)}\right).$$

C.2   Auxiliary Functions

1. Auxiliary Functions $w(x,t)$ for 1D Hyperbolic and Parabolic Equations

In the case of nonhomogeneous boundary conditions

$$P_1[u] \equiv \alpha_1\frac{\partial u}{\partial x} + \beta_1 u\Big|_{x=0} = g_1(t), \quad |\alpha_1| + |\beta_1| \neq 0,$$

$$P_2[u] \equiv \alpha_2\frac{\partial u}{\partial x} + \beta_2 u\Big|_{x=l} = g_2(t), \quad |\alpha_2| + |\beta_2| \neq 0 \tag{C.1}$$

the solution to the BVP can be expressed as the sum of two functions

$$u(x,t) = v(x,t) + w(x,t), \tag{C.2}$$

where $w(x,t)$ is an auxiliary function satisfying the boundary conditions and $v(x,t)$ is a solution of the boundary value problem with zero boundary conditions.

We seek an auxiliary function $w(x,t)$ in a form

$$w(x,t) = g_1(t)\overline{X}(x) + g_2(t)\overline{\overline{X}}(x), \tag{C.3}$$

where $\overline{X}(x)$ and $\overline{\overline{X}}(x)$ are polynomials of 1st or 2nd order. The coefficients of these polynomials are adjusted to satisfy the boundary conditions.

Functions $\overline{X}(x)$ and $\overline{\overline{X}}(x)$ should be chosen in such a way that

$$P_1\left[\overline{X}(0)\right] = 1, \quad P_2\left[\overline{X}(l)\right] = 0,$$

$$P_1\left[\overline{\overline{X}}(0)\right] = 0, \quad P_2\left[\overline{\overline{X}}(l)\right] = 1.$$

If $\beta_1 \neq 0$ or $\beta_2 \neq 0$ then functions $\overline{X}(x)$ and $\overline{\overline{X}}(x)$ are polynomials of the 1st order

$$\overline{X}(x) = \gamma_1 + \delta_1 x, \quad \overline{\overline{X}}(x) = \gamma_2 + \delta_2 x. \tag{C.4}$$

Coefficients $\gamma_1, \delta_1, \gamma_2, \delta_2$ of these polynomials are defined uniquely and depend on the types of boundary conditions:

$$\gamma_1 = \frac{\alpha_2 + \beta_2 l}{\beta_1 \beta_2 l + \beta_1 \alpha_2 - \beta_2 \alpha_1}, \quad \delta_1 = \frac{-\beta_2}{\beta_1 \beta_2 l + \beta_1 \alpha_2 - \beta_2 \alpha_1},$$

$$\gamma_2 = \frac{-\alpha_1}{\beta_1 \beta_2 l + \beta_1 \alpha_2 - \beta_2 \alpha_1}, \quad \delta_2 = \frac{\beta_1}{\beta_1 \beta_2 l + \beta_1 \alpha_2 - \beta_2 \alpha_1}. \tag{C.5}$$

If $\beta_1 = \beta_2 = 0$ then functions $\overline{X}(x)$ and $\overline{\overline{X}}(x)$ are polynomials of the 2nd order

$$\overline{X}(x) = x - \frac{x^2}{2l}, \quad \overline{\overline{X}}(x) = \frac{x^2}{2l}. \tag{C.6}$$

Below we present auxiliary functions for different types of boundary conditions.

1. Boundary conditions: $\begin{cases} u(0,t) = g_1(t), \\ u(l,t) = g_2(t). \end{cases}$

   Auxiliary function: $w(x,t) = \left[1 - \dfrac{x}{l}\right] \cdot g_1(t) + \dfrac{x}{l} \cdot g_2(t).$

2. Boundary conditions: $\begin{cases} u(0,t) = g_1(t), \\ u_x(l,t) = g_2(t). \end{cases}$

   Auxiliary function: $w(x,t) = g_1(t) + x g_2(t).$

3. Boundary conditions: $\begin{cases} u(0,t) = g_1(t), \\ u_x(l,t) + h_2 u(l,t) = g_2(t). \end{cases}$

   Auxiliary function: $w(x,t) = \left[1 - \dfrac{h_2}{1 + h_2 l} x\right] \cdot g_1(t) + \dfrac{x}{1 + h_2 l} \cdot g_2(t).$

4. Boundary conditions: $\begin{cases} u_x(0,t) = g_1(t), \\ u(l,t) = g_2(t). \end{cases}$

   Auxiliary function: $w(x,t) = (x - l)g_1(t) + g_2(t).$

5. Boundary conditions: $\begin{cases} u_x(0,t) = g_1(t), \\ u_x(l,t) = g_2(t). \end{cases}$

   Auxiliary function: $w(x,t) = \left[x - \dfrac{x^2}{2l}\right] \cdot g_1(t) + \dfrac{x^2}{2l} \cdot g_2(t).$

6. Boundary conditions: $\begin{cases} u_x(0,t) = g_1(t), \\ u_x(l,t) + h_2 u(l,t) = g_2(t). \end{cases}$

   Auxiliary function: $w(x,t) = \left[x - \dfrac{1 + h_2 l}{h_2}\right] \cdot g_1(t) + \dfrac{1}{h_2} \cdot g_2(t).$

7. Boundary conditions: $\begin{cases} u_x(0,t) - h_1 u(0,t) = g_1(t), \\ u(l,t) = g_2(t). \end{cases}$

Auxiliary function: $w(x,t) = \dfrac{x-l}{1+h_1 l} \cdot g_1(t) + \dfrac{1+h_1 x}{1+h_1 l} \cdot g_2(t).$

8. Boundary conditions $\begin{cases} u_x(0,t) - h_1 u(0,t) = g_1(t), \\ u_x(l,t) = g_2(t). \end{cases}$

Auxiliary function: $w(x,t) = -\dfrac{1}{h_1} \cdot g_1(t) + \left[x + \dfrac{1}{h_1} \right] \cdot g_2(t).$

9. Boundary conditions $\begin{cases} u_x(0,t) - h_1 u(0,t) = g_1(t), \\ u_x(l,t) + h_2 u(l,t) = g_2(t). \end{cases}$

Auxiliary function: $w(x,t) = \dfrac{h_2(x-l) - 1}{h_1 + h_2 + h_1 h_2 l} \cdot g_1(t) + \dfrac{1+h_1 x}{h_1 + h_2 + h_1 h_2 l} \cdot g_2(t).$

2. Auxiliary Functions $w(x,y)$ for Poisson Equation in Rectangular Domain

Consider the boundary value problem for Poisson equation with nonhomogeneous boundary conditions

$$P_1[u] \equiv \alpha_1 \frac{\partial u}{\partial x} + \beta_1 u \bigg|_{x=0} = g_1(y), \qquad P_2[u] \equiv \alpha_2 \frac{\partial u}{\partial x} + \beta_2 u \bigg|_{x=l_x} = g_2(y),$$

$$P_3[u] \equiv \alpha_3 \frac{\partial u}{\partial y} + \beta_3 u \bigg|_{y=0} = g_3(x), \qquad P_4[u] \equiv \alpha_4 \frac{\partial u}{\partial y} + \beta_4 u \bigg|_{y=l_y} = g_4(x). \tag{C.7}$$

The solution to the BVP can be expressed as the sum of two functions

$$u(x,y) = v(x,y) + w(x,y), \tag{C.8}$$

where $w(x,y)$ is an auxiliary function satisfying the boundary conditions and $v(x,y)$ is a solution of the boundary value problem with zero boundary conditions.

We seek an auxiliary function $w(x,y,t)$ in a form

$$w(x,y,t) = g_1(y)\overline{X} + g_2(y)\overline{\overline{X}} + g_3(x)\overline{Y} + g_4(x)\overline{\overline{Y}}$$
$$+ A \cdot \overline{XY} + B \cdot \overline{X}\overline{\overline{Y}} + C \cdot \overline{\overline{X}}\overline{Y} + D \cdot \overline{XY}, \tag{C.9}$$

where $\overline{X}(x)$, $\overline{\overline{X}}(x)$ and $\overline{Y}(y)$, $\overline{\overline{Y}}(y)$ are polynomials of 1st or 2nd order. The coefficients of these polynomials are adjusted to satisfy the boundary conditions.

We will choose the functions $\left\{ \overline{X}, \overline{\overline{X}} \right\}$ in such a way that the function $\overline{X}(x)$ satisfies homogeneous boundary condition at $x = l_x$ and the function $\overline{\overline{X}}(x)$ satisfies homogeneous boundary condition at $x = 0$,

$$P_2\left[\overline{X}(l_x)\right] = 0, \qquad P_1\left[\overline{\overline{X}}(0)\right] = 0.$$

Also, it is convenient to normalize the functions $\overline{X}(x)$ and $\overline{\overline{X}}(x)$ so that

$$P_1\left[\overline{X}(0)\right] = 1, \qquad P_2\left[\overline{\overline{X}}(l_x)\right] = 1.$$

The final choice of functions $\{\overline{X},\ \overline{\overline{X}}\}$ depends on the type of boundary conditions for the function $u(x,y)$.

Suppose β_1 and β_2 are not both zero. In this case we can search for $\overline{X}(x)$ and $\overline{\overline{X}}(x)$ as polynomials of the first order:

$$\overline{X}(x) = \gamma_1 + \delta_1 x \quad \text{and} \quad \overline{\overline{X}}(x) = \gamma_2 + \delta_2 x. \tag{C.10}$$

This choice yields the system of equations

$$\begin{cases} P_1\left[\overline{X}\right] \equiv \alpha_1\,\overline{X}_x + \beta_1\overline{X}\big|_{x=0} = \beta_1\gamma_1 + \alpha_1\delta_1 = 1, \\ P_2\left[\overline{X}\right] \equiv \alpha_2\,\overline{X}_x + \beta_2\overline{X}\big|_{x=l_x} = \beta_2\gamma_1 + (\alpha_2 + \beta_2 l_x)\,\delta_1 = 0, \end{cases}$$

and

$$\begin{cases} P_1\left[\overline{\overline{X}}\right] \equiv \alpha_1\overline{\overline{X}}_x + \beta_1\overline{\overline{X}}\big|_{x=0} = \beta_1\gamma_2 + \alpha_1\delta_2 = 0, \\ P_2\left[\overline{\overline{X}}\right] \equiv \alpha_2\overline{\overline{X}}_x + \beta_2\overline{\overline{X}}\big|_{x=l_x} = \beta_2\gamma_2 + (\alpha_2 + \beta_2 l_x)\,\delta_2 = 1. \end{cases}$$

From the above conditions we may find a unique solution for coefficients γ_1, δ_1, γ_2, δ_2:

$$\gamma_1 = \frac{\alpha_2 + \beta_2 l_x}{\beta_1\beta_2 l_x + \beta_1\alpha_2 - \beta_2\alpha_1}, \quad \delta_1 = \frac{-\beta_2}{\beta_1\beta_2 l_x + \beta_1\alpha_2 - \beta_2\alpha_1},$$

$$\gamma_2 = \frac{-\alpha_1}{\beta_1\beta_2 l_x + \beta_1\alpha_2 - \beta_2\alpha_1}, \quad \delta_2 = \frac{\beta_1}{\beta_1\beta_2 l_x + \beta_1\alpha_2 - \beta_2\alpha_1}. \tag{C.11}$$

If $\beta_1 = \beta_2 = 0$ then functions $\overline{X}(x)$ and $\overline{\overline{X}}(x)$ are polynomials of the 2nd order

$$\overline{X}(x) = x - \frac{x^2}{2l}, \quad \overline{\overline{X}}(x) = \frac{x^2}{2l}. \tag{C.12}$$

Similarly, we will choose the functions $\{\overline{Y},\ \overline{\overline{Y}}\}$ in such a way that

$$P_3\left[\overline{Y}(0)\right] = 1, \quad P_3\left[\overline{Y}(l_y)\right] = 0,$$
$$P_4\left[\overline{\overline{Y}}(0)\right] = 0, \quad P_4\left[\overline{\overline{Y}}(l_y)\right] = 1.$$

If $\beta_3 \neq 0$ or $\beta_4 \neq 0$ then the functions $\overline{Y}(y)$ and $\overline{\overline{Y}}(y)$ are polynomials of the 1st order and we may write

$$\overline{Y}(y) = \gamma_3 + \delta_3 y, \quad \overline{\overline{Y}}(y) = \gamma_4 + \delta_4 y. \tag{C.13}$$

The coefficients γ_3, δ_3, γ_4, δ_4 of these polynomials are defined uniquely and depend on the types of boundary conditions:

$$\gamma_3 = \frac{\alpha_4 + \beta_4 l_y}{\beta_3\beta_4 l_y + \beta_3\alpha_4 - \beta_4\alpha_3}, \quad \delta_3 = \frac{-\beta_4}{\beta_3\beta_4 l_y + \beta_3\alpha_4 - \beta_4\alpha_3},$$

$$\gamma_4 = \frac{-\alpha_3}{\beta_3\beta_4 l_y + \beta_3\alpha_4 - \beta_4\alpha_3}, \quad \delta_4 = \frac{\beta_3}{\beta_3\beta_4 l_y + \beta_3\alpha_4 - \beta_4\alpha_3}. \tag{C.14}$$

If $\beta_3 = \beta_4 = 0$, then $\overline{Y}(y)$ and $\overline{\overline{Y}}(y)$ can be taken as polynomials of the second order:

$$\overline{Y}(y) = y - \frac{y^2}{2l_y}, \quad \overline{\overline{Y}}(y) = \frac{y^2}{2l_y}. \tag{C.15}$$

Coefficients A, B, C, and D of the auxiliary function $w(x,y)$ are defined from the boundary conditions:

At the edge: $x = 0$: $P_1[w]_{x=0} = g_1(y) + (P_1[g_3(0)] + A)\overline{Y} + (P_1[g_4(0)] + B)\overline{\overline{Y}}$,

At the edge: $x = l_x$: $P_2[w]_{x=l_x} = g_2(y) + (P_2[g_3(l_x)] + C)\overline{Y} + (P_2[g_4(l_x)] + D)\overline{\overline{Y}}$,

At the edge: $y = 0$: $P_3[w]_{y=0} = g_3(x) + (P_3[g_1(0)] + A)\overline{X} + (P_3[g_2(0)] + C)\overline{\overline{X}}$,

At the edge: $y = l_y$: $P_4[w]_{y=l_y} = g_4(x) + (P_4[g_1(l_y)] + B)\overline{X} + (P_4[g_2(l_y)] + D)\overline{\overline{X}}$.

3. Auxiliary Functions, $w(x, y, t)$, for 2D Hyperbolic or Parabolic Equations in Rectangular Domain

Consider the boundary value problem for hyperbolic or parabolic equation with nonhomogeneous boundary conditions

$$P_1[u] \equiv \alpha_1 \frac{\partial u}{\partial x} + \beta_1 u \bigg|_{x=0} = g_1(y, t), \quad P_2[u] \equiv \alpha_2 \frac{\partial u}{\partial x} + \beta_2 u \bigg|_{x=l_x} = g_2(y, t),$$

$$P_3[u] \equiv \alpha_3 \frac{\partial u}{\partial y} + \beta_3 u \bigg|_{y=0} = g_3(x, t), \quad P_4[u] \equiv \alpha_4 \frac{\partial u}{\partial y} + \beta_4 u \bigg|_{y=l_y} = g_4(x, t). \quad \text{(C.16)}$$

The solution to the boundary value problem can be expressed as the sum of two functions

$$u(x, y, t) = v(x, y, t) + w(x, y, t), \quad \text{(C.17)}$$

where $w(x, y, t)$ is an auxiliary function satisfying the boundary conditions (C.17) and $v(x, y, t)$ is a solution of the boundary value problem with zero boundary conditions.

We seek an auxiliary function $w(x, y, t)$ in a form

$$w(x, y, t) = g_1(y, t)\overline{X} + g_2(y, t)\overline{\overline{X}} + g_3(x, t)\overline{Y} + g_4(x, t)\overline{\overline{Y}}$$

$$+ A(t)\overline{XY} + B(t)\overline{X\overline{Y}} + C(t)\overline{\overline{X}Y} + D(t)\overline{\overline{XY}}, \quad \text{(C.18)}$$

where $\overline{X}(x)$, $\overline{\overline{X}}(x)$ and $\overline{Y}(y)$, $\overline{\overline{Y}}(y)$ are polynomials of 1st or 2nd order. The coefficients of these polynomials are adjusted to satisfy the boundary conditions. Formulas (C.10)–(C.12) are used to construct functions $\overline{X}(x)$, $\overline{\overline{X}}(x)$, and formulas (C.13)–(C.15) – to construct functions $\overline{Y}(y)$, $\overline{\overline{Y}}(y)$.

Using these results we can find the coefficients $A(t)$, $B(t)$, $C(t)$ and $D(t)$ in the auxiliary function $w(x, y, t)$.

At the boundary $x = 0$ we have

$$P_1[w]_{x=0} = g_1(y, t) + (P_1[g_3(0, t)] + A)\overline{Y} + (P_1[g_4(0, t)] + B)\overline{\overline{Y}}.$$

At the boundary $x = l_x$ we have

$$P_2[w]_{x=l_x} = g_2(y, t) + (P_2[g_3(l_x, t)] + C)\overline{Y} + (P_2[g_4(l_x, t)] + D)\overline{\overline{Y}}.$$

At the boundary $y = 0$ we have

$$P_3[w]_{y=0} = g_3(x, t) + (P_3[g_1(0, t)] + A)\overline{X} + (P_3[g_2(0, t)] + C)\overline{\overline{X}}.$$

At the boundary $y = l_y$ we have

$$P_4[w]_{y=l_y} = g_4(x, t) + (P_4[g_1(l_y, t)] + B)\overline{X} + (P_4[g_2(l_y, t)] + D)\overline{\overline{X}}.$$

To simplify the above we may choose

$$A(t) = -P_3[g_1(y, t)]_{y=0}, \quad B(t) = -P_4[g_1(y, t)]_{y=l_y},$$

$$C(t) = -P_3[g_2(y, t)]_{y=0}, \quad D(t) = -P_4[g_2(y, t)]_{y=l_y}.$$

D

The Sturm-Liouville Problem for Circular and Rectangular Domains

D.1 The Sturm-Liouville Problem for a Circle

Let us consider the following Sturm-Liouville problem for a circle:

$$\nabla^2 u + \lambda y = 0, \quad 0 \leq r < l, \ 0 \leq \varphi < 2\pi, \tag{D.1}$$

$$\alpha \frac{\partial u(r, \varphi)}{\partial r} + \beta u(r, \varphi) \bigg|_{r=l} = 0, \tag{D.2}$$

$$u(r, \varphi) = u(r, \varphi + 2\pi), |\alpha| + |\beta| \neq 0.$$

In polar coordinates, (r, φ), the Laplacian has the form

$$\nabla^2 u = \frac{1}{r} \frac{\partial}{\partial r} \left(r \frac{\partial u}{\partial r} \right) + \frac{1}{r^2} \frac{\partial^2 u}{\partial \varphi^2}.$$

Separating the variables

$$u(r, \varphi) = R(r)\Phi(\varphi) \tag{D.3}$$

we obtain from Equation (D.1):

$$\frac{r \frac{\partial}{\partial r} \left(r \frac{\partial R}{\partial r} \right) + \lambda r^2 R}{R(r)} \equiv -\frac{\Phi''(\varphi)}{\Phi(\varphi)} = \nu. \tag{D.4}$$

For periodic in φ function $\Phi(\varphi)$ we have the following Sturm-Liouville problem:

$$\Phi'' + \nu\Phi = 0, \ 0 \leq \varphi < 2\pi,$$
$$\Phi(\varphi) \equiv \Phi(\varphi + 2\pi).$$

Its solutions are

$$\Phi = \Phi_n(\varphi) = \begin{cases} \cos n\varphi, \\ \sin n\varphi, \end{cases} \quad \nu = \nu_n = n^2, \quad n = 0, 1, 2, \ldots \tag{D.5}$$

For each $\nu = n^2$ we have equation for $R(r)$:

$$r \frac{\partial}{\partial r} \left(r \frac{\partial R}{\partial r} \right) + \left(\lambda r^2 - n^2 \right) R = 0, \quad 0 \leq r < l. \tag{D.6}$$

Function $R(r)$ should satisfy the boundary condition

$$\alpha \frac{\partial R}{\partial r} + \beta R \bigg|_{r=l} = 0, |\alpha| + |\beta| \neq 0,$$

which follows from Equation (D.2); also this function should be bounded $r = 0$:

$$|R(0)| < \infty.$$

Therefore, we have the following Sturm-Liouville for $R(r)$:

$$r^2 R''(r) + r R'(r) + \left(\lambda r^2 - n^2\right) R(r) = 0, \quad 0 \le r < l. \tag{D.7}$$

$$\alpha \frac{\partial R}{\partial r} + \beta R \bigg|_{r=l} = 0, \quad |\alpha| + |\beta| \ne 0, \tag{D.8}$$

$$|R(0)| < \infty. \tag{D.9}$$

Equation (D.7) by a substitution $x = r\sqrt{\lambda}$ becomes the Bessel equation of n-th order:

$$x^2 y'' + x y' + \left(x^2 - n^2\right) y = 0.$$

It has two particular solutions $J_n(\sqrt{\lambda}r)$ and $N_n(\sqrt{\lambda}r)$, but because the second one is unbounded at $r \to 0$ it has to be dropped for the internal problem, thus the eigenfunction of the BVP (D.7)-(D.9) is

$$R_n(r) = J_n\left(\sqrt{\lambda}r\right). \tag{D.10}$$

From the homogeneous boundary condition given in Equation (D.8) we have

$$\alpha\sqrt{\lambda}\, J_n'\left(\sqrt{\lambda}l\right) + \beta J_n\left(\sqrt{\lambda}l\right) = 0.$$

Setting $\sqrt{\lambda}l \equiv \mu$ we obtain a transcendental equation defining μ

$$\alpha\mu J_n'(\mu) + \beta l J_n(\mu) = 0, \tag{D.11}$$

which has an infinite number of roots which we label as

$$\mu_0^{(n)}, \quad \mu_1^{(n)}, \quad \mu_2^{(n)}, \ldots$$

The corresponding values of λ are thus

$$\lambda_{nm} = \left(\frac{\mu_m^{(n)}}{l}\right)^2, \quad n, m = 0, 1, 2, \ldots \tag{D.12}$$

From here we see that we need only *positive roots*, $\mu_m^{(n)}$, because negative roots do not give new values of λ_{nm}.

The eigenfunctions are

$$R_{nm}(r) = J_n\left(\frac{\mu_m^{(n)}}{l}r\right). \tag{D.13}$$

The index $m = 0$ corresponds to the first root of Equation (D.11). (It should be noted that very often the roots are labeled with the starting value $m = 1$ in the literature.).

Eigenfunctions $R_{nm}(r)$ belonging to different eigenvalues λ_{nm} for some fixed value of n are orthogonal with weight r:

$$\int_0^l r R_{nm_1}(r) R_{nm_2}(r) dr = 0, \tag{D.14}$$

or

$$\int_0^l r J_n\left(\mu_{m_1}^{(n)} r/l\right) J_n\left(\mu_{m_2}^{(n)} r/l\right) dr = 0 \quad \text{for} \quad m_1 \neq m_2. \tag{D.15}$$

Combining the results for the functions $\Phi_n(\varphi)$ and $R_{nm}(r)$ we see that for each eigenvalue λ_{nm} there are two linearly independent eigenfunctions:

$$V_{nm}^{(1)}(r, \varphi) = J_n\left(\frac{\mu_m^{(n)}}{l} r\right) \cos n\varphi \quad \text{and} \quad V_{nm}^{(2)}(r, \varphi) = J_n\left(\frac{\mu_m^{(n)}}{l} r\right) \sin n\varphi. \tag{D.16}$$

Since

$$\int_0^{2\pi} d\varphi = 2\pi, \quad \int_0^{2\pi} \cos^2 n\varphi \, d\varphi = \pi, \quad \int_0^{2\pi} \sin^2 n\phi \, d\phi = \pi \, (n > 0)$$

the norms of the eigenfunctions $V_{nm}^{(1)}(r, \varphi)$ and $V_{nm}^{(2)}(r, \varphi)$ are

$$\left\| V_{0m}^{(1)} \right\|^2 = 2\pi \left\| R_{nm}(r) \right\|^2, \quad \left\| V_{nm}^{(1)} \right\|^2 = \left\| V_{nm}^{(2)} \right\|^2 = \pi \left\| R_{nm}(r) \right\|^2 \quad \text{for} \quad n > 0. \tag{D.17}$$

The squared norm $\| R_{nm} \|^2 = \int_0^l r J_n^2\left(\frac{\mu_m^{(n)}}{l} r\right) dr$ is:

1) For the *Dirichlet* boundary condition $\alpha = 0$ and $\beta = 1$, in which case eigenvalues are obtained from the equation

$$J_n(\mu) = 0$$

and we have

$$\| R_{nm} \|^2 = \frac{l^2}{2} \left[J_n'\left(\mu_m^{(n)}\right) \right]^2. \tag{D.18}$$

2) For the *Neumann* boundary condition $\alpha = 1$ and $\beta = 0$, in which case eigenvalues are obtained from the equation

$$J_n'(\mu) = 0$$

and we have

$$\| R_{nm} \|^2 = \frac{l^2}{2\left(\mu_m^{(n)}\right)^2} \left[\left(\mu_m^{(n)}\right)^2 - n^2 \right] J_n^2\left(\mu_m^{(n)}\right). \tag{D.19}$$

3) For the *mixed* boundary condition $\alpha = 1$ and $\beta = h$, in which case eigenvalues are obtained from the equation

$$\mu J_n'(\mu) + hl J_n(\mu) = 0$$

and we have

$$\| R_{nm} \|^2 = \frac{l^2}{2\left(\mu_m^{(n)}\right)^2} \left[\left(\mu_m^{(n)}\right)^2 + l^2 h^2 - n^2 \right] J_n^2\left(\mu_m^{(n)}\right). \tag{D.20}$$

D.2   The Sturm-Liouville Problem for the Rectangle

Let us consider the following Sturm-Liouville problem for the rectangle:

$$V_{xx}(x,y) + V_{yy}(x,y) + \lambda V(x,y) = 0, \quad 0 < x < a,\ 0 < y < b, \tag{D.21}$$

$$P_1[V] \equiv \alpha_1 V_x + \beta_1 V|_{x=0} = 0, \qquad P_2[V] \equiv \alpha_2 V_x + \beta_2 V|_{x=l_x} = 0,$$
$$P_3[V] \equiv \alpha_3 V_y + \beta_3 V|_{y=0} = 0, \qquad P_4[V] \equiv \alpha_4 V_y + \beta_4 V|_{y=l_y} = 0. \tag{D.22}$$

To solve Equation (D.21) for $V(x,y)$ we make the assumption that the variables are independent and attempt to separate them using the substitution

$$V(x,y) = X(x)Y(y).$$

From here we obtain two separate one dimensional BVP:

$$X''(x) + \lambda_x X(x) = 0,$$

$$\alpha_1 X'(0) + \beta_1 X(0) = 0, \quad \alpha_2 X'(l_x) + \beta_2' X(l_x) = 0 \tag{D.23}$$

and

$$Y''(x) + \lambda_y Y(y) = 0,$$

$$\alpha_3 Y'(0) + \beta_3 Y(0) = 0, \quad \alpha_4 Y'(l_y) + \beta_4 Y(l_y) = 0, \tag{D.24}$$

where

$$\lambda_x + \lambda_y = \lambda.$$

The boundary conditions for $X(x)$ and $Y(y)$ follow from the corresponding conditions for the function $V(x,y)$. For example, from the condition

$$\alpha_1 V_x(0,y) + \beta_1 V(0,y) = \alpha_1 X'(0)Y(y) + \beta_1 X(0)Y(y)$$

$$- \left[\alpha_1 X'(0) + \beta_1 X(0) \right] Y(y) = 0$$

it follows (since $Y(y) \neq 0$) that

$$\alpha_1 X'(0) + \beta_1 X(0) = 0.$$

Solutions to Equations (D.23) and (D.24) (given in Appendix C) depend on the boundary conditions and have the generic forms

$$X(x) = C_1 \cos \sqrt{\lambda_x} x + C_2 \sin \sqrt{\lambda_x} x,$$

and

$$Y(y) = D_1 \cos \sqrt{\lambda_y} y + D_2 \sin \sqrt{\lambda_y} y.$$

Boundary conditions for problem (D.23) lead to the system for defining coefficients C_1 and C_2

$$\begin{cases} C_1 \beta_1 + C_2 \alpha_1 \sqrt{\lambda_x} = 0, \\ C_1 \left[-\alpha_2 \sqrt{\lambda_x} \sin \sqrt{\lambda_x} l_x + \beta_2 \cos \sqrt{\lambda_x} l_x \right] \\ \quad + C_2 \left[\alpha_2 \sqrt{\lambda_x} \cos \sqrt{\lambda_x} l_x + \beta_2 \sin \sqrt{\lambda_x} l_x \right] = 0. \end{cases} \tag{D.25}$$

This system of linear homogeneous algebraic equations has a nontrivial solution only when its determinant equals zero

$$(\alpha_1\alpha_2\lambda_x + \beta_1\beta_2)\tan\sqrt{\lambda_x}x - \sqrt{\lambda_x}(\alpha_1\beta_2 - \beta_1\alpha_2) = 0.$$

It is easy to determine (for instance by using graphical methods) that this equation has an infinite number of roots $\{\lambda_{xn}\}_1^\infty$, which conforms to the general Sturm-Liouville theory. For each root λ_{xn}, we obtain a nonzero solution of Equations (D.25)

$$C_1 = C\frac{\alpha_1\sqrt{\lambda_{xn}}}{\sqrt{\lambda_{xn}\alpha_1^2 + \beta_1^2}}, \quad C_2 = -C\frac{\beta_1}{\sqrt{\lambda_{xn}\alpha_1^2 + \beta_1^2}},$$

where $C \neq 0$ is an arbitrary constant.

Similarly boundary conditions for problem (D.24) lead to the system for defining coefficients D_1 and D_2:

$$\begin{cases} D_1\beta_3 + D_2\alpha_3\sqrt{\lambda_y} = 0, \\ D_1\left[-\alpha_4\sqrt{\lambda_y}\sin\sqrt{\lambda_y}l_y + \beta_4\cos\sqrt{\lambda_y}l_y\right] \\ \quad +D_2\left[\alpha_4\sqrt{\lambda_y}\cos\sqrt{\lambda_y}l_y + \beta_4\sin\sqrt{\lambda_y}l_y\right] = 0. \end{cases} \tag{D.26}$$

This system of linear homogeneous algebraic equations has a nontrivial solution only when its determinant equals zero:

$$(\alpha_3\alpha_4\lambda_y + \beta_3\beta_4)\tan\sqrt{\lambda_y}y - \sqrt{\lambda_y}(\alpha_3\beta_4 - \beta_3\alpha_4) = 0.$$

This equation has an infinite number of roots $\{\lambda_{ym}\}_1^\infty$. For each root λ_{ym} we obtain a nonzero solution of Equations (D.26):

$$D_1 = D\frac{\alpha_3\sqrt{\lambda_{ym}}}{\sqrt{\lambda_{ym}\alpha_3^2 + \beta_3^2}}, \quad D_2 = -D\frac{\beta_3}{\sqrt{\lambda_{ym}\alpha_3^2 + \beta_3^2}},$$

where $D \neq 0$ is an arbitrary constant.

Collecting the above results we have eigenfunctions for the Sturm-Liouville problem defined by Equations (D.23) and (D.24) given by

$$X_n(x) = \frac{1}{\sqrt{\alpha_1^2\lambda_{xn} + \beta_1^2}}\left[\alpha_1\sqrt{\lambda_{xn}}\cos\sqrt{\lambda_{xn}}l_x - \beta_1\sin\sqrt{\lambda_{xn}}l_x\right],$$

$$Y_m(y) = \frac{1}{\sqrt{\alpha_3^2\lambda_{ym} + \beta_3^2}}\left[\alpha_3\sqrt{\lambda_{ym}}\cos\sqrt{\lambda_{ym}}l_y - \beta_3\sin\sqrt{\lambda_{ym}}l_y\right] \tag{D.27}$$

(square roots should be taken with positive signs).

The eigenvalues of the problem are

$$\lambda_{xn} = \left[\frac{\mu_{xn}}{l_x}\right]^2 \quad \text{and} \quad \lambda_{ym} = \left[\frac{\mu_{ym}}{l_y}\right]^2, \tag{D.28}$$

where μ_{xn} is the n-th root of the equation

$$\tan\mu_x = \frac{(\alpha_1\beta_2 - \alpha_2\beta_1)l_x\mu_x}{\mu_x^2\alpha_1\alpha_2 + l_x^2\beta_1\beta_2}, \tag{D.29}$$

and μ_{ym} is the m-th root of the equation

$$\tan \mu_y = \frac{(\alpha_3 \beta_4 - \alpha_4 \beta_3) \, l_y \mu_y}{\mu_y^2 \alpha_3 \alpha_4 + l_y^2 \beta_3 \beta_4}. \tag{D.30}$$

The norms of the eigenfunctions are given by

$$\|X_n\|^2 = \int_0^{l_x} X_n^2(x) dx = \frac{1}{2} \left[l_x + \frac{(\alpha_1 \beta_2 - \alpha_2 \beta_1)(\lambda_{xn} \alpha_1 \alpha_2 - \beta_1 \beta_2)}{(\lambda_{xn} \alpha_1^2 + \beta_1^2)(\lambda_{xn} \alpha_2^2 + \beta_2^2)} \right],$$

$$\|Y_m\|^2 = \int_0^{l_y} Y_m^2(y) dy = \frac{1}{2} \left[l_y + \frac{(\alpha_3 \beta_4 - \alpha_4 \beta_3)(\lambda_{ym} \alpha_3 \alpha_4 - \beta_3 \beta_4)}{(\lambda_{ym} \alpha_3^2 + \beta_3^2)(\lambda_{ym} \alpha_4^2 + \beta_4^2)} \right] \tag{D.31}$$

in which case functions $X_n(x)$ and $Y_m(y)$ are bounded by the values ± 1.

If λ_{xn} and $X_n(x)$ are eigenvalues and eigenfunctions of Equation (D.23), and λ_{ym} and $Y_m(y)$ are eigenvalues and eigenfunctions of Equation (D.24), then

$$\lambda_{nm} = \lambda_{xn} + \lambda_{ym} \tag{D.32}$$

and

$$V_{nm}(x, y) = X_n(x) Y_m(y) \tag{D.33}$$

are the eigenvalues and eigenvectors, respectively, of the problem in Equation (D.21).

The functions $V_{nm}(x, y)$ are orthogonal and their norms are

$$\|V_{nm}\|^2 = \|X_n\|^2 \|Y_m\|^2. \tag{D.34}$$

Any twice differentiable function $F(x, y)$ obeying the same boundary conditions as the functions V_{nm}, can be resolved in absolutely and uniformly converging to $F(x, y)$ series in these functions.

E

The Heat Conduction and Poisson Equations for Rectangular Domains – Examples

E.1 The Laplace and Poisson Equations for a Rectangular Domain with Nonhomogeneous Boundary Conditions – Examples

Consider the general boundary value problem for Poisson equation given by

$$\nabla^2 u = \frac{\partial^2 u}{\partial x^2} + \frac{\partial^2 u}{\partial y^2} = -f(x, y), \tag{E.1}$$

with nonhomogeneous boundary conditions

$$P_1[u] \equiv \alpha_1 \frac{\partial u}{\partial x} + \beta_1 u \Big|_{x=0} = g_1(y), \quad P_2[u] \equiv \alpha_2 \frac{\partial u}{\partial x} + \beta_2 u \Big|_{x=l_x} = g_2(y),$$

$$P_3[u] \equiv \alpha_3 \frac{\partial u}{\partial y} + \beta_3 u \Big|_{y=0} = g_3(x), \quad P_4[u] \equiv \alpha_4 \frac{\partial u}{\partial y} + \beta_4 u \Big|_{y=l_y} = g_4(x). \tag{E.2}$$

To deal with nonhomogeneous boundary conditions we introduce an auxiliary function, $w(x, y)$, and seek a solution of the problem in the form

$$u(x, y) = v(x, y) + w(x, y),$$

where $v(x, y)$ is a new unknown function, and the function $w(x, y)$ is chosen so that it satisfies the given non-homogeneous boundary conditions (E.2). Then function $v(x, y)$ satisfies the homogeneous boundary conditions – the solution to this problem was discussed in Section 7.13.

We shall seek an auxiliary function, $w(x, y)$, in the form

$$w(x, y) = g_1(y)\overline{X} + g_2(y)\overline{\overline{X}} + g_3(x)\overline{Y} + g_4(x)\overline{\overline{Y}}$$

$$+ A\overline{XY} + B\overline{X}\overline{\overline{Y}} + C\overline{\overline{X}}\overline{Y} + D\overline{\overline{XY}}, \tag{E.3}$$

where $\overline{X}(x)$, $\overline{\overline{X}}(x)$, $\overline{Y}(y)$ and $\overline{\overline{Y}}(y)$ are polynomials of 1st or 2nd order. The coefficients of these polynomials will be adjusted in such a way that function $w(x, y)$ satisfies the boundary conditions given in Equations (E.2). Notice, that function $w(x, y)$ depends only on the boundary conditions (E.2), i.e. is the same for Poisson and Laplace equations.

Consistent Boundary Conditions

Let the boundary functions *satisfy consistency conditions* (i.e. the boundary functions take the same values at the corners of the domain), in which case we have

$$P_1[g_3(x)]_{x=0} = P_3[g_1(y)]_{y=0}, \quad P_1[g_4(x)]_{x=0} = P_4[g_1(y)]_{y=l_y},$$

$$P_2[g_3(x)]_{x=l_x} = P_3[g_2(y)]_{y=0}, \quad P_2[g_4(x)]_{x=l_x} = P_4[g_2(y)]_{y=l_y}. \tag{E.4}$$

In this case functions $\overline{X}(x)$, $\overline{\overline{X}}(x)$ are constructed via formulas (C.10)–(C.12), and functions $\overline{Y}(y)$, $\overline{\overline{Y}}(y)$ – via formulas (C.13)–(C.15) (see Appendix C part 2).

It is easy to verify that the auxiliary function, $w(x, y)$, satisfies the given boundary conditions:

$$P_1[w]_{x=0} = g_1(y), \quad P_2[w]_{x=l_x} = g_2(y),$$
$$P_3[w]_{y=0} = g_3(x), \quad P_4[w]_{y=l_y} = g_4(x).$$

Inconsistent Boundary Conditions

Suppose the boundary functions *do not satisfy consistency conditions*. We shall search for an auxiliary function as the sum of two functions:

$$w(x, y) = w_1(x, y) + w_2(x, y). \tag{E.5}$$

Function $w_1(x, y)$ is an auxiliary function satisfying the consistent boundary conditions

$$P_1[w_1]_{x=0} = g_1(y), \quad P_2[w_1]_{x=l_x} = g_2(y),$$
$$P_3[w_1]_{y=0} = -A\overline{X}(x) - C\overline{\overline{X}}(x),$$
$$P_4[w_1]_{y=l_y} = -B\overline{X}(x) - D\overline{\overline{X}}(x), \tag{E.6}$$

where

$$A = -P_3[g_1(y)]_{y=0}, \quad B = -P_4[g_1(y)]_{y=l_y},$$
$$C = -P_3[g_2(y)]_{y=0}, \quad D = -P_4[g_2(y)]_{y=l_y}. \tag{E.7}$$

Such a function was constructed above. In this case it has the form

$$w_1(x, y) = g_1(y)\overline{X} + g_2(y)\overline{\overline{X}}. \tag{E.8}$$

The function $w_2(x, y)$ is a particular solution of the *Laplace problem* with the following boundary conditions

$$P_1[w_2]_{x=0} = 0, \quad P_2[w_2]_{x=l_x} = 0,$$
$$P_3[w_2]_{y=0} = g_3(x) + A\overline{X}(x) + C\overline{\overline{X}}(x),$$
$$P_4[w_2]_{y=l_y} = g_4(x) + B\overline{X}(x) + D\overline{\overline{X}}(x). \tag{E.9}$$

This problem was considered in detail in Section 7.12. The solution of this problem has a form

$$w_2(x, y) = \sum_{n=1}^{\infty} [A_n Y_{1n}(y) + B_n Y_{2n}(y)] X_n(x), \tag{E.10}$$

where λ_{xn} and $X_n(x)$ are eigenvalues and eigenfunctions of the Sturm-Liouville problem

$$X'' + \lambda X = 0 \quad (0 < x < l_x),$$
$$P_1[X]|_{x=0} = P_2[X]|_{x=l_x} = 0. \tag{E.11}$$

The coefficients A_n and B_n are defined by the formulas

$$A_n = \frac{1}{||X_n||^2} \int_0^{l_x} \left[g_4(x) + B\overline{X}(x) + D\overline{\overline{X}}(x) \right] X_n(x) dx,$$
$$B_n = \frac{1}{||X_n||^2} \int_0^{l_x} \left[g_3(x) + A\overline{X}(x) + C\overline{\overline{X}}(x) \right] X_n(x) dx. \tag{E.12}$$

Thus, we see that eigenvalues and eigenfunctions of this boundary value problem depend on the types of boundary conditions (see Appendix C part 1 for a detailed account).

Having the eigenvalues, λ_{xn}, we obtain a similar equation for $Y(y)$ given by

$$Y'' - \lambda_{xn}Y = 0 \quad (0 < y < l_y). \tag{E.13}$$

We shall choose fundamental system $\{Y_1,\ Y_2\}$ of solutions in such a way that

$$P_3\left[Y_1(0)\right] = 0, \quad P_3\left[Y_1(l_y)\right] = 1,$$

$$P_4\left[Y_2(0)\right] = 1, \quad P_4\left[Y_2(l_y)\right] = 0. \tag{E.14}$$

Two particular solutions of the previous Equation (E.13) are $\exp(\pm\sqrt{\lambda_n}y)$ but for future analysis it is more convenient to choose two linearly independent functions $Y_1(y)$ and $Y_2(y)$ in the form

$$Y_1(y) = a \sinh\sqrt{\lambda_n}y + b \cosh\sqrt{\lambda_n}y,$$

$$Y_2(y) = c \sinh\sqrt{\lambda_n}(l_y - y) + d \cosh\sqrt{\lambda_n}(l_y - y). \tag{E.15}$$

The values of coefficients a, b, c, and d depend on the types of boundary conditions $P_3[u]_{y=0}$ and $P_4[u]_{y=l_y}$. It can be verified that the auxiliary function

$$w(x,y) = w_1(x,y) + w_2(x,y)$$

$$= w_1(x,y) + \sum_{n=1}^{\infty} \left[A_n Y_{1n}(y) + B_n Y_{2n}(y)\right] X_n(x) \tag{E.16}$$

satisfies the given boundary conditions when $n \to \infty$.

Below we present auxiliary functions for different types of boundary condition.

1. Boundary conditions $\begin{cases} P_3[u] \equiv u|_{y=0} = g_3(x) & \text{(Dirichlet condition)}, \\ P_4[u] \equiv u|_{y=l_y} = g_4(x) & \text{(Dirichlet condition)}. \end{cases}$

   Fundamental system:

   $$Y_{1n}(y) = \frac{\sinh\sqrt{\lambda_{xn}}y}{\sinh\sqrt{\lambda_{xn}}l_y}, \quad Y_{2n}(y) = \frac{\sinh\sqrt{\lambda_{xn}}(l_y - y)}{\sinh\sqrt{\lambda_{xn}}l_y}.$$

   If $\lambda_{x0} = 0$, $X_0(x) \equiv 1$ then

   $$Y_{1n}(y) = \frac{y}{l_y}, \quad Y_{2n}(y) = 1 - \frac{y}{l_y}.$$

2. Boundary conditions $\begin{cases} P_3[u] \equiv u|_{y=0} = g_3(x) & \text{(Dirichlet condition)}, \\ P_4[u] \equiv \dfrac{\partial u}{\partial y}\bigg|_{y=l_y} = g_4(x) & \text{(Neumann condition)}. \end{cases}$

   Fundamental system:

   $$Y_{1n}(y) = \frac{\sinh\sqrt{\lambda_{xn}}y}{\sinh\sqrt{\lambda_{xn}}l_y}, \quad Y_{2n}(y) = \frac{\sinh\sqrt{\lambda_{xn}}(l_y - y)}{\sinh\sqrt{\lambda_{xn}}l_y}.$$

   If $\lambda_{x0} = 0$, $X_0(x) \equiv 1$ then

   $$Y_{1n}(y) = \frac{y}{l_y}, \quad Y_{2n}(y) = 1 - \frac{y}{l_y}.$$

3. Boundary conditions
$$\begin{cases} P_3[u] \equiv u|_{y=0} = g_3(x) & \text{(Dirichlet condition)}, \\ P_4[u] \equiv \dfrac{\partial u}{\partial y} + h_4 u\bigg|_{y=l_y} = g_4(x) & \text{(mixed condition)}. \end{cases}$$

Fundamental system:

$$Y_{1n}(y) = \frac{\sinh\sqrt{\lambda_{xn}}\,y}{h_4\sinh\sqrt{\lambda_{xn}}l_y + \sqrt{\lambda_{xn}}\cosh\sqrt{\lambda_{xn}}l_y},$$

$$Y_{2n}(y) = \frac{h_4\sinh\sqrt{\lambda_{xn}}(l_y - y) + \sqrt{\lambda_{xn}}\cosh\sqrt{\lambda_{xn}}(l_y - y)}{h_4\sinh\sqrt{\lambda_{xn}}l_y + \sqrt{\lambda_{xn}}\cosh\sqrt{\lambda_{xn}}l_y}.$$

If $\lambda_{x0} = 0$, $X_0(x) \equiv 1$ then

$$Y_{1n}(y) = \frac{y}{1 + h_4 l_y}, \quad Y_{2n}(y) = 1 - \frac{h_4}{1 + h_4 l_y}y.$$

4. Boundary conditions
$$\begin{cases} P_3[u] \equiv \dfrac{\partial u}{\partial y}\bigg|_{y=0} = g_3(x) & \text{(Neumann condition)}, \\ P_4[u] \equiv u|_{y=l_y} = g_4(x) & \text{(Dirichletcondition)}. \end{cases}$$

Fundamental system:

$$Y_{1n}(y) = \frac{\cosh\sqrt{\lambda_{xn}}\,y}{\cosh\sqrt{\lambda_{xn}}l_y}, \quad Y_{2n}(y) = -\frac{\sinh\sqrt{\lambda_{xn}}(l_y - y)}{\sqrt{\lambda_{xn}}\cosh\sqrt{\lambda_{xn}}l_y}.$$

If $\lambda_{x0} = 0$, $X_0(x) \equiv 1$ then

$$Y_{1n}(y) = 1, \quad Y_{2n}(y) = y - l_y.$$

5. Boundary conditions
$$\begin{cases} P_3[u] \equiv \dfrac{\partial u}{\partial y}\bigg|_{y=0} = g_3(x) & \text{(Neumann condition)}, \\ P_4[u] \equiv \dfrac{\partial u}{\partial y}\bigg|_{y=l_y} = g_4(x) & \text{(Neumann condition)}. \end{cases}$$

Fundamental system:

$$Y_{1n}(y) = \frac{\cosh\sqrt{\lambda_{xn}}\,y}{\sqrt{\lambda_{xn}}\sinh\sqrt{\lambda_{xn}}l_y}, \quad Y_{2n}(y) = -\frac{\cosh\sqrt{\lambda_{xn}}(l_y - y)}{\sqrt{\lambda_{xn}}\sinh\sqrt{\lambda_{xn}}l_y}.$$

If $\lambda_{x0} = 0$, $X_0(x) \equiv 1$ then

$$Y_{1n}(y) = \frac{1}{2l_y}y^2, \quad Y_{2n}(y) = y - \frac{1}{2l_y}y^2.$$

6. Boundary conditions
$$\begin{cases} P_3[u] \equiv \dfrac{\partial u}{\partial y}\bigg|_{y=0} = g_3(x) & \text{(Neumann condition)}, \\ P_4[u] \equiv \dfrac{\partial u}{\partial y} + h_4 u\bigg|_{y=l_y} = g_4(x) & \text{(mixed condition)}. \end{cases}$$

Fundamental system:

$$Y_{1n}(y) = \frac{\cosh\sqrt{\lambda_{xn}}\,y}{\sqrt{\lambda_{xn}}\sinh\sqrt{\lambda_{xn}}l_y + h_4\cosh\sqrt{\lambda_{xn}}l_y},$$

$$Y_{2n}(y) = -\frac{h_4 \sinh \sqrt{\lambda_{xn}}(l_y - y) + \sqrt{\lambda_{xn}} \cosh \sqrt{\lambda_{xn}}(l_y - y)}{\sqrt{\lambda_{xn}} \left[\sqrt{\lambda_{xn}} \sinh \sqrt{\lambda_{xn}} l_y + h_4 \cosh \sqrt{\lambda_{xn}} l_y\right]}.$$

If $\lambda_{x0} = 0$, $X_0(x) \equiv 1$ then

$$Y_{1n}(y) = \frac{y}{h_4}, \quad Y_{2n}(y) = y - \frac{1 + h_4 l_y}{h_4}.$$

7. Boundary conditions
$$\begin{cases} P_3[u] \equiv \dfrac{\partial u}{\partial y} - h_3 u \Big|_{y=0} = g_3(x) & \text{(mixed condition)}, \\ P_4[u] \equiv u\big|_{y=l_y} = g_4(x) & \text{(Dirichlet condition)}. \end{cases}$$

Fundamental system:

$$Y_{1n}(y) = \frac{h_3 \sinh \sqrt{\lambda_{xn}} y + \sqrt{\lambda_{xn}} \cosh \sqrt{\lambda_{xn}} y}{h_3 \sinh \sqrt{\lambda_{xn}} l_y + \sqrt{\lambda_{xn}} \cosh \sqrt{\lambda_{xn}} l_y},$$

$$Y_{2n}(y) = -\frac{\sinh \sqrt{\lambda_{xn}}(l_y - y)}{h_3 \sinh \sqrt{\lambda_{xn}} l_y + \sqrt{\lambda_{xn}} \cosh \sqrt{\lambda_{xn}} l_y}.$$

If $\lambda_{x0} = 0$, $X_0(x) \equiv 1$ then

$$Y_{1n}(y) = \frac{1 + h_3 y}{1 + h_3 l_y}, \quad Y_{2n}(y) = \frac{y - l_y}{1 + h_3 l_y}.$$

8. Boundary conditions
$$\begin{cases} P_3[u] \equiv \dfrac{\partial u}{\partial y} - h_3 u \Big|_{y=0} = g_3(x) & \text{(mixed condition)}, \\ P_4[u] \equiv \dfrac{\partial u}{\partial y} \Big|_{y=l_y} = g_4(x) & \text{(Neumann condition)}. \end{cases}$$

Fundamental system:

$$Y_{1n}(y) = \frac{h_3 \sinh \sqrt{\lambda_{xn}} y + \sqrt{\lambda_{xn}} \cosh \sqrt{\lambda_{xn}} y}{\sqrt{\lambda_{xn}} \left[\sqrt{\lambda_{xn}} \sinh \sqrt{\lambda_{xn}} l_y + h_3 \cosh \sqrt{\lambda_{xn}} l_y\right]},$$

$$Y_{2n}(y) = -\frac{\cosh \sqrt{\lambda_{xn}}(l_y - y)}{\sqrt{\lambda_{xn}} \sinh \sqrt{\lambda_{xn}} l_y + h_3 \cosh \sqrt{\lambda_{xn}} l_y}.$$

If $\lambda_{x0} = 0$, $X_0(x) \equiv 1$ then

$$Y_{1n}(y) = \frac{1 + h_3 y}{1 + h_3 l_y}, \quad Y_{2n}(y) = \frac{y - l_y}{1 + h_3 l_y}.$$

9. Boundary conditions
$$\begin{cases} P_3[u] \equiv \dfrac{\partial u}{\partial y} - h_3 u \Big|_{y=0} = g_3(x) & \text{(mixed condition)}, \\ P_4[u] \equiv \dfrac{\partial u}{\partial y} + h_4 u \Big|_{y=l_y} = g_4(x) & \text{(mixed condition)}. \end{cases}$$

Fundamental system:

$$Y_{1n}(y) = \frac{h_3 \sinh \sqrt{\lambda_{xn}} y + \sqrt{\lambda_{xn}} \cosh \sqrt{\lambda_n} y}{(\lambda_{xn} + h_3 h_4) \sinh \sqrt{\lambda_{xn}} l_y + \sqrt{\lambda_{xn}}(h_3 + h_4) \cosh \sqrt{\lambda_{xn}} l_y},$$

$$Y_{2n}(y) = -\frac{h_4 \sinh \sqrt{\lambda_{xn}}(l_y - y) + \sqrt{\lambda_{xn}} \cosh \sqrt{\lambda_{xn}}(l_y - y)}{(\lambda_{xn} + h_3 h_4) \sinh \sqrt{\lambda_{xn}} l_y + \sqrt{\lambda_{xn}}(h_3 + h_4) \cosh \sqrt{\lambda_{xn}} l_y}.$$

If $\lambda_{x0} = 0$, $X_0(x) \equiv 1$ then

$$Y_1(y) = \frac{1 + h_3 y}{h_3 + h_4 + h_3 h_4 l_y}, \quad Y_2(y) = \frac{h_4(y - l_y) - 1}{h_3 + h_4 + h_3 h_4 l_y}.$$

Examples

Example E.1 Find the stationary distribution of temperature within a very long (infinite) parallelepiped of rectangular cross section $(0 \leq x \leq \pi, 0 \leq y \leq \pi)$ if the faces $y = 0$ and $y = \pi$ follow the temperature distributions

$$u(x, 0) = \cos x \quad \text{and} \quad u(x, \pi) = \cos 3x$$

respectively and the constant heat flows are supplied to the faces $x = 0$ and $x = \pi$ from outside:

$$\frac{\partial u}{\partial x}(0, y) = \sin y \quad \text{and} \quad \frac{\partial u}{\partial x}(\pi, y) = \sin 5y.$$

Generation (or absorption) of heat by internal sources is absent.

Solution. The problem is expressed as Laplace equation

$$\frac{\partial^2 u}{\partial x^2} + \frac{\partial^2 u}{\partial y^2} = 0$$

under the conditions

$$P_1[u]_{x=0} \equiv \frac{\partial u}{\partial x}(0, y) = \sin y \qquad (\alpha_1 = 1,\ \beta_1 = 0 - \text{Neumann condition}),$$

$$P_2[u]_{x=\pi} \equiv \frac{\partial u}{\partial x}(\pi, y) = \sin 5y \qquad (\alpha_2 = 1,\ \beta_2 = 0 - \text{Neumann condition}),$$

$$P_3[u]_{y=0} \equiv u(x, 0) = \cos x \qquad (\alpha_3 = 0,\ \beta_3 = 1 - \text{Dirichlet condition}),$$

$$P_4[u]_{y=\pi} \equiv u(x, \pi) = \cos 3x \qquad (\alpha_4 = 0,\ \beta_4 = 1 - \text{Dirichlet condition}).$$

The solution to this problem can be expressed as the sum of two functions

$$u(x, y) = w(x, y) + v(x, y),$$

where $v(x, y)$ is a new unknown function and $w(x, y)$ is an auxiliary function satisfying the boundary conditions.

The boundary functions *satisfy the consistent conditions*, that is, (see formulas (E.4))

$$P_1[\cos x]_{x=0} = \left.\frac{\partial(\cos x)}{\partial x}\right|_{x=0} = 0 \ = \ P_3[\sin y]_{y=0} = \sin y|_{y=0} = 0,$$

$$P_1[\cos 3x]_{x=0} = \left.\frac{\partial(\cos 3x)}{\partial x}\right|_{x=0} = 0 \ = \ P_4[\sin y]_{y=\pi} = \sin y|_{y=\pi} = 0,$$

$$P_2[\cos x]_{x=\pi} = \left.\frac{\partial(\cos x)}{\partial x}\right|_{x=\pi} = 0 \ = \ P_3[\sin 5y]_{y=0} = \sin 5y|_{y=0} = 0,$$

$$P_2[\cos 3x]_{x=\pi} = \left.\frac{\partial(\cos 3x)}{\partial x}\right|_{x=\pi} = 0 \ = \ P_4[\sin 5y]_{y=\pi} = \sin 5y|_{y=\pi} = 0.$$

Let construct the auxiliary function (E.3) satisfying the given boundary condition. In our problem $\beta_1 = \beta_2 = 0$, so (see formulas (C.12))

$$\overline{X}(x) = x - \frac{x^2}{2l_x}, \quad \overline{\overline{X}}(x) = \frac{x^2}{2l_x}.$$

From formulas (C.13), (C.14) we have $\gamma_3 = 1,\ \delta_3 = -1/l_y,\ \gamma_4 = 0,\ \delta_4 = 1/l_y$, so

$$\overline{Y}(y) = 1 - \frac{y}{\pi}, \quad \overline{\overline{Y}}(y) = \frac{y}{\pi}.$$

Coefficients (C.16) of the auxiliary function (E.3) are zero:

$$A = -P_3[\sin y]_{y=0} = -\sin y|_{y=0} = 0,$$
$$B = -P_4[\sin y]_{y=\pi} = -\sin y|_{y=\pi} = 0,$$
$$C = -P_3[\sin 5y]_{y=0} = -\sin 5y|_{y=0} = 0,$$
$$D = -P_4[\sin 5y]_{y=\pi} = -\sin 5y|_{y=\pi} = 0.$$

So, the auxiliary function (E.3) has the form:

$$w(x,y) = \sin y \cdot \overline{X} + \sin 5 \cdot \overline{\overline{X}} + \cos x \cdot \overline{Y} + \cos 3x \cdot \overline{\overline{Y}}$$
$$= \sin y \left(x - \frac{x^2}{2\pi} \right) + \sin 5y \frac{x^2}{2\pi} + \cos x \left(1 - \frac{y}{\pi} \right) + \cos 3x \frac{y}{\pi}.$$

Function $v(x,y)$ is the solution to the Poisson problem with zero boundary conditions

$$\frac{\partial^2 u}{\partial x^2} + \frac{\partial^2 u}{\partial y^2} = \tilde{f}(x,y),$$

where

$$\tilde{f}(x,y) = \frac{\partial^2 w}{\partial x^2} + \frac{\partial^2 w}{\partial y^2} = \frac{1}{2\pi} \left[\sin y \cdot (x^2 - 2\pi x - 2) + \sin 5y \cdot (2 - 25x^2) \right.$$
$$\left. + 2\cos x \cdot (y - \pi) - 18y \cdot \cos 3x \right].$$

Eigenvalues and eigenfunctions are (see Appendix C part 1)

$$\lambda_{nm} = \lambda_{xn} + \lambda_{ym} = n^2 + m^2, \quad n = 0,1,2,\ldots, \quad m = 1,2,\ldots,$$
$$V_{nm}(x,y) = X_n(x) \cdot Y_m(y) = \cos nx \cdot \sin my,$$
$$\|V_{nm}\|^2 = \|X_n\|^2 \cdot \|Y_m\|^2 = \begin{cases} \pi^2/2, & \text{if } n = 0, \\ \pi^2/4, & \text{if } n > 0. \end{cases}$$

The solution $v(x,y)$ is defined by the series

$$v(x,y) = \sum_{n=0}^{\infty} \sum_{m=1}^{\infty} C_{nm} \cos nx \cdot \sin my,$$

where $C_{nm} = \frac{f_{nm}}{\lambda_{nm}}$, $f_{nm} = \frac{1}{\|V_{nm}\|^2} \int_0^\pi \int_0^\pi \tilde{f}(x,y) \cdot \cos nx \cdot \sin my \, dx dy$.
Thus,

$$C_{01} = \frac{f_{01}}{\lambda_{01}} = -\frac{1}{\pi} \left(\frac{\pi^2}{3} + 1 \right), \quad C_{n1} = \frac{f_{n1}}{\lambda_{n1}} = \frac{2}{\pi n^2 (n^2 + 1)},$$
$$C_{05} = \frac{f_{05}}{\lambda_{05}} = \frac{1}{\pi} \left(\frac{1}{25} - \frac{\pi^2}{6} \right), \quad C_{n5} = \frac{f_{n5}}{\lambda_{n5}} = (-1)^{n+1} \frac{50}{\pi n^2 (n^2 + 25)},$$
$$C_{1m} = \frac{f_{1m}}{\lambda_{1m}} = -\frac{2}{\pi m (m^2 + 1)}, \quad C_{3m} = \frac{f_{3m}}{\lambda_{3m}} = (-1)^m \frac{18}{\pi m (m^2 + 9)},$$
$$C_{nm} = 0 \text{ in other cases.}$$

The final solution is (see Figure E.1):

$$u(x,y) = w(x,y) + v(x,y) = \sin y \cdot \left(x - \frac{x^2}{2\pi} \right) + \sin 5y \cdot \frac{x^2}{2\pi} + \cos x \cdot \left(1 - \frac{y}{\pi} \right) + \cos 3x \cdot \frac{y}{\pi}$$
$$+ \sum_{n=0}^{\infty} [C_{n1} \sin y + C_{n5} \sin 5y] \cdot \cos nx + \sum_{n=1}^{\infty} [C_{1m} \cos x + C_{3m} \cos 3x] \cdot \sin my.$$

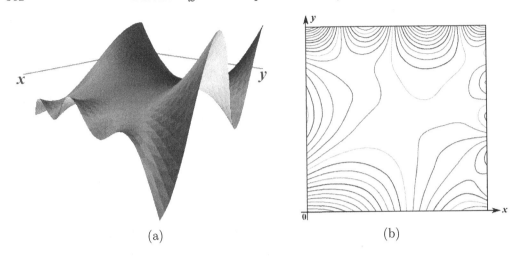

FIGURE E.1
Surface plot (a) and lines of equal temperature (b) for Example E.1.

Example E.2 A heat-conducting thin uniform rectangular membrane $(0 \leq x \leq l_x, 0 \leq y \leq l_y)$ is thermally insulated over its lateral faces. The bounds $x = 0$ and $y = 0$ are thermally insulated, the other bounds, $x = l_x$ and $y = l_y$, are held at fixed temperature $u = u_0$. One constant internal source of heat acts at the point (x_0, y_0) of the membrane. The value of this source is $Q = $ const. Find the steady-state temperature distribution in the membrane.

Solution. The problem may be explored in the solution of the Poisson equation

$$\frac{\partial^2 u}{\partial x^2} + \frac{\partial^2 u}{\partial y^2} = -Q \cdot \delta(x - x_0)\delta(y - y_0)$$

under the conditions

$$P_1[u]_{x=0} \equiv \frac{\partial u}{\partial x}(0, y) = 0 \quad (\alpha_1 = 1, \ \beta_1 - 0 - \text{Neumann condition}),$$

$$P_2[u]_{x=l_x} \equiv u(l_x, y) = u_0 \quad (\alpha_2 = 0, \ \beta_2 = 1 - \text{Dirichlet condition}),$$

$$P_3[u]_{y=0} \equiv \frac{\partial u}{\partial y}(x, 0) = 0 \quad (\alpha_3 = 1, \ \beta_3 = 0 - \text{Neumann condition}),$$

$$P_4[u]_{y=l_y} \equiv u(x, l_y) = u_0 \quad (\alpha_4 = 0, \ \beta_4 = 1 - \text{Dirichlet condition}).$$

The solution to this problem can be expressed as the sum of two functions

$$u(x, y) = v(x, y) + w(x, y),$$

where $v(x, y)$ is a new unknown function and $w(x, y)$ is an auxiliary function satisfying the boundary conditions.

The boundary functions *satisfy the consistent conditions*, that is, (see formulas (E.4))

$$P_1[g_3]_{x=0} = \frac{\partial(g_3 \equiv 0)}{\partial x}\bigg|_{x=0} = 0 = P_3[g_1]_{y=0} = \frac{\partial(g_1 \equiv 0)}{\partial y}\bigg|_{y=0} = 0,$$

$$P_1[g_4]_{x=0} = \frac{\partial(g_4 \equiv u_0)}{\partial x}\bigg|_{x=0} = 0 = P_4[g_1]_{y=l_y} = g_1|_{y=l_y} = 0,$$

$$P_2[g_3]_{x=l_x} = g_3|_{x=l_x} = 0 = P_3[g_2]_{y=0} = \frac{\partial(g_2 \equiv u_0)}{\partial y}\bigg|_{y=0} = 0,$$

$$P_2[g_4]_{x=l_x} = g_4|_{x=l_x} = u_0 = P_4[g_2]_{y=l_y} = g_2|_{y=l_y} = u_0.$$

Let us construct the auxiliary function (E.3) satisfying the given boundary condition. In our problem $\gamma_1 = -l_x$, $\delta_1 = 1$, $\gamma_2 = 1$, $\delta_2 = 0$ and $\gamma_3 = -l_y$, $\delta_3 = 1$, $\gamma_4 = 1$, $\delta_4 = 0$ (see formulas (C.10), (C.11) and (C.13), (C.14)), so

$$\overline{X}(x) = x - l_x, \quad \overline{\overline{X}}(x) = 1, \quad \overline{Y}(y) = y - l_y, \quad \overline{\overline{Y}}(y) = 1.$$

Coefficients (C.16) of the auxiliary function (E.3) are zeros:

$$A = -P_3[g_1]_{y=0} = 0, \quad B = -P_4[g_1]_{y=l_y} = 0,$$
$$C = -P_3[g_2]_{y=0} = 0, \quad D = -P_4[g_2]_{y=l_y} = u_0.$$

So, the auxiliary function (E.3) has the form:

$$w(x,y) = u_0\overline{\overline{X}} + u_0\overline{\overline{Y}} + u_0\overline{\overline{XY}} = 3u_0.$$

This function $w(x,y)$ in this example is a harmonic function, that is $\nabla^2 w(x,y) = 0$.

Function $v(x,y)$ is the solution to the Poisson problem with zero boundary conditions

$$\frac{\partial^2 v}{\partial x^2} + \frac{\partial^2 v}{\partial y^2} = \tilde{f}(x,y),$$

where $\tilde{f}(x,y) = f(x,y) + \frac{\partial^2 w}{\partial x^2} + \frac{\partial^2 w}{\partial y^2} = -Q \cdot \delta(x - x_0)\delta(y - y_0)$.

Eigenvalues and eigenfunctions are (see Appendix C part 1)

$$\lambda_{nm} = \lambda_{xn} + \lambda_{ym} = \pi^2\left(\frac{(2n-1)^2}{4l_x^2} + \frac{(2m-1)^2}{4l_y^2}\right), \quad n,m = 1,2,\ldots$$

$$V_{nm}(x,y) = X_n(x) \cdot Y_m(y) = \cos\frac{\pi(2n-1)x}{2l_x}\cos\frac{\pi(2m-1)y}{2l_y},$$

$$\|V_{nm}\|^2 = \|X_n\|^2 \cdot \|Y_m\|^2 = \frac{l_x l_y}{4}.$$

The solution $v(x,y)$ is defined by the series

$$v(x,y) = \sum_{n=1}^{\infty}\sum_{m=1}^{\infty} C_{nm}\cos\frac{(2n-1)\pi x}{2l_x}\cos\frac{(2m-1)\pi y}{2l_y},$$

where $C_{nm} = \frac{f_{nm}}{\lambda_{nm}}$ with

$$f_{nm} = \frac{1}{\|V_{nm}\|^2}\int_0^{l_x}\int_0^{l_y} f(x,y)\cdot\cos\frac{(2n-1)\pi x}{2l_x}\cos\frac{(2m-1)\pi y}{2l_y}\,dxdy$$

$$= \frac{4Q}{l_x l_y}\cos\frac{(2n-1)\pi x_0}{2l_x}\cos\frac{(2m-1)\pi y_0}{2l_y},$$

$$C_{nm} = \frac{f_{nm}}{\lambda_{nm}} = \frac{16Q l_x l_y}{\pi^2\left[(2n-1)^2 l_y^2 + (2m-1)^2 l_x^2\right]}\cos\frac{(2n-1)\pi x_0}{2l_x}\cos\frac{(2m-1)\pi y_0}{2l_y}.$$

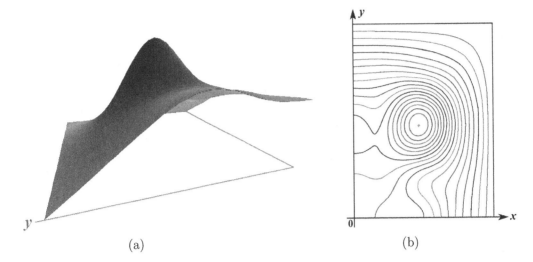

FIGURE E.2
Surface plot (a) and lines of equal temperature (b) for Example E.2.

The final solution is:

$$u(x,y) = w(x,y) + v(x,y) = 3u_0 + \sum_{n=1}^{\infty}\sum_{m=1}^{\infty} C_{nm} \cos \frac{(2n-1)\pi x}{2l_x} \cos \frac{(2m-1)\pi y}{2l_y}.$$

This solution was obtained for the case $x_0 = 2$, $y_0 = 3$, $u_0 = 10$, $Q = 100$ (see Figure E.2).

Example E.3 Find the electrostatic potential in a change-free rectangular domain ($0 \leq x \leq l_x$, $0 \leq y \leq l_y$), if one part of the bound ($x = 0$ and $y = 0$) is held at fixed potential $u = u_1$ and the other part ($x = l_x$ and $y = l_y$) at fixed potential $u = u_2$.

Solution. The problem is the Laplace equation

$$\frac{\partial^2 u}{\partial x^2} + \frac{\partial^2 u}{\partial y^2} = 0,$$

under the conditions

$$
\begin{aligned}
P_1[u]_{x=0} &\equiv u(0,y) = u_1 &&(\alpha_1 = 0,\ \beta_1 = 1 - \text{Dirichlet condition}), \\
P_2[u]_{x=l_x} &\equiv u(l_x,y) = u_2 &&(\alpha_2 = 0,\ \beta_2 = 1 - \text{Dirichlet condition}), \\
P_3[u]_{y=0} &\equiv u(x,0) = u_1 &&(\alpha_3 = 0,\ \beta_3 = 1 - \text{Dirichlet condition}), \\
P_4[u]_{y=l_y} &\equiv u(x,l_y) = u_2 &&(\alpha_4 = 0,\ \beta_4 = 1 - \text{Dirichlet condition}).
\end{aligned}
$$

The solution to this problem can be expressed as the sum of two functions

$$u(x,y) = v(x,y) + w(x,y),$$

where $v(x,y)$ is a new unknown function and $w(x,y)$ is an auxiliary function satisfying the boundary conditions.

The boundary functions *do not satisfy the consistent conditions* at points $(0, l_y)$ and $(l_x, 0)$:

$$
\begin{aligned}
P_1[g_3]_{x=0} = g_3(0) = u_1 &= P_3[g_1]_{y=0} = g_1(0) = u_1, \\
P_1[g_4]_{x=0} = g_4(0) = u_2 &\neq P_4[g_1]_{y=l_y} = g_1(l_y) = u_1, \\
P_2[g_3]_{x=l_x} = g_3(l_x) = u_1 &\neq P_3[g_2]_{y=0} = g_2(0) = u_2, \\
P_2[g_4]_{x=l_x} = g_4(l_x) = u_2 &= P_4[g_2]_{y=l_y} = g_2(l_y) = u_2.
\end{aligned}
$$

Thus, we seek function $w(x, y)$ as a sum of two functions

$$
w(x, y) = w_1(x, y) + w_2(x, y),
$$

where $w_1(x, y)$ is the auxiliary function satisfying the consistent boundary conditions (see (E.6))

$$
\begin{aligned}
P_1[w_1]_{x=0} = g_1(y), \quad P_2[w_1]_{x=l_x} = g_2(y), \\
P_3[w_1]_{y=0} = -A\overline{X}(x) - C\overline{\overline{X}}(x), \\
P_4[w_1]_{y=l_y} = -B\overline{X}(x) - D\overline{\overline{X}}(x),
\end{aligned}
$$

and has the form (E.8)

$$
w_1(x, y) = g_1(y)\overline{X} + g_2(y)\overline{\overline{X}}.
$$

Like in the previous situations, we find functions $\overline{X}(x)$ and $\overline{\overline{X}}(x)$

$$
\overline{X}(x) = 1 - \frac{x}{l_x}, \quad \overline{\overline{X}}(x) = \frac{x}{l_x},
$$

and coefficients A, B, C, D (see (E.7))

$$
\begin{aligned}
A = -P_3[g_1(y)]_{y=0} = -u_1, \quad B = -P_4[g_1(y)]_{y=l_y} = -u_1, \\
C = -P_3[g_2(y)]_{y=0} = -u_2, \quad D = -P_4[g_2(y)]_{y=l_y} = -u_2.
\end{aligned}
$$

Thus, the auxiliary function $w_1(x, y)$, which satisfies the consistent boundary conditions

$$
\begin{aligned}
P_1[w_1]_{x=0} = u_1, & \qquad P_2[w_1]_{x=l_x} = u_2, \\
P_3[w_1]_{y=0} = u_1 + (u_2 - u_1)\frac{x}{l_x}, & \quad P_4[w_1]_{y=l_y} = u_1 + (u_2 - u_1)\frac{x}{l_x},
\end{aligned}
$$

is

$$
w_1(x, y) = u_1 + (u_2 - u_1)\frac{x}{l_x}.
$$

Function $w_1(x, y)$ is harmonic, i.e. $\nabla^2 w_1(x, y) = 0$.

The function $w_2(x, y)$ is a particular solution of the *Laplace problem* with the following boundary conditions (see (E.9)):

$$
P_1[w_2]_{x=0} = 0, \quad P_2[w_2]_{x=l_x} = 0,
$$

$$
P_3[w_2]_{y=0} = g_3(x) + A\overline{X}(x) + C\overline{\overline{X}}(x) = \frac{x}{l_x}(u_1 - u_2),
$$

$$
P_4[w_2]_{y=l_y} = g_4(x) + B\overline{X}(x) + D\overline{\overline{X}}(x) = (u_2 - u_1)\left(1 - \frac{x}{l_x}\right).
$$

The solution of this problem has the form (E.10):

$$
w_2(x, y) = \sum_{n=1}^{\infty} \{A_n Y_{1n}(y) + B_n Y_{2n}(y)\} \cdot X_n(x),
$$

where λ_{xn}, $X_n(x)$ – eigenvalues and eigenfunctions of the respective Sturm-Liouville problem:

$$\lambda_{xn} = \left[\frac{n\pi}{l_x}\right]^2, \quad X_n(x) = \sin\frac{n\pi x}{l_x}, \quad \|X_n\|^2 = \frac{l_x}{2}, \quad n = 1, 2, \ldots$$

and functions $Y_{1n}(y)$ and $Y_{2n}(y)$ for the given boundary functions are

$$Y_{1n}(y) = \frac{\sinh\sqrt{\lambda_n}y}{\sinh\sqrt{\lambda_n}l_y}, \quad Y_{2n}(y) = \frac{\sinh\sqrt{\lambda_n}(l_y - y)}{\sinh\sqrt{\lambda_n}l_y}.$$

Coefficients A_n and B_n are determined with formulas (E.12):

$$A_n = \frac{1}{\|X_n\|^2}\int_0^{l_x}\left[g_4(x) + B\overline{\overline{X}} + D\overline{\overline{X}}\right]X_n(x)dx$$

$$= \frac{2}{l_x}\int_0^{l_x}(u_2 - u_1)\left(1 - \frac{x}{l_x}\right)\sin\frac{n\pi x}{l_x}dx = \frac{2}{n\pi}(u_2 - u_1),$$

$$B_n = \frac{1}{\|X_n\|^2}\int_0^{l_x}\left[g_3(x) + A\overline{\overline{X}} + C\overline{\overline{X}}\right]X_n(x)dx$$

$$= \frac{2}{l_x}\int_0^{l_x}\frac{x}{l_x}(u_1 - u_2)\sin\frac{n\pi x}{l_x}dx = (-1)^n\frac{2}{n\pi}(u_2 - u_1).$$

So,

$$w_2(x, y) = 2(u_2 - u_1)\sum_{n=1}^{\infty}\frac{1}{n\pi \cdot \sinh\sqrt{\lambda_n}l_y}\left\{\sinh\sqrt{\lambda_n}y + (-1)^n\sinh\sqrt{\lambda_n}(l_y - y)\right\}\cdot\sin\frac{n\pi x}{l_x}.$$

Function $w_2(x, y)$ is a particular solution of the *Laplace problem* $\nabla^2 w_2(x, y) = 0$.

Function $v(x, y)$ is the solution to the *Poisson problem with zero boundary conditions*, where

$$\tilde{f}(x, y) = \frac{\partial^2 w_1}{\partial x^2} + \frac{\partial^2 w_1}{\partial y^2} + \frac{\partial^2 w_2}{\partial x^2} + \frac{\partial^2 w_2}{\partial y^2} \equiv 0,$$

so, in this example

$$v(x, y) \equiv 0.$$

The solution to the problem is:

$$u(x, y) = w_1(x, y) + w_2(x, y) = u_1 + (u_2 - u_1)\frac{x}{l_x}$$

$$+ 2(u_2 - u_1)\sum_{n=1}^{\infty}\frac{1}{n\pi \cdot \sinh\sqrt{\lambda_n}l_y}\left\{\sinh\sqrt{\lambda_n}y + (-1)^n\sinh\sqrt{\lambda_n}(l_y - y)\right\}\cdot\sin\frac{n\pi x}{l_x}.$$

This solution was obtained for the case $u_1 = 10$, $u_2 = 20$ (see Figure E.3).

E.2 The Heat Conduction Equations with Nonhomogeneous Boundary Conditions – Examples

Consider the general boundary value problem for heat equation given by

$$\frac{\partial u}{\partial t} = a^2\left[\frac{\partial^2 u}{\partial x^2} + \frac{\partial^2 u}{\partial y^2}\right] - \gamma u + f(x, y, t), \tag{E.17}$$

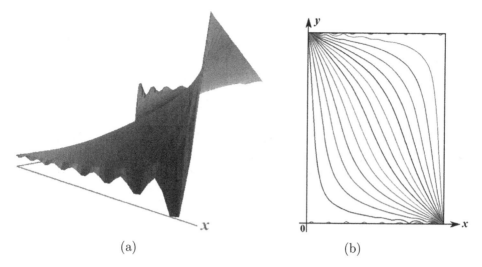

(a) (b)

FIGURE E.3
Surface plot (a) and lines of equal potential (b) for Example E.3.

with nonhomogeneous initial and boundary conditions

$$u(x, y, t)|_{t=0} = \varphi(x, y),$$

$$P_1[u] \equiv \alpha_1 \frac{\partial u}{\partial x} + \beta_1 u \bigg|_{x=0} = g_1(y, t), \quad P_2[u] \equiv \alpha_2 \frac{\partial u}{\partial x} + \beta_2 u \bigg|_{x=l_x} = g_2(y, t),$$

$$P_3[u] \equiv \alpha_3 \frac{\partial u}{\partial y} + \beta_3 u \bigg|_{y=0} = g_3(x, t), \quad P_4[u] \equiv \alpha_4 \frac{\partial u}{\partial y} + \beta_4 u \bigg|_{y=l_y} = g_4(x, t). \tag{E.18}$$

To deal with nonhomogeneous boundary conditions we introduce an auxiliary function, $w(x, y, t)$, and seek a solution of the problem in the form

$$u(x, y, t) = v(x, y, t) + w(x, y, t),$$

where $v(x, y, t)$ is a new unknown function, and the function $w(x, y, t)$ is chosen so that it satisfies the given nonhomogeneous boundary conditions (E.18). Then function $v(x, y, t)$ satisfies the homogeneous boundary conditions – the solution to this problem was discussed in Section 9.1.2.

We shall seek an auxiliary function, $w(x, y)$, in the form

$$w(x, y, t) = g_1(y, t)\overline{X} + g_2(y, t)\overline{\overline{X}} + g_3(x, t)\overline{Y} + g_4(x, t)\overline{\overline{Y}}$$
$$+ A(t)\overline{XY} + B(t)\overline{X\overline{Y}} + C(t)\overline{\overline{X}Y} + D(t)\overline{\overline{XY}}, \tag{E.19}$$

where $\overline{X}(x)$, $\overline{\overline{X}}(x)$, $\overline{Y}(y)$ and $\overline{\overline{Y}}(y)$ are polynomials of 1st or 2nd order. The coefficients of these polynomials will be adjusted in such a way that function $w(x, y, t)$ satisfies the boundary conditions given in Equations (E.18).

Consistent Boundary Conditions

Let the boundary functions *satisfy consistency conditions* (i.e. the boundary functions take the same values at the corners of the domain), in which case we have

$$P_1[g_3(x,t)]_{x=0} = P_3[g_1(y,t)]_{y=0}, \quad P_1[g_4(x,t)]_{x=0} = P_4[g_1(y,t)]_{y=l_y},$$

$$P_2[g_3(x,t)]_{x=l_x} = P_3[g_2(y,t)]_{y=0}, \quad P_2[g_4(x,t)]_{x=l_x} = P_4[g_2(y,t)]_{y=l_y}. \tag{E.20}$$

In this case functions $\overline{X}(x)$, $\overline{\overline{X}}(x)$ are constructed via formulas (C.10)–(C.12), and functions $\overline{Y}(y)$, $\overline{\overline{Y}}(y)$ – via formulas (C.13)–(C.15) (see Appendix C part 2).

It is easy to verify that the auxiliary function, $w(x,y,t)$, satisfies the given boundary conditions:

$$P_1[w]_{x=0} = g_1(y,t), \quad P_2[w]_{x=l_x} = g_2(y,t),$$

$$P_3[w]_{y=0} = g_3(x,t), \quad P_4[w]_{y=l_y} = g_4(x,t).$$

Inconsistent Boundary Conditions

Suppose the boundary functions *do not satisfy consistency conditions*. Then we shall search for an auxiliary function as the sum of two functions:

$$w(x,y,t) = w_1(x,y,t) + w_2(x,y,t). \tag{E.21}$$

Function $w_1(x,y,t)$ is an auxiliary function satisfying the consistent boundary conditions

$$P_1[w_1]_{x=0} = g_1(y,t), \quad P_2[w_1]_{x=l_x} = g_2(y,t),$$

$$P_3[w_1]_{y=0} = -A(t)\overline{X}(x) - C(t)\overline{\overline{X}}(x),$$

$$P_4[w_1]_{y=l_y} = -B(t)\overline{X}(x) - D(t)\overline{\overline{X}}(x), \tag{E.22}$$

where

$$A(t) = -P_3[g_1(y,t)]_{y=0}, \quad B(t) = -P_4[g_1(y,t)]_{y=l_y},$$

$$C(t) = -P_3[g_2(y,t)]_{y=0}, \quad D(t) = -P_4[g_2(y,t)]_{y=l_y}. \tag{E.23}$$

Such a function was constructed above (for the case of consistent boundary conditions). In this case it has the form

$$w_1(x,y,t) = g_1(y,t)\overline{X} + g_2(y,t)\overline{\overline{X}}. \tag{E.24}$$

Reading Exercise. Verify Equation (E.24).

The function $w_2(x,y,t)$ is a particular solution of the *Laplace problem*

$$\frac{\partial^2 w_2}{\partial x^2} + \frac{\partial^2 w_2}{\partial y^2} = 0$$

with the following boundary conditions

$$P_1[w_2]_{x=0} = 0, \quad P_2[w_2]_{x=l_x} = 0,$$

$$P_3[w_2]_{y=0} = g_3(x,t) + A(t)\overline{X}(x) + C(t)\overline{\overline{X}}(x),$$

$$P_4[w_2]_{y=l_y} = g_4(x,t) + B(t)\overline{X}(x) + D(t)\overline{\overline{X}}(x). \tag{E.25}$$

This problem was considered in detail in Section 7.12. The solution of this problem has a form

$$w_2(x, y, t) = \sum_{n=1}^{\infty} \left[A_n(t) Y_{1n}(y) + B_n(t) Y_{2n}(y) \right] X_n(x), \tag{E.26}$$

where λ_{xn} and $X_n(x)$ are eigenvalues and eigenfunctions of the Sturm-Liouville problem

$$\begin{aligned} X'' + \lambda X = 0 \quad (0 < x < l_x), \\ P_1[X]|_{x=0} = P_2[X]|_{x=l_x} = 0. \end{aligned} \tag{E.27}$$

The coefficients $A_n(t)$ and $B_n(t)$ are defined by the formulas

$$A_n(t) = \frac{1}{||X_n||^2} \int_0^{l_x} \left[g_4(x, t) + B(t)\overline{X}(x) + D(t)\overline{\overline{X}}(x) \right] X_n(x) dx,$$

$$B_n(t) = \frac{1}{||X_n||^2} \int_0^{l_x} \left[g_3(x, t) + A(t)\overline{X}(x) + C(t)\overline{\overline{X}}(x) \right] X_n(x) dx. \tag{E.28}$$

Thus, we see that eigenvalues and eigenfunctions of this boundary value problem depend on the types of boundary conditions (see Appendix C, part 1 for a detailed account).

Having the eigenvalues, λ_{xn}, we obtain a similar equation for $Y(y)$ given by

$$Y'' - \lambda_{xn} Y = 0 \quad (0 < y < l_y). \tag{E.29}$$

We shall choose fundamental system $\{Y_1, Y_2\}$ of solutions in such a way that

$$\begin{aligned} P_3[Y_1(0)] = 0, \quad P_3[Y_1(l_y)] = 1, \\ P_4[Y_2(0)] = 1, \quad P_4[Y_2(l_y)] = 0. \end{aligned} \tag{E.30}$$

Two particular solutions of the previous Equation (F.13) are $\exp(\pm\sqrt{\lambda_n}y)$ but for future analysis it is more convenient to choose two linearly independent functions $Y_1(y)$ and $Y_2(y)$ in the form

$$\begin{aligned} Y_1(y) = a \sinh \sqrt{\lambda_n} y + b \cosh \sqrt{\lambda_n} y, \\ Y_2(y) = c \sinh \sqrt{\lambda_n}(l_y - y) + d \cosh \sqrt{\lambda_n}(l_y - y). \end{aligned} \tag{E.31}$$

Reading Exercise. Prove that the two functions in Equation (E.31) are the solutions to Equation (E.29).

The values of coefficients a, b, c, and d depend on the types of boundary conditions $P_3[u]_{y=0}$ and $P_4[u]_{y=l_y}$. It can be verified that the auxiliary function

$$w(x, y, t) = w_1(x, y, t) + w_2(x, y, t)$$

$$= w_1(x, y, t) + \sum_{n=1}^{\infty} \left[A_n(t) Y_{1n}(y) + B_n(t) Y_{2n}(y) \right] X_n(x) \tag{E.32}$$

satisfies the given boundary conditions when $n \to \infty$.

Fundamental systems of functions $Y_1(y)$ and $Y_2(y)$ for different types of boundary conditions are presented at the first part of this Appendix.

Examples

Example E.4 A heat-conducting thin uniform rectangular plate $(0 \leq x \leq l_x, 0 \leq y \leq l_y)$ is thermally insulated over its lateral faces. The edge at $y = 0$ of the plate is kept at the constant temperature $u = u_1$, the edge $y = l_y$ at constant temperature $u = u_2$, the remaining boundary is thermally insulated. The initial temperature distribution within the plate is

$$u(x, y, 0) = u_0 = \text{const.}$$

Find the temperature $u(x, y, t)$ of the plate at any later time, if generation (or absorption) of heat by internal sources is absent.

Solution. The problem may be resolved by solving the equation

$$\frac{\partial u}{\partial t} = a^2 \left[\frac{\partial^2 u}{\partial x^2} + \frac{\partial^2 u}{\partial y^2} \right],$$

under the conditions

$$u(x, y, 0) = \varphi(x, y) = u_0,$$

$$\frac{\partial u}{\partial x}(0, y, t) = \frac{\partial u}{\partial x}(l_x, y, t) = 0, \quad u(x, 0, t) = u_1, \quad u(x, l_y, t) = u_2.$$

The solution to this problem can be expressed as the sum of two functions, as explained above,

$$u(x, y, t) = w(x, y, t) + v(x, y, t).$$

The boundary value functions *satisfy the conforming conditions*, that is,

$$g_1\big|_{y=0} = \frac{\partial g_3}{\partial x}\bigg|_{x=0} = 0, \quad g_1\big|_{y=l_y} = \frac{\partial g_4}{\partial x}\bigg|_{x=0} = 0,$$

$$g_2\big|_{y=0} = \frac{\partial g_3}{\partial x}\bigg|_{x=l_x} = 0, \quad g_2\big|_{y=l_y} = \frac{\partial g_4}{\partial x}\bigg|_{x=l_x} = 0.$$

An auxiliary function satisfying the given boundary condition is giving by formula (F.3) with

$$\overline{X}(x) = x - \frac{x^2}{2l_x}, \quad \overline{\overline{X}}(x) = \frac{x^2}{2l_x}, \quad \overline{Y}(y) = 1 - \frac{y}{l_y}, \quad \overline{\overline{Y}}(y) = \frac{y}{l_y},$$

$$A(t) = B(t) = C(t) = D(t) = 0 \text{ (because boundary functions are zero)},$$

in which case we have

$$w(x, y, t) = u_1 + (u_2 - u_1)\frac{y}{l_y}.$$

Given this expression for $w(x, y, t)$ we see that the separation of the function $u(x, y, t)$ into functions $w(x, y, t)$ and $v(x, y, t)$ is a separation into a stationary solution corresponding to the boundary conditions and the solution describing the relaxation of the temperature to the stationary state.

The relaxation process to a steady state described by the function $v(x, y, t)$ is the solution to the boundary value problem with zero boundary conditions where the stationary solution is described by

$$\tilde{f}(x, y, t) = -\frac{\partial w}{\partial t} + a^2 \left(\frac{\partial^2 w}{\partial x^2} + \frac{\partial^2 w}{\partial y^2} \right) = 0,$$

$$\tilde{\varphi}(x, y) = u_0 - u_1 - (u_2 - u_1)\frac{y}{l_y}.$$

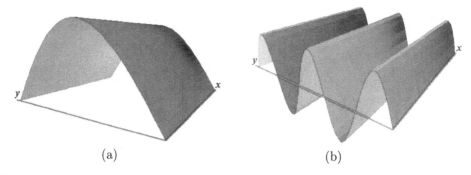

(a)   (b)

FIGURE E.4
Eigenfunctions (a) $V_{01}(x, y)$ and (b) $V_{05}(x, y)$ for the free surface in Example E.4.

Eigenvalues and eigenfunctions of this problem can be easily obtained:

$$\lambda_{nm} = \lambda_{xn} + \lambda_{ym} = \pi^2 \left[\frac{n^2}{l_x^2} + \frac{m^2}{l_y^2} \right], \quad n = 0, 1, 2, \ldots \quad m = 1, 2, 3, \ldots$$

$$V_{nm}(x, y) = X_n(x)Y_m(y) = \cos\frac{n\pi x}{l_x} \sin\frac{m\pi y}{l_y},$$

$$\|V_{nm}\|^2 = \|X_n\|^2 \, \|Y_m\|^2 = \begin{cases} l_x l_y/2, & n = 0 \\ l_x l_y/4, & n > 0. \end{cases}$$

The three-dimensional picture shown in Figure E.4 depicts the two eigenfunctions, $V_{01}(x, y)$ and $V_{05}(x, y)$, chosen as examples.

Applying Equation (9.30), we obtain

$$C_{nm} = \frac{1}{\|v_{nm}\|^2} \int_0^{l_x} \int_0^{l_y} \left[u_0 - u_1 - (u_2 - u_1)\frac{y}{l_y} \right] \cos\frac{n\pi x}{l_x} \sin\frac{m\pi y}{l_y} dx dy.$$

From this formula we have

$$C_{0m} = \frac{2}{l_y} \int_0^{l_y} \left[u_0 - u_1 - (u_2 - u_1) \cdot \frac{y}{l_y} \right] \sin\frac{m\pi y}{l_y} dx dy$$

$$= \frac{2}{m\pi} \left\{ (u_0 - u_1) [1 - (-1)^m] + (u_2 - u_1)(-1)^m \right\},$$

for $n > 0$, $C_{nm} = 0$. And as we obtained, the temperature distribution does not depend on x at all. This result could be anticipated from the very beginning, since the initial and boundary conditions do not depend on x. In other words, the solution is actually one-dimensional for this problem. Hence, the distribution of temperature inside the rectangular plate for some instant of time is described by the series

$$u(x, y, t) = u_1 + (u_2 - u_1)\frac{y}{l_y} + \sum_{m=1}^{\infty} C_{0m} e^{-\lambda_{0m} a^2 t} \sin\frac{m\pi y}{l_y}.$$

Figure E.5 shows two snapshots of the solution at the times $t = 0$ and $t = 10$. This solution was obtained in the case when $a^2 = 0.25$, $l_x = 4$, $l_y = 6$, $u_0 = 10$, $u_1 = 20$, $u_2 = 50$.

It is interesting to compare the first and second pictures in Figure F.2. The first one is very rough, the second is smooth. The explanation of the roughness is that at $t = 0$ the initial and boundary conditions do not match (the solution is non-physical). But at any

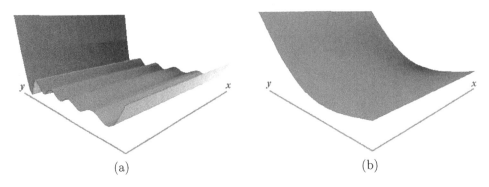

FIGURE E.5
Surface graph of temperature at (a) $t = 0$ and (b) $t = 10$ for Example E.4.

nonzero time the temperature distribution very quickly becomes smooth, as it should be for real, physical situations. Here we see that this method of solution works quite well in approximating a real, physical situation.

Example E.5 A heat-conducting thin uniform rectangular plate $(0 \leq x \leq l_x, 0 \leq y \leq l_y)$ is thermally insulated over its lateral faces. The edge at $y = 0$ is thermally insulated, the edge at $y = l_y$ is kept at constant zero temperature, the edge at $x = 0$ is kept at constant temperature $u = u_1$ and the edge at $x = l_x$ is kept at the temperature

$$u(l_y, x, t) = \cos \frac{3\pi y}{2l_y} e^{-t}.$$

The initial temperature distribution within the plate is $u(x, y, 0) = u_0 = \text{const}$. Find the temperature $u(x, y, t)$ of the plate at any later time, if generation (or absorption) of heat by internal sources is absent.

Solution. The problem is described by the equation

$$\frac{\partial u}{\partial t} = a^2 \left[\frac{\partial^2 u}{\partial r^2} + \frac{\partial^2 u}{\partial y^2} \right]$$

under the conditions

$$u(x, y, 0) = \varphi(x, y) = u_0,$$

$$u(0, y, t) = u_1, \quad u(l_x, y, t) = \cos \frac{3\pi y}{2l_y} \cdot e^{-t}, \quad \frac{\partial u}{\partial y}(x, 0, t) = 0, \quad u(x, l_y, t) = 0.$$

The boundary functions *do not satisfy the conforming conditions* at point $(0, l_y)$, that is,

$$g_1\big|_{y=l_y} = u_1 \quad \neq \quad g_4\big|_{x=0} = 0.$$

So, in this case we shall search for an auxiliary function as the sum of two functions

$$w(x, y, t) = w_1(x, y, t) + w_2(x, y, t),$$

where $w_1(x, y, t)$ is an auxiliary function satisfying the conforming boundary value functions

$$P_1[w_1]_{x=0} = g_1(y, t) = u_1, \quad P_2[w_1]_{x=l_x} = g_2(y, t) = \cos \frac{3\pi y}{2l_y} e^{-t},$$

$$P_3[w_1]_{y=0} = \frac{\partial w_1}{\partial y}(x, 0) = g_3(x, t) = 0, \quad P_4[w_1]_{y=l_y} = w_1(x, l_y) = u_1 \left(1 - \frac{x}{l_x}\right),$$

FIGURE E.6
Surface graph of the particular solution $w_2(x, y, t)$ to Example E.5.

and $w_2(x, y, t)$ is a particular solution of the Laplace equation

$$\nabla^2 w_2(x, y, t) = \left[\frac{\partial^2 w_2}{\partial x^2} + \frac{\partial^2 w_2}{\partial y^2} \right] = 0$$

with the following boundary conditions

$$P_1[w_2]_{x=0} = w_2(0, y, t) = 0, \quad P_2[w_2]_{x=l_x} = w_2(l_x, y, t) = 0,$$

$$P_3[w_2]_{y=0} = \frac{\partial u}{\partial y}(x, 0, t) = 0, \quad P_4[w_2]_{y=l_y} = w_2(x, l_y, t) = u_1\left(\frac{x}{l_x} - 1\right).$$

The auxiliary function, $w_1(x, y, t)$, is

$$w_1(x, y, t) = u_1\left(1 - \frac{x}{l_x}\right) + \cos\frac{3\pi y}{2l_y}e^{-t}\frac{x}{l_x} = u_1 + \frac{x}{l_x}\left(\cos\frac{3\pi y}{2l_y}e^{-t} - u_1\right).$$

The particular solution, $w_2(x, y, t)$, of the problem has the form

$$w_2(x, y, t) = \sum_{n=1}^{\infty} \left\{ A_n(t) Y_{1n}(y) + B_n(t) Y_{2n}(y) \right\} X_n(x),$$

where λ_{xn} and $X_n(x)$ are eigenvalues and eigenfunctions of the respective Sturm-Liouville problem:

$$\lambda_{xn} = \left[\frac{n\pi}{l_x}\right]^2, \quad X_n(x) = \sin\frac{n\pi x}{l_x}, \quad \|X_n\|^2 = \frac{l_x}{2}, \quad n = 1, 2, 3, \ldots$$

The functions $Y_{1n}(y)$ and $Y_{2n}(y)$ for the given boundary conditions are (see formulas (C.12))

$$Y_{1n}(y) = \frac{\cosh\sqrt{\lambda_n}y}{\cosh\sqrt{\lambda_n}l_y}, \quad Y_{2n}(y) = -\frac{\sinh\sqrt{\lambda_n}(l_y - y)}{\sqrt{\lambda_n}\cosh\sqrt{\lambda_n}l_y}.$$

Coefficients $A_n(t)$ and $B_n(t)$ are given by

$$A_n = \frac{2}{l_x}\int_0^{l_x} u_1\left(\frac{x}{l_x} - 1\right)\sin\frac{n\pi x}{l_x}dx = -\frac{2u_1}{n\pi}, \quad B_n = 0.$$

Thus,

$$w_2(x, y, t) = -\frac{2u_1}{\pi}\sum_{n=1}^{\infty}\frac{1}{n}\frac{\cosh\sqrt{\lambda_n}y}{\cosh\sqrt{\lambda_n}l_y}\sin\frac{n\pi x}{l_x}$$

as graphed in Figure E.6.

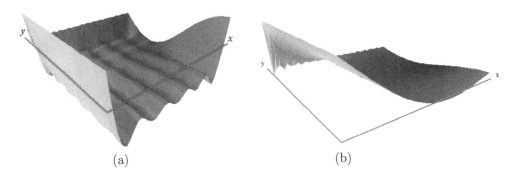

FIGURE E.7
Surface graph of the plate temperature at (a) $t = 0$ and (b) $t = 2$ for Example E.5.

The function $v(x, y, t)$ is the solution to the boundary value problem with zero boundary conditions where

$$\tilde{f}(x, y, t) = \frac{x}{l_x} \cos \frac{3\pi y}{2l_y} e^{-t} \left[1 - a^2 \left(\frac{3\pi}{2l_y} \right)^2 \right],$$

$$\tilde{\varphi}(x, y) = u_0 - u_1 - \frac{x}{l_x} \left(\cos \frac{3\pi y}{2l_y} e^{-t} - u_1 \right) + \frac{2u_1}{\pi} \sum_{n=1}^{\infty} \frac{1}{n} \frac{\cosh \sqrt{\lambda_n} y}{\cosh \sqrt{\lambda_n} l_y} \sin \frac{n\pi x}{l_x}.$$

Therefore, the eigenvalues and eigenfunctions of problem are

$$\lambda_{nm} = \lambda_{xn} + \lambda_{ym} = \pi^2 \left[\frac{n^2}{l_x^2} + \frac{(2m-1)^2}{4l_y^2} \right], \quad n, m = 1, 2, 3, \ldots$$

$$V_{nm}(x, y) = X_n(x) Y_m(y) = \sin \frac{n\pi x}{l_x} \cos \frac{(2m-1)\pi y}{2l_y}, \|V_{nm}\|^2 = \|X_n\|^2 \cdot \|Y_m\|^2 = \frac{l_x l_y}{4}.$$

(As before, these eigenvalues and eigenfunctions can be found in Appendix C part 1 or easily derived by the reader.)

Applying the Equations (9.30) and (9.35), we obtain

$$C_{nm} = \frac{4}{l_x l_y} \int_0^{l_x} \int_0^{l_y} \tilde{\varphi}(x, y) \sin \frac{n\pi x}{l_x} \cos \frac{(2m-1)\pi y}{2l_y} dx dy,$$

$$f_{nm}(t) = \frac{4}{l_x l_y} \int_0^{l_x} \int_0^{l_y} \tilde{f}(x, y, t) \sin \frac{n\pi x}{l_x} \cos \frac{(2m-1)\pi y}{2l_y} dx dy.$$

Thus, we have

$$T_{nm}(t) = \int_0^t f_{nm}(\tau) e^{-a^2 \lambda_{nm}(t-\tau)} d\tau$$

and

$$u(x, y, t) = [w_1(x, y, t) + w_2(x, y, t)] + v(x, y, t)$$

$$= u_1 + \frac{x}{l_x} \left(\cos \frac{3\pi y}{2l_y} e^{-t} - u_1 \right) - \frac{2u_1}{\pi} \sum_{n=1}^{\infty} \frac{1}{n} \frac{\cosh \sqrt{\lambda_n} y}{\cosh \sqrt{\lambda_n} l_y} \sin \frac{n\pi x}{l_x}$$

$$+ \sum_{n=1}^{\infty} \sum_{m=1}^{\infty} \left[T_{nm}(t) + C_{nm} e^{-\lambda_{nm} a^2 t} \right] \sin \frac{n\pi x}{l_x} \cos \frac{(2m-1)\pi y}{2l_y}.$$

Figure E.7 shows two snapshots of the solution at the times $t = 0$ and $t = 2$. This solution was obtained for the case when $a^2 = 1$, $l_x = 4$, $l_y = 6$, $u_0 = -1$ and $u_1 = 1$.

Bibliography

[1] J.W. Brown and R.V. Churchill, Fourier Series and Boundary Value Problems, McGraw-Hill, Inc., New York et al., 1993.

[2] R. Haberman, Applied Partial Differential Equations with Fourier Series and Boundary Value Problems, Pearson Education, Inc., Upper Saddle River, NJ, 2013.

[3] Y. Pinchover and J. Rubinstein, An Introduction to Partial Differential Equations, Cambridge University Press, Cambridge, 2005.

[4] L.D. Landau and E.M. Lifshitz, Fluid Mechanics, Pergamon Press, Oxford, 1987.

[5] R.K. Dodd, J.C. Eilbeck, J.D. Gibbon and H.C. Morris, Solitons and Nonlinear Wave Equations, Academic Press, London et al., 1982.

[6] M.J. Ablowitz and H. Segur, Solitons and the Inverse Scattering Transform, SIAM, Philadelphia, 1981.

[7] V. Henner, T. Belozerova, and K. Forinash, Mathematical Models in Physics: Partial Differential Equations, Fourier Series, and Special Functions, CRC Press, Taylor and Francis Group, Boca Raton, 2009.

[8] V. Henner, T. Belozerova, M. Khenner, Ordinary and Partial Differential Equations, CRC Press, Taylor and Francis Group, Boca Raton, 2013.

Index

A

Airy functions, 335–337
Analytical function, 141
Analytic formula for functions, 292
Angular frequency, 283
Antinodes, 208
Arbitrary constant, 4
Arbitrary differentiable functions, 6, 9
Arbitrary function, 4
Arbitrary length, Fourier expansions on
    intervals of, 289–290
Associated Legendre Functions,
    331–334
    Fourier-Legendre series in, 334–335
Auxiliary functions, 118
    for 1D hyperbolic and parabolic
        equations, 343–345
    for 2D hyperbolic or parabolic
        equations in rectangular domain,
        347
    for Poisson equation in rectangular
        domain, 345–347
Axisymmetric case for three-dimensional
    Laplace equation, 164–165
Axisymmetric oscillations of membrane,
    209–213

B

Bessel equation, 309–312
Bessel functions, 207, 309, 315
    of order, 310
    properties, 312–315
    spherical Bessel functions, 320–322
Bessel's inequality, 329
Bilinear formulation of KdV equation,
    271–272
Boundary conditions, 2, 45–48, 101–103,
    142, 227, 229
    eigenvalues and eigenfunctions of 1D
        Sturm-Liouville problem, 341–343
    Ill-posed problem, 142–143

maximum principle and consequences,
    144–146
in two-dimensional hyperbolic
    equations, 189–191
well-posed boundary value problems,
    143–144
Boundary value problems (BVP), 38, 99,
    101, 315–319
    diffusion equation, 100
    heat conduction, 99–100
    for Laplace equation in rectangular
        domain, 167–169
Breathers, 280
Bright solitons, 279–280
Burgers equation, 261; *see also* Nonlinear
    Schrödinger equation (NSE)
    kink solution, 261–262
    symmetries, 262–263
BVP, *see* Boundary value problems

C

Canonical forms, 22
    elliptic equations, 24–26
    hyperbolic equations, 23–24
    parabolic equations, 26–27
Cartesian coordinates, 139
Cauchy problem, 52, 142
    D'Alembert's formula, 57–58
    general solution of, 264–267
    Green's function, 58–59
    in infinite region, 88–91
    interaction of kinks, 265–267
    for nonhomogeneous wave equation,
        57–60
    well-posedness of, 59–60
Characteristic curves, 12–19
Circle, Sturm-Liouville problem for,
    349–351
Circular domain; *see also* Finite rectangular
    domain
    heat conduction within, 237

Circular domain (*Continued*)
  homogeneous heat equation, Fourier method for, 238–241
  nonhomogeneous heat equation, Fourier method for, 241–247
Cnoidal waves, 268–270
Completeness
  equation, 297
  property, 33, 62
Consistency conditions, 83–84
Consistent boundary conditions, 355–356, 368
Continuity equation, 99
Convergence of Fourier series, 286–288
Cosine functions, Fourier series in, 290–292
Cosine series, 292–293
Cylindrical coordinates, 178

D

D'Alembert method, 52–58, 88
  characteristic triangle, 53, 57
  propagation of initial displacement, 54
  propagation of initial pulse, 55
Dark solitons, 280–281
Defocusing NSE, 278
Delay theorem, 305
Differential equation, 29, 45–46
Diffusion equation, 100, 140, 142
Dirac delta function, 300
Directional derivative, 12
Dirichlet boundary condition, 30, 47, 190, 317, 351
Dirichlet boundary value problem, 164
Dirichlet condition, 228, 237
Dirichlet homogeneous boundary conditions, 234
Dirichlet problem, 143, 151, 167
Dirichlet theorem, 288
Dirichlet type, 240
Dispersion of waves, 88, 91
  cauchy problem in infnite region, 88–91
  propagation of wave train, 91–93
Dissipative processes, 2
Double Fourier series for function, 296
Driven edge boundary condition, 190
Duhamel's principle, 114–118
Dynamic equations, 177

E

Eigenfunctions, 30, 231, 309, 350–351
  of 1D Sturm-Liouville problem, 341–343
Eigenvalues, 31, 231, 309
  of 1D Sturm-Liouville problem, 341–343
  problem, 30
Elastic boundary, 47
Elastic end boundary condition, 50
Electrical oscillations in circuit, 50–51
Electric energies, 86
Electro-motive force (emf), 51
Elliptic differential equations, 139–140
Elliptic equations, 22, 24–26, 60, 139, 141
  application of Bessel functions for solution of Poisson equations in circle, 156–160
  ball, three-dimensional Laplace equation for, 164–166
  boundary conditions, 142–146
  BVP for Laplace equation in a rectangular domain, 167–169
  cylinder, three-dimensional Laplace equation for, 160–164
  elliptic differential equations, 139–140
  Green's function for Poisson equations, 171–176
  harmonic functions, 141
  Helmholtz equation, 177–179
  Laplace equation in polar coordinates, 146–147
  Poisson equation, 151–153, 169–171
  Poisson integral, 154–156
  related physical problems, 139–140
  Schrödinger equation, 180–181
  spherical coordinates, 140
emf, *see* Electro-motive force
Envelope function, 277
Equations of motion, 187–191
Error function, 250
Euler's formula, 295

F

f-fold degeneracy, 31
Fick's law, 100
Finite rectangular domain; *see also* Circular domain
  heat conduction within, 227–236

homogeneous heat equation, Fourier method for, 230–232

nonhomogeneous heat equation, Fourier method for, 233–236

types of boundary conditions, 228–229

First-order equations
linear, 7–12
quasilinear, 12–19

First-order partial differential equations, 23

First harmonic function, 66

Fixed edge boundary condition, 190

Fixed end boundary conditions, 46

Focusing NSE, 278–279

Force, 189

Forced axisymmetric oscillations, 216–218

Forced oscillations, 71
of uniform rod, 50

Fourier-Bessel Series, 40, 315–319

Fourier-Legendre series
in associated Legendre functions, 334–335
in Legendre polynomials, 328–331

Fourier coefficients, 286

Fourier cosine transform, 301

Fourier expansion
of function, 286
on intervals of arbitrary length, 289–290
method, 29

Fourier formulas, 284–286

Fourier method, 146, 229
graphical solution of eigenvalue equation, 70, 109
for homogeneous equations, 60–71, 103–111
for nonhomogeneous equations, 71–76, 118–126
in oscillations of rectangular membrane, 192–205
small transverse oscillations of circular membrane, 206–209, 214–216
in two-dimensional parabolic equations, 230–236, 238–247

Fourier series, 286, 293
convergence of, 286–288
in cosine or sine functions, 290–292
for functions of several variables, 295–296
generalized Fourier series, 296–298
for non-periodic functions, 288–289

Fourier transforms, 90, 299–303; *see also* Laplace transforms
of even or odd functions, 301

Free edge boundary conditions, 190

Free end boundary conditions, 46

Free heat exchange, 105, 230–232

Frequency spectrum, 68

Friction forces, 189

Fundamental harmonic function, 66

G

Galilean symmetry, 262–263

Gamma function, 310–311, 322–324

Gaussian beam, 180

Generalized Fourier series, 33, 296–298

Generalized solution, 107

General solution, 7–10

Gibbs phenomenon, 298–299

Green's formula, 188

Green's function, 58–59, 114–118, 249
homogeneous boundary conditions, 171–175
nonhomogeneous boundary conditions, 175–176
for Poisson equations, 171

Green's theorem, 57

Gross-Pitaevskii equation, 277

Group velocity, 92–93

H

Harmonic analysis, 283–284

Harmonic functions, 141, 154

Harmonic polynomials, 141

Heat
exchange, 227
terminology, 101

Heat conduction, 99–100
within circular domain, 237–247
equations with nonhomogeneous boundary conditions, 366–374
within finite rectangular domain, 227–236
in infinite medium, 248–249

Heat equation, 1, 21, 145, 264
in infinite region, 131–133

Heaviside function, 301

Helmholtz equation, 177–179

Hermitian operator, 31

Hirota's method, 272–274

Homogeneous boundary conditions, 29, 61, 71, 103, 112
    in finite rectangular domain, 233–236
    in oscillations of rectangular membrane, 192–203
    in small transverse oscillations of circular membrane, 206–209, 214–216
Homogeneous equations, 30, 112
Homogeneous heat equation, 229
    Fourier method for, 230–232, 238–241
    homogeneous heat-conduction equation, 248
Hook's law, 50
Hopf-Cole transformation, 264
Hyperbolic equations, 21, 23–24, 142

I

Ill-posed problem, 142–143
Incomplete elliptic integral, 269
Inconsistent boundary conditions, 355–366, 368–374
Infinite medium, heat conduction in, 248–249
Infinite trigonometric series, 283
Initial-boundary value problem, 2
Initial conditions, 10–12, 45–48, 101–103, 227
    in two-dimensional hyperbolic equations, 189–191
Initial value problems (IVPs), 10, 48, 306 307
Integral surface, 12
Integrating factor, 5
Interior Dirichlet problem, 148
Interior Neumann's problem, 144
Inverse Laplace transform, 303
Inverse scattering transform, 276
IVPs, *see* Initial value problems

J

Jacobi amplitude, 269
Jacobian of transformation, 15, 22
Jacobi elliptic cosine, 270

K

KdV equation, *see* Korteweg-de Vries equation

Kinetic energy, 84–87
Kink
    interaction, 265–267
    solution, 261–262
    trajectories of kink centers, 267
Korteweg-de Vries equation (KdV equation), 267; *see also* Poisson equation
    bilinear formulation, 271–272
    cnoidal waves, 268–270
    Hirota's method, 272–274
    multisoliton solutions, 274–277
    solitons, 270–271
    symmetry properties, 267–268

L

Laplace equation, 3, 26, 100, 139–144, 152, 173
    and exterior BVP for circular domain, 151
    and interior BVP for circular domain, 147–150
    in polar coordinates, 146–147
    for rectangular domain with nonhomogeneous boundary conditions, 355–366
Laplace integral, 303
Laplace transforms (LT), 76–79, 110–111, 303
    applications for ODE, 306–307
    of functions, 304
    properties, 304 306
Legendre equation, 165–166, 324–328
Legendre function, 309
    associated Legendre Functions, 331–334
Legendre polynomials, 324–328
    Fourier-Legendre series in, 328–331
Linear equation, 3, 8, 14, 128
Linear Euclidian space, 33
Linear heat flow equation, 99
Linear mass density, 44
Linear operator, 30
Linear periodic waves, 270
Longitudinal free oscillations of uniform rod, 49
Longitudinal vibrations, 48
    electrical oscillations in circuit, 50–51
    rod oscillations, 48–50
LT, *see* Laplace transforms

M

Magnetic energies, 86
Maximum principle, 129–131
   and consequences, 144–146
Measurement error, 2
Membranes, 187
Minimum principle, 130
Mixed boundary condition, 30, 47,
   317, 351
Mixed condition, 102, 228, 237
Mixed problem, 143
Mnemonic rule, 24
Modulated wave, 92
Monochromatic traveling wave, 91
Multisoliton solutions, 274–277

N

Navier-Stokes equation, 127
Neumann boundary condition, 30, 47,
   190–191, 317, 351
Neumann conditions, 101–102, 171,
   228, 237
Neumann functions, 315
Neumann problem, 143, 148, 151
Neumann type, 247
Newton's law, 103, 108
   of cooling, 144
Newton's second law, 1
   for motion, 188
Non-axisymmetric case for
   three-dimensional Laplace
   equation, 165–166
Non-free heat exchange, 229, 233
Non-periodic functions, Fourier series for,
   288–289
Non-viscous Burgers equation, 3, 18, 261
Nonhomogeneous boundary conditions,
   118–126, 175–176
   in circular domain, 246–247
   equations, 78–83
   heat conduction equations with,
     366–374
   Laplace and Poisson equations
     for rectangular domain with,
     355–366
   in oscillations of rectangular
     membrane, 203–205
   in small transverse oscillations of
     circular membrane, 218–220

Nonhomogeneous equations, 111–114,
   152, 233
Nonhomogeneous heat equation, 229
   Fourier method for, 233–236,
     241–245
   with nonhomogeneous boundary
     conditions, 246–247
Nonhomogeneous linear equation, 71, 111
Nonhomogeneous wave equation, 44, 57–60
Nonlinear equations, 261
   Burgers equation, 261–263
   general solution of Cauchy problem,
     264–267
   KdV equation, 267–277
   NSE, 277–281
Nonlinear medium, 277
Nonlinear Schrödinger equation (NSE), 277;
   see also Burgers equation
   solitary waves, 278–281
   symmetry properties, 277–278
Nonlinear spatially periodic waves, 270
Nonnegative real discrete eigenvalues, 325
NSE, *see* Nonlinear Schrödinger equation

O

ODE, *see* Ordinary differential equation
One-dimension (1D), 38, 144
   auxiliary functions for 1D hyperbolic
     and parabolic equations, 343–345
   eigenvalues and eigenfunctions of 1D
     Sturm-Liouville problem, 341–343
   heat conduction equation, 101
   problem, 127
   wave equations, 51
One-dimensional hyperbolic equations, 61
   boundary and initial conditions, 45–48
   consistency conditions and generalized
     solutions, 83–88
   dispersion of waves, 88–93
   finite intervals, 60–71
   fourier method for nonhomogeneous
     equations, 71–76
   longitudinal vibrations of rod and
     electrical oscillations, 48–51
   LT method, 76–79
   nonhomogeneous boundary conditions
     equations, 78–83
   traveling waves, 52–57
   tsunami effect, 93–94
   wave equation, 43–45

One-dimensional parabolic equations,
    101–103
    fourier method for homogeneous
        equations, 103–111
    fourier method for nonhomogeneous
        equations, 118–126
    Green's function and Duhamel's
        principle, 114–118
    heat conduction and diffusion,
        99–100
    heat equation in infinite region,
        131–133
    and initial and boundary conditions,
        101–103
    large time behavior of solutions,
        126–129
    maximum principle, 129–131
    nonhomogeneous boundary conditions,
        118–126
    nonhomogeneous equations, 111–114
One-parametric family, 13
Ordinary differential equation (ODE), 1, 8,
        18
    applications of Laplace transform for,
        306–307
    problems, 29
Orthogonal equation, 31
Orthogonal functions, 296–297
Orthonormal eigenfunctions, 35
Oscillations of rectangular membrane,
        191
    Fourier method for homogeneous
        equations, 192–199
    Fourier method for nonhomogeneous
        equations, 199–205

P

Parabolic equations, 21, 26–27, 100,
        142
Parity property, 333
Parsevale's equality, 297, 317, 329
Partial differential equation (PDE), 1,
        4, 46
    equations, 29
PDE, *see* Partial differential equation
Periodic functions, 283–284
Periodic processes, 283–284
Physical laws, 29–30
Piecewise continuous function, 288
Point of discontinuity, 288

Poisson equation, 3, 21, 139–140, 142–143,
        151–153, 156, 171–173; *see also*
        Korteweg-de Vries equation (KdV
        equation)
    auxiliary functions for 2D hyperbolic or
        parabolic equations in, 347
    auxiliary functions in rectangular
        domain, 345–347
    with homogeneous boundary
        conditions, 169–171
    for rectangular domain with
        nonhomogeneous boundary
        conditions, 355–366
Poisson integral, 154
Poisson kernel, 154
Polar coordinates equation, 152
Potential energy, 85, 87

Q

Quadratic equation, 1
Quasilinear equation, 4

R

Rectangle, Sturm-Liouville problem for,
        352–354
Rectangular domain
    auxiliary functions for Poisson equation
        in, 345–347
    Laplace and Poisson equations with
        nonhomogeneous boundary
        conditions, 355–366
Recurrence relations, 321
Reflection symmetry, 263
Riemann's lemma, 287–288
Rod oscillations, 48–50
Rodrigues' formula, 327

S

Scalar product, 40
    functions, 33
Scaling
    symmetry, 263
    transformation, 263
Schrödinger equation, 180–181
Second-order equations
    canonical forms, 22–27
    classification of, 21–22
Second boundary value problems, 148

Second harmonic function, 66
Second order linear PDEs, 29
Self-adjoint operator, 31
Semi-infinite medium, heat conduction in, 250–254
Shift theorem, 78, 305
Shock wave, 19
Sine function, 283
   Fourier series in, 290–292
Small transverse oscillations of circular membrane, 205
   axisymmetric oscillations of membrane, 209–213
   boundary conditions, 205–206
   equations, Fourier method for, 218–220
   forced axisymmetric oscillations, 216–218
   homogeneous equations, Fourier method for, 206–209
   nonhomogeneous equations, Fourier method for, 214–216
Solitary waves, 270–271, 278–281
Solitons, 270–271
Spectral density, 300
Spherical Bessel functions, 320–322
Spherical coordinates, 164
Spherical Neumann functions, 320
Standing waves, 66
Static distributions of temperature, 3
Stretched edge boundary condition, 190
Sturm-Liouville boundary problem, 104–105
Sturm-Liouville problem, 29, 61–63, 157, 170, 356
   for circle, 349–351
   examples, 34–40
   for rectangle, 352–354
   theorem, 32–34
Superposition principle, 25
Symmetry
   of Burger's equation, 262–263
   properties of KdV equation, 267–268
   properties of NSE, 277–278

Third boundary value problems, 148
Third harmonic function, 66
Three-dimensional Green's function, 249
Three-dimensional Helmholtz equation, 180
Three-dimensional Laplace equation
   axisymmetric case, 164–165
   for ball, 164
   for cylinder, 160–164
   graph of function, 161
   non-axisymmetric case, 165–166
   surface plot, 161
Three-dimensional space, 12–13
Three-soliton solution, 276
Time-independent Schrödinger equation, 181
Transcendental equation, 316
Translational symmetries, 263
Traveling wave form, 261–262
Trigonometric Fourier expansion, 38, 284
Trigonometric series, complex form of, 294–295
Tsunami effect, 93–94
Two-dimension (2D)
   hyperbolic or parabolic equations in rectangular domain, auxiliary functions for, 347
   Laplace equation, 141
Two-dimensional hyperbolic equations
   derivation of equations of motion, 187–191
   oscillations of rectangular membrane, 191–205
   small transverse oscillations of circular membrane, 205–220
Two-dimensional parabolic equations
   circular domain, heat conduction within, 237–247
   finite rectangular domain, heat conduction within, 227–236
   infinite medium, heat conduction in, 248–249
   semi-infinite medium, heat conduction in, 250–254
Two-fold degeneracy, 37
Two-soliton solution, 276

T

Telegraph equations, 51
Thermal conductivity, 99
Thermal diffusivity, 99

U

Uncertainty relation, 92
Uniform convergence, 286
Uniform mass density, 189

W

Wave
    equation, 3, 21, 24, 43–45, 180
    number, 89
    numbers, 299
    propagation of wave train, 91–93

propagation on inclined bottom,
        93–94
    speed, 45
Weierstrass criterion, 288
Well-posed boundary value problems,
        143–144

Printed in the United States
by Baker & Taylor Publisher Services